BASIC
GENETICS

Other Books of Interest...

Bacterial Phage and Molecular Genetics
  U. Winkler
Concepts in Photobiology: Photosynthesis and Photomorphogenesis
  G.S. Singhal et al
Cyanobacterial and Algal Metabolism and Environmental Biotechnology
  T. Fatma
Cyanobacterial Nitrogen Metabolism and Environmental Biotechnology
  A.K. Rai
Essentials of Molecular Biology
  D. Freifelder
Experiments in Plant Physiology: A Laboratory Manual
  D. Bajracharya
Flowering Plants: Taxonomy and Phylogeny
  B. Bhattacharyya and B.M. Johri
Genetics
  A.V.S.S. Sambamurty
Genome Analysis in Eukaryotes: Developments and Evolutionary Aspects
  R.N. Chatterjee and L. Sanchez
Glossary of Genetics: Classical and Molecular
  R.K. Sharma and M.B. Sharma
The Gymnosperms
  C. Biswas and B.M. Johri
Hybrid Cultivar Development
  S.S. Banga and S.K. Banga
Microbial Genetics
  D. Freifelder
Molecular Biology, 2/e
  D. Freifelder
Plant Cell, Tissue and Organ Culture: Fundamental Methods
  O.L. Gamborg and G.C. Phillips
Plant Molecular Biology and Biotechnology
  K.K. Tewari and G.S. Singhal
Plant Tissue Culture and Molecular Biology: Applications and Prospects
  P.S. Srivastava
Pollen Biology: A Laboratory Manual
  K.R. Shivanna and N.S. Rangaswamy
Transport Phenomena in Plants
  K. Raman

Forthcoming

Advanced Genetics
  G.S. Miglani
Ecology
  N.S. Subrahmanyam and A.V.S.S. Sambamurty
Introductory Practical Biochemistry
  S.K. Sawhney and Randhir Singh

BASIC
GENETICS

Gurbachan S. Miglani

NAROSA PUBLISHING HOUSE
New Delhi   Madras   Bombay   Calcutta
London

Gurbachan S. Miglani
Professor of Genetics
College of Basic Sciences & Humanities
Punjab Agricultural University
Ludhiana-141 004, INDIA

Copyright © 2000 Narosa Publishing House

NAROSA PUBLISHING HOUSE

6 Community Centre, Panchsheel Park, New Delhi 110 017
22 Daryaganj, Prakash Deep, Delhi Medical Association Road, New Delhi 110 002
35–36 Greams Road, Thousand Lights, Chennai 600 006
306 Shiv Centre, D.B.C. Sector 17, K.U. Bazar P.O., Navi Mumbai 400 705
2F–2G Shivam Chambers, 53 Syed Amir Ali Avenue, Calcutta 700 019
3 Henrietta Street, Covent Garden, London WC2E 8LU, UK

All rights reserved. No part of this publication may be reproduced, stored in a retrieval system, or transmitted in any form or by any means, electronically, mechanically, photocopying, recording or otherwise, without the prior written permission of the publishers.

All export rights for this book vest exclusively with Narosa Publishing House. Unauthorised export is a violation of Copyright Law and is subject to legal action.

ISBN 81-7319-300-2

Published by N.K. Mehra for Narosa Publishing House, 6 Community Centre, Panchsheel Park, New Delhi 110 017 and printed at Rajkamal Electric Press, Delhi 110 033 (India).

IN THE MEMORY OF

MY FATHER

S. HARNAM SINGH MIGLANI
(1916-1988)

WHO WAS A TEACHER *PAR EXCELLENCE*

PREFACE

An invitation from the University Grants Commission, New Delhi prompted me to write this text book **Basic Genetics**. After teaching genetics at undergraduate and postgraduate levels to the students of various constituent colleges of the Punjab Agricultural University for over 28 years. I felt myself confident enough to write a text book and the result is in your hands. Thus my students' contribution in the book is tremendous.

Objective of any text book is to make concepts and principles of the subject clear to the reader in a simple language; sometimes this may lead to over-simplification. Principles of genetics operate at cell, individual, and population level. Attempt has been made to explain various phenomenon of genetics at molecular level, wherever possible.

In light of the fact that genetics is a fast developing subject, no text book can be complete in itself. Although it has been my effort to include all basic terms, concepts, phenomenon and principles in this book but still many things might have been left out. Responsibility of including or not including certain items is entirely mine. A phenomenon in genetics may sometimes be explained with the help of one or more out of several known examples. Selection of examples and style of presentation is entirely my choice.

This book has been divided into 27 chapters. This text book can be used by university level students all over India. Postgraduate students, teachers and researchers can use it as a reference book. Chapter-wise details are given in the beginning of the book. Each chapter begins with detailed contents. Text is followed by hints on some useful reading materials, questions, numerical problems (wherever possible). A section on glossary includes more than 1,000 terms explained in simple words.

Bibliography has been given special attention and forms a separate section. Subject and Author indexes given at the end of the book will have special value for the reader.

I have consulted a large number of available old and new text books, original research papers, review articles and dictionaries in genetics. My head bows with respect before all these authors.

All my colleagues in the department have been very helpful in the writing of this book. Dr Ravi, Professor of Genetics (Retd.), made all his materials readily available to me. He also gave many useful suggestions for improvement of the manuscript. Moral support of Dr. R.G. Saini, Senior Geneticist-cum-Head, Department of Genetics, PAU was always a source of inspiration to me. I am thankful to Dr Ashok K. Sharma for sparing his valuable time during library consultations.

My mother Smt. Apar Kaur, wife Harjit Kaur, daughter Simmi, and son Jemmy were very eager to see the book completed.

Manuscript of the book was type set and all the figures were drawn on computer with utmost devotion and efficiency by Mr. Gagandeep Singh. Without the help of my nephew, Mr Sandeep Singh, M.Sc. (Entomology), compilation of Subject and Author indexes would not have been possible.

The financial aid provided by the University Grants Commission, New Delhi for preparing the manuscript of the book is gratefully acknowledged.

I am thankful to M/s Narosa Publishing House, New Delhi for bringing out this book in excellent format and design.

Gurbachan S. Miglani

CONTENTS

Preface vii

1 **Introduction** 1
    Heredity and Variation 1
    Certain Facts about Life 1
    Basis of Science of Genetics 2
    Theories of Inheritance 2
    Continuity of Life 4
    Qualitative and Quantitative Traits 4
    Genotype and Phenotype 5
    Genotypic and Environmental Variation 5
    Genetic Perspective 5
    Brief History of Genetics 6
    Genetic Analysis 7
    Some Useful Reading Hints 8
    Questions 8

2 **Principles of Inheritance** 9
    Brief Life Sketch of Mendel 9
    Mendel's Working Method 10
    Monohybrid Crosses 10
    Incomplete Dominance and Codominance 14
    Dihybrid Crosses 14
    Rediscovery of Mendelism 19
    Impact of Mendelism and its Rediscovery 19
    Concept of Dominance 19
    Scale of Dominance 22
    Some Useful Reading Hints 22
    Questions 22
    Numericals 23

3 **Cell Division and its Significance** 25
    Cell Cycle 25
    Cell Division 31
    Some Useful Reading Hints 37
    Questions 37

4 Chromosome Theory of Inheritance — 38
Background 38
Parallelism Between Behaviour of Chromosomes 39
and Mendelian Factors (Genes)
Chromosome Theory of Inheritance 42
Some Useful Reading Hints 43
Questions 43

5 Chromosomes — 44
Prokaryotic Chromosomes 44
Eukaryotic Chromosomes 44
Unusual Chromosomes 55
Some Useful Reading Hints 62
Questions 63

6 Sex-linked, Sex-limited and Sex-influenced Inheritance — 64
Sex-linked Inheritance 64
Holandric Genes 68
X- and Y-linked Inheritance 68
Sex-limited Characters 69
Sex-influenced Traits 73
Some Useful Reading Hints 75
Questions 75
Numericals 76

7 Multiple Alleles — 77
Multiple Allelism 77
Conclusions from Multiple Allelism 85
Some Useful Reading Hints 85
Questions 85
Numericals 86

8 Gene-gene Interactions — 87
Dominance and Epistasis 87
Epistatic and Hypostatic Genes 87
Interaction Between Nuclear Genes 88
Nuclear and Cytoplasmic Gene Interaction 97
Conclusions from Gene-gene Interactions 100
Totipotency 100
Pleiotropy 101
Homeostasis 102
Some Useful Reading Hints 105
Questions 105
Numericals 106

9 Systems Used in Genetic Analysis 109
Experimental Method 109
Life Cycles 110
Some Useful Reading Hints 130
Questions 130

10 Genetic Analysis in Diploid Eukaryotes 132
Time of Segregation 132
Endosperm Genetics 132
Linkage and Genetic Recombination 133
Terminology 135
Linkage in Plant Systems 136
Linkage in Animal Systems 145
Complete Linkage 154
Cytological Basis of Crossing-over 154
Crossing-over Occurs at Four-strand Stage 154
Machanism of Recombination 157
Molecular Models of Recombination 159
Gene Conversion 168
Significance of Crossing-over 170
Some Useful Reading Hints 171
Questions 172
Numericals 173

11 Genetic Analysis in Haploid Eukaryotes 176
Gene Mapping in *Neurospora* 176
Gene Mapping in Yeast and *Chlamydomonas* 186
Some Useful Readinging Hints 186
Questions 186
Numericals 187

12 Genetic Analysis in Bacteria and Viruses 190
Recombination in Bacteria 190
Recombination in Viruses 203
Some Useful Reading Hints 207
Questions 208
Numericals 208

xii *Basic Genetics*

13 Genetic Material — 211
Nature of Genetic Materials 211
Structure of DNA 216
Structure of RNA 221
Properties of Genetic Material 222
Replication of DNA in Prokaryotes 226
Replication of DNA in Eukaryotes 231
Some Useful Reading Hints 232
Questions 233

14 Gene Structure — 234
Concept of Gene 234
What is a Gene? 245
Colinearity 251
Overlapping Genes 252
Some Useful Reading Hints 252
Questions 253
Numericals 254

15 Gene Function — 257
Relationship Between Gene and Enzyme 257
Genes Control Biochemical Reactions 257
Relationship Between Genotype and Phenotype 259
One Gene-one Enzyme 262
One-gene-one Polypeptide 265
One Cistron-one Polypeptides 267
One Gene-one Primary Cellular Function 267
Gene as a Functional Entity 268
Some Useful Reading Hints 268
Questions 268

16 Transcription — 270
Central Dogma and its Reversal 270
Gene Expression 271
RNAs, their Processing and Functions 275
Genetic Code 284
Hormones and Gene Expression 289
Penetrance 290
Expressivity 290
Phenocopies 291
Some Useful Reading Hints 292
Questions 293

17 Translation — 295
Translation in Prokaryotes 295
Translation in Eukaryotes 305
Some Useful Reading Hints 310
Questions 310

18 Gene Regulation — 311
The Meaning of Gene Regulation 311
Terminology of Gene Regulation 311
Components of Gene Regulation in Prokaryotes 312
Lactose operon in *E. coli* 314
Is Operon Concept Applicable to Eukaryotes? 326
Short-term Versus Long-term Regulation 326
Mechanisms of Gene Regulation in Eukaryotes 326
Hormones and Gene Regulation 330
Antisense Inhibition of Gene Expression 336
Antisense RNA in Gene Regulation 336
Molecular Zippers in Gene Regulation 337
Some Useful Reading Hints 339
Questions 330

19 Mutations — 341
Definition 341
Characteristics of Mutations 341
Classification of Mutations 342
Spontaneous Mutations 350
Induced Mutations 355
Detection of Mutations 371
Applications of Induced Mutations 379
Environmental Mutagenic Hazards 379
Some Useful Reading Hints 381
Questions 381

20 Genes in Populations 383
Laboratory and Natural Populations 383
Binomial Distribution of Genotype Frequencies 383
Hardy-Weinberg Law 383
Determination of Gene Frequency 387
Some Useful Reading Hints 393
Questions 393
Numericals 394

21 Quantitative Inheritance 395
Quantitative Traits 395
Multiple Factor Hypothesis 396
Polygenes 402
Genetic and Environment Components of Continuous Variation 403
Heritability 403
Some Useful Reading Hints 406
Questions 406
Numericals 407

22 Extranuclear and Extrachromosomal Inheritance 409
Extranuclear Inheritance 409
Yeast Mitochondrial Genome 420
Universality of Code 422
Mitochondrial Genes and Diseases 422
Maternal Effects 423
Nucleo-ctoplasmic Interactions 424
Extrachromosomal Inheritance 428
Importance of Plasmids 432
Some Useful Reading Hints 433
Questions 434

23 Sex-determination and Dosage Compensation 435
Sex-determination 435
Dosage Compensation 441
Some Useful Reading Hints 444
Questions 445

24 Human Genetics — 446
Genetics and Man 446
Genetic Basis of Disorders 447
Changes in Chromosome Number or Structure 449
Diagnosis and Carrier Detection 45449
Somatic Cell Hybridization 455
Treatment for Inherited Disease 459
Gene Therapy 460
DNA Fingerprinting 462
LOD Score 462
Some Useful Reading Hints 463
Questions 464

25 Genetic Engineering — 465
Restriction Endonucleases 465
Gene/DNA Cloning 467
Gene Transfer 469
Chromosome 471
Microsatellites 473
Polymerase Chain Reaction 474
Transposons 475
High Density Molecular Maps 476
Variable Number of Tandem Repeat Loci 476
Applications of Genetic Engineering 476
Hazards of Genetic Engineering 478
Some Useful Reading Hints 478
Questions 479

26 Evolution — 481
Lamarck's Theory of Evolution 481
Theory of Natural Selection 482
Mutation Theory of Evolution 484
Modern Synthetic Theory of Evolution 485
Molecular Evolution 497
A Model of Speciation 498
Some Useful Reading Hints 500
Questions 501

xvi   *Basic Genetics*

27   Probability and Chi-square   503
    Probability   503
    Chi-square   505
    Some Useful Reading Hints   507
    Questions   507

Glossary   508

Bibliography   570

Subject Index   595

Author Index   625

1

INTRODUCTION

Genetics is a branch of biology that precisely understands the mechanisms responsible for transmission of biological properties from one generation to the next. Thus genetics deals with the understanding of the nature, molecular structure, organization, biological function, regulation and manipulation of hereditary particles called genes which play a central role in processes of life. It also cover the development of ways and means to use the knowledge of genetics for the welfare of mankind.

Heredity and Variation

In short, genetics is science of heredity and variation. This term was coined by Basteson in 1905. **Heredity** is the cause of similarities between individuals. For this reason brothers and sisters with the same parents resemble each other. **Variation** is the cause of the differences between individuals. This is reason that brothers and sisters who do resemble each other are still unique individuals.

Certain Facts about Life

(a) Life comes only from pre-existing life. (b) Life exists is diverse forms: bacteria, fungi, plants (gymnosperms, angiosperms), animals (protozoa, fishes, amphibians, reptiles, birds, mammals). (c) Like begets like, e.g., dogs produce pups and cats produce kitten. (d) Traits, i.e., biological characteristics, show recombination. We generally say that the child has nose like his father and eyes like his mother. (e) Variation exists in living forms. Members of a species in a natural population differ from one an other. (f) Living things have individually. Biological individuality has pattern, limitations and characteristics that depends atleast in part on the parents of an individual. (g) Similarities and differences are heritable.

Basis of Science of Genetics

Basis of science of genetics is variation. If there were no variation there would have been no science of genetics. Principles of genetics apply only to sexually reproducing organisms having biparental parentage. The hereditary transmission of biological properties is obviously an aspect of **sexual reproduction** and is important for transmission of biological properties from one generation to the next. Genes play central role in different processes of life.

Theories of Inheritance

Preformation

Swammerdam in 1694 claimed that he could see a miniature man called **"homunculus"** inside the human sperm head. Thus it was believed that human body was preformed and an individual was considered to be a simple enlargement of this preformed structure. With the improvement of microscopy it was shown that "homunculus" was an imagination. This view is of historic importance only.

Encapsulation

According to Bonnet (1720-1793), female contains all hereditary information of all her immediate and remote progeny. But with discovery of fertilization by Koelreuter (1733-1793), it became clear that hereditary information is transmitted both through male and female gametes, in plants as well as in animals.

Pangenesis

Charles Darwin in 1868 believed that very small exact but invisible copies of each body organ and component, called **"gemmules"**, were transported by the blood stream to the sex organs and these were assembled into gametes. Upon fertilization, "gemmules" of the opposite sex mixed up. During development these "gemmules" then separated out to different parts of the body. Thus the individuals constituted a mixture of maternal and paternal organs. Pangenesis hypothesis was rejected on the basis of an experiment carried out by Galton (1822-1911). He transfused the blood between white and black rabbits but did not get any rabbit with intermediate skin colour in their progeny.

Inheritance of Acquired Traits

Lamarck in 1889 tried to explain extraordinary ability of hereditary factors to respond directly to environment. He assumed that each hereditary particle had a spiritual conscious like property that could absorb and interpret messages from outside. He believed that variations in organisms are induced in response to urgent need and he regarded these variations as heritable. This theory could not be proven experimentally.

Blood/blending Inheritance

Heredity was supported to be transmitted as though it were a **miscible fluid** like blood in animals and man. The "bloods" of the parents were assumed to mix, fuse or blend in their offspring.

Epigenesis

Adult structures of plants and animals develop from uniform embryonic tissues. Thus, development of an individual is epigenetic and genes act in **epigenetic systems**. Process of development consists of more than growth. The sex cells are more or less drops of structureless liquid containing nothing whatsoever resembling the body that it is to develop from them. Development of an individual does not stop at some arbitrary stage in life. An individual that has stopped developing is considered dead.

Germplasm Theory

Weismann (1834-1914) established through his famous experiment on mice that "pangenesis" and "inheritance of acquired traits" could not be verified. According to this theory, multicellular organisms give rise to two types of tissues - somatoplasm and germplasm. **Somatoplasm** consists of tissues that are essential for functioning of the organism but lack the property of entering into sexual reproduction. Changes that occur in somatic cells are not passed on from parents to offspring and are thus not heritable. **Germplasm** is set aside for reproductive purposes as the cells of these tissues divide meiotically to form gametes. Any change in the germplasm is transmitted from parents to offspring and could lead to changed inheritance. McLaren (1981) has provided a new look at the problem of germ cells and soma.

Continuity of Life

There is continuity of germplasm between all descendent generations. Continuity of life results from its inherent characteristic of reproduction. **Reproduction** is the basic necessity of a species to continue its existence on this planet. The continuity of life is perpetuation of a species to repeat its genetic possession in every subsequent generation. A species stores biological information in **deoxyribonucleic acid (DNA)**. Each cell of a given species has a constant amount of DNA which is doubled by replication and cell division ensures equal distribution of DNA of parent cell to daughter cells. The gametes have half the amount of DNA while the zygote possesses the amount of DNA characteristic of the species. Replication of DNA is the **molecular basis of reproduction** and hence the continuity of life. Replication faithfully keeps the genetic information intact over generations. Continuity of germplasm between all the descendent generations of a species explains many biological similarities that are inherited.

Qualitative and Quantitative Traits

Traits are some biological attributes of individuals within a species for which various heritable differences can be defined. Biological property, characteristics or simply character are terms synonymously used for trait. Traits may be measured in terms of colour, shape, pattern, molecular structure or statistical parameters. Qualitative is a trait for which there are a fewer number of discrete phenotypes that can be distinguished by visual observation. A fewer number of genes determine such traits. Data can be put into a few discrete classes. Traits are less affected by environment.

Quantitative is a trait for which there is a range of phenotypes differing by degree. Data can not be put into a few discrete classes. Statistical parameters, viz., mean, mode, median, range, variance, standard deviation, coefficient of variation are used to measure quantitative traits. A large number of genes with small additive effect are generally involved in determination of such traits. Expression of these traits is affected by environment. Qualitative traits have been found to be more useful for understanding the mechanics of inheritance.

Genotype and Phenotype

Johannsen (1911) gave concept of genotype and phenotype. **Genotype** is sum total of heredity, i.e., the genetic constitution that an organism receives from its parents. **Phenotype** is appearance of the organism i.e., a sum total of all its characteristics such as colour, form, size, behaviour, chemical composition, structure. The phenotype is external manifestation of a genotype of an organism and is a result of the structure of a genotype with an environment. Both are absolutely necessary. The equation phenotype = genotype X environment describes relationship between genotype and environment. Genotype is relatively stable through out the life of an individual whereas phenotype changes continuously during development.

The interplay of the effects of genetic and non-genetic factors on development is called the genotype X environment interaction. Genes determines the potentials and environment determines whether that potential is to be reached or not.

Genotypic and Environmental Variation

When two or more individuals develop in the same environment and come to possess different phenotypes, the individuals then have different genotypes. This type of variation is known as **genotypic variation**. When individuals with the same genotypes develop in different environment, their phenotypes may be quite different, e.g., identical twins raised in different environments. This type of variation is known as **environmental variation**. The variation that is observed in different individuals in a natural population at phenotypic level is known as phenotypic variation which is sum total of genotypic variation and environmental variation and their **interaction**. Mathematically, $V_P = V_G + V_E + V_{GE}$, where V_P = phenotypic variance, V_G = genotypic variance, V_E = environmental variance and V_{GE} = variance due to interaction between genotypic and environmental factors. Biological properties of an individuals are governed by genes, the units of heredity. Genotypic variations arise due to mutations in genes.

Genetic Perspective

There are two basic approaches to study genetics. **Approach I** deals with study of subject matter of genetics starting from atoms and molecules of biological significance and then studying the structure of macromolecules, viz., deoxyribonucleic acid (DNA) and ribonucleic acid

6   Basic Genetics

(RNA), proteins and enzymes. Next step is the study of structure and properties of gene, namely, replication, recombination and mutation. This is followed by the study of structure and functions of cells. A stage is now set for defining genetics and study of mechanics of inheritance, i.e., classical genetics, also known as transmission genetics. This is followed by population genetics, evolutionary genetics and recent trends in genetics. Lastly, applications of genetics are studied in relation to human welfare. Approach I is not in its historical perspective

Approach II first defines genetics and then principles of inheritance are studied. This is followed by studying the nature, structure, organization, function and properties of genetic material. This is followed by the study of population genetics and evolutionary genetics. Lastly, applications of genetics are study in relation to human welfare. Approach II is more or less historical in its perspective. Genetic perspective may be defined as the study of various progresses made in the field of genetics in its historical perspective.

Brief History of Genetics

Genetics began in mid nineteenth (1866) century when Gregor John Mendel carefully analysed mechanism of inheritance. His experiments were simple which brought forth the most significant principles that determine how traits are passed from one generation to the next. His experiments set a stage where the subject could be understood and the rules could clearly be formed to detect the presence of genes through hybridization experiments without knowing the nature of genes or gene products. Mendelian principles of heredity are equally applicable to most by higher organisms, e.g., wheat, rice, animals and even human beings.

Mendel's experiments were not appreciated for 34 years because nobody could believe that genes were discrete objects. In 1900, Mendelism was rediscovered and birth of genetics took place. In mid-1900s when biochemistry highly developed, geneticists began to think about biochemical nature of genes and the causes of genetic variation. The observations made between 1920s and 1940s pointed out DNA as the genetic material in most of the viruses. Critical analysis by J.D. Watson and F.H.C. Crick in 1953 gave the molecular structure of DNA and with this genetics entered the DNA age. This was a very exciting phase in the development of genetics. By the end of 1960s it was known how a gene is copied, how a gene is expressed, how a mutation arises, how genes are turned on and turned off according to the needs of the cell or organism. It became possible to identify the products of thousands of genes. These developments constituted a part of important branch of genetics called molecular genetics. In 1970s genetics under went its most recent revolution, i.e., the development of **recombinant DNA technology**. This technique which forms the backbone of genetic engineering involves procedures to identify, sequence, isolate, clone and transfer genes from one organism to another at the will of molecular geneticist bypassing all reproductive barriers. This development had enormous effect on genetic research, particularly to understand gene structure, gene expression and its regulation in plants and animals. Principles of genetics are being used not only in basic studies but also are being applied in the field of agriculture, animal husbandry, forestry, medicine, law, gene therapy and genetic counselling.

Genetics is a **science of potentials**. Human intervention in genetic mechanisms have used these potentials for his own benefit. It is the genes and not the traits that are inherited. No trait is simple enough to be governed by a single gene. When we talk about inheritance of a trait, in fact we refer to inheritance of differences(s) existing between two parental lines for a particular trait. A trait is result of complex interactions involving many genes.

Genetic Analysis

In genetics we deal with similarities and differences. To understand how certain differences are inherited, genetic analysis in conducted. In genetic analysis we explore whether a particular contrasting trait is governed by one or more gene difference and whether gene in question is nuclear or cytoplasmic. In animals, it is to be found out whether a gene is sex-linked or autosomal. Genetic analysis also includes assignment of a gene to a particular chromosome and to find specific position of the gene in a chromosome in relation to the other known genes in the nuclear or extranuclear genome. Genetic analysis in certain cases may even be extended to find out nucleotide sequence of a gene. Several prokaryotes and haploid and diploid eukaryotes have been found to be very useful organisms in genetic analysis. These organisms and the procedures used in genetic analysis will be dealt with in Chapters 10-13. The organism used in genetics experiments should fulfil certain conditions: (1) should have a number of well-defined contrasting differences, (2) should be sexually reproducing and have biparental parentage, (3) should undergo inherent recombination, (4) controlled matings should be possible, (5) should have short life cycle, (6) should yield a large number of offspring, (7) convenience in handling, and (8) relatively inexpensive, etc.

8 Basic Genetics

Some Useful Reading Hints

Story of origin of genetics has been written by Stern and Sherwood (1966). Interesting account of mechanism of Mendelian heredity has been given by Morgan (1919), Morgan et al. (1922), Ribisson (1979), Sinnott et al. (1958). Classical Papers in Genetics complied by Peters (1959) are worth reading.Pontecorvo (1958) has dealt with variuos trends in genetic analysis.

A detailed history of genetics was provided by Dunn (1965), Sturtevant (1965), and Stubbe (1972). Selected Papers which mark the beginning of molecular genetics were compiled by Taylor (1965). Ashburner and Wright edited various volumes on the genetics and biology of the best known multicellular eukaryote, *Drosophila*.

Some of the authors who wrote text books in general genetics, cytogenetics, human genetics, and molecular genetics, whose landmark will always be useful to the students of genetics, are: Auerbach (1961), Burnham (1962), Dobzhansky (1962), Burdette (1963), Taylor (1963), Brewbaker (1964), Darlington (1965), Herskowitz (1965) Brown (1972), Jenkins (1975), King (1975), Schulz-Schaeffer (1980), Mackean (1981), Swaminathan et al. (1983), Sarin (1985), Strickberger (1986), Dulbecco (1987), Freifelder (1987), Zubay (1987), Farnsworth (1988), Burns and Bottino (1989), Brown (1992), Tamarin (1996) Lewin (1996).

A few dictionaries which students, teachers and research workers will find useful are by King and Stansfield (1990), Rieger et al. (1991), Somani (1995), Miglani (1998).

Questions

1. Define: genetics, trait, preformation, encapsulation, pangenesis, inheritance of acquired traits, blood/blending inheritance, epigenesis, germplasm theory, genetic perspective.
2. Differentiate between: heredity and variation, genotype and phenotype, qualitative and quantitative traits, genotype and phenotype, genotypic and environmental variation.
3. State certain facts about life.
4. Would science of genetics have born if there would have been no variation?
5. Describe various theories of Inheritance put forth time to time. Which one of them is close to the present known mechanism of inheritance?
6. What do you understand by continuity of life?
7. Write brief history of genetics.
8. What do you understand by genetic analysis? Why is genetic analysis considered a pre-requisite for crop and animal improvement?.
9. What are certain conditions the organism used in genetics experiments should fulfil?
10. What are the areas which science of genetics touches upon?
11. What do you understand by genetic perspective?

2

PRINCIPLES OF INHERITANCE

A satisfactory answer to the mechanics of inheritance came from the work of an Austrian monk Gregor Johann Mendel. He provided first proof for a theory that explained heredity by transmission of units in the reproductive cells. Thus he put an end to false notions such as those concerning bloods. Two principles deduced from his work are known as (1) principle of "segregation" and (2) principle of "independent assortment" (now preferably called "independent segregation"). The significance of his work was not appreciated until 34 years after his work was published in 1866, when his work was rediscovered independently by three workers in 1900. Because of his work, Mendel is known as **"Father of Genetics"**.

Brief Life Sketch of Mendel (1822-1884)

Mendel was born in Silisoan, a village in Heinzendorf in Austria. Mendel joined an Augustinian monastry of St. Thomas at Brunn (now in Czechoslovakia) in 1843. In 1847 he was made a priest in the monastry. He was sent to the University of Vienna as a teacher of science in 1854 where he served for 14 years. In 1856, he observed occurrence of two types of seeds in peas growing in his monastry. He became interested in them and set out to study the effect of hybridization between different varieties. He took 2 years to select different varieties of the plant. He concentrated on only 7 pairs of varieties. All the selected varieties had pure lines. After completion of his experiments he formulated generalizations which were presented at two meetings of the "Natural history Society of Brunn" in 1865. His findings were published in the "Proceedings of Brunn Natural Science Society" in 1866. For further details on life of Mendel, readers may refer to Iltis (1932).

Mendel's Working Method

Mendel avoided complexities that had troubled earlier workers by studying each trait and record of the progeny and parents individually. He focussed his attention to a single trait at a time - flower colour, for example. When behaviour of each single trait was established he then studied two traits simultaneously - such as flower colour and vine length (height). He established **pure lines** of the plants for the characters he studied. Mendel crossed two plants artificially different in a pair of contrasting traits and observed the appearance of the first hybrid, or "F_1", generation. He then crossed the hybrid plants together and raised a second filial generation, "F_2". In F_2 he counted the number of plants possessing each of the contrasting traits in which the parents differed. From his observations in F_1 and F_2 generations, he formulated his hypothesis. He tested his hypothesis experimentally by selfing all the F_2 plants to obtain third filial generation, "F_3". He studied seven traits. The contrasting pairs of phenotypes are given in **Fig. 2.1**. He studied these traits singly through **monohybrid crosses** and two traits together through **dihybrid crosses**.

Monohybrid Crosses

The results of Mendel's experiments involving monohybrid crosses in garden Pea (*Pisum sativum*) upto F_2 generation are given in **Table 2.1**. The following **generalizations** were drawn from the results of the monohybrid crosses: (1) When a pea plant showing one phenotype is crossed with a plant having another contrasting phenotype for the same trait, the F_1 generation consisted of seeds all having phenotype same as one of the two parents. The phenotype that appears in F_1 generation is termed as **dominant** while the other that does not appear in F_1 generation is termed as **recessive**. F_1 individual is called **hybrid**. (2) The **reciprocal crosses** between the parents produced the same results. (3) When the plants of F_1 were allowed to undergo self-pollination, the recessive trait which had remained unexpressed in F_1 generation reappeared in F_2 generation. The F_2 progeny consisted of dominant and recessive phenotype in ratio 3:1. This is known as **monohybrid phenotypic ratio**. (4) When plants of F_2 generation were self pollinated, the individuals with recessive phenotypes would always breed true. However, only one third among the dominant were found to be pure and the remaining two thirds repeated the ratio of 3 dominant: 1 recessive in F_3 generation.

Principles of Inheritance 11

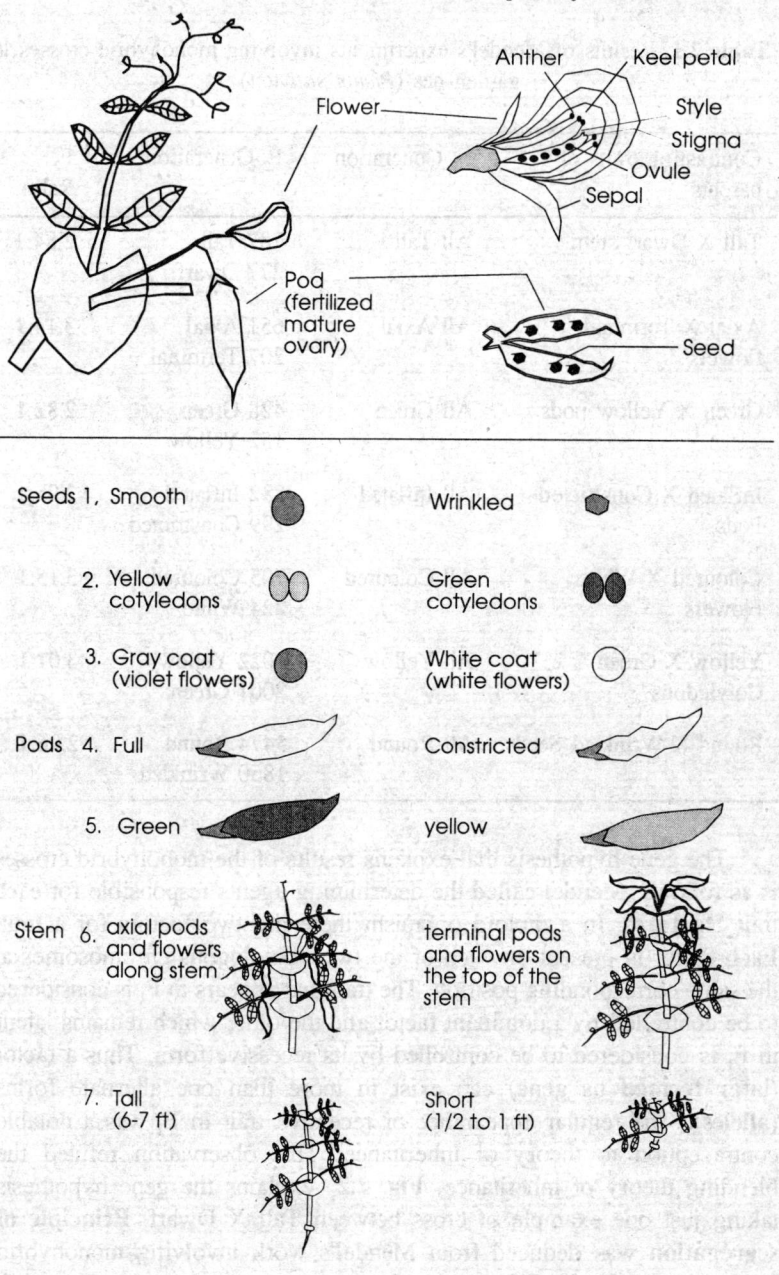

Seeds	1. Smooth	Wrinkled	
	2. Yellow cotyledons	Green cotyledons	
	3. Gray coat (violet flowers)	White coat (white flowers)	
Pods	4. Full	Constricted	
	5. Green	yellow	
Stem	6. axial pods and flowers along stem	terminal pods and flowers on the top of the stem	
	7. Tall (6-7 ft)	Short (1/2 to 1 ft)	

Fig. 2.1. Seven characters in Peas studied by Mendel

Basic Genetics

Table 2.1. Results of Mendel's experiments involving monohybrid crosses in garden pea (*Pisum sativum*)

Contrasting traits of parents	F_1 Generation	F_2 Generation	F_2 Ratio
Tall X Dwarf stem	All Tall	787 Tall 277 Dwarf	2.84:1
Axial X Terminal flowers	All Axial	651 Axial 207 Terminal	3.14:1
Green X Yellow pods	All Green	428 Green 152 Yellow	2.82:1
Inflated X Constricted Pods	All Inflated	882 Inflated 299 Constricted	2.95:1
Coloured X White Flowers	All Coloured	705 Coloured 224 White	3.15:1
Yellow X Green Cotyledons	All Yellow	6022 Yellow 2001 Green	3.01:1
Round X Wrinkled Seeds	All Round	5474 Round 1850 Wrinkled	2.96:1

The gene hypothesis that explains results of the monohybrid crosses is as follows: Mendel called the determining agents responsible for each trait **"factors"**. In a diploid organism there are two factors for a trait. Each factor is present on each of the two homologous chromosomes at the same corresponding position. The trait that appears in F_1 is considered to be controlled by a dominant factor and the other which remains latent in F_1 is considered to be controlled by its recessive form. Thus a factor (**later termed as gene**) can exist in more than one alternate forms (**alleles**). The regular appearance of recessive trait in F_2 was a notable contraception to theory of inheritance. This observation refuted the blending theory of inheritance. **Fig. 2.2** explains the gene hypothesis taking just one example of cross between Tall X Dwarf. **Principle of segregation** was deduced from Mendel's work involving monohybrid crosses. Accordingly, The two members of a gene pair separate from each other during meiosis, so that one-half of the gametes carry one member

Principles of Inheritance 13

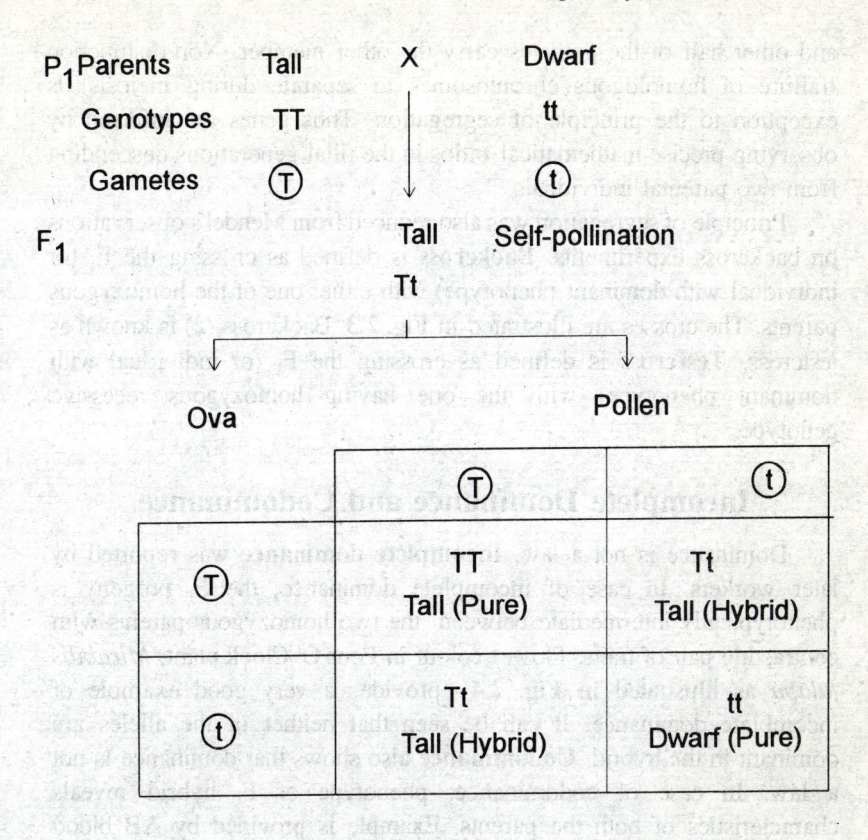

P₁ Parents   Tall   X   Dwarf
Genotypes   TT   tt
Gametes   (T)   (t)

F₁   Tall   Self-pollination
                      Tt

Ova   Pollen

	(T)	(t)
(T)	TT Tall (Pure)	Tt Tall (Hybrid)
(t)	Tt Tall (Hybrid)	tt Dwarf (Pure)

F₂   Monohybrid genotypic ratio

1 TT   :   2 Tt   :   1 tt

Monohybrid phenotypic ratio

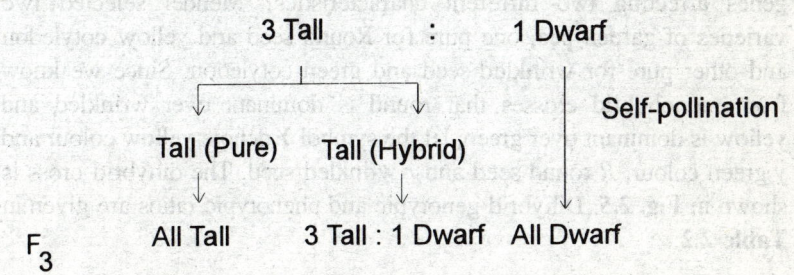

3 Tall   :   1 Dwarf

                                        Self-pollination

Tall (Pure)   Tall (Hybrid)

F₃   All Tall   3 Tall : 1 Dwarf   All Dwarf

Fig. 2.2. Gene hypothesis for a monohybrid cross

and other half of the gametes carry the other member. Non-disjunction (failure of homologous chromosomes to separate during meiosis) is exception to the principle of segregation. Thus genes are inferred by observing precise mathematical ratios in the filial generations descending from two parental individuals.

Principle of segregation was also reduced from Mendel's observations on backcross experiments. **Backcross** is defined as crossing the F_1 (or individual with dominant phenotype) with either one of the homozygous parents. The crosses are illustrated in **Fig. 2.3**. Backcross (2) is known as testcross. **Testcross** is defined as crossing the F_1 (or individual with dominant phenotype) with the one having homozygous recessive genotype.

Incomplete Dominance and Codominance

Dominance is not a law. **Incomplete dominance** was reported by later workers. In case of incomplete dominance, the F_1 progeny is phenotypically intermediate between the two homozygous parents with contrasting pair of traits. Flower colour in Four O' Clock plant, *Mirabilis jalapa* as illustrated in **Fig. 2.4** provides a very good example of incomplete dominance. It can be seen that neither of the alleles are dominant in the hybrid. **Codominance** also shows that dominance is not a law. In case of codominance, phenotype of F_1 hybrid reveals characteristics of both the parents. Example is provided by AB blood group of ABO system.

Dihybrid Crosses

What happens when a dihybrid cross is made involving two pairs of genes affecting two different characteristics? Mendel selected two varieties of garden pea, one pure for Round seed and yellow cotyledon and other pure for wrinkled seed and green cotyledon. Since we know from monohybrid crosses that round is dominant over wrinkled and yellow is dominant over green, let the symbol Y denote yellow colour and y green colour, R round seed and r wrinkled seed. The dihybrid cross is shown in **Fig. 2.5**. Dihybrid genotypic and phenotypic ratios are given in **Table 2.2**.

Principles of Inheritance

F₁ Tall   Tall
    Tt   TT
    ↓   ↓
    Ova   Pollen
  (T)   (t)   (T)

                TT   Tt
                Tall   Tall

Genotypic ratio   1 TT   :   1 Tt
Phenotypic ratio   All Tall

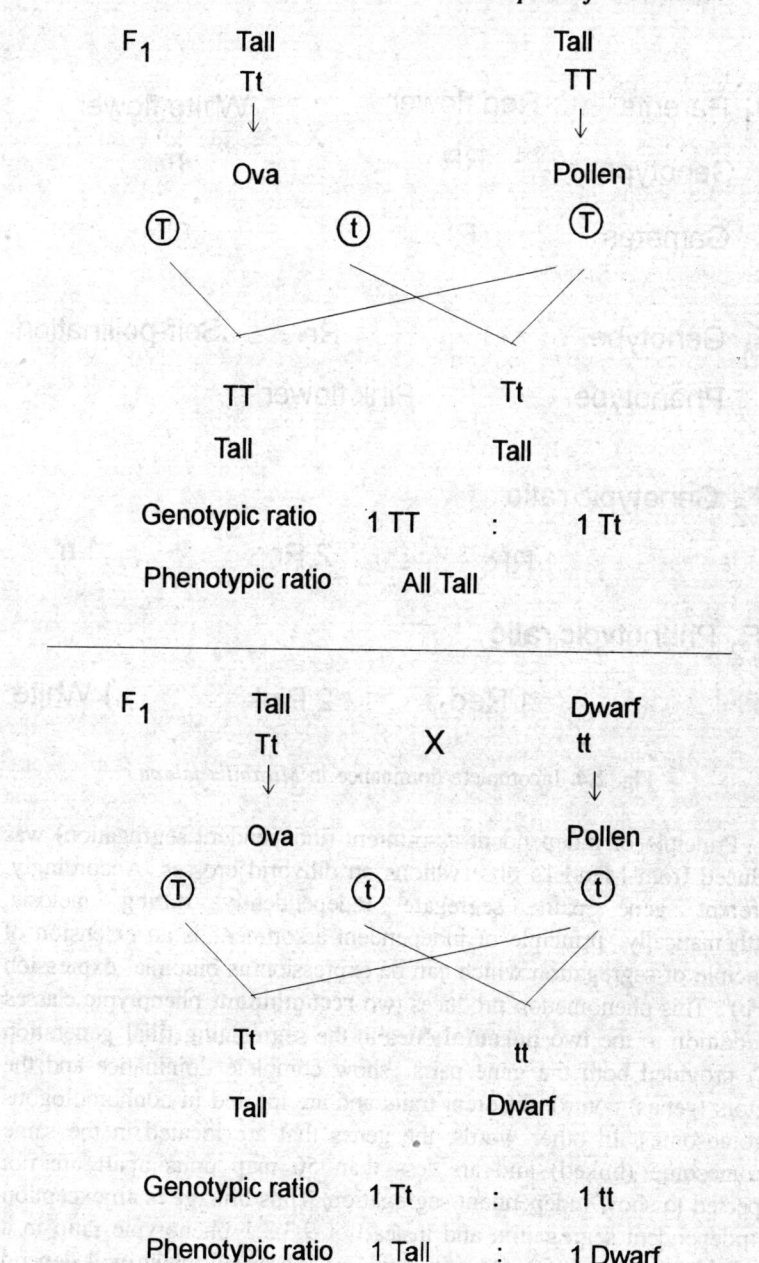

F₁ Tall   X   Dwarf
    Tt   tt
    ↓   ↓
    Ova   Pollen
  (T)   (t)   (t)

                Tt   tt
                Tall   Dwarf

Genotypic ratio   1 Tt   :   1 tt
Phenotypic ratio   1 Tall :   1 Dwarf

Fig. 2.3. Gene hypothesis for a monohybrid back cross and test cross

16 Basic Genetics

P₁ Parents   Red flower   X   White flower

Genotypes   RR   rr

Gametes   ®   ⓡ

F₁ Genotype   Rr   Self-pollination

Phenotype   Pink flower

F₂ Genotypic ratio

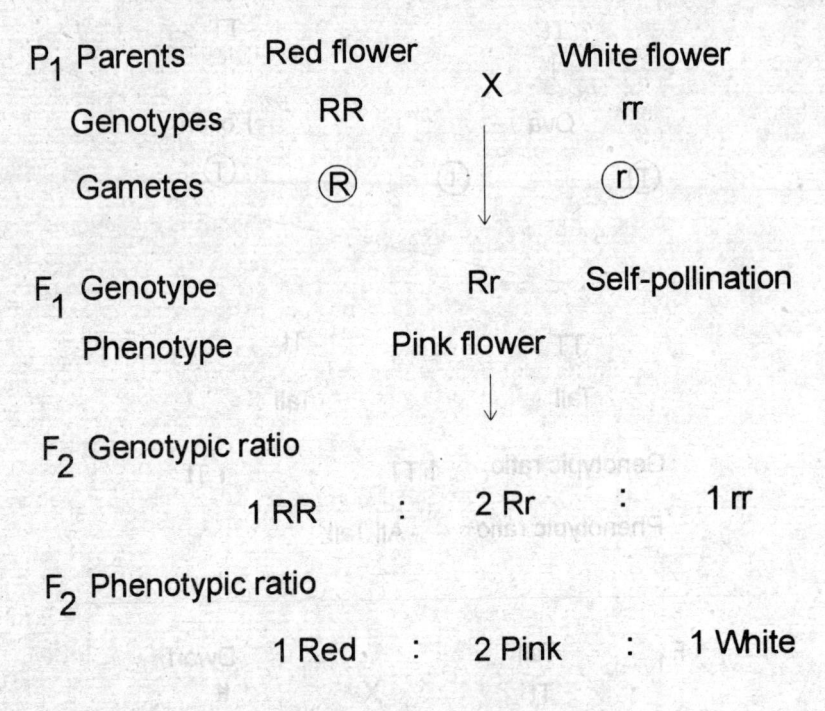

1 RR   :   2 Rr   :   1 rr

F₂ Phenotypic ratio

1 Red   :   2 Pink   :   1 White

Fig. 2.4. Incomplete dominance in *Mirabilis jalapa*

Principle of independent assortment (independent segregation) was deduced from Mendel's observations on dihybrid crosses. Accordingly, different gene pairs segregate independently during meiosis. Mathematically, principle of independent assortment is an extension of principle of segregation which can be expression as binomial expression $(a+b)^n$. This phenomenon produces two **recombinant** phenotypic classes in addition to the two **parental** ones in the segregating filial generation (F₂) provided both the gene pairs show complete dominance and the factors (genes) control different traits and are located in nonhomologous chromosomes. In other words, the genes that are located in the same chromosome (linked) and are less than 50 map units apart are not expected to show independent segregation. Thus linkage is an exception to independent segregation and instead of 9:3:3:1 phenotypic ratio in a dihybrid cross a different ratio will be observed which will depend

Principles of Inheritance 17

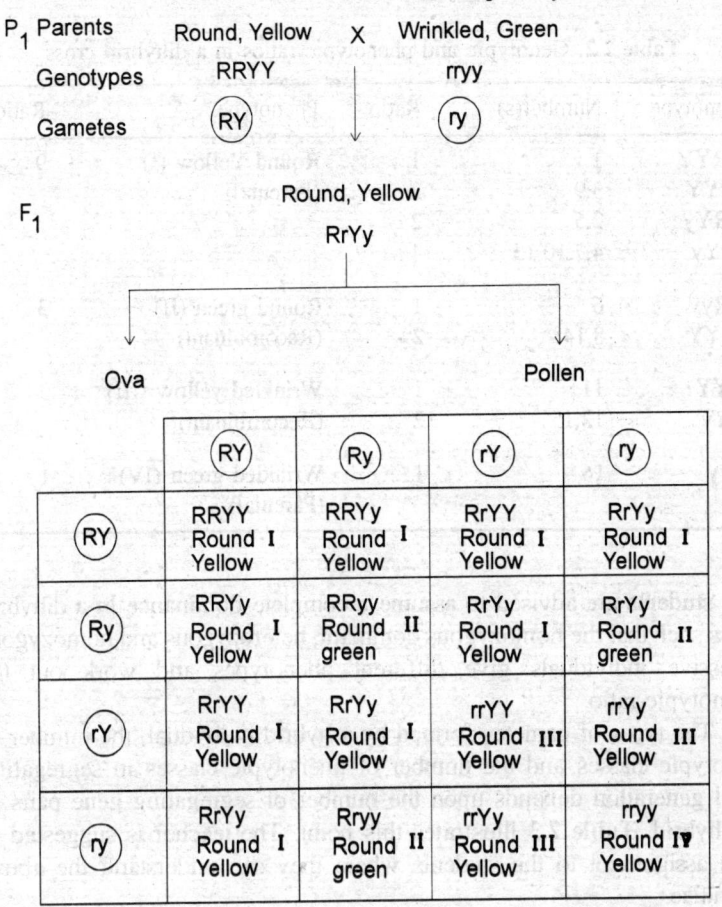

P₁ Parents	Round, Yellow	×	Wrinkled, Green
Genotypes	RRYY		rryy
Gametes	(RY)		(ry)

F₁ Round, Yellow
RrYy

Ova / Pollen

	RY	Ry	rY	ry
RY	RRYY Round I Yellow	RRYy Round I Yellow	RrYY Round I Yellow	RrYy Round I Yellow
Ry	RRYy Round I Yellow	RRyy Round II green	RrYy Round I Yellow	RrYy Round II green
rY	RrYY Round I Yellow	RrYy Round I Yellow	rrYY Round III Yellow	rrYy Round III Yellow
ry	RrYy Round I Yellow	Rryy Round II green	rrYy Round III Yellow	rryy Round IV Yellow

Dihybrid F₂ genotypic and phenotypic ratios are summarised in Table 2.2.

Fig. 2.5. Gene hypothesis for a dihybrid cross

upon degree of linkage. Principle of independent segregation can also be understood from backcrosses on the dihybrids. Independent segregation is important means for generation of new genotypic and hence phenotypic combinations. Test cross is generally used to determine (a) whether an individual of dominant phenotype phenotype is homozygous or heterozygous and (b) degree of linkage. Even when different gene pairs show incomplete dominance or codominance, the segregation is independent and the genotypic ratio remains the same but the number of phenotypes, however, increases.

18   Basic Genetics

Table 2.2. Genotypic and phenotypic ratios in a dihybrid cross

Genotype	Number(s)	Ratio	Phenotype	Ratio
RRYY	1	1	Round Yellow (I)	9
RrYY	3,9	2	(Parental)	
RRYy	2,5	2		
RrYy	4,7,10,13	4		
RRyy	6	1	Round green (II)	3
RrYY	8,14	2	(Recombinant)	
rrYY	11	1	Wrinkled yellow (III)	3
rrYy	12,15	2	(Recombinant)	
rryy	16	1	Wrinkled green (IV) (Parental)	1

The students are advised to assume incomplete dominance in a dihybrid cross such that the homozygous dominant, heterozygous and homozygous recessive individuals give different phenotypes and work out the phenotypic ratio.

The types of gametes formed by a hybrid individual, the number of genotypic classes and the number of phenotypic classes in segregating filial generation depends upon the number of segregating gene pairs in the hybrid. **Table 2.3** illustrates this point. The teacher is suggested to give assignment to the students where they can understand the above formulae.

Table 2.3. Types of gametes formed by a hybrid individual, the number of genotypic classes and the number of phenotypic classes when 'n' number of gene pairs are segregating

No. of segregating loci	Types of gametes formed by F_1 hybrid	No. of genotypic classes in F_2	No. of phenotypic classes in F_2 in case of	
			Complete dominance	Incomplete dominance
n	2^n	3^n	2^n	3^n

Rediscovery of Mendelism

Mendelism was rediscovered independently in the year 1900 by three persons, namely, C. Correns of Germany, Hugo deVries of Holland and E. von Tschermak of Austria. This event went a long way in understanding Mendelism in its real sense as by that time details of cell division and chromosome structure were known. This rediscovery marks the beginning of science of formal genetics, also known as classical or transmission genetics.

Impact of Mendelism and its Rediscovery

An important consequence of Mendel's work and its subsequent rediscovery was to replace the blending theory of inheritance with a **particulate theory** of inheritance. Mendel's observation that one of the parental characteristic was absent in F_1 hybrid and reappeared in unchanged form in the F_2 generation was inconsistent with blending theory. It was concluded from Mendel's work that traits from the parental lines were transmitted as two different element of particulate nature that retained their purity in the F1 hybrids. This thus gave rise to most believable particulate theory of inheritance.

Concept of Dominance

Earlier a gene was considered dominant if it was expressed in the heterozygous condition and recessive if it was masked. Sometimes heterozygous individuals for a particular locus reveals an intermediate phenotype. For example, bar-eyed heterozygote *Drosophila* individuals have a kidney-shaped eye which is different from either of the homozygous parents. Thus heterozygous individuals in this case exhibit incomplete dominance rather than a simply dominant or recessive situation. The following example will provide futher insight on concept of dominance.

Inheritance of a Temperature-sensitive Lethal Gene

Inheritance of a temperature-sensitive lethal gene at different temperatures shows that dominance-recessiveness relationship is not a property of a gen, it is its expression in a particular environment (**Fig. 2.6**). At 18°C, the heterozygous in identical to both parents and no dominance relationship exists. At 20°C, a^+ is dominant a and at 27°C,

P₁   X   P₂
a^+a^+ $a\,a$

This genotype is functional at 18, 20 and 27 °C

This genotype is functional at 18, but dies at 20 and 27 °C

F₁
$A\,a$

This genotype is functional at 18 and 20 but dies at 27 °C

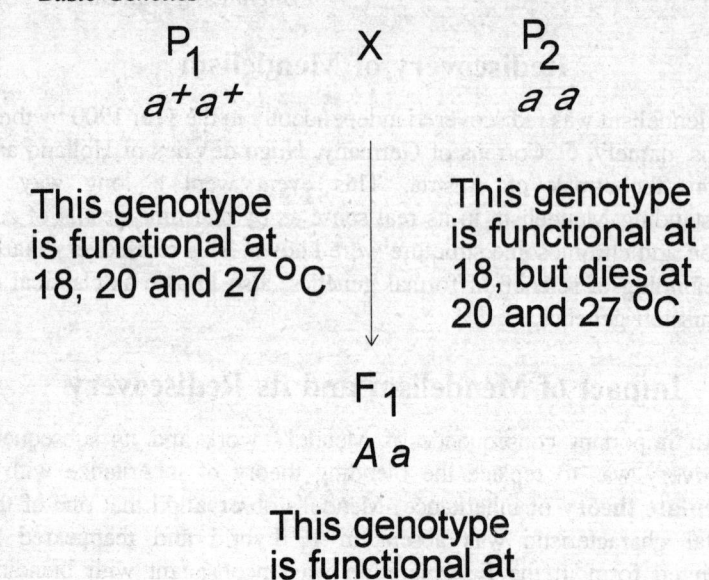

Fig. 2.6. Inheritance of a temperature-sensitive lethal gene

a is dominant over a^+. Thus we see that we can't classify these genes simply dominant or recessive without referring to the condition(s) at which the relationships exist.

It we consider alleles from a functional perspective, a clearer understanding of dominant and recessiveness emerges. Assume that a gene $a^+ \rightarrow$ an enzyme that converts a substrate "X" to a product "Y". Suppose a mutation changes $a^+ \rightarrow a$, various consequences may be: (i) gene a produces a nonfunctional enzyme or none at all. In heterozygous condition (a^+a), if $a+$ can still produce enough of normal enzyme, the organism may show normal phenotype (as done by a^+a^+). The organism of genotype aa, the reaction "X" $\rightarrow$ "Y" will not occur and mutant phenotype will result. Such a mutant gene (a) will be classified as recessive, with a^+ as completely dominant over a. (ii) gene a produces a non-functional enzyme or none at all. In heterozygous condition (a^+a), a^+ gene may not produce enough of normal enzyme. Here reaction "X" to "Y" may proceed slowly. Here an intermediate phenotype may be produced. Thus a^+ will be incompletely dominant over a gene. (iii) gene *a* produces an enzyme that has a greater affinity for substrate "X" than does the a^+ gene product. Enzyme produced by gene *a* converts "X" to a new product "Y_2" which causes a new phenotype to appear in the heterozygote that is identical to that of homozygous a phenotype. Therefore, gene *a* may be regarded as dominant. Dominance of a over a^+ may also arise when the a gene product inactivates the normal a^+ gene product.

white Gene in *Drosophila*

Classification of alleles as dominant or recessive ultimately depends on the function of the gene products produced and level of analysis applied to the phenotypes in question. For example, in the analysis of a white-eyed heterozygous female *Drosophila* (w^+/w), a number of conclusions are possible, depending on how the phenotype is analysed: (a) When eye colour is visually inspected, w^+/w heterozygous female has phenotype same as wildtype (w^+/w^+) female fly. Here w^+ allele is dominant over w. (b) Pteridine pigment analysis reveals that w^+/w ♀ has quantitatively more pigments than wildtype (w^+/w^+) and w/w females. This shows overdominance. (c) Spermatheca analysis shows different dominance-recessiveness relationship. w^+/w female reveals structural abnormalities less extreme than those of white-eyed female. This indicates incomplete dominance. Basically, then, dominance is often also a function of level of analysis employed. The point here is that dominance and recessiveness have meaning only when discussed in terms of gene products.

Locus A has two alleles A1 and A2 and three genotypes - two homozygotes, A1A1 and A2A2, and one heterozygote, A1A2.

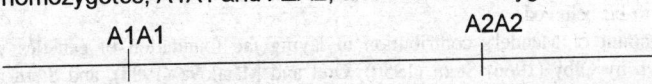

```
    A1A1                      A2A2
─────┼─────────────────────────┼─────
```

Assume that phenotype of A1A1 is P1 and that of A2A2 is P2.
(1) A1 will be dominant over A2 if phenotype of A1A2 is P1.
(2) A2 will be dominant over A1 if phenotype of A1A2 is P2.
(3) The two alleles will show incomplete dominance if phenotype of A1A2 ranges between P1 and P2.
(4) The two alleles will show codominance if phenotype of A1A2 shows expression of both the alleles.
(5) The two alleles will show overdominance if phenotype of A1A2 is beyond the range of P1 and P2.

Fig. 2.7. Scale of dominance

Scale of Dominance

Scale of dominance is represented in **Fig. 2.7** Let us consider a locus A with two alleles A_1 and A_2. Let us assume that phenotype of homozygote A_1A_1 is P_1 and that of A_2A_2 If phenotype of heterozygote A_1A_2 is (a) same as P_1, A_1 is dominant over A_2, (b) same as P_2, A_2 is dominant over A_1, (c) ranges between P_1 and P_2, A_1 and A_2 show incomplete dominance, (d) exhibiting properties of both P_1 and P_2 phenotypes, codominance is the result, (e) beyond the range $P_1 - P_2$, overdominance is said to be existing between alleles A_1 and A_2.

Some Useful Reading Hints

Mendel's paper (1886) originally published in the Proceedings of the Brünn Natural History Society were translated into English under the title "Experiments in Plant Hybridization" and reprinted in the collections of Peters (1959). Correns (1909) reporteds his hybridisation experiments in *Mirabilis jalapa*. Punnett (1911) explained as to how by use of a checker board we can find out diploid genotypic combinations resulting from random union of male and female gametes.

Complete life sketch of Mendel was given by Iltis (1932). Bateson (1909) dealt with Mendel's principles of heredity as they seemed to be after rediscovery of Mendel's work. Fisher (1936) questions whether Mendel's work was rediscovered. Tschermak-Seysenegg (1951) gaves an details of the rediscovery of Mendel's work. Corcos and Monaghan (1985) discussed role of deVries in the rediscovery of Mendel's work and questions whether deVries was an indepenent disoverer of his work. Corcos and Monaghan (1987) debateed whether Correns was an independent discoverer of Mendelism? Some workers cast doubt on the authencity of Mendel's work. Douglas and Novitski (1977) dealt with the question: What chance did Mendel's experiments give him of notice linkage? Pilgrim (1986) commented on Mendel's work as a solution to the too-good-to-be-true paradox. Weiling (1971) considered Mendel's data in *Pisum* experiments "too good" to be believed.

An account of Mendel's contribution in laying the foundation of genetics was penned down by Olby (1966) Stern (1950), Orel and Matalova (1983), and Stern and Sherwood (1966). Rodgers (1991) pointed out linkage as a mechanism Mendel never knew.

East (1936) and Gowen (1952) dealt with genetic basis of heterosis. Trehan and Gill (1987) concluded sub-unit interation as a molecular mechanism of heterosis.

Questions

1. Why is Mendel known as Father of genetics?
2. Give a brief life sketch of Mendel.
3. Describe the methodology Mendel used to work out the mechanism of inheritance.

Principles of Inheritance

4. Define: monohybrid cross, dihybrid cross, incomplete dominance, codominance, principle of segregation, principle of independent assortment (independent segregation).
5. Describe Mendel's observations on the monohybrid crosses.
6. Do incomplete dominance and codominance provide exceptions to Mendel's generalizations? Give reasons for your answer.
7. What generalizations did Mendel draw from his experiments involving dihybrid crosses?
8. Who rediscovered Mendelism? What was the impact of Mendelism and its rediscovery?
9. Principle of independent assortment (independent segregation) is just an expansion of principle of segregation. Comment.
10. Is dominance a law? Give experimental evidences to support your answers.
11. What insight does the experiment on temperature-sensitive lethal mutations provide about dominance?
12. Dominance-recessive relationship between two alleles may vary with the level of analysis. Illustrate this point through the example of *white* gene in *Drosophila*.
13. What do you understand by scale of dominance?

Numericals

1. In peas, tall plant is dominant over dwarf. (a) If a plant homozygous for tall is crossed with one homozygous for dwarf, what will be the appearance of F_1; of the F_2; of the offspring of a cross of F_1 with its tall parent; with its dwarf parent? (b) Let the allele for tall be represented by T and the allele for dwarfness by t. What will be the gametes produced by the parents and the height of the offspring (tall or dwarf) from each of the following crosses: $Tt \times tt$; $TT \times tt$?.
2. In four'O clock flowers, red flower colour R is incompletely dominant over white r, the heterozygous plant being pink flowered. (a) In the following crosses, in which the genotypes of the parents are given, what are the gametes produced by each parent and what will be the flower colour of the offspring from each cross $Rr \times RR$; $rr \times Rr$; $RR \times rr$; $Rr \times Rr$? (b) How would you produce four'O clock seeds all of which would yield pink flowered plants when sown?
3. In cattle hornless condition P is dominant over the normal p. A certain hornless bull is bred to true cows. With cow A, which is horned, a hornless calf is produced; with cow B, also horned, a normal calf is produced; with cow C, which is hornless, a horned calf is produced. What are the genotypes of the four parents, and what further offspring would you expect from these matings.
4. A woman has a rare abnormality of the eyelids called ptosis, which makes it impossible for her to open her eyes completely. The condition has been found to depend on a single dominant gene (P). The woman's father had ptosis, but her mother had normal eyelids. Her father's mother had normal eye lids. (a) What are the probable genotypes of the woman, her father, and the mother? (b) What proportion of her children will be expected to have ptosis if she marries a man with normal eye-lids.

24 Basic Genetics

5. Albinism in humans is controlled by a recessive gene (c) From marriages between normally pigmented people known to be carriers (Cc) and albinos (cc), (a) what proportion of the children would be expected to be albinos? (b) what is the chance that any pregnancy would result in an albino child? (c) what is the chance in a family of three that one would be normal and two albino?

6. If both the partners were known to be carriers (Cc) for albinism, what is the chance of following combinations in families of four: (a) all four normal (b) three normal and one albino (c) two normal and two albino (d) one normal and three albino.

7. Phenylkelonuria in humans is caused by a recessive allele p. If both partners are known to be carriers (Pp), what is the chance in the following combinations with five children that (a) all are normal (b) four are normal and one is affected (c) three are normal and two are affected (d) two are normal and three are affected (e) one is normal and four are affected (f) all are affected.

8. In poultry, feathered legs (F) are dominant over clear legs (f) and peacomb (P) over single comb (p). Two cocks A and B were bred to two hens C and D. All four birds were feathered legs and pea combed. The crosses A X C and A X D yielded all feathered and pea combed birds, cross B X C produced feathered and clean legged but all pea combed birds. Cross B X D produced all feathered birds but segregation for pea combed and single combed. Find the genotypes of four birds.

9. In rabbits, black fur is dependent on a dominant gene (B) and brown on the recessive allele (b). Normal length fur is determined by dominant gene (R) and short (rex) by the recessive allele (r). (a) Diagram and summarise the F_1 and F_2 result of a cross between a homozygous black rabbit with normal length fur and a brown rex rabbit. (b) What proportion or the normal black F_2 rabbits from the above cross may be expected to be homozygous for both gene pairs? (c) Diagram and summarize a testcross between the F_1 and the fully recessive brown rex parent.

10. Hornless, or polled, condition in cattle is due to a completely dominant gene, P normally horned cattle being pp. The gene for red colour (R) shows no dominance to that for white (R), the hybrid (RR') being roan colour. Assuming independent segregation, give the genotypic and phenotypic expectations from the following matings: (a) PP RR X pp R' R', (b) PP RR' X pp R'R', (c) PP RR' X PP R R' and (d) hornless roan (whose mother was horned) X horned white.

3

CELL DIVISION AND ITS SIGNIFICANCE

Cell Cycle

Cell cycle is a cyclic process in which a cell at the interphase stage passes through a sequence of events to divide into two daughter cells to enter again into interphase stage. Hence cell cycle consists of two periods - interphase and period of divisions which alternate with each other.

For many years, the cytologists were primarily concerned with the period of divisions, because they were able to observe the changes in the behaviour of chromosomes under the light microscope. The **interphase** was considered a resting phase. Later on it was realized that cells spend most of their life span in interphase which is a period of intense biosynthetic activity in which cell doubles in size and replicates its chromosomes precisely.

The basic outline of the cell cycle is the same in all the organisms. Time taken for each stage and entire cell cycle differ in different species. The most important biochemical activity that a cell undertakes in preparation for cell division is the replication of its DNA. DNA replication does not occur throughout the interphase, but at a specific time of the interphase - the **S phase**. The time interval between the end of telophase and the beginning of S phase is termed is **G_1 phase**. The interval between the completion of DNA replication (S phase) and beginning of prophase is termed as **G_2 phase**.

Duration of Cell Cycle and its Phases

The duration of cell cycle and its phases can be measured by the use of radioactive **tritiated (^{3}H) thymidine** and autoradiography. Growing roots are shown in the presence of ^{3}H thymidine for a short period (~30 minutes). Only those cells which are at S phase will incorporate the ^{3}H-26 thymidine in their DNA. It is known as **pulse labelling**. After short pulse labelling (30 minutes) roots are grown in absence of ^{3}H thymidine and root tip samples are studied after different intervals of time. Mitotic preparations are made from these root tips.

Duration of the different phases of cell cycle is worked out by counting the number of labelled dividing mitotic cells out of total number of cells. For an interval of time, no labelled mitotic figures are seen. The time interval after which the labelled prophase cells appear will be equal to the duration of G_2.

Duration
↓
Pulse labelling → Labelled appearance of Prophase

After some duration all the dividing will appear labelled. The time interval in which all the dividing cells will appear labelled will be equal to the duration of division (M).

Beginning of labelled → All dividing cells labelled
Prophase cells (Prophase, metaphase, anaphase and telophase cells)

Gradually the percentage of labelled were at G_1 phase at the time of pulse labelling enter into mitosis. Thus the time interval between the appearance of labelled prophase and their subsequent disappearance is equivalent to the duration of S phase. Gradually all the dividing cells appear unlabelled.

Subsequently labelled mitotic cells will be seen when earlier labelled cells complete interphase and reenter mitosis. Thus the duration of complete cell cycle is estimated by measuring the time interval between successive appearance of labelled cells in mitosis (**Fig. 3.1**). Duration of G_1 phase is determined by the formula G_1 = cell cycle - $(G_2 + M + S)$.

The duration of complete cell cycle and its component phases vary from cells type to cell type as shown in **Table 3.1**. G_1 have the most-variable period of cell cycle. Cells have **2C amount of DNA** during G_1 and each chromosome has only one chromatid. During G_1 preparation for chromosome duplication takes place. Quantity of nucleotides, histones, tubulins and DNA polymerase increases.

S phase lasts for 6-10 h Chromosomes replicate. Euchromatin replicates during early S phase and heterochromatin during later half of

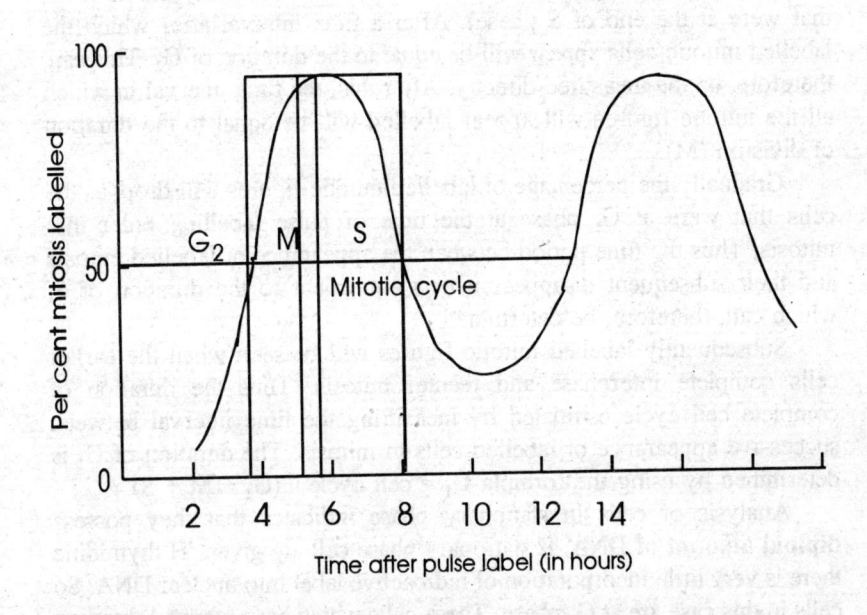

Fig. 3.1. Measuring of duration of complete cell cycle

S phase. Histones are also synthesized. Whereas a typical cells of adult *Drosophila* will take ∼ 10 h to replicate its DNA, the same amount of DNA in the nucleus of an early embryonic cell will be replicated in only 3 minutes. All the replicases are active at the same time in embryonic cells while in adult cells only a small percentage of **replicons** are active

Table 3.1. Duration of complete cell cycle and its component phases in different cell types

Cell type	Average cell cycle	G_1	S	G_2	Mitosis
Lateral root-tips of *Vicia faba*	∼ 14 h at 22°C	2½ h	6 h	3½ h	2 h
Mouse fibroblasts	∼ 22.5 h	9 h	10 h	2½ h	1 h

at any given time. For a period equal to the duration of G_2 no labeled mitotic figures will be seen. First labelled cells to reach mitosis will be those cells completing DNA synthesis in presence of ^{3}H-thymidine (cells that were at the end of S phase). After a time interval after which the labelled mitotic cells appear will be equal to the duration of G_2. This can, therefore, be the measured directly. After this, the time interval in which all the mitotic figures will appear labelled will be equal to the duration of division (M).

Gradually the percentage of labelled mitotic figures will drop, as the cells that were at G_1 phase at the time of pulse labelling enter into mitosis. Thus the time period between the appearance of labelled mitosis and their subsequent disappearance is equivalent to the duration of S, which can, therefore, be determined.

Subsequently labelled mitotic figures will be seen when the earlier cells complete interphase and reenter mitosis. Thus the duration of complete cell cycle estimated by measuring the time interval between successive appearance of labelled cells in mitosis. The duration of G_1 is determined by using the formula G_1 = cell cycle - (G_2 + M + S).

Analysis of cells in stationary phase indicates that they possess **diploid amount of DNA**. If stationary phase cell are given ^{3}H thymidine there is very little incorporation of radioactive label into nuclear DNA. So cells in this case are at G_1 phase. These cells which have stopped dividing and are looked up in this state are said to lack in G_0 state to distinguish them from typical G_1 cells that well soon enter S phase G_0 and G_1 have different physiological properties.

Thus the most important decision concerning whether or not a cell will continue to divide is made in G_1 phase, and that division is whether or not to start DNA synthesis. Cells in G_1 must receive a **positive signal** to proceed into S phase.

Control of Cell Cycle

Nuclear transplantation and cell fusion experiments indicate that the cytoplasm from a cell actively engaged in DNA synthesis (amphibian egg or the HeLa cell) is capable of activating DNA synthesis in a nucleus that was no longer synthesizing DNA. Thus activation of DNA synthesis results from the positive action of some type of cytoplasmic factor(s).

Similarly, the transition from G_2 to M requires the presence of some type of factor(s). This conclusion is based on the studies in which mitotic cells are fused with cells in other stages of the cell cycle.

Cells at G_1 fused with cells at S → nucleus of G_1 cells is activated to begin DNA replication in advance

Cells at G_2 fused with cells at S → each nucleus continued with its previous activity

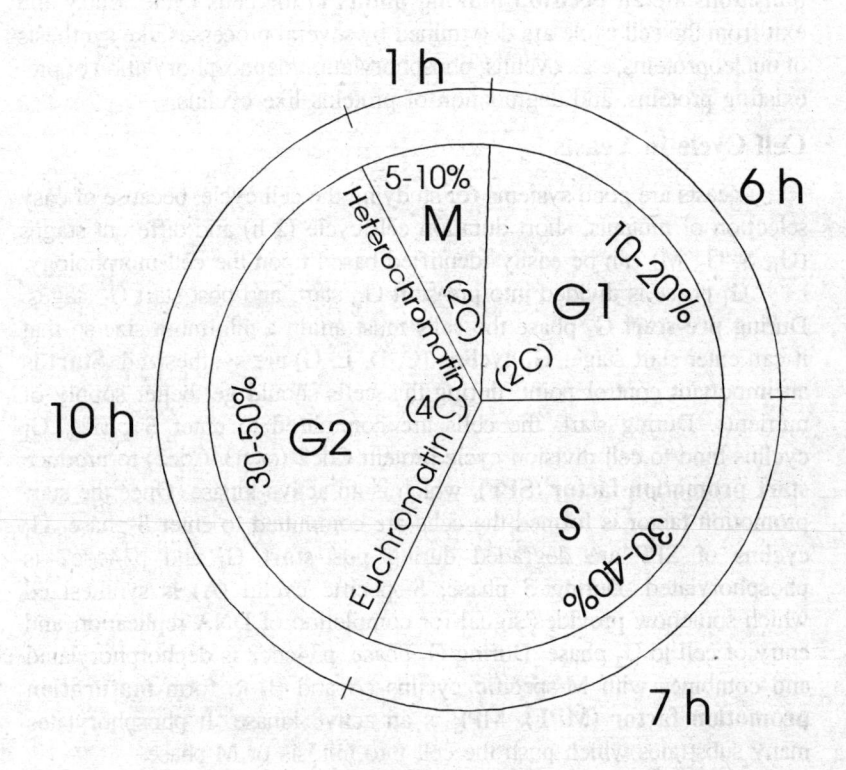

Fig. 3.2. Representative cell cycle

In multicellular eukaryotes, a single celled zygote through process of cell division and cellular differentiation gives rise to a multicellular organism. This holds or all plants and animals, including man. Cells arise from pre-exiting cells by a process known as **cell reproduction**. During multiplication, every cell passes through a cycle of changes referred to as cell cycle. Cell cycle is represented in **Fig. 3.2**. A cell cycle consists of two main states - interphase and mitotic phase.

Biochemistry and Genetics of Cell Cycle

Eukaryotes from yeast to humans have similar cell division cycles. Four cell cycle stages have been identified: G_1, S, G_2 and M. Differentiated mature and non-dividing cells are said to be at G_0 stage. Earlier G_0 to G_1 transition was found to be a crucial stage in cell cycle. However, later on it was found that G_0/G_1, G_1/S, S/G_2, G_2/M, M/G1 transitions are all **decision making points** in the cells cycle. Entry and exit from the cell cycle are determined by several processes like synthesis of nucleoproteins, e.g., cyclins, phosphorylation/dephosphorylation of pre-existing proteins, and degradation of proteins like cyclins.

Cell Cycle in Yeasts

Yeasts are good systems for studying the cell cycle, because of easy selection of mutants, short duration cell cycle (2 h) and different stages (G_1, S, G_2, M) can be easily identified based upon the cell morphology.

G_1 phase is divided into pre-start G_1, start, and post-start G_1 stages. During **pre-start G_1** phase the cells must attain a minimum size so that it can enter start stage. **G_1 cyclins** (C, D, E, G) are synthesized. **Start** is an important control point, during this cells should get better supply of nutrients. During start, the cells are committed to enter S phase. G_1 cyclins bind to cell division cycle protein Cdc2 (or p34Cdc2) to produce **start promotion factor (SPF)**, which is an active kinase. Once the start promotion factor is formed the cells are committed to enter S-phase. G_1 cyclins of SPF are degraded during **post-start** G_1 and p34cdc2 is phosphorylated. During S phase, S-specific cyclin (A) is synthesized which somehow provides signal for completion of DNA replication and entry of cell to G_2 phase. During G_2 phase, p34cdc2 is dephorphorylated and combines with M-specific cyclins (A and B) to form **maturation promotion factor (MPF)**. MPF is an active kinase. It phosphorylates many substrates which push the cell into mitosis or M phase.

In yeast co-ordinated action of Cdc2 and Cdc13 cyclin is also regulated by Cdc25 and Wee1. Cdc25 is a phosphatase, which dephosphorylates Cdc2 to make it functional. Gene *Wee1* produces a kinase which phosphorylates Cdc2 exercising a negative control. Similarly, another gene *nim1* (never in mitosis) exercises a negative control on Wee1 by phosphorylation. Thus both *nim1* and *Wee1* are believed to produce kinases and *Cdc25* produces phosphatase. The phosphatase activity removes phosphate from tyrosine residue (tyr-15) of

Cdc2. An increased amount of Cdc25 causes premature entry into mitosis resulting into lethality. Increase in the ratio of Cdc25 and Wee1 produced increases the cell size required for the entry into mitosis, whereas decrease in the ratio result into decrease in critical size.

MPF initial metaphase by breaking up the nuclear envelop. It phosphorylates the **lamina proteins** of nuclear envelop. Phosphorylated lamina subunits are dissociated leading to the breakdown of nuclear envelop. MPF also induces chromosome condensation and assembly of the spindle. During anaphase, cyclin component of MPF is phosphorylated and becomes substrate for protease. Finally, the cyclin gets destroyed and Cdc2 returns to initial state. This resets the cyclin to prestart G1.

Cell Division

Interphase

This phase is 10-23 h long and it comprises of first gap (G_1), synthetic (S) and second gap (G_2) subphases. Duration of these subphases varies in different organisms. Interphase is wrongly described as resting phase as the cell is metabolically very active during this phase. G_1 phase includes initial growth of the cell and cell prepares itself for synthesis of DNA enzymes, other proteins and RNA molecules. No change occurs in the amount of DNA in the cell. During S phase, the chromosomes replicate by new DNA synthesis. Each chromatids joined together by a centromere. A diploid (2n) cell thus has double the amount (4n) of DNA. RNA and some proteins (histones) are synthesized in this phase. During G_2 phase the cell further grows and prepares itself for division. Cytoplasmic organelles (centrioles, mitochondria and **Golgi apparatus** are doubled. Proteins for apparatus are doubled. Proteins for spindle apparatus are synthesized. Cell is metabolically active and stores energy required for division of the cell.

Mitosis

As a result of mitosis, one somatic cell produces two identical daughter cells. It is an equational division. It comprises of karyokinesis (a division of nucleus) and cytokinesis (division of cytoplasm). Different events of mitosis are represented in **Fig. 3.3**.

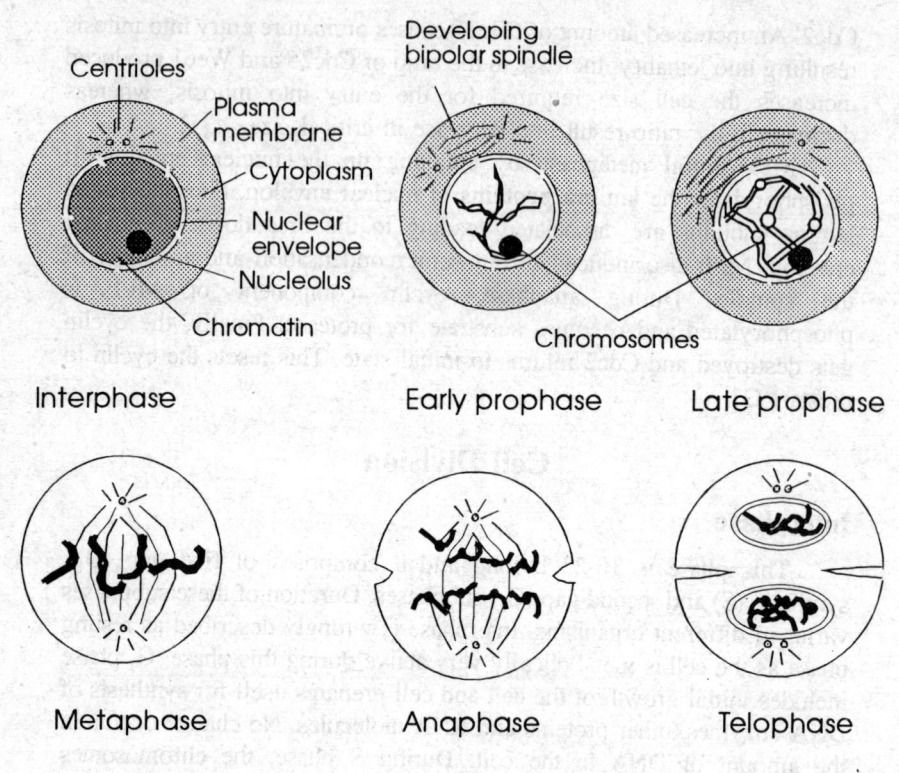

Fig. 3.3. Diagrammatic summary of mitosis in animal cells

Karyokinesis

Prophase. Chromosomes become visible. Each chromosome consists of two sister chromatids joined together by a centromere. Chromosomes shorten and thicken. Nucleoli start decreasing in size. Nuclear envelope begins to breakdown.

Metaphase. Centrioles push each other apart to lie at opposite poles. Chromosomes get arranged on the equatorial or metaphase plate. Nucleolus and nuclear membrane are not visible. Spindle apparatus is very prominent.

Anaphase. Centromere of each chromosome splits into two by replication of its DNA. Now each chromatid becomes a separate single-stranded chromosome. Sister chromatids move to opposite poles. Anaphase ends when all the chromatids reach the opposite poles.

Telophase. Nuclear envelop slowly appears round each group of chromosomes. Chromosomes become thin and long and finally they become indistinguishable. Two nuclei are seen in a single cell. Nucleoli appear again.

Cytokinesis

During division of cytoplasm equal distribution of cytoplasmic organelles is ensured. Animal cells divide their cytoplasm by **furrowing**. A circular constriction appears round the middle of the cell. Plant cells divide their cytoplasm by **cell plate** formation. It starts from the centre of the cell and grows towards the periphery.

Significance of Mitosis

It helps maintain size of the cell. A cell when full grown divides by mitosis instead of growing further. A fertilized egg produces an embryo which develops into an adult by repeated mitotic divisions. It keeps the number of chromosomes equal in all the cells of an individual. Mitosis provides new cells to replace the old worn out and dying cells. Mitosis produces new cells for healing of wounds and regeneration. It plays role in asexual as well as sexual reproduction. Mitosis being essentially similar in many kind of organisms provides evidence of basic relationship of organisms.

Meiosis

It is confined to certain types of cells (oocytes, spermatocytes and sporocytes) and takes place at a particular time. It consists of two divisions (meiosis I and meiosis II) that take place in succession. A parental cell produces four daughter cells, each having half the number of chromosomes (n) and half the amount of nuclear DNA present in the parental cell. Four product cells are genetically non-identical. Meiosis is thus a reductional division. Different phases and subphases of meiosis are shown in **Fig. 3.4** and are described here briefly.

Meiosis I

During meiosis I, the two homologous chromosomes of each pair separate from each other and go to separate daughter cells thus reduces the number of chromosomes from diploid to haploid condition. Meiosis I is, therefore, known as reductional division Meiosis I comprises of four phases - prophase I, metaphase I, anaphase I and telophase I.

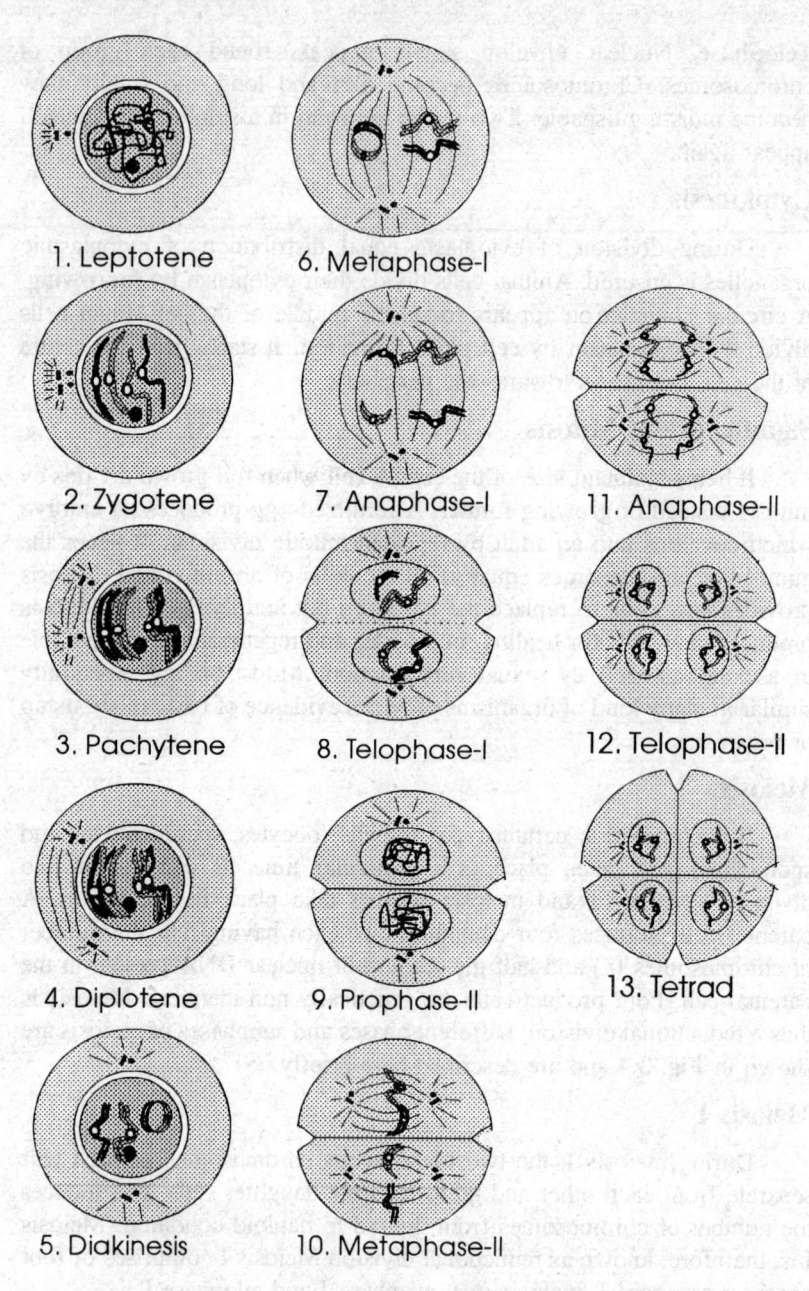

1. Leptotene
2. Zygotene
3. Pachytene
4. Diplotene
5. Diakinesis
6. Metaphase-I
7. Anaphase-I
8. Telophase-I
9. Prophase-II
10. Metaphase-II
11. Anaphase-II
12. Telophase-II
13. Tetrad

Fig. 3.4. Diagrammatic summary of different phases of meiosis

Prophase I. It may be divided into five substages - leptotene, zygotene, pachytene, diplotene and diakinesis.

Leptotene. **Centrioles** move apart. Chromosomes are thin and long. They are distinguishable individually.

Zygotene. The homologous chromosomes come to lie side-by-side. Pairing starts. The chromosomes become shorter and thicker.

Pachytene. The homologous chromosomes are completely paired. This pairing of homologous chromosomes is known as **synapsis**. A pair of homologous chromosome is known as a **bivalent**. Two chromatids of a chromosome are known as **sister chromatids** and those of **homologous chromosomes** are termed as **nonsister chromatids**. **Crossing over** occurs here.

Diplotene. Homologous chromosomes start separating. The phenomenon is known as disjunction. Separation of chromosomes does not take place at all the points as the sites where crossing over (exchange of segments of nonsister chromatids of homologous chromosomes) has already occurred are showing cross-type structure known as **chiasmata** (singular **chiasma**).

Diakinesis. The chromosomes condense further and shorter in size and become thicker. The nucleoli begin to disappear. The nuclear envelop starts breaking down.

Metaphase I. Spindle apparatus is completely formed. The pairs of homologous chromosomes are arranged in an equatorial line in the spindle apparatus, forming a **metaphase plate**. Each centromere is joined by a single spindle fibre instead of two as in mitosis.

Anaphase I. From each pair of homologous chromosomes, two chromatids joined by centromere move as a unit to one pole of a spindle and the remaining two chromatids joined by a centromere move to the opposite pole. So during anaphase I, homologous chromosomes of each pair are separated. Here real reduction in the number of chromosomes takes place.

Telophase I. The chromosomes at each pole of spindle elongate and form a nucleus with a nuclear envelope and nucleoli. The spindle and astral rays gradually disappear.

Cytoplasm divides at its middle but the two daughter cells formed from one parental cell remain intact. A brief interphase may appear between meiosis I and meiosis II in some cases. During this interphase there is no replication of chromosomes.

Meiosis II

It is an equational division. The four stages are this division are known as prophase II, metaphase II, anaphase II and telophase II. These divisions are shown in **Fig. 3.4** and are described below briefly.

Prophase II. Centrioles move apart in pairs. Spindle is layed down. Nuclear envelop breaks down. Nucleoli disappear.

Metaphase II. The chromosomes take positions at the centre of the spindle forming a metaphase plate. Each centromere is joined by two spindle fibres, one from each pole of the spindle pole.

Anaphase II. The centromere split into two each. The two chromatids of each chromosome start moving away from each other. Finally, they reach the poles of the spindle. Here they are called the chromosomes. Reduction of haploidy in terms of DNA content takes place.

Telophase II. The group of chromosomes at each end of the spindle get enclosed by a nuclear envelop. Nucleoli are layed down. Spindle are lost.

Often cytokinesis is delayed until both nuclear divisions Meiosis I and Meiosis II are completed so that four cells are simultaneously formed. The four haploid cells produced by division of a single diploid cell are genetically nonidentical.

Significance of Meiosis

Meiosis produces gametes for sexual reproduction. Meiosis halves the chromosome number in gametes and fertilization restores the original chromosome number. Crossing-over during meiosis results formation of chromosomes having new combination of hereditary units, genes. Random distribution of paternal and maternal chromosomes during anaphase I gives new combinations of chromosomes. Meiosis being similar in all sexually reproducing organisms offers further evidence of basic evolutionary relationship of living organisms.

Some Useful Reading Hints

Swanson (1957) and Swanson et al. (1981) gave a comprehensive account of cell division. Nicklas (1971) gave a general account of mitosis. Murray and Kirschner (1989) gave a unified look at dominoes and clocks, the two views of the cell cycle. Murray (1989) considered the cell cycle as cdc2 cycle. Murray and Kirschner (1991) described mechanism of control of the cell cycle. Nurse (1990) describeed a universal control of mechanism regulating onset of M-phase. Zimmerman (1981) dealt with mitosis, particularly cytokinesis. Broek et al. (1991) dealt with involvement of $p34^{cdc2}$ in establishing the dependency of S phase on mitosis. King et al. (1994) dwelt on mitosis in transition. Koshland (1994) dealt with the basics of mitosis. Strange (1992) explained how cell cycle advances. Murray (1992) and Li and Nicklas (1995) argued that mitosis forces control as a cell-cycle checkpoint and Maddin et al. (1995) discuss role of MCM3 complex in cell cycle regulation of DNA replication in vertebrate cells. Rhoades (1961) gave a general account of Meiosis. Wolfe (1981) dealt with biology of the Cell.

Questions

1. Describe: Cell cycle, cell division, cell multiplication, interphase, G_1 phase, S phase, G_2 phase, mitosis, meiosis, karyokinesis, cytokinesis, prophase, metaphase, anaphase, telophase, leptotene, zygotene, pachytene, diplotene, diakinesis, bivalent, tetrad, crossing-over, chiasma.
2. Differentiate between:
    (a) prophase-I and prophase-II
    (b) metaphase-I and metaphase-II
    (c) anaphase-I and anaphase-II
    (d) telophase-I and telophase-II
    (e) mitosis and meiosis.
3. How is the duration of different phases of cell-cycle worked out?
4. What does lead to activation of DNA synthesis?
5. Represent diagrammetically a typical cell cycle.
6. Why are yeasts considered suitable model for the study of cell cycle?
7. What are the processes that determine entry and exit from cell cycle?
8. Write short notes on:
    (a) The genes involved in cell cycle
    (b) The proteins involved in cell cycle.
9. How is the duration of complete cell cycle determined?
10. Assuming 2n=8, draw different stages of mitosis.
11. Assuming 2n=8, draw different stages of meiosis.
12. What is the significance of mitosis in life of an organism?
13. What is the significance of meiosis in life of an organism?
14. How is meiosis important in generation of genotypic and phenotypic variability at population level?
15. Mitosis and meiosis are essentially similar in all the organisms. What is the evolutionary significance of this observation?

4

CHROMOSOME THEORY OF INHERITANCE

Background

Living cells of most organisms (eukaryotic) are characterized by the presence of **nucleus**. One of the components of nucleus is fine, threadlike strands, the **chromatin** which consists of **deoxyribonucleic acid (DNA)** and **basic proteins** (histones or protamines). By the beginning of 20th century it was realised that the hereditary material consisting of genes might be carried by the chromosomes. Chromosome could be seen under the microscope. During nuclear division the chromatin strands contract and appear thicker to become **chromosomes**. The chromosomes can be seen under an ordinary light microscope. The number and morphology of the chromosomes are specific, distinct and ordinarily constant for each species.

Pairing and disjunction of homologous chromosomes at meiosis explains the mechanism of gene segregation. The independent assortment of genes can like-wise be deduced from knowledge of chromosome behaviour as observed in microscopic preparations. Diploid cells contain two sets of chromosomes that are derived from two different parents. At meiosis various pairs of chromosomes are assorted and distributed to gametes independently of each other. Gametes may thus contain any mixture of **maternal and paternal chromosomes**. Thus is precisely the manner in which the genes behave in inheritance (**Fig. 4.1**).

Thus chromosomes constitute physical inheritance. In other words, genes are located in the chromosomes. Sutton (1902) and Boveri (1902) independently discovered parallelism between behaviour of chromosomes and Mendelian factors (genes) during meiosis and fertilization.

(a) Segregation of chromosomes and genes

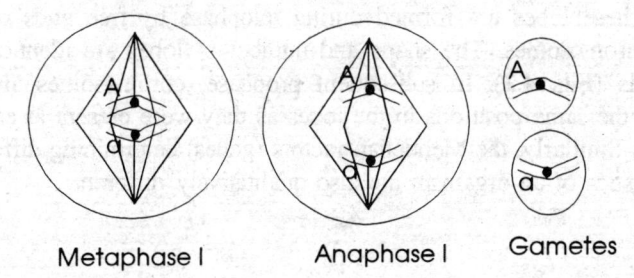

Metaphase I   Anaphase I   Gametes

(b) Independent assortment of chromosomes and genes

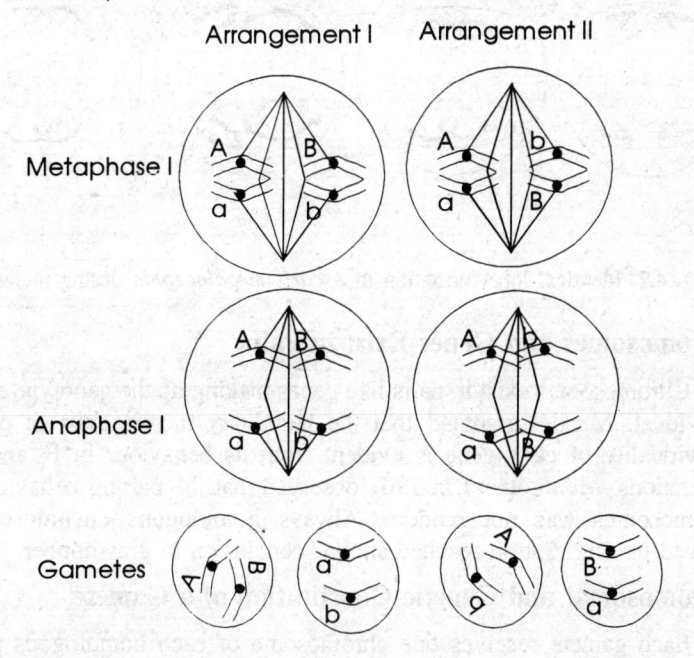

Arrangement I   Arrangement II

Metaphase I

Anaphase I

Gametes

Fig. 4.1. (a) Segregation and (b) independent assortment of chromosomes

Parallelism Between Behaviour of Chromosomes and Mendelian Factors (Genes)

Individuality of Chromosomes and Genes

Individuality of each chromosome is maintained like the individuality of the genes from generation to generation. It was seen by Boveri (1889)

that in round worm *Ascaris megalocephala*, specific number of lobes appear. These lobes are formed during telophase by free ends of V-shaped chromosomes. The shape and number of lobes are identical in sister cells (**Fig. 4.2**). In subsequent prophase, chromosomes always appear at the same positions in the lobes as they were present at earlier telophase. Similarly, the Mendelian factors (genes) determining different characteristics of an organism are also qualitatively different.

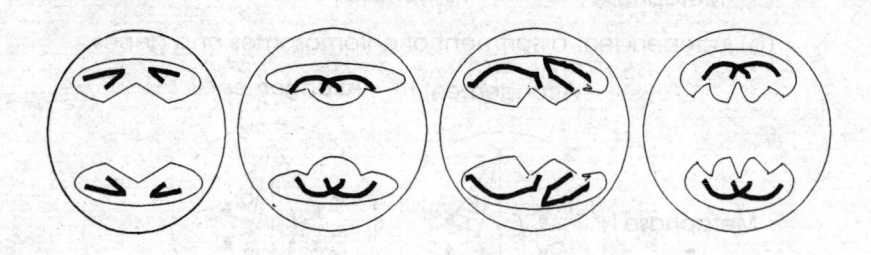

Fig. 4.2. Identical lobe formation in *Ascaris megalocephla* during meiosis

Chromosomes and Genes Exist in Pairs

Chromosomes exist in pairs like genes making up the genotype of an individual. Mendel assumed that the hereditary factors exist in pairs. Individuality of each gene is evident from its behaviour in F_1 and F_2 generations. Montgomery in 1901 observed that the pairing behavior of chromosomes was not random. Always **homologous chromosomes** showed pairing. Sutton reached similar conclusion in grasshopper.

Chromosomal and Genetic Constitution of a Gamete

Each gamete receives one chromosome of each homologous pair. Mendel postulated the segregation of the paired genes during gamete formation, each gamete receiving only one factor of each pair.

Independent Segregation of Chromosomes and Genes

The chromosomes of various homologous pairs are assorted at random in meiosis segregate independently of the chromosomes of every other pair. Crothers gave evidence for random orientation of chromosomes from the study of meiosis on grasshopper having 2n = 3. In one pair of chromosomes, one homologue was larger than the other while there was another chromosome that had no homologue. She was able to observe two metaphase-I arrangements in equal frequency (**Fig. 4.3**). Mendel held that the factors of each pair are independently of every other pair in their distribution of gametes.

Metaphase I   Anaphase I   Anaphase I

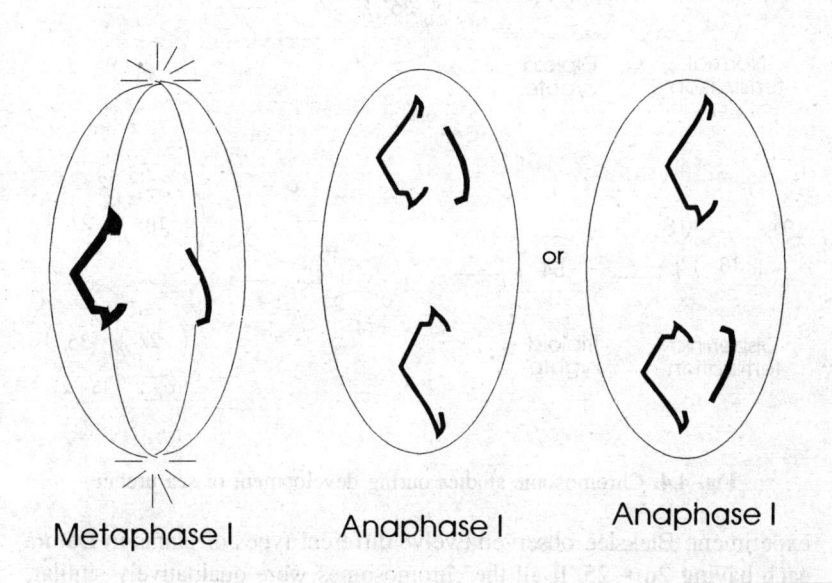

Fig. 4.3. Meiosis in grasshopper having 2n = 3

Chromosome and Genetic Constitution of a Zygote

The chromosomes from two parents come together in the zygote, restoring the diploid number of chromosomes for the offspring. Mendel held that the paternal and maternal characters mix up in the progeny.

Qualitative Differences Between Chromosomes and Genes

There are qualitative differences between different chromosome pairs. Every cell must get at least one chromosome of each kind for normal development. Boveri studied development of Sea urchin (2n = 36) (**Fig. 4.4**). It was observed that two cells with identical chromosome number do not necessarily develop in a similar fashion. Every cell must get at least one chromosome of each type for normal development. In a later

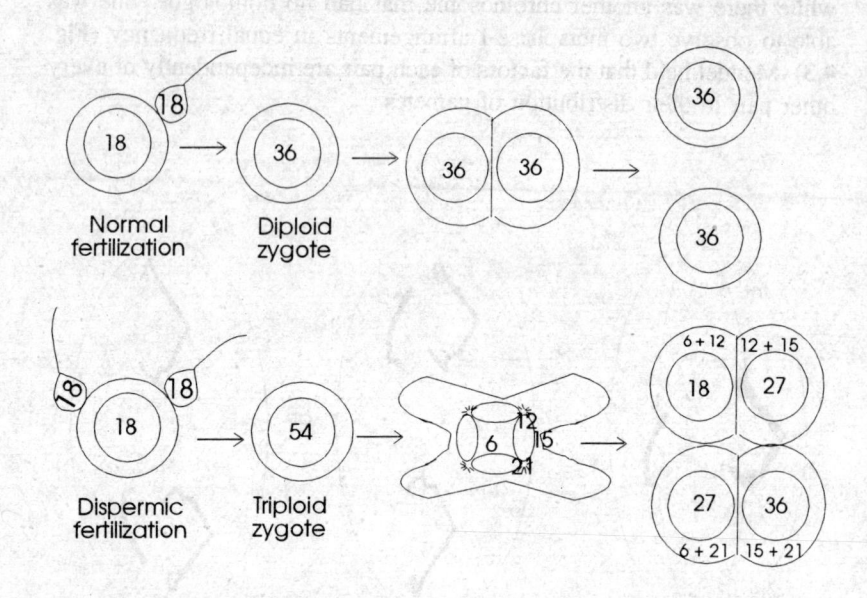

Fig. 4.4. Chromosome studies during development of sea urchin

experiment, Blakslee observed twelve different types of plants in Datura each having 2n = 25. If all the chromosomes were qualitatively similar, all plants having one extra chromosome should appear more or less alike. Blakslee observed that shape and size of the capsule were different in each of twelve types of plants. This indicated the qualitative differences among different chromosomes.

Chromosome Theory of Inheritance

Firm basis for Mendel's laws of heredity exists in the behaviour of chromosomes during meiosis and fertilization. This paved the way for chromosome theory of inheritance. Considering the above parallel behavior, Sutton and Boveri independently suggested that genes are the actual physical hereditary units located in the chromosomes in a linear order. Since there are two chromosomes of each kind, there must be two genes of each kind, one in each of the homologous chromosomes. Thus the genes which actually determine the traits are carried by chromosomes from parents to offspring. This led to formulation of chromosome theory of inheritance. The chromosomes behaviour during meiosis is seen under the microscope whereas the existence of genes is inferred from observation of behaviour of traits in hybridization experiments. The **chromosome theory of inheritance** states that chromosomes constitute the physical basis of the inheritance and that the genes are located in the chromosomes.

Some Useful Reading Hints

Sutton (1903) and Boveri (1904) drew parallelism between behaviour of genes and chromosomes during meiosis and fertilization. Bridges (1916) reported non-disjunction as proof of the chromosome theory of heredity.

Questions

1. How do the pairing and disjunction of homologous chromosomes at meiosis explain the mechanism of gene segregation?
2. What is the contribution of the following scientists in development of chromosome theory of inheritance: Sutton, Boveri, Crothers, Blakslee, and Montgomery?
3. Describe various points of parallelism between behaviour of chromosomes and Mendelian factors (genes) during meiosis and fertilization?
4. Describe chromosomal theory of inheritance.

5

CHROMOSOMES

Prokaryotic Chromosomes

Genetic material in viruses and prokaryotic and eukaryotic cells exists in the form of chromosomes. In **viruses** chromosomes are in the form of RNA or DNA molecules. Their size ranges from 3.5 to 260 kb and contain 2 to 250 genes. For example, Tobaco Mosaic Virus (TMV) is a single stranded 3.5 kb long RNA virus having 4 genes; ϕX174 is a circular single stranded DNA virus having 5 kb length of chromosome; SV40 is a circular double stranded DNA virus of the same length and has 5 genes; λ phage (50 kb), T even phages (165 kb) and smallpox virus (267 kb) have linear double stranded DNA chromosome and all of them have about 250 genes.

In **bacteria**, the chromosomes are in the form of circular DNA molecules. In *E. coli*, chromosome is 4,700 kb (1500 μm) long and comprises of 1500-3000 genes which are packed in a cell $\sim$2μm long and 1μm in diameter. This chromosome is highly condensed and is called a **nucleoid** or folded chromosome. DNA is organised into a large number of **loops** with the help of certain proteins. Loops of DNA are further supercoiled.

Eukaryotic Chromosomes

In eukaryotes, chromosomes are diffused throughout the nucleus during interphase. These appear in form of darkly stained bodies during cell division. These were observed for the first time by Nageli in 1842. Linear cell nuclei are made up of proteins, RNA, DNA and phospholipids (**Table 5.1**). Basic proteins are **protamines** (in spermatozoa) and **histones. Acidic proteins** (non-histones) perform structural and regulatory functions. Enzymatic proteins include DNA polymerases, RNA polymerase, **DNA ligase, endonucleases and exonucleases,**

Table 5.1. Chemical structure of nucleus

Proteins	RNA	DNA	Phospholipids
78%	19.5%	1.0%	1.5%
15% Basic	hnRNA		
65% Others	mRNA, snRNA		
	tRNA, rRNA		

Table 5.2. Amount of DNA (C value) in nucleus of some species

Species	Amount of DNA
Human diploid cell	5.6 pg DNA (174 cm)
Brassica sp.	1.6 to 3.2 pg
6x Triticales	42.4 pg
8x Triticales	52 pg
Trillium	120 pg (37 meters)

1 pg = 10^{-12} g, 1 dalton = 1.66×10^{-12} pg

transcription factors. Amount of DNA in nucleus varies from 1.6 pg (*Brassica* spp.) to 120 pg (*Trillium* spp.) **(Table 5.2)**.

Nucleus

Structurally, the nucleus is composed of the **nuclear envelop**, the chromatin, the nucleolus, the nuclear sap and small nucleoribonucleoproteins (snRNPs). The nuclear envelop consists of two concentric membranes separated by a **perinuclear space** of 10-15 nm width. Outer surface of the outer membrane has many ribosomes. Inner surface of the inner membrane is closely associated with condensed chromatin, i.e., **heterochromatin**.

Nuclear envelop is not a continuous structure but is interrupted at several places by **nuclear pores** (50-80 nm). About 5-15% nuclear envelop area is occupied by nuclear pores in mammals. Nuclear pores are octagonal in shape. Plants and protozoa have 20-36% nuclear area. **Nuclear membranes** perform various functions. They allow free exchange of ions between the nucleus and the cytoplasm, involved in

Proteins synthesis and the act as sites of attachment for heterochromatin. Exchange of macromolecules such as **ribonucleoprotein (RNP) granules** and **fibrils** and proteins take place through nuclear pores. In fungal cells where there is less amount of **endoplasmic reticulum (ER)**, its function is performed by nuclear envelop. It is believed that **evagination** of double nuclear envelop give rise to initials of cell organelles like mitochondria and plastids.

Nucleolus

The nucleus contains at least one large, dense, spherical structure known as nucleolus. Size of the nucleolus is related with the synthetic activity of the cell. Therefore, the cells with little or no synthetic activities, e.g., sperms, blastomeres, muscle cells, etc. are found to contain smaller nucleoli whereas the cells which are actively involved in protein synthesis, e.g., oocytes, nerve cells, secretary cells, contain large sized nucleoli. Number of nucleoli per cell may be one, two or more, which is characteristic of a species.

Nucleoli are rich in RNA having 5-20% of the total nuclear RNA. Nucleolus contains large amount of proteins mainly phosphoproteins. It contains certain enzymes required for RNA formation. These ribosomal proteins are acid phosphatase, nucleoside phosphorylase, NAD^+ synthesizing enzymes, RNA methylase, RNA polymerase, exo- and endo-nucleases.

Nucleolus is found around the **nucleolar organizer** a region of a chromosome which has many copies of genes for rRNAs. The nucleolar organizer is usually located in a secondary constriction on the chromosome. Nucleolus is composed of following four zones:

Fibrillar zone. Central region of nucleolus is occupied by RNA fibrils of 5-10 nm diameter. Ribosomal proteins are bound to rRNAs in this zone (primary rRNA transcript).

Granular zone. Fibrillar zone is surrounded by granular zone consisting of dense RNP granules of 15-20 nm diameter. Processing and maturation of pre-ribosomal particles occur in this region.

Amorphous zone. Matrix in which fibrils and granules are suspended constitutes this zone. It consists of different enzymes and phosphoproteins.

Nucleolus associated chromatin. It is composed of a chromatin fibre situated around the nucleolus and extending into it. Ribosomal RNA is transcribed from this chromatin. This chromatin contains multiple copies of genes coding for 18S, 28S and 5.8S rRNA (900 in *Xenopus*). The amount of nuclear DNA, chromosomal DNA, rDNA, number of rDNA genes and number of nucleoli in *Xenopus laevis* are given in **Table 5.3**.

Table 5.3. Nuclear DNA, chromosomal DNA, rDNA, number of rDNA genes and number of nucleoli in somatic cells and oocytes of *Xenopus laevis*

Cell type	Nuclear DNA (pg)	Chromosomal DNA (pg)	rDNA (pg)	No. of rDNA genes	No. of nucleoli
Somatic cell	6	6	0.012	900	2
Oocyte	37	12	25	2,000,000	1000-1500

Synthesis of a **eukaryotic ribosome** is a complex phenomenon in which several regions of the cell are involved. The 18S, 5.8S and 28S rRNAs are synthesized as a part of a much longer precursor RNA molecules in the nucleolus - 5S RNA is synthesized on the chromosome outside the nucleolus, and the 80 ribosomal proteins are synthesized in the cytoplasm. rRNA genes are tandemly arranged and are separated by spacer DNA which is not transcribed. Repititive rDNA cistrons in various organisms are given in **Table 5.4**.

Table 5.4. Repetitive rDNA cistrons in various organisms

Species	No. of rRNA cistrons per haploid genome
E. coli	8 - 22
B. subtilis	9 - 10
HeLa cells	160 - 640
Drosophila	130
Xenopus	450

The nucleoli undergo cyclic changes during cell division. These disappear at late prophase stage and reappear at late telophase. Nucleolus

Basic Genetics

remains associated with a region of the chromosome known as nucleolar organizer. **Nucleoli organizing regions (NOR)** the last to undergo condensation, thus producing a secondary constriction on the chromosome.

Several functions assigned to nucleolus include biogenesis of ribosomes and central component of **synaptonemal complex**. Ribosomal DNA is highly amplified in amphibian oocytes. **Gene amplification** is a process by which at early stages of meiotic prophase of the set of genes is replicated selectively while the rest of the genome remains constant, e.g., rDNA of oocytes of *Xenopus* where eggs are very large cells (1.3 nm is diameter) that accumulate large number of ribosomes (10^{12}) for use during early development. So many ribosomes are synthesized by the amplification of rDNA genes 1000 fold. Somatic cells have two nucleoli per cell, whereas each oocytes of *Xenopus laevis* has 1,000-1,500 nucleoli. The amplified DNA is lost during the course of development.

Chromosome Size

The chromosome size in eukaryotes varies. **Yeast haploid genome has 14,000 kb DNA and individual chromosomes contain 200-300 kb DNA. Human haploid genome** comprises of 3,000,000 kb DNA and individual chromosomes contain 60,000-2,50,000 kb DNA. In nondividing eukaryotic cells, chromosomal DNA is complexed with histone proteins to form a beaded chromatin called **nucleosomes**. Nucleosome is a basic repeating subunit of **chromatin**, consisting of a core particle composed of two molecules of each of 4 different **histones** (H2a, H2b, H3 and H4) around which ∼ 160 bp DNA is wound. Beaded chromatin is 110 Å in diameter and there is 7-fold reduction in length of DNA in it. This chromatin is organized into shorter thicker chromatin fiber of 300 Å diameter with the help of H1 histone. This chromatin fibre (30 nm) condenses into a **chromatid** of compact metaphase chromosome with the help of non-histone proteins. Two chromatids when join together by a centromere form a complete chromosome. There are four essential components of a eukaryotic chromosomes - centromere, telomeres, replication origin, and genes.

The size of the chromosome varies from species to species and relatively remains constant for a particular chromosome. The length of chromosomes may vary from 0.2 μm (in fungi and birds) to 50 μm (in some plants) with average length of 6 μm. Some species have all the chromosomes of almost similar size whereas others have chromosomes differing greatly in size. The chromosome size ratio of largest to smallest chromosome is 1000:1 in *Droceraceae*, 40:1 in *Commelinaceae*, 5:1 in *Crepis* species. Birds, lizards and **Liliaceae** species have two type of chromosomes **macro-chromosomes** and **micro-chromosomes**.

When the cell is not dividing chromosomes are in the form of very long, thin, fine thread like structures called chromatin, enclosed within a nucleus. Chromosomes can be easily seen (through the microscope) during cell division, because these are compactly organised. Each species has a constant number of chromosomes in somatic cells of its body. It is called as **somatic chromosome number** (2n) or **diploid chromosome number**. Chromosome number does not have any relation to the complexity of the organism. The chromosomes of somatic cells are usually present in pairs. One chromosome of each pair is contributed by maternal parent and other by paternal parent. Gametic or gametophytic cells are **haploid** having only one copy of each chromosome. In diploid state members of the same chromosome pair are genetically corresponding and are called as **homologous chromosomes** or **homologues**. Each homologous pair is made up of identical partners, except sex chromosomes. Each chromosome has a specific shape and size.

Chromosome Shape

Shape of the chromosomes depends upon the position of the centromere. Centromere has the attachment point for spindle fibres. Attachment points are called as **kinetochores**. Shape of chromosome is determined by the position of the centromere. Centromere is a located in a thinner segment of the chromosome called as **primary constriction**. The regions adjacent to centromere frequently have highly repetitive DNA sequence and may stain more darkly with basic dyes. The centromere divides the chromosome vertically into two parts called as **chromosome arms**. Chromosomes are classified according to the position of the centromere as follows:

Metacentric. When the centromere is in the middle of the chromosome. During metaphase such chromosomes appear as V-shaped in structure.

Telocentric. When the centromere is at the end of the chromosome. During metaphase such chromosomes acquire rod-shaped structure.

Acrocentric. When the centromere is near the end of the chromosome. Such chromosomes appear hook-shaped at metaphase.

Submetacentric. When the centromere is slightly displaced from the middle. During metaphase such chromosomes are seen as J-shaped structures.

Centromere position, shape of metaphase and anaphase chromosomes and arm ratio of different types of chromosomes is shown in **Fig. 5.1**. For any particular chromosome the position of the centromere is fixed. During

Centromere may be Metaphase	Median	Submedian	Subterminal	Terminal
Chromosome type	Metacentric	Submetacentric	Acrocentric	Telocentric
Anaphase	V-shaped	L-shaped	J-shaped	I-shaped
Arm ratio (long : short)	1:1	2:1	3:1	1:0

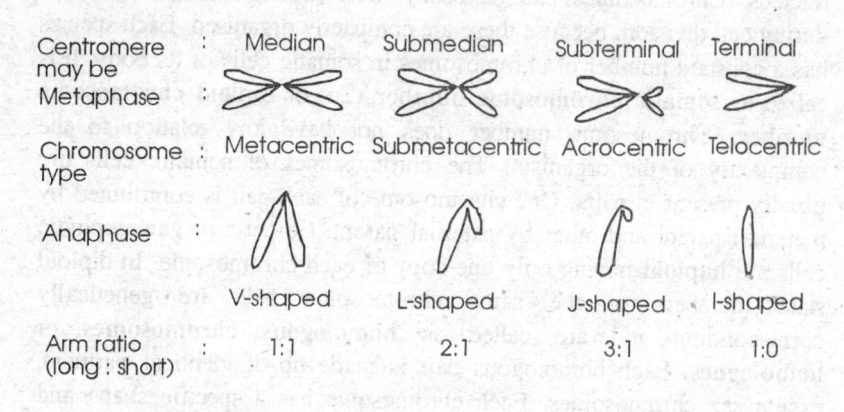

Fig. 5.1. Chromsome types based on position of centromere

cell division, chromatin becomes compactly organised into darkly staining bodies called as chromosomes. These were observed for the first time by Strasburger (1857). Term chromosome was given by Waldeyer (1888) Sutton and Boveri (1902) suggested that chromosomes were the physical basis of hereditary information which is transmitted from one to next generation. Morgan (1910) gave experimental evidence for the chromosomal theory of inheritance. Chromosomes are best seen in squash preparations. Morphology and number of chromosomes are studied during metaphase and anaphase stages of mitosis and pachytene stage of meiosis. The pachytene stage of meiosis in many species is ideal for the study of **chromosome morphology** because at this stage morphological markers other than centromere, telomeres and nucleolus organizing region are **chromomeres** and **knobs**.

Chromosome Number

The number of chromosomes is constant and specific for a particular species. These are of great importance in the determination of phylogeny and taxonomy of the species. The chromosome number of some important animal and plant species is given in **Table 5.5**. The number and set of chromosomes present in the gametic cells (such as sperm and egg) is known as gametic or haploid set of chromosomes. The haploid set of chromosomes is also known as **genome**. The somatic or body cells of most of organisms contain two haploid sets or genomes and are known as diploid. This chromosome number is referred as somatic or diploid chromosome number.

Some generalizations have been drawn from the studies of chromosomes from different species. The organisms with less number of chromosomes contain comparatively large-sized chromosomes than the organisms having many chromosomes. The diameter of the chromosomes may vary from 0.2 μm to 2 μm. Plants, in general, have larger chromosomes than animals. Monocots in general have large sized chromosomes than dicots. Annuals have small chromosomes than perennials.

Some Parts of a Chromosome

Kinetochore. Kinetochore is composed of protein and RNA and is a disc-shaped protein structure that is attached to the **centromeric chromatin** (**Fig. 5.2**). Small fibres are attached to the kinetochore during all division and help in movement of chromosomes. Mostly each chromosomes has

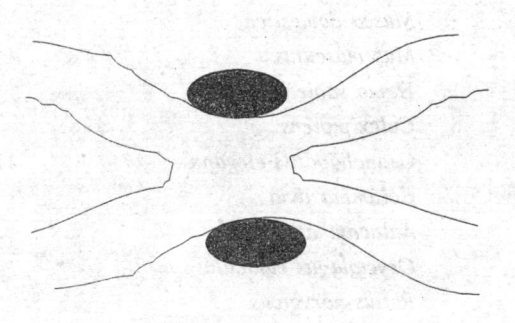

Fig. 5.2. Kinetochore attached to the centromere chromatin

52 Basic Genetics

Table 5.5. Diploid number of chromosomes in some common animals and plants

Common Name	Species	Diploid No.
ANIMALS		
Alligator	*Alligator mississipiensis*	32
Carp	*Cyprinus carpio*	104
Cat	*Felis catus*	38
Cattle	*Bos taurus*	60
Chicken	*Goillus domesticus*	78
Chimpanzee	*Pan troglodytes*	48
Cockroach	*Blatta germanica*	23♂, 24♀
Dog	*Canis familiaris*	78
Donkey	*Equus asinus*	62
Flatworm	*Planaria torva*	16
Freshwater hydra	*Hydra vulgaris attenuata*	32
Frog	*Rana pipiens*	26
Fruit fly	*Drosophila melanogaster*	8
Golden hamster	*Mesocricetus auratus*	44
Gorilla	*Gorilla gorilla*	48
Grasshoper	*Melanoplus differentialis*	24
Guinea pig	*Cavia cobaya*	64
Honeybee	*Apid mellifera*	32♀, 16♂
Horse	*Equus calibus*	64
House fly	*Musca domestica*	12
House mouse	*Mus musculus*	40
Man	*Homo sapiens*	46
Mosquito	*Culex pipiens*	6
Nematode	*Ceanohabditis elegans*	11♂, 12♀
Pigeon	*Columbia livia*	ca. 80
Protozoa	*Aulacantha species*	~1600
Rabbit	*Oryctolagus cuniculus*	44
Rat	*Rattus norvgicus*	42
Red ant	*Formica sanguinea*	48
Rhesus monky	*Macaca mulatta*	42
Round warm	*Ascaris melagocephala*	2

Table 5.5 contd...

Common Name	Species	Diploid No.
Sheep	Ovis aries	54
Silkworm	Bombyx mori	56
Starfish	Asterias forbesi	36
Toad	Bufo americanus	22
PLANTS		
Amercian cotton	Gossypium hirsutum	52
Asiatic cotton	Gossypium herbaceum	26
Barley	Hordeum vulgare	14
Black gram	Cicer arietinum	16
Bread wheat	Triticum aestivum	42
Broad bean	Vicia faba	12
Cabbage	Brassica oleracea	18
Compositae	Haplopappus gracilis	4
Corn	Zea mays	20
Durum wheat	Triticum durum	28
Evening primrose	Oenothera biennis	14
Garden onion	Allium cepa	16
Garden pea	Pisum sativum	14
Green algae	Acetabularia mediterranea	ca. 20
Indian fern	Ophioglossum reticulatum	1262
Kidney bean	Phaseolus vulgaris	22
Lentil	Leus culinaris	14
Mash	Vigna mungo	22
Metha	Trigonella species	16
Moong	Vigna radiata	22
Moth	Vigna aconitifolics	22
Pine	Pinus species	24
Potato	Solanum tuberosum	48
Raddish	Raphanus sativus	18
Rape	Brassica Napus	38
Raya	Brassica juncea	36
Rice	Oryza sativa	24
Rye	Secle cereale	14

Table 5.5 contd...

Common Name	Species	Diploid No.
Snapdragon	*Antirrhinum majus*	16
Sorghum	*Sorghum vulgare*	14
Spiderwort	*Tradescantia virginiana*	24
Squash	*Cucurbita pepo*	40
Sugarcane	*Saccharum officinarum*	80
Tobacco	*Nitcotiana tabacum*	48
Tomato	*Lycopersicon esculentum*	24
White clover	*Trifolium repens*	32
White oak	*Quercus alba*	24
HAPLOIDS		
Amoeba	*Amoeba proteus*	ca. 250
Green algae	*Chlamydononas reinhardi*	16
Mold (fungus)	*Aspergillus nidulans*	8
Penicillin mold	*Penicillium species*	4
Pink bread mold	*Neurospora crassa*	7
Slime mold	*Dictyostelium discoideum*	7
Yeast	*Saccharomyces cerevisiae*	ca. 17

one centromere, such chromosomes are known as **monocentric**. Certain chromosomes may becomes **dicentric** (as in Maize) or **polycentric** (Luzula plants, *Ascais megalocephalus*). In some cases under certain conditions, **spindle fibers** may be attached to some region of the chromosomes other than the centromere. This region acts as secondary centre of **chromosome movement** and is called as **Neocentromere** (as in maize).

Secondary constriction. In addition to a primary constriction formed by centromere, certain chromosomes have secondary constriction also. This region is responsible for the formation of nucleolus during telophase and is associated with nucleolus during interphase and prophase. It is therefore called as **nucleolus organizing region** (NOR) (McClintock, 1934). NOR is the site of multiple repeats of DNA unit coding for rRNA which is the main component of the nulceolus). Segment of chromosome beyond secondary constriction is called as **satellite**. Mostly each species possess at least one pair of nucleolus organizing chromosomes.

Telomeres. Terminal regions of the chromosomes are called as telomers. Telomeres maintain the integrity or identity of the chromosomes. These have **constitutive heterochromatin** (highly repetitive DNA sequences).

The chromomeres are the lead-like structures along the entire length of a pachytene chromosome at specific regions. These are formed due to local condensation or tight folding of the chromatin. Heterochromatin chromomeres stain darker and are larger than **euchromatic chromomeres**. Number, size and position of the chromomeres on each chromosome are reasonably constant and can be used as reliable morphological markers of chromosomes. Chromomeres are also formed in **polytene chromosomes** and **lamp brush chromosomes**.

Unusual Chromosomes

B-chromosomes

B-chromosomes are a kind of **supernumerary chromosomes** that may be missing in an organism or may be found as extra chromosomes over and above the standard diploid or polyploid chromosomes complement of **A-chromosome**. These are formed in natural populations of may plant and animal species. These have been reported in more than 1000 species of plants belonging to bryophytes, pteridophytes, gymnosperms and angiosperms. These chromosomes have the following characteristics: They are dispensable and are not found in all individuals of a species. They are absent in developed agronomic strains. These may not be formed in all type of cells. Sometimes they are restricted only to aerial parts and absent in roots and may be eliminated during meiosis. These are not homologous with any of the A-chromosomes and during meiosis pair among themselves. Their inheritance is non-Mendelian, sometimes due to **non-disjunction** during pollen mitosis. They are smaller than A chromosomes and have their own unique pattern of heterochromatin distribution. In general they are genetically inert but may rarely organize nucleoli and carry functional genetic material. When present in high number, they suppress vigour and fertility. They effect on homologous pairing in species hybrids. They replicate their DNA during late S-phase. Their origin and function are unknown.

Largest B chromosome is two-thirds the shortest A-chromosome. Each B-chromosome is composed of a short and long arm. Short arm is visible when two B- chromosomes are paired. Long arm is composed of proximal euchromatic region which is divided into two parts and distal heterochromatic region is divided into four parts or segments (**Fig. 5.3**).

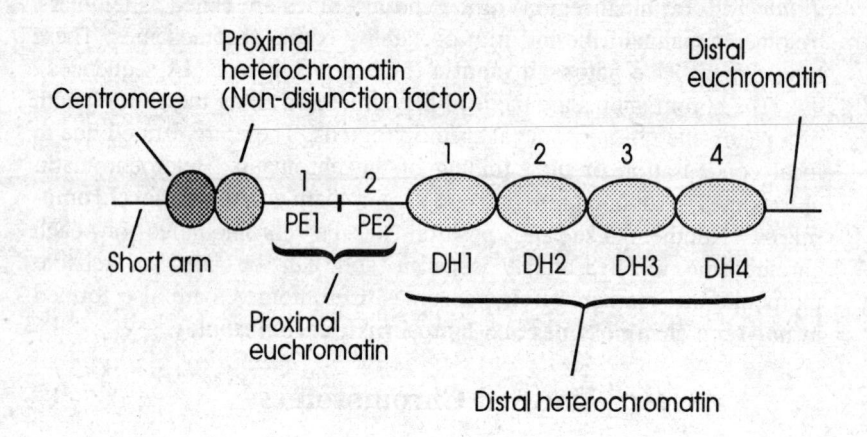

Fig. 5.3. Structure of a B-chromosome

Genes located on B-chromosomes. Usually B-chromosomes are genetically silent. They lack major functional genes. But certain B-chromosomes do exhibit qualitative effects. For example, in *Plantago coronapus*, genes for male-sterility and pigmentation. In many cases, B-chromosomes carry nucleolus organising region (NOR) as reported in amphibians, insects, plants. Genes for achene colour in *Haplopappus gracilis*, for leaf-striping in *Zea mays*, for meiotic pairing in *Aegilops*, and esterase in *Scillia autumnalis* have been reported. Formation of **puff** on polytene B-chromosomes of blackfly is interpreted as the site of heavy transcription. Lampbrush B-chromosomes in amphibians also provide evidence of transcriptional activity. Thus supernumerary chromosomes are not devoid of genetic activity, although they may carry only a few genes.

Effects of B-chromosomes. Numerous effects of B-chromosomes have been observed. There is gradual decrease in pollen-fertility and seed set with the increase in the number of B-chromosome (0-8) in rye and 30-34B in maize reduce fertility. Flowering is delayed by B-chromosomes and other plant characters, like height, weight and tiller number, are adversely affected. Addition of rye B-chromosomes to wheat also has similar effects. In rye, more than 6 B-chromosomes drastically affect the chiasma frequency and in maize, B-chromosomes affect the recombination frequency at specific loci. B chromosomes suppress or promote **homoeologous pairing** in a number of intergenic or interspecific hybrids like Wheat X *Aegilops* spp, Wheat X rye, *Lolium* X *Festuca*, *Lolium* X *Lolium*.

Use of B-chromosomes. B chromosomes are used in gene location or chromosome mapping. **B-A translocations** involving interchanges between normal A-chromosomes and B-chromosomes are used for gene location. (**Fig. 5.4**). B-A translocations have been reported in maize, rye, *Lolium* and pearl millet. In maize, a number of B-A translocations involving almost every maize chromosome have been reported and used for locating genes on specific A segments and on B-chromosomes also.

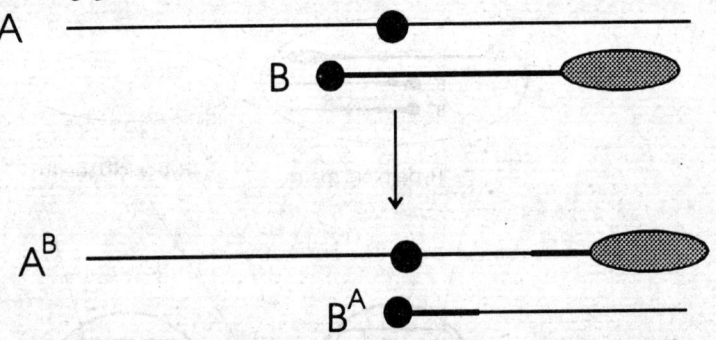

Fig. 5.4. B-A translocation involving interchange between normal A-type and B chromosome

Cross B-A translocations homozygote (A^BA^B, B^AB^A) carrying dominant marker with a normal plant carrying a recessive marker. While the results of the cross are given below the details of the cross are given in **Fig. 5.5**.

♀ A^BA^B, B^AB^A (RR) X ♂ rr → F_1 seed - All coloured aleurone and embryo

♀ rr X ♂ A^BA^B, B^AB^A (RR) → colourless aleurone : coloured Aleurone, and coloured embryo : colourless embryo

On the other hand, if a dominant allele of a gene is not located on a particular B^A chromosome, **hypoploidy** for that chromosome will not uncover the relevant recessive phenotype. By using B^A translocation, *su* (sugary) was located on the 4; O_2 (opaque) and gl^1 (glossy) on chromosome 7, in maize.

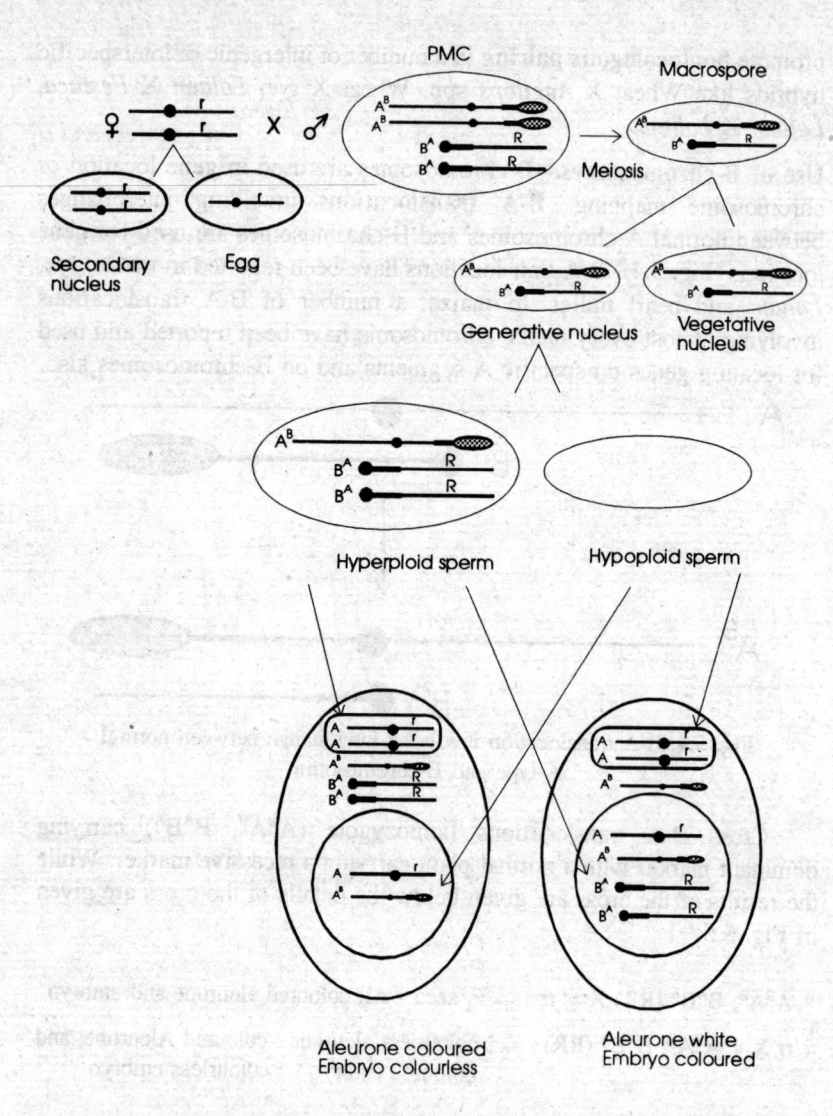

Fig. 5.5. Cross involving B-A translocation homozygote (AA, BB) carrying a dominant marker with a normal plant carrying a recessive marker

Polytene or Polynemic Chromosomes

These chromosomes, reported by Balbiani (1881), are very large sized and have more than one thousand chromatin (DNA) fibres aligned side by side. These chromosomes have been observed in some salivary glands cells, Malpighian tubules, ovarian nurse cells and gut epithelial cells of the larval of dipteran species (*Drosophila, Chironomus, Sciara* and *Rhyncosciara*). These have also been reported from **suspensor cells** of developing embryos in leguminous plants. These have been studied in detail in salivary gland cells of third instar larvae of *D. melanogaster*. Here the chromosomes are 2000 μm in length (7.5 μm in somatic cells). Bridges (1936) was first to construct a salivary chromosome map of *Drosophila melanogaster* (**Fig. 5.6**).

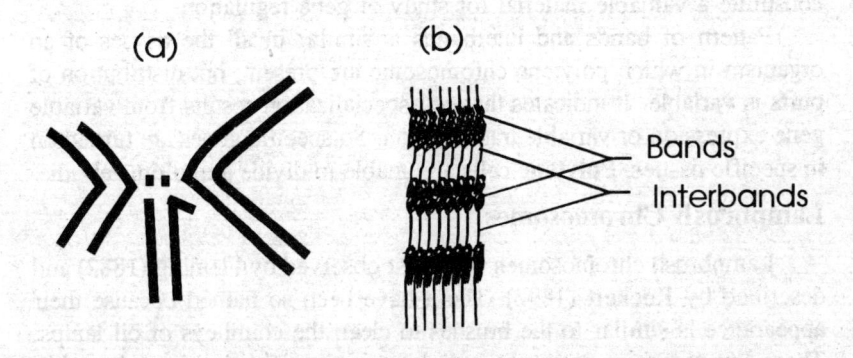

Fig. 5.6. (a) Chromosome complement of brain ganglia cells and (b) bands and interbands of salivary chromosomes of *Drosophila melanogaster*

These giant chromosomes are formed by the replication of the chromosomal DNA many times (9-10 times or more) without the separation of **daughter chromatids**, the process is known as **endomitosis**. So each chromosome is multistranded or polytenic consisting of ∼ 1000 chromatin strand in *Drosophila* and ∼ 16000 in *Chironomus*. Also the homologous chromosomes are paired along the entire length (**somatic synapsis**). This permits the identification of deletions, inversions and duplications as regions looped out the chromosomes. Pericentromeric heterochromatic regions of all the chromosomes are tightly paired, so that all the chromosomes appear to be fused at this region. This region is called as **chromocenter**. The **satellite DNAs** of chromocenter are under replicated with respect to the rest of the chromosome. These chromosomes have many dark **bands** alternating with lightly stained **interbands** throughout their length. There are ∼ 5000 bands in *Drosophila* genome. They have characteristic morphology and positions. Each band is formed due to the alignment of chromomeres of all the chromatin strands of two homologous chromosomes. Alignment of uncoiled chromatin fibers constitutes interband.

During certain stages of development diameter, of specific sites of chromosomes is greatly increased. These swellings are called as **chromosome puffs (Balbiani rings)**. In the puff region the chromomeric DNA unfolds into open loops and is involved in gene transcription. The remarkable feature about polytene chromosome is that gene transcription can be visualized directly under the microscope. These chromosomes thus constitute a valuable material for study of gene regulation.

Pattern of bands and interbands is similar in all the tissues of an organism in which polytene chromosome are present, but distribution of puffs is variable. It indicates that cell-specialization results from variable gene expression or variable transcription. So specific genes are turned on in specific tissues. **Polytene** cells are unable to divide and ultimately die.

Lampbrush Chromosomes

Lampbrush chromosomes were first observed by Fleming (1882) and described by Ruckert (1892). These have been so named because their appearance is similar to the brushes to clean the chimneys of oil lamps. These chromosomes occur at the diplotene stage of meiotic prophase-I in oocytes of all animal species, in spermatocytes of several species, and in giant nucleus of unicellular algae *Acetabularia*. These have been extensively studied in **amphibian oocytes**. Since these chromosomes are observed during prolonged **diplotene** stage (200 days to 12 years), these are in the from of bivalents. During diplotene, the homologous chromosomes begin to separate from each other and are held together at chiasmata. Each chromosome of a bivalent pair consists of a large number of chromosomes of variable sizes, which are distributed throughout its length. From each of vast majority of chromosomes, a pair of lateral loops extends in opposite directions, perpendicular to the main axis of the chromosome. In some cases more than one pair of loops may emerge from a single chromomere. These lateral loops give these chromosomes the appearance of a lampbrush (**Fig. 5.7**).

There are approximately 10,000 loops per oocyte. These chromosomes are extremely long and in some cases and range form 400-800 μm in length. Total length of 11 bivalents in *Triturus viridescens* in 5,900 μm and of DNA being 10 meters (1 μm DNA $\sim$ 3 kbp.).

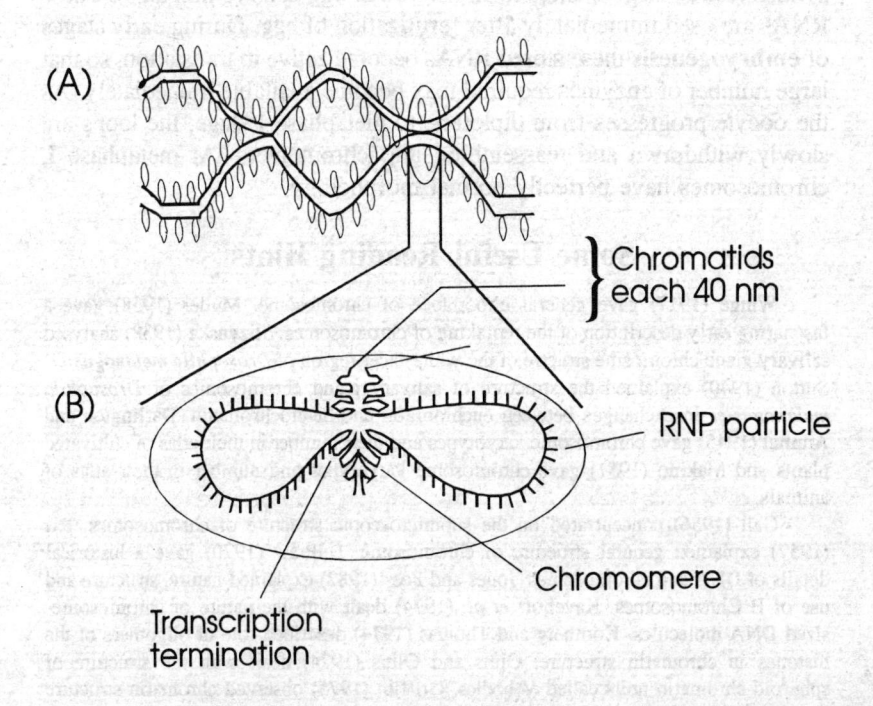

Fig. 5.7. (A) Lampbrush chromosome bivalent showing two chiasmata and (B) Mangified view of two lateral loops arising from the chromosome

Chromomeres represent the regions in which the chromatids are tightly folded. Loops are formed in pairs because each chromosome has two sister chromatids. Length of the loops is variable in different species and increases with genomic size. Loops are formed due to unfolding of some of chromomeric DNA. Loops are the active site of transcription. Mostly each loop represents a single unit transcription (one gene). Some loops may contain more than one gene. Each loop is asymmetrical in appearance. It is thinner at one end and gradually becomes thicker towards other end. Thin end is the site of **transcription initiation (promoter** site) and thick end is the site of **transcription termination**. **Lampbrush loops** are an adaptation for intensive transcription. The process of egg formation or oogenesis takes 200 days (*Salamander*) to several years (upto 12 years in human female), upto puberty sexual maturity in animals. During oogenesis, many types of RNA are synthesized in large quantities and stored in egg cells (1 mm size). These RNAs are used immediately after fertilization of egg. During early stages of **embryogenesis** these stored RNAs become active in translation, so that large number of enzymes required may become available immediately. As the oocyte progresses from diplotene to metaphase-I stage, the loops are slowly withdrawn and reassembled into chromomers. At metaphase I, chromosomes have perfectly normal morphology.

Some Useful Reading Hints

Winge (1917) gave general importance of chromosome. Muller (1938) gave a fascnating early description of the remaking of chromosomes. Slizynska (1938) analysed salivary gland chromosme structre of the white-facet region of *Drosophila melanogaster*. Sutton (1940) explained the structure of salivary gland chromosomes of *Drosophila melanogaster* in exchanges between euchromatin and heterochromatin. Darlington and Ammal (1945) gave chromosome karyotypes and their number in their atlas of cultivated plants and Makino (1951) gave chromosome karyotypes and number in their atlas of animals.

Gall (1956) concentrated on the submicroscopic structure of chromosomes. Ris (1957) explained general structure of chromosome. DuPraw (1970) gave a historical details of DNA and chromosomes. Jones and Pees (1982) explained nature, structure and use of B Chromosomes. Kavenoff *et al.* (1974) dealt with the nature of chromosome-sized DNA molecules. Kornberg and Thomas (1974) described role of oligomers of the histones in chromatin structure. Olins and Olins (1974) dealt with the structure of spheroid chromatin units called V bodies. Griffith (1975) observed chromatin structure as deduced from a minochromosome.

Noll (1977) explained how DNA folding occurs in the nucleosome. Thomas (1967) and Thomas (1984) dealt with the higher order structure of chromatin and histone H1.

Adoplh *et al.* (1977). isolated a protein scaffold from mitotic HeLa cell chromosomes. Bak and Jeuthen (1977) dealt with high order structure of mitotic chromosomes, in general while Bak *et al.* (1977) dealt with high order structure of human mitotic chromosomes, in particular. Laemmli *et al.* (1977) described the role of non-histone proteins in metaphase chromosome structure. Taniguchi and Tukayama (1986) gave high order structure of metaphase chromosomes as an evidence for a multiple coiling model. Verma (1988) dealt with molecular and structural aspects of heterochromatin. Callan (1963) and Callan (1986) gave description of the nature of lampbrush chromosomes.

Aalalen *et al.* (1980) described structure of histone H1 and its location in chromatin. Kornberg and Klug (1981) dealt with the structure of nucleosome. Laskey and Earnshaw (1980) explained the steps involved in nucleosome assembly. Graziano *et al.* (1994) reported that histone H1 is located in the interior of the chromatin 30-nm filament. Eissenberg *et al.* (1985) edited in a book form some selected topics in chromatin structure. Garza *et al.* (1989) described the procedure of mapping the *Drosophila* genome with yeast artificial chromosomes. Manuelidis (1990) expressed a view of interphase chromosomes.

Chromosomes 63

Murray and Szostak (1983) and Murray and Szostak (1987) dealt with construction of artifical chromosomes in yeast. Pederson et al. (1986) dealt with structure of transcriptionally active chromatin. Touchette (1990) presents an evolutionary aspect of chromatin structure. Wagner et al. (1993) gave complete picture of synthesis of chromosomes.

Questions

1. Describe the characteristics of prokaryotic and eukaryotic chromosomes.
2. Describe structure and characteristics of nucleus and nucleolus.
3. What is the significance of chromosome number for a species?
4. On what factors does chromosome size and shape depend?
5. Describe various parts of a chromosome.
6. Describe the two types of chromatin - heterochromatin and euchromatin.
7. Explain structure of interphase chromatin with special reference to DNA and histones
8. What type of DNA-histone interactions are found in chromosome.
9. What do you understand by chromatosomes, nucleosomes and nucleosome phasing.
10. What are various characteristics of metaphase chromosomes?
11. Describe scaffold or radial-loop model of chromosomes structure?
12. What do you know about the higher order organization of chromatid?
13. How do the metaphase chromosomes provide evidence for a multiple-coiling model of chromosome structure?
14. Describe the supersolenoid or coiled-coil model of chromosome structure.
15. Write a short note on the higher order packing of chromatin fibres
16. What are unusual chromosomes? How do they differ from ordinary somatic and germ cell chromosomes?
17. What are the B-chromosomes? What are the functions and effects of these chromosomes?
18. What are the polytene or polynemic chromosomes? How are these chromosomes formed? What is the utility of these chromosomes in cytogenetical studies?
19. Describe the structure, origin and usefulness of lampbrush chromosomes.
20. What are the artificial chromosomes? What are different structural components of yeast artificial chromosome? How was the first yeast artificial chromosome first synthesised?

6

SEX-LINKED, SEX-LIMITED AND SEX-INFLUENCED INHERITANCE

There are several instances when we do not get expected Mendelian phenotypic ratios in a segregating progeny. Sex-linked, sex-limited and sex-influenced inheritance provide some of the examples of phenotypic ratio departures in males and females.

Sex-linked Inheritance

Sex-linked traits show following characteristics: There are differences in **reciprocal crosses** giving different phenotypic ratios. Males that carry (inherit) X-linked recessive character due to **hemizygous** condition, because Y-chromosome does not contain counterpart allele. Affected male transmits the X-linked gene to all their daughters but none to their sons. **Heterozygous** females are **carriers** because their sons may express the recessive trait, even when their father is normal. An opposite pattern of non-reciprocal inheritance is found in organisms in which male is **homogametic** and female is **heterogametic**. This pattern is found in birds, some reptiles, fishes, moths and butterflies. The following examples have been taken only from those organisms where male is heterogametic.

white Eyes in *Drosophila*

Morgan (1910) discovered **sex-linked genes** in *Drosophila*. *white* gene was located in the X chromosome. Sex-linked genes follow a **criss-cross** or **skip-generation** inheritance where male transmits male transmits his sex-linked genes to his grandsons through his daughters, never to or through his sons. This is the mode **X chromosome** is transmitted (**Fig. 6.1**). This experiment provided evidence to chromosome theory of inheritance and **white** gene in *Drosophila* was the first gene to have been

Sex-linked, Sex-limited and Sex-influenced Inheritance

assigned to a chromosome. Morgan was awarded Nobel Prize for this pioneering work in 1934.

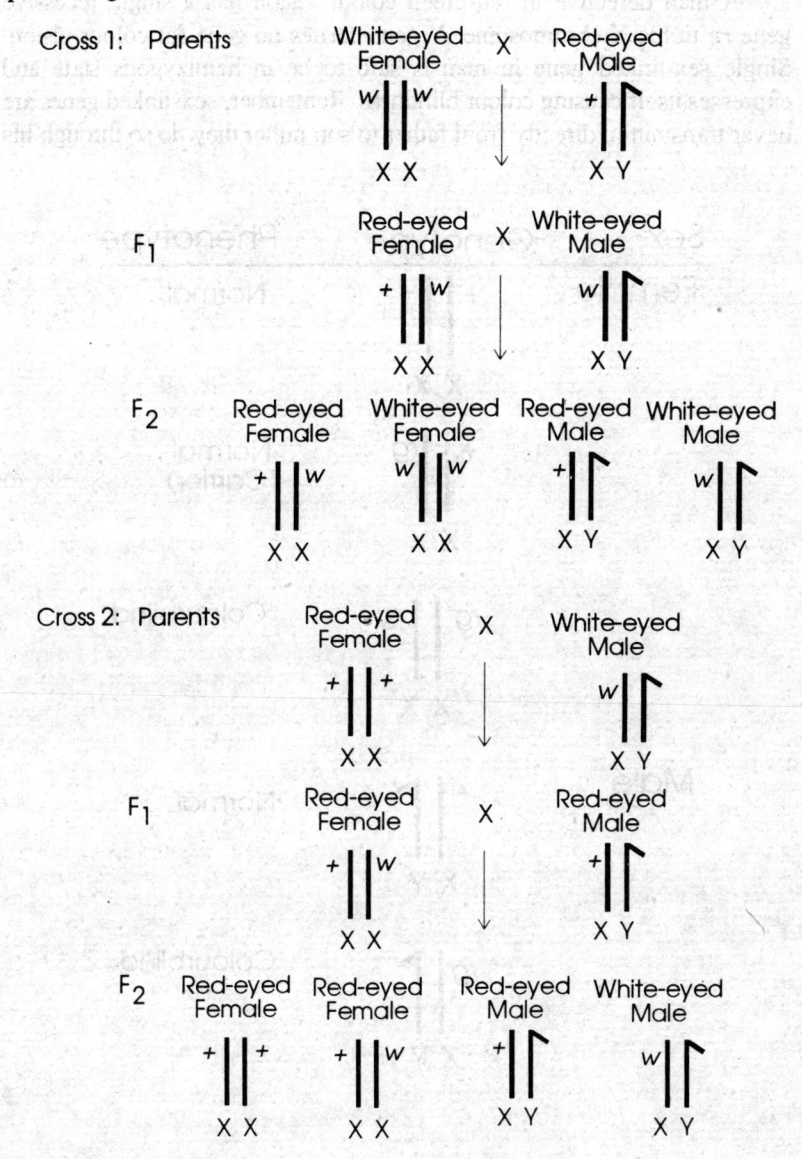

Cross 1: Parents: White-eyed Female (ww, XX) × Red-eyed Male (+, XY)

F_1: Red-eyed Female (+/w, XX) × White-eyed Male (w, XY)

F_2: Red-eyed Female (+/w, XX), White-eyed Female (ww, XX), Red-eyed Male (+, XY), White-eyed Male (w, XY)

Cross 2: Parents: Red-eyed Female (+/+, XX) × White-eyed Male (w, XY)

F_1: Red-eyed Female (+/w, XX) × Red-eyed Male (+, XY)

F_2: Red-eyed Female (+/+, XX), Red-eyed Female (+/w, XX), Red-eyed Male (+, XY), White-eyed Male (w, XY)

Fig. 6.1. White eye colour in *Drosophila melanogaster* showing sex-linked inheritance

Red-green Colour Blindness in Man

A man defective in red-green colour vision has a single recessive gene *rg* in his X chromosome. Y gene carries no gene for colour vision. Single sex-linked gene in man is said to be in hemizygous state and expresses itself causing colour blindness. Remember, sex-linked genes are never transmitted directly from father to son rather they do so through his

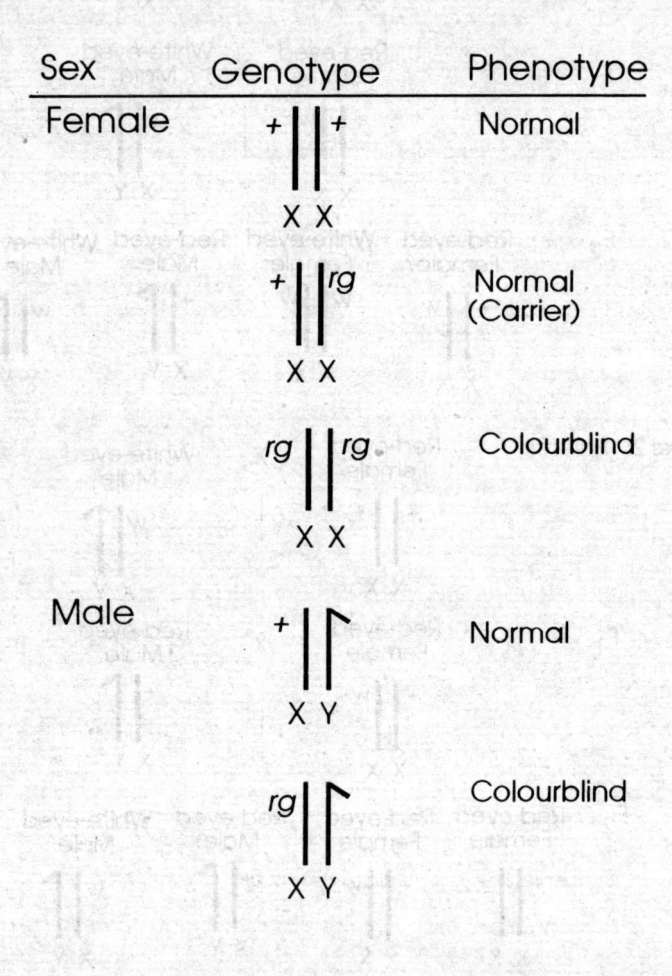

Sex	Genotype	Phenotype
Female	+ + (X X)	Normal
	+ rg (X X)	Normal (Carrier)
	rg rg (X X)	Colourblind
Male	+ (X Y)	Normal
	rg (X Y)	Colourblind

Fig. 6.2. Genotypes and phenotypes for sex-linked trait red-green colour blindness in man

Sex-linked, Sex-limited and Sex-influenced Inheritance 67

daughter. Thus *rg* gene also shows criss-cross pattern of inheritance and there is higher incidence of this condition in males than in females. In humans, sex-linked genes are discovered through **pedigree analysis**. The pedigree (**Fig. 6.2**) shows how X-linked recessive genes are expressed in males, then carried unexpressed through females (known as carriers) in the next generation, to be expressed in their sons.

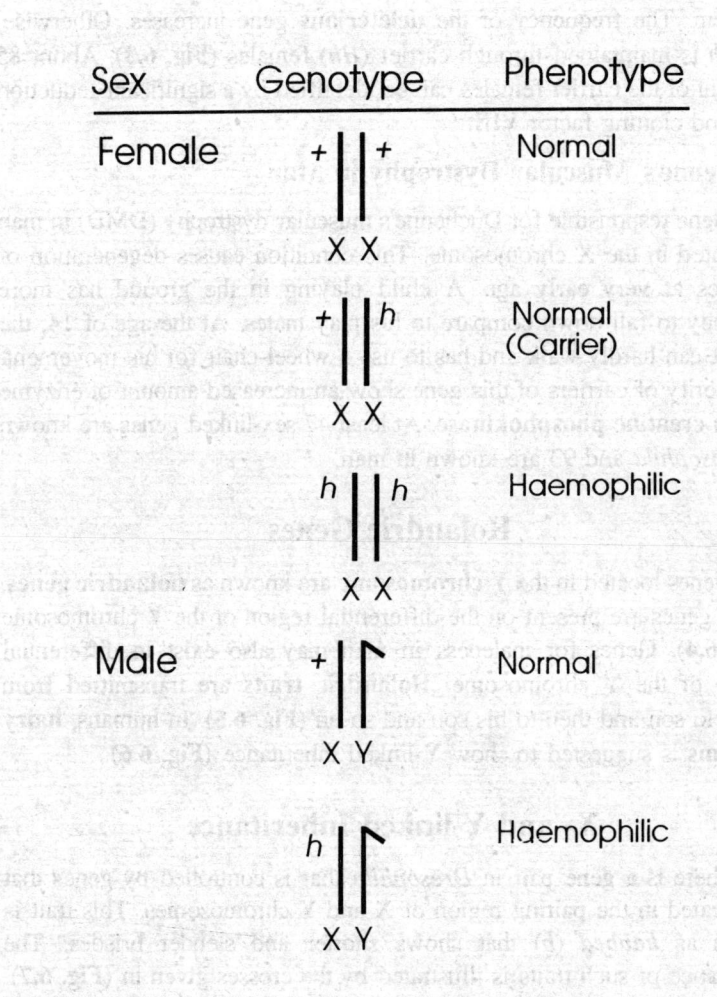

Sex	Genotype	Phenotype
Female	+/+ XX	Normal
	+/h XX	Normal (Carrier)
	h/h XX	Haemophilic
Male	+/– XY	Normal
	h/– XY	Haemophilic

Fig. 6.3. Genotypes and phenotypes for sex-linked trait haemophilia in man

Hemophilia in Man

It is a condition where blood fails to clot. The individuals possessing genotype *hh* do not survive up to the age of maturity and hence do not reproduce. An injection has been developed that contains **blood clotting factor (Factor VIII)** which when given periodically, a hemophilic individual can lead a normal life and reproduce. This, however, poses a problem. The frequency of the **deleterious gene** increases. Otherwise, gene *h* is maintained through carrier (*Hh*) females (**Fig. 6.3**). About 85 per cent of the carrier females can be identified by a significant reduction in blood clotting factor VIII.

Duchenne's Muscular Dystrophy in Man

Gene responsible for Duchenne's muscular dystrophy (DMD) in man is located in the X chromosome. This condition causes degeneration of muscles at very early age. A child playing in the ground has more tendency to fall down compare to his play mates. At the age of 14, the patient can hardly walk and has to use a wheel-chair for his movement. A majority of carriers of this gene show an increased amount of enzyme **serum creatine phosphokinase**. At least 47 sex-linked genes are known in *Drosophila* and 93 are known in man.

Holandric Genes

Genes located in the **Y chromosome** are known as **holandric genes**. These genes are present on the differential region of the Y chromosome (**Fig. 6.4**). Genes for maleness in man may also exist in differential region of the Y chromosome. **Holandric traits** are transmitted from father to son and then to his son and so on (**Fig. 6.5**). In humans, **hairy ear rims** is suggested to show Y-linked inheritance (**Fig. 6.6**).

X- and Y-linked Inheritance

There is a gene pair in *Drosophila* that is controlled by genes that are located in the pairing region of X and Y chromosomes. This trait is known as ***bobbed*** (*b*) that shows shorter and slender bristles. The inheritance of such traits is illustrated by the crosses given in (**Fig. 6.7**). In F_1 generation, results in the reciprocal crosses are same. X- and Y-linkage is revealed only in the F_2 generation.

Region containing the genes	Type of inheritance shown	Example
Differential region of X chromosome	Criss-cross pattern	Haemophilia in Man
Differential region of Y chromosome	Holandric	Hairy ear rims in Man
Pairing region of X and Y chromosomes	Pseudoautosomal	Bobbed trait in *Drosophila*

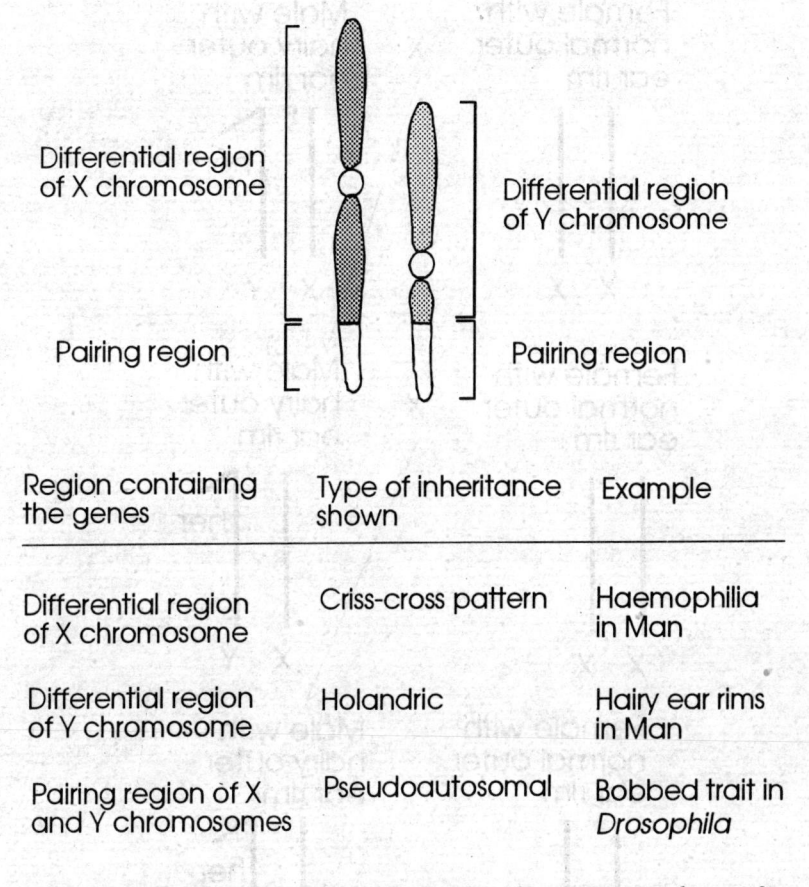

Fig. 6.4. Different regions of human X and Y chromosomes and type of inheritance shown by the genes located in them

Sex-limited Characters

Genes for a particular character may be present in both the sexes but they are expressed only in one of the sexes. Sex-limited characters are those whose expression is determined by the presence or absence of one of the the sex-hormones. The phenotypic effect is thus limited to one sex or the other. Thus one sex is uniform in expression of a particular trait and yet the same genes produce a different phenotype or do not express at all in the other sex in its progeny.

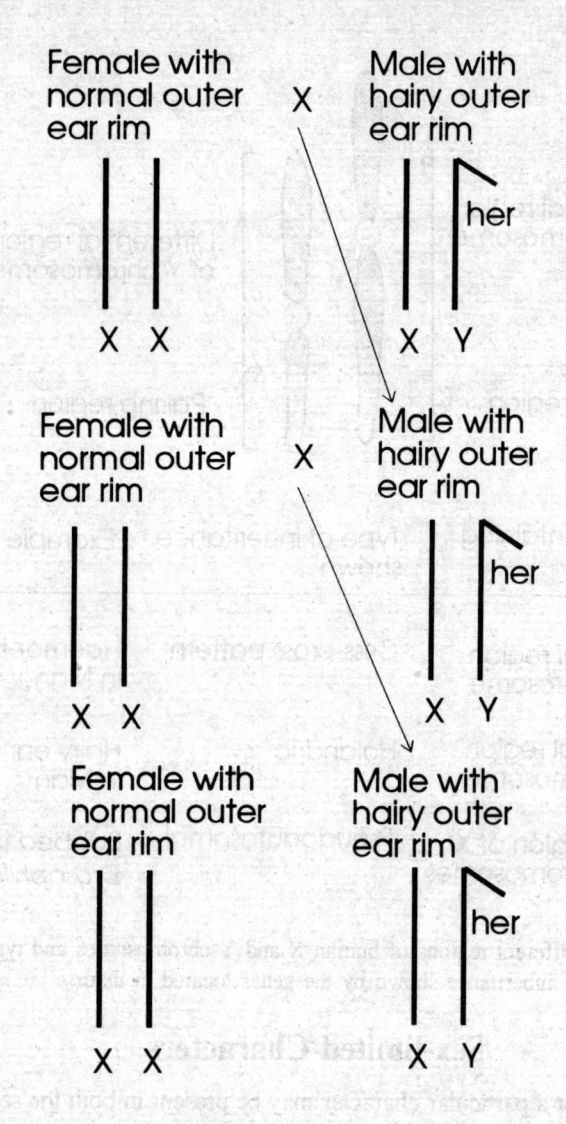

Fig. 6.5. Inheritance of a holandric trait hairy outer ear rim in man

Plumage in Birds

Feathering in domestic fowl is of two types: hen-feathering and cock-feathering. **Cock-feathering** is characterized by long, pointed, curving

Sex-linked, Sex-limited and Sex-influenced Inheritance 71

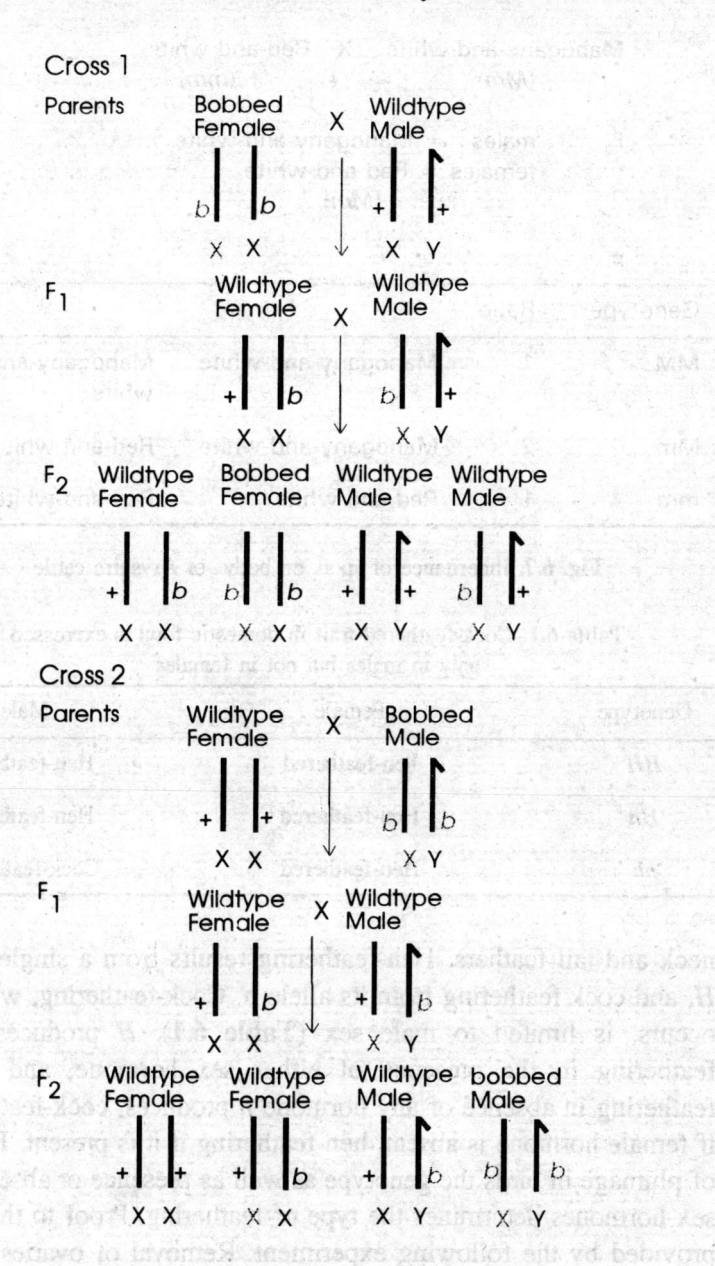

Fig. 6.6. Inheritance of a X- and Y-linked trait bobbed in *Drosophila melanogaster*

```
        Mahogany-and-white    X   Red-and-white
        ♂   (Mm)              ↓   ♀ (mm)

    F₁      males   = Mahogany-and-white
            females = Red-and-white
                      (Mm)
                        ↓
    F₂
```

Genotype	Ratio	♂	♀
MM	1	Mahogany-and-white	Mahogany-and-white
Mm	2	Mahogany-and-white	Red-and-white
mm	1	Red-and-white	Red-and-white

Fig. 6.7. Inheritance of spots on body of Aryshire cattle

Table 6.1. Cock-feathered trait in domestic fowl is expressed only in males but not in females

Genotype	Female	Male
HH	hen-feathered	Hen-feathered
Hh	Hen-feathered	Hen-feathered
hh	Hen-feathered	Cock-feathered

neck and tail feathers. Hen-feathering results from a single gene, *H*, and cock feathering from its allele h. Cock-feathering, where it occurs, is limited to male sex (**Table 6.1**). *H* produces hen-feathering in the presence of either sex hormone, and cock-feathering in absence of any hormone *h* produces, cock-feathering if female hormone is absent, hen-feathering if it is present. In case of plumage in birds the genotype as well as presence or absence of sex-hormones determines the type of feathering. Proof to this was provided by the following experiment. Removal of ovaries in *hh* hen produces cock-feathering. Thus female hormones inhibits expression of *hh* genotype, i.e., cock-feathering in females. Removal of testes in $h^+\_$ males produces cock-feathering because male hormones inhibit hen-feathering. Thus hormones of both sexes limit the gene action.

Colour of Clover Butter Fly

Males are always yellow; females may be yellow or white. Gene for white (*W*) is dominant over gene for yellow (*w*) colour. Gene *W* does not express in males (**Table 6.2**).

Some More Examples

Milk production in mammals, horns in Rambowillet sheep, egg production in chickens, age of onset of menstruation, width of pelvis, stuttering and distribution of body hairs are other examples of sex-limited traits. Various types of spots are seen on the body of Ayrshire cattle. Expression of this trait is also limited only to one sex. Genes for the sex-limited traits are autosomal and are present in both the sexes but their expression is limited to only one sex.

Table 6.2. Expression of white colour in clover butterfly is limited only to females

Genotypes	Female	Male
WW	White	Yellow
Ww	White	Yellow
ww	White	Yellow

Sex-influenced Traits

Those characters in which the dominance of an autosomal gene depends on the sex of the individual. In case of sex-influenced traits the heterozygotes express differently in males and females because of differences in the male and female hormones. Thus a trait may be dominant in one sex and recessive in the other. Sex influenced traits is also known as **sex-controlled traits**. This phenomenon is also known as **sex-influenced dominance**.

Baldness in Man

Baldness is more common in man than in women. Baldness is due to a dominant gene, *B*. Expression of baldness is associated with sex as illustrated in **Table 6.3**.

Table 6.3. Expression of a dominant gene for baldness in heterozygoous is associated with sex

Genotype	Female	Male
BB	Bald	Bald
Bb	Nonbald	Bald
bb	Nonbald	Nonbald

Horns in Sheep

In sheep autosomal genes responsible for horns behave differently in the presence of male and female hormones (Fig. 6.8).

```
           Dorset sheep      X    Suffolk sheep
           ♀ (h⁺h⁺)          ↓    ♂ (hh)
     F₁    females hornless  X    males horned
           (h⁺h)             ↓    (h⁺h)
     F₂
```

Males	Females
1 h^+h^+ horned	1 h^+h^+ horned
2 h^+h horned	2 h^+h hornless
1 hh hornless	1 hh hornless

Phenotypic ratios:

3 horned : 1 hornless   3 hornless : 1 horned

Fig 6.8. Inheritance of horns in sheep

Some More Examples

Gout is another sex-influenced in man which is more frequent in males. **Spina bifida** (forked spine with open spinal cord), another sex-influenced trait in man, is more frequent in females. **White forelock** in humans is due to gene *w'*. Normal gene is *w*. Heterozygous (*w'w*) males show white forelock whereas females are normal. **Short index finger** is also a sex-influenced which is due to a dominant gene *S* which in heterozygous state (*Ss*) expresses in males but not in females. **Fissure of upper lip, cleft palate, absence of upper lateral incisor, stuttering and enlargement of terminal joints of fingers** or some other well known sex-influenced traits in man.

Some Useful Reading Hints

Morgan (1910) was first to assign a gene (white) to a particular chromosome (sex chromosome) in *Drosophila melanogaster* thus providing evidence to chromosomal theory of inheritance. Morgan (1911) attempted to analyse the constitution of the chromosomes on the basis of sex-limited inheritance in *Drosophila* Bridges (1925) discussed sex in Relation to chromosomes and genes. Burgoyne (1982) describeed genetic homology and crossing over in the X and Y chromosomes in mammals. Corcos (1983) provided a fresh look at pattern baldness, a sex-influenced trait.

Questions

1. Define sex-liked inheritance. Why is it also known as criss-cross inheritance or skip-generation inheritance? Describe various characteristics of sex-linked inheritance?
2. Explain with the help of labelled diagrams inheritance pattern of the following sex-linked traits: *white* eyes in *Drosophila*, red-green colour blindness in man, hemophilia in man, and Duchenne's muscular dystrophy in Man.
3. Enlist atleast ten more sex-linked traits.
4. What are holandric genes? How does the inheritance of holandric traits differ from the sex-linked traits? Explain with the help of a labelled diagram inheritance of holandric trait hairy outer ear rim. Do you know of some other holandric trait?
5. What are the characteristics of X- and Y-linked inheritance? Explain inheritance of bobbed trait in *Drosophila*. Do you know of any other X- and Y-linked trait?
6. What are the characteristics of sex-limited characters? Genes for such traits are present in both the sexes but their expression is limited only to one of them, why?
7. Explain with the help of diagrams inheritance of the following sex-limited traits: plumage in birds, colour of clover butter fly, and spots in Ayrshire cattle.
8. Enlist atleast five more sex-limited traits.
9. What are sex-influenced traits? How do they differ from sex-limited traits?
10. Explain the following sex-influenced traits with the help of labelled diagrams: (i) baldness in man and (ii) horns in sheep.
11. Enlist atleast five more examples of sex-influenced traits.

Numericals

1. In *Drosophila*, white eyes w are recessive to red W (sex-linked) and vestigial wings vg are recessive to normal long wings Vg (autosomal) of a homozygous white, long female is crossed with a homozygous red, vestigial male, what will be the appearance of F_1 of F_2 of the offspring of a cross of F_1 with each parent type.
2. In *Drosophila*, two red eyed, long winged flies bred together produce the following offspring: Females: 3/4th red long; 1/4th red vestigial; Males : 3/8th red long; 3/8th white long, 1/8th red vestigial; 1/8th white vestigial. What are the genotypes of the parents?
3. A colourblind man marries a woman of normal vision. They have sons and daughters of normal vision. All of them marry persons of normal vision. Where is the colourblindness expected to appear in grand children?
4. A man and a woman of normal vision have one and two daughters. Son is colourblind and has a daughter of normal vision. First daughter of normal vision has one colourblind and one normal son. Second daughter of normal vision has 5 sons, all normal. what are probable genotypes of parents, children and grand children?
5. A man's maternal grandmother had normal vision, his maternal grandfather was colourblind, his mother is colourblind, his father is of normal vision. What are the genotypes of two parents and grand-parents? What of vision has this man? What type have his sisters. If he marries a woman genotypically like one of his sister, what type of vision is expected in their children?
6. A girl of normal vision whose father was colourblind marries a man of normal vision whose father was also colourblind. What type of vision is expected in their children?
7. A colourblind man marries a woman of normal vision. They have sons and daughters of normal vision. What are the genotypes of man and woman?
8. A man and a woman of normal vision have one son and one daughter. Son is colourblind and has a daughter of normal vision. Daughter has normal vision, but one of her sons is colourblind and other normal. What are the genotypes of man, woman, their son and daughter?

7

MULTIPLE ALLELES

Multiple Allelism

Multiple allelism is a condition where two or more alternate alleles represent the same locus in a given pair of chromosomes. Although only two actual alleles of a given gene can exist in a diploid cell, the total number of different allelic forms that might exist in a population is often quite large. Number of possible diploid combinations depends upon number of allelic in a series. If n is the no. of alleles in a series, number of diploid combinations (genotypes) is given by the expression n(n+1)/2. For example, if number of alleles is 5, number of diploid genotypes in a population will be 1/2{5(5+1)} = 1/2 X 5 X 6 = 15.

ABO Blood Groups in Man

K. Landsteiner in 1900 reported ABO blood group system in man. This system consisted of 3 alleles I^A, I^B and i. These 3 alleles from 6 different genotypes and yield 4 phenotypes. I^A is dominant over i. I^B is dominant over i. I^A and I^B are codominant (**Table 7.1**). **Codominance** is the condition where expression of both the alleles in a heterozygote can be observed. Red blood cells contain antigens and serum contains antibodies A particular blood type contains particular types of antigens and antibodies (**Table 7.2**).

Table 7.1. Genotypes and phenotypes of human blood under ABO system

Genotypes(s)	Blood type
$I^A I^A$, $I^A i$	A
$I^B I^B$, $I^B i$	B
$I^A I^B$	AB
ii	O

Table 7.2. Antigens and antibodies present in different bloods types

Phenotype	Antigens	Antibodies
A	A	Anti- B
B	B	Anti- A
AB	A + B	None
O	None	Anti-A + Anti-B

Clumping of antigens and antibodies. Antigen A clumps with **Antibody-A** Antigen B Clumps with **Antibody-B**. These reactions, called **agglutination**, form basis of testing blood types. Thus it is possible to **cross-match** blood and determine compatibility before transfusion.

Universal donor and universal acceptor. Persons with O type of blood lack both antigens A and B and can give blood to members of any group. Thus persons having O type of blood are called **universal donors**. Persons with 'AB' type of blood possess both antigens A and B and can receive blood from any person. Thus the persons having 'AB' type of blood are called **universal acceptor**. Now a days blood type of donor and recipient is crossed checked. If there is no clumping then transfusion is done. Blood typing is helpful in deciding disputed cases of paternity and in criminal cases.

Self-incompatibility Alleles in Plants

Some plants just donot self-pollinate. A single flowers bearing male and female flowers may not produce any seed. The same plant may, however, cross with certain other plants, so obviously they are not sterile. This phenomenon is known as **self-incompatibility**. In case of self-incompatibility, ovum and pollen produced by the same plant do not produce seed. Self-incompatibility has a genetic basis. If a pollen grain bears a self-incompatibility (S) allele that is also present in the maternal parent, then it will not germinate. However, if that allele is not present in the maternal tissue, then the pollen grain produces a pollen tube containing the male nucleus, and this tube affects fertilization (**Table 7.3**). Large number of self-incompatiblity alleles are known to be present in different plant species. For example, the number of alleles in evening primrose and clover is 50 each while in some species it is 100.

Rhesus (Rh) Alleles in Man

The **Rh factor** was discovered in 1940 by Landsteiner and Wiener from rabbits immunized with blood of a monkey. the resulting antibodies

Table 7.3. Genetic control of self-incompatibility

Parents	S_1, S_2 X S_1, S_2	S_1, S_2 X S_2, S_3	S_1, S_2 X S_3, S_4
Pollen	S_1, S_2	S_2, S_3	S_3, S_4
Egg Cells	S_1, S_2	S_1, S_2	S_1, S_2
Progeny	None	S_1, S_3	S_1, S_3 S_1, S_4
		S_2, S_3	S_2, S_3 S_2, S_4
Results	Fully incompatible	Semi-compatible	Fully compatible

agglutinated not only RBCs of the monkey but also those of a high percentage of Caucasian people of New York. A test for **Rh Incompatibility** is accomplished by placing a drop of blood from the subject on a slide and introducing **anti-Rh** serum. Agglutination of erythrocytes indicates incompatibility, when as even distribution of erythrocytes indicates no reaction. Rh is a multiple allelic system. There are several Rh⁺ and Rh⁻ alleles: Rh⁺ alleles - R^1, R^2, R^0, R^z and Rh⁻ alleles - r, r′, r″, r^y.

Cross-matching of the Rh factor as well as of the ABO system is now used for establishing compatibility of blood types for transfusion. **Erythroblastosis fetalis** is a condition found in new born infants. These infants are Rh⁺ and so are their fathers; but their mothers are Rh⁻. Rh⁺ foetus developing in the uterus of an Rh⁻ mother causes formation of anti-Rh antibodies in mother's blood stream. These antibodies may attack the RBCs of the foetus which may cause death of the child.

Eye Colour in *Drosophila melanogaster*

Genes for eye colour exist in all the major chromosomes of *Drosophila melanogaster*. These are large number of alleles at the *white* locus of *Drosophila* (**Table 7.4**). This multiple allelic series and some other eye colour mutants have contributed immensely in understanding structure and function of gene (Chapter 14).

Coat Colour in Rabbits

Multiple alleles in this series are c (full colour), c^{ch} (Chinchilla, a light greyish colour), c^h (Himalyan, albino with black extremities, and c (albino). In this series dominance is in the order of alleles listed in **Table 7.5**.

Table 7.4. Alleles at the *white* locus of *Drosophila melanogaster*

Allele	Symbol	Allele	Symbol
Wildtype	w^+	apricot	w^a
White	w	cherry	w^{ch}
ivory	w^i	eosin	w^e
pearl	w^p	blood	w^{bl}
tinged	w^t	coral	w^{co}
honey	w^h		

Histocompatibility Alleles in Man

Histocompatibility antigens are those antigens that determine the acceptance or rejection of a tissue graft. These antigens are produced by histocompatibiulity genes. **Histocompatibility genes** are the genes that code for histocompatibility antigens. **Human leukocyte antigen (HLA) complex** comprises of genes that are responsible for synthesis of antigens that determine acceptance or rejection of a tissue graft. Histocompatibility is also known as **tissue compatibility.** Tissues can be transplanted between genetically identical animals without concern for immunologic rejection, whereas transplants between genetically non-identical animals are usually rejected with time. The timing and intensity of the rejection reaction are functions of the genetic differences between donor and recipient. Each of these genes have a series of alleles. This allelic series serves as an excellent example of multiple allelism in humans.

The HLA type of an individual is determined by testing his or her lymphocytes with a battery of standard antibodies from donors of known HLA type. In a transplant the recipient's **immune system** will recognize the graft as "foreign" and reject it if an allele is present in the donor tissue that is not present in the recipient. In the typing test, this rejection is observed as **lymphocyte death.** The immune system will not reject tissue that lacks some alleles present in the recipient-it only rejects the **"strange" alleles. Table 7.6** shows examples using different alleles of A and B genes assuming that all alleles of other genes between graft and recipient are identical.

Table 7.6. Strange alleles of the donor are rejected by the recipient

Transplant No.	Recipient genotype	Donor genotype	Result
1	A_1A_2, B_5B_5	A_1A_1, B_5B_7	Rejected (B_7)
2	A_2A_3, B_7B_{12}	A_1A_2, B_7B_7	Rejected (A_1)
3	A_1A_2, B_7B_5	A_1A_2, B_7B_7	Accepted
4	A_2A_3, B_7B_5	A_3A_3, B_5B_5	Accepted

HLA complex. Genes involved in the production of cell-surface antigens that are recognized by the rejection or tolerance of tissue transplants are collectively called Histocompatibility genes or loci. The action of antigen-producing alleles at many of these loci seems to be codominant. Individuals will reject the tissue of donors that carry alleles which they themselves donot carry. **HLA system** is based on three classes of HLA genes on chromosome number 6 which determine **immunological acceptability** (Fig. 7.1). Each gene may possess 8 to 40 alleles (**Table 7.7**), each allele specifying a particular antigen. A particular number 6 chromosome may have one of 75,000 theoretically possible combinations of HLA alleles; each particular combination is called a haplotype. An individual usually possesses two different haplotypes, one from each parent. This provides millions of different possible HLA diploid

Class	Genes
Class I	B, C, A
Class II	SB DR DC BR ∧ ∧ ∧ ∧ α β α β α β α β
Class III	C2, BF, C4A, C4B.

SB, DR, DC and BR are regions, every one containing two genes α and β

Fig. 7.1. Three classes (I-III) of genes in human leucocyte antigen (HLA) complex

82 Basic Genetics

Table 7.7. Multiple alleles of HLA complex in Man

Gene	# alleles
HLA-A	17
HLA-B	32
HLA-C	8
HLA-D	12
HLA-DR	10
HLA-C$_2$	5
HLA-C$_4$A	6
HLA-C$_4$B	4
HLA-BF	11

genotypes. This enormous populational variability helps explain the difficulties in successful transplantation of tissues and organs, since most unrelated people will differ markedly in their HLA genotypes. Tests to obtain as close an HLA match as possible between transplant recipient and donor are necessary. Hypothetical pedigree for the inheritance of HLA-A and HLA-B antigen alleles in given in **Fig. 7.2**. Each parent carries two haplotypes and each haplotype in commonly transmitted unchanged. Rare crossovers do occur and occasional siblings may show

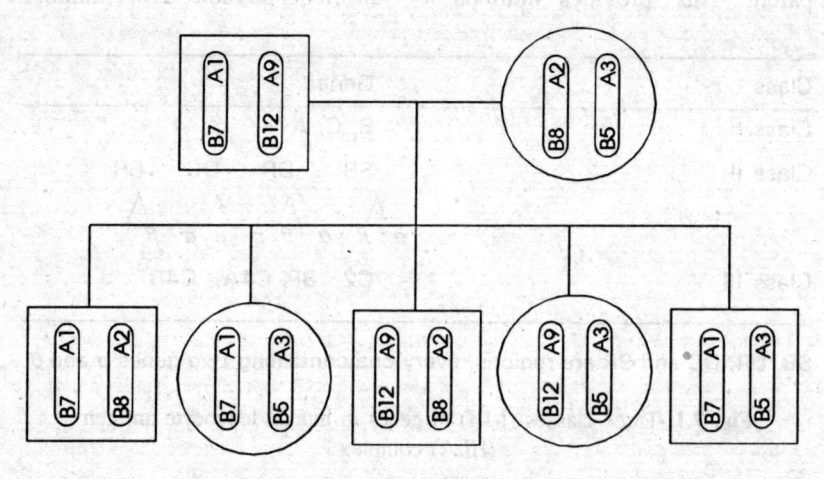

Fig. 7.2. Hypothetical pedigree for inheritance of HLA antigen alleles

similar antigens. HLA tests can also be used to help decide question of **genetic relatedness** (such as paternity problems). There is a correlation between particular HLA antigens and the incidence of particular diseases. In **autoimmune disease**, individual's own tissues and organs are rejected.

Functions of HLA genes. Tissue transplants have not been a normal aspect of human evolution. **Graft rejection** in man-made phenomenon. Hence HLA antigens must be performing some normal functions in man. Some HLA alleles seem to confer resistance or susceptibility to certain specific diseases (**Table 7.8**). These alleles may have played some role in our evolutionary history in response to some environmental selection pressure. HLA genes may be intimately involved in cancer. Presumably one of the normal functions of the HLA genes is to recognize cancer cells as a kind of "foreign" agent within the body. The development of cancer cells may be a fairly common phenomenon in healthy individuals with these cells being recognized and destroyed before they cause noticeable damage. A cancerous tumor may be the result of a failure of this detection and destruction system at some stage. The HLA genes also play an important role in normal immune response.

Table 7.8. Possible relationship of some HLA antigens with diseases

Disease	HLA antigen	Frequency of antigen in patients (with disease)	Frequency of antigen in controls (without disease)	Relative risk
Joint diseases				
Ankylosing sondylitis	B27	90	9.4	87.0
Reiter,s disease	B27	79	9.4	37.0
Rhematoid arthritis	D4	50	19.4	4.2
Skin diseases				
Psoriasis	Cw6	87	33.1	13.3
Pemphigus	D4	87	32.1	14.4
Systemic lumpus erythematosus	D3	70	28.2	5.2
Gland diseases				
Hashimoto's thyroiditis	D5	19	6.9	3.2
Grave's disease	D3	56	26.3	3.7

Table 7.8 Contd...

Addison's disease	D3	69	26.3	6.3
Insulin-dependent diabetes	D4	75	32.2	6.4
Gastrointestinal diseases				
Celiac disease	D3	79	26.3	10.8
Hemochromatosis	A3	76	28.2	8.2
Ulcerative colitis	B5	80	30.8	9.3
Malingnancies				
Acute lymphatic leucemia	A2	60	53.6	1.3
Hogdkin's disease	A1	40	32.0	1.4
Other diseases				
Multiple scelrosis	D2	59	25.8	4.1
Pernicious anemia	D5	25	5.8	5.4

Lethal Genes

A gene that renders inviable an organism or a cell possessing it in proper arrangement for expression, is known as a lethal gene. Cuenot (1940) studied mice having yellow coat colour (**Fig. 7.3**). This hypothesis

```
            Yellow          X              wildtype
            AʸA             ↓              AA

F₁          1 yellow:       1              wildtype
            AʸA                            AA

F₁          Yellow X        F₁             Yellow
            AʸA             ↓              AʸA

            1 AʸAʸ :        2 AʸA :        1 AA
            died            yellow         wildtype
```

Fig. 7.3. Results of Cuenot's experiment involving lethal gene in mice

that individuals of genotype A^yA^y die was confirmed by removal of uteri from pregnant females of F_1 yellow X F_1 yellow; one-fourth of the embryos were found dead.

Conclusions from Multiple Allelism

The conclusions that can be drawn from studies on multiple alleles are: a gene can exist several alternate forms. There are several mutable sites within a genes. Members of allelic series can show any type of dominance relationship with one another.

Some Useful Reading Hints

Bateman (1947) dealt with self-incompatibility alleles in a population. Thompson and Kirch (1992) discussed that S locus of flowering plants provided a system where **self-rejection** is self-interest. Matton et al. (1994) discussed that self-incompatibility is a mechanism to avoid **illegitimate offspring** in plants. Nasrallah et al. (1994) dealt with the mechanism signalling the arrest of pollen tube development in self-incompatible plants. Wihelmi and Preiss (1996) discovered that self-sterility in *Arabidopsis* due to defective pollen tube guidance. Ikoda et al. (1997) discovered that an aquaporin-like gene required for the *Brassica* self-incompatibility response.

Yamamoto (1990) dealt with molecular genetic basis of the histo-blood group ABO system. Brown et al. (1993) gave three-dimensional structure of the human class II histocompatibility antigen HLA-DRI. Strickberger (1986), Jenkins (1975), Sarin (1985), Zubay (1987), Farnsworth (1988), Burns and Bottino (1989), Brown (1992), Tamarin (1996) provided detailed account of various examples of multiple alleles.

Questions

1. What is the relationship between number of alleles at a locus with number of diploid genotypes in a population?
2. What is the genetic basis of blood types in ABO system in man?
3. What types of antigens and antibodies are present in different blood groups of ABO system in man and which antigens and antibodies show clumping with each other?
4. What is the basis and method of blood group testing?
5. What do you understand by concept of universal donor and universal acceptor?
6. Why some of the plants just do not self-pollinate even when the male and female flowers are perfectly normal?
7. What is the genetic basis of self-incompatibility?
8. Why is it necessary to test the blood for Rh type before transfusion? What is erythroblastosis fetalis?
9. Describe multiple alleles at the *white* locus of *Drosophila*?

10. Dominance in the alleles of multiple allelic series may be different. Illustrate this point with the help of example of coat colour in rabbits.
11. What is the genetic basis of rejection or acceptance of a graft by the host?
12. What is HLA complex? Mention different classes of genes of the HLA locus?
13. What is the extent of genetic diversity of antigens of the histocompatibility system?
14. What is the relationship between histocompatibility antigens with resistance or susceptibility of various diseases?
15. What are normal functions of the histocompatibility system?
16. In what way are the histocompatibility genes involved in cancer?
17. What conclusions can be drawn from the studies on multiple allelism?
18. What is the significance of study of lethal genes?
19. What has been contribution of this multiple allelic series in understanding the gene.

Numericals

1. In humans, a series of alleles has been associated with the blood typing groups as follows: I^A-, A type; I^B-, B type; ii, O type; I^A and I^B are co-dominant; $I^A I^B$ heterozygotes have AB type blood; i is recessive to both I^A and I^B. What phenotypes and ratios might be expected from the following mating: (a) $I^A I^A$ X $I^B I^B$; (b) $I^A I^B$ X ii (c) $I^A I^A$ X $I^B I^B$ and (d) $I^A i$ X ii?
2. A case was brought before certain judge in which a woman of blood group O presented a baby of blood group O, which she claimed as her child, and brought suit against a man of group AB whom she claimed was the father of the child. What bearing might the blood-type information have on the case?
3. In another case, a woman of blood group AB presented a baby of group O, which she claimed as her baby. What bearing might the blood type information have on the case. In the following three problems, determine the genotypes.
(a) One parent is AB and the other B, but of the children one-fourth are A, one fourth AB and one half B.
(b) One parent is group A and the other group B, but all four groups are represented among the children.
(c) In the two following cases of disputed paternity, determine the probable father of the child: (a) the mother belongs to group B, the child to O, one possible father to A and the other to AB. (b) the mother belongs to group B, the child to AB, one possible father to A and the other to B.
4. A baby has blood type AB. What can you tell about the genotypes of its parents? What would you predict about the blood types of children it will later produce?
5. If one parent is A blood type and the other is B, give their respective genotypes if they produce a large number of children whose blood types are: (a) All AB, (b) Half AB and half B, (c) Half AB and half A, (d) 1/4 AB, 1/4 A, 1/4 AB, 1/4 O.

8

GENE-GENE INTERACTIONS

In characters studied by Mendel, variation for each trait was controlled by a single gene. Actually, variation in many characters may be governed by two or more genes. Genes are not merely separate elements producing distinct individual effects but they could interact with each other to give completely **novel phenotypes**. Gene interactions may give **modified F_2 ratios**. This so happens because many characters are influenced by two or more gene pairs which interact. This is explained by the a few examples here. In all these cases it is assumed that both the gene pairs show complete dominance. In case of independent assortment, we deal with two gene pairs or more all governing different characters whereas in gene-gene interactions we deal with two or more gene pairs affecting one and the same character.

Dominance and Epistasis

In case of dominance one member of allelic pair marks the expression of the other member (interallelic interaction) whereas in case of epistasis there is suppression of action of a gene or genes by a gene or genes not allelomorphic to those expressed (intergenic interaction).

Epistatic and Hypostatic genes

Epistatic are the genes that suppress the gene expression whereas hypostatic are the genes whose action is suppressed. Gene interactions are classified on the basis of the manner in which the concerned genes influence or modify the expression of each other. Some of the common interactions are described below.

Interactions Between Nuclear Genes

Interaction Between Two Dominant Genes

A Typical Dihybrid Ratio - Novel Phenotype. An example of interaction between two dominant independently inherited genes is provided by comb shape in fowls (Bateson and Punnet, 1905). The crosses are shown in **Fig. 8.1**. Cross I shows that difference between rose and single comb is due to one gene difference. Similarly, cross II shows that difference between pea and single comb is due to one gene difference. Cross III shows that two dominant genes R and P give a

```
Cross I           Rose       X      Single
                   RR                  rr
                              ↓
           F₁      Rose       X      F₁ Rose
                    Rr                  Rr
                              ↓
           F₂     3/4 Rose  :  1/4 Single
                  1 RR
                  2 Rr              1 rr

Cross II          Pea        X      Single
                   pp                  pp
                              ↓
           F1      Pea        X      Pea
                   Pp         ↓      Pp

           F2     3 Pea      :  1 Single
                  1 PP
                  2 Pp              1 rr

Cross III         Rose       X      Pea
                  RRpp              rrPP
                              ↓
           9 Walnut  :  3 Rose  :  3 Pea  :  1 Single
             R-P-         R-pp        rrP-       rrpp
```

Fig. 8.1. Crosses showing that difference between rose and single comb shape in poultry is due to one gene difference. Difference between pea and single comb is due to one gene difference. Two dominant genes interact to produce a novel phenotype

```
                    Rose
                   ↗
              R  ╱
    rrpp      ╱
    Single   ╳──────→ Walnut
             ╲
              P  ╲
                   ↘
                    Pea
```

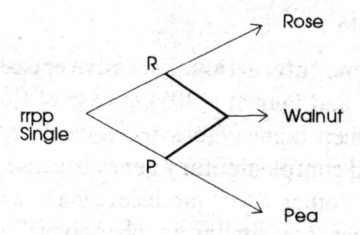

Fig. 8.2. Biochemical pathway suggested to explain interaction between two dominant genes for comb shape in poultry

typical dihybrid ratio (9:3:3:1) but interact to produce a novel phenotype. Single comb is produced due to interaction of two different recessive genes. Possible metabolic pathway is shown in **Fig. 8.2**. Molecular mechanism of this interaction is based on the interaction of products of different genes as explained in **Fig. 8.3**.

Gene	Product	
R	Chain A	———
r	No Chain A	
P	Chain B	– – –
p	No Chain B	

Genotype(s)	Product	Phenotype	Ratio
R_P_	= = =	Walnut	9
R_pp	═══	Rose	3
rrP_	= = =	Pea	3
rrpp	No protein	Single	1

Fig. 8.3. Molecule model suggested to explain interaction between dominant and recessive genes for comb shape in pultry

Recessive Epistasis

Complementary Gene Interaction. Recessive epistasis is exemplified by the work of Bateson and Punnet (1905) in case of flower colour in sweet peas. Either gene when homozygous recessive is epistatic to the other. Such genes are called **complementary genes** because two dominant genes complement each other to produce the wildtype phenotype. Complementary genes are similar in phenotypic effect when present separately. When together they interact to produce a wild-type phenotype. When two different inbred strains, both with white flowers, are crossed the results obtained are shown in **Fig. 8.4**. The genes showing this type of interaction are known as complementary genes. Because two dominant genes interact to produce the wild type phenotype (purple in above example). Mechanism of complementary gene interaction is illustrated in **Fig. 8.5**.

Strain 1		Strain 2
White	X	White
CCpp	↓	*ccpp*
F_1 Purple	X F_1	Purple
CcPp	↓	*CcPp*
F_2 9 *C-P-*		3 *C-pp*
		3 *ccP-*
		1 *ccpp*
9 Purple	:	7 White

Fig. 8.4. Complementary gene interaction for flower colour in sweet pea. This is one type of recessive epistasis

Another example of complementary gene interaction is provided by HCN content in clovers in which case two dominant genes *Ac* and *Li* are required for production of HCN. Write a biochemical pathway to explain the above gene interaction.

Supplementary Gene Interaction. Coat colour in mice is an other example of recessive epistasis. Black, albino and **agouti patterns** of coat colour in mice are shown in **Fig. 8.6**. This type of interaction is known as **supplementary gene interaction**. Possible metabolic pathway is given

Gene	Product	Representation
C	An enzyme that synthesizes purple pigment	● ● ● ● ●
c	No enzyme synthesized, therefore, o pigment formed.	
P	A protein that forms a foundation for the deposition of the pigment	 ⊓⊔⊓⊔⊓⊔⊓⊔⊓⊔
p	No protein synthesized, therefore, no foundation formed.	

Genotype (s)	Product(s)	Phenotype	Ratio
C_P_	● ● ● ● ● Foundation + Pigment	Purple	9
C_pp	● ● ● ● ● No Foundation + Pigment	White	
ccP_	 ⊓⊔⊓⊔⊓⊔⊓⊔⊓⊔ Foundation + No Pigment	White	7
ccpp	No Foundation + No Pigment	White	

Fig. 8.5. Molecular model explaning complementary gene interaction in case of flower colour in sweet peas

	Albino ccAA	X	Black CCaa
F₁	CcAa Agouti	X F1	CcAa Agouti
F₂	9 C-A-	3 C-aa	3 ccA- 1 ccaa

| 9 Agouti | 3 Black | 4 Albino |

Fig. 8.6. Recessive epistasis exhibited by coat colour genes in mice is supplementary gene interaction

in **Fig. 8.7**. Molecular mechanism of supplementary gene interaction assumes that gene *C* controls a step for the production of melanin pigment. The mutation *c* represents a defect in biosynthetic pathway for melanin pigment. The mutation *a* specifies that the pigment be dispersed uniformly throughout the length of the hair whereas its dominant allele *A* specifies agouti pattern (**Fig. 8.8**).

92 Basic Genetics

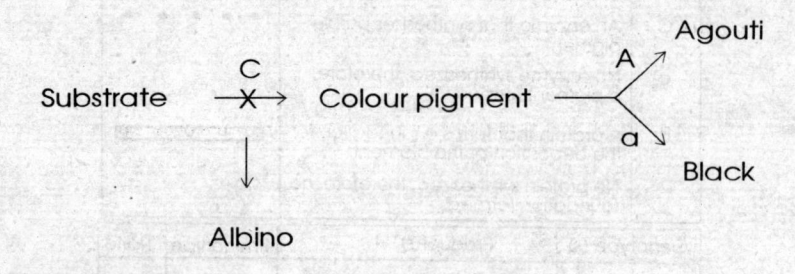

```
                    C                           A ↗ Agouti
    Substrate  ──✕──→  Colour pigment  ──────<
                    ↓                           a ↘ Black
                  Albino
```

Fig. 8.7. Possible metabollic pathway for supplementary gene interaction shown by recessive genes for coat colour in mice

Gene	Action
C	Controls a step in the formation of melanin pigment
c	Represents a defect in pathway for melanin pigment
A	Specifies agouti pattern when pigment is formed
a	Specifies that pigment be dispersed througout the length of the hair

Genotype(s)	Product(s)	Phenotype	Ratio
C_A_	Pigment + Agouti pattern	Agouti	9
C_aa	Pigment + Pigment dispersal	Black	3
ccA_	No Pigment + Agouti pattern	Albino	4
ccaa	No pigment + Pigment dispersal	Albino	

Fig. 8.8. Molecular model explaining supplementary gene interaction shown by recessive genes for coat colour in mice

Dominant Epistasis - Duplicate Genes

When two gene pairs seem to be identical in function, either dominant gene or both dominant gene together give the same effect. Such genes are called **duplicate genes** and the type of epistasis is called **dominant epistasis**. Capsule shape in Bursa (Shepherd's purpose) provides an example (**Fig. 8.9**). Possible metabolic pathway involved in above gene-gene interaction is given in **Fig. 8.10**.

$$
\begin{array}{lcl}
\text{Triangular} & & \text{Ovoid} \\
T_1T_1\ T_2T_2 & \times & t_1t_1\ t_2t_2 \\
& \downarrow & \\
F_1\ \text{Triangular} & F_1 & \text{Triangular} \\
T_1t_1\ T_2t_2 & \times & T_1t_1\ T_2t_2 \\
& \downarrow & \\
F_2\ 9\ T_1\text{-}\ T_2\text{-} & & 1\ t_1t_1\ t_2t_2 \\
\ \ \ 3\ T_1\text{-}\ t_2t_2 & & \\
\ \ \ 3\ t_1t_1\ T_2\text{-} & & \\
\end{array}
$$

---

15 Triangular   1 Ovoid

Fig. 8.9. Dominant epistasis exhibited by duplicate genes for fruit shape in Shepherd's purse.

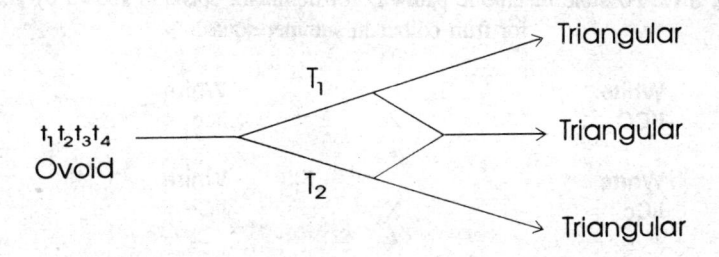

$$
t_1t_2t_3t_4\ \text{Ovoid} \longrightarrow
\begin{cases}
T_1 \longrightarrow \text{Triangular} \\
\quad\quad \longrightarrow \text{Triangular} \\
T_2 \longrightarrow \text{Triangular}
\end{cases}
$$

Fig. 8.10. Possible metabollic pathway for duplicate epistasis shown by genes for capsule shape in Bursa (Shepherd's purse)

Fruit colour in summer squash provides another example of dominant epistasis (**Fig. 8.11**). In this case there is complete dominance but one gene when dominant is epistatic to the other. Metabolic pathway given in **Fig. 8.12** satisfies the observations made in this cross.

94   Basic Genetics

```
         White              X              Green
         WWYY                               wwyy
                            ↓
    F₁   White              X         F₁   White
         WwYy               ↓              WwYy

    F2   9 W_Y_    3 W_yy  +  3 wwY_    1 wwyy
```

12 White	6 Yellow	1 Green

Fig. 8.11. Fruit colour in summer squash provides example of dominant epistasis

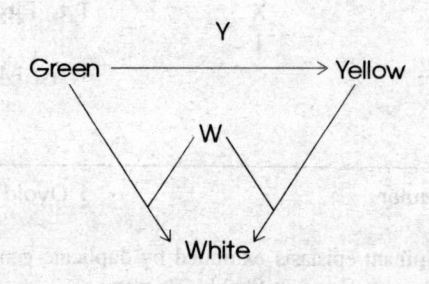

```
                         Y
           Green ───────────────→ Yellow
                   ↘            ↙
                       W
                       ↓
                     White
```

Fig. 8.12. Possible metabolic pathway for dominant epistasis shown by genes for fruit colour in summer squash

```
              White                              White
              IICC              X                iicc
                                ↓
        F₁    White                        F₁    White
              IiCc              X                IiCc
                                ↓
        F₂    9 I-C-                              3 iiC-
              3 I-cc
              1 iicc
```

13 White	:	3 Coloured

Fig. 8.13. Dominance recessive epistasis exhibited by inhibitory genes for feather colour in fowls. This type of effect is also called suppression gene interaction

Dominant-Recessive Epistasis

Inhibitory Genes. In case of inhibitory genes, one gene when dominant is epistatic to the other but other when recessive homozygous is epistatic to the first. Thus there is dominant recessive epistasis. Example is provided by feather colour in fowls. Gene I inhibits production of colour even when gene for colour is present and cc prevent colour because it cannot provide substrate for colour (**Fig. 8.13**). The type of interaction described above is called **suppression gene interaction**.

Suppression of one gene by another is also illustrated by inheritance production of a chemical called "malvidin" in the plant of genus *Primula*. Dominant gene *K* is required for production of pigment. However, action of this dominant gene K may be suppressed by a nonallelic dominant gene *D*. Selfing of F_1 dihybrid (*KkDd*) produces in F_2 13 malvadin minus : 3 malvidin plus individuals. Confirm the results through suppression gene interaction hypothesis with the help of a cross.

	Spherical AAbb	X		Spherical aaBB	
F_1	Disc AaBb	X	F_1	Disc AaBb	
F_2	9 A_B_	3 A_bb +	3 aaB_		1 aabb
	9 Disc	6 Spherical			1 Elongate

Fig. 8.14. Polymeric effect in case of fruit shape in summer squash

Polymeric Effect

Polymeric effect is shown between dominant genes in case of fruit shape in summer squash (**Fig. 8.14**). Either dominant gene shows the same effect but interaction betweenn dominant genes produced a novel phenotype. Metabolic pathway for polymeric effect is given in **Fig. 8.15**.

The relationship between different types of gene-gene interactions can be understood by studying **Table 8.1**.

Basic Genetics

Table 8.1. Relationship between different types of gene interactions

Types of epistasis	Genotypes			
	A-B-	A-bb	aaB-	aabb
No epistasis: classical ratio	9	3	3	1
Supplementary interaction Recessive epistasis aa epistatic to B, b	9	3	4 (3+1) Homozygous recessive at one locus produces certain phenotype regardless of allelic condition at other locus	
Dominant epistasis A epistatic to B, b	12 (9+3) Dominant allele at one locus. A producers certain phenotype regardless of allelic condition at other locus		3	1
Dominant & Recessive epistasis (Inhibitory factors) A epistatic to B, b bb epistatic to a, A	13 (9+3+1)		3	
Duplicate recessive epistasis (complementary interaction) aa epistatic to B, b bb epistatic to A, a	9	7 (3+3+1) Homozygous recessive at either or both loci produces same phenotype.		
Duplicate dominant epistasis (Duplicate factors) A epistatic to B, b B epistatic to a, A	15 (9+3+3) Dominant allele at either or both loci produces same phenotype.			1
Duplicate genes with cumulative effect (Polymerism) A-bb and aaB-wild identical phenotype	9	6 (3+3) Dominant allele at either locus produces some phenotype		1

```
                                    → Spherical
                           A  ╱
               _____╱  ╲
    aabb                   ╲  ╱  → Disc
    Elongate              B  ╲
                              ╲
                               → Spherical
```

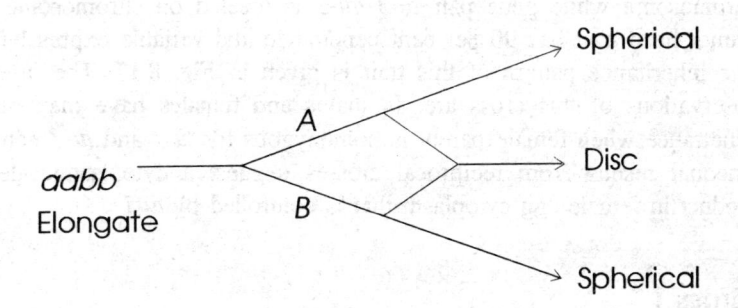

Fig. 8.15. Possible metabollic pathway for polymeric effect shown by genes for fruit shape in summer squash

Modifier Genes

A modifier gene affects or modifies the expression of an other gene. Coat colour in horse and mice illustrates action of a modifier gene (**Fig. 8.16**). *DD* and *Dd* genotypes permit full expression of black colour produced by another gene *B* and of brown colour produced by another allele *b*. Genotype *dd* dilutes the colour pigment to a milky appearance. Black colour is formed when both dominant genes *B* and *D* are present.

	Dilute black	X	Brown
	BBDD		bbDD
		↓	
F1	Black	X	Black
	BbDd		BbDd
		↓	
F₂	9 B_D_	3 B_dd 3 bbD_	1 bbdd
	Black	Dilute Brown	Dilute
		black	brown

Fig. 8.16. Coat colour in horse and mice illustrates action of a modifier gene

Nuclear and Cytoplasmic Gene Interaction

An interesting example of both gene-gene and gene-environment interaction is the inheritance of tumor-head growth in *Drosophila*. The tumor-head (tu-h) phenotype is a function of interaction of one autosomal and one sex linked gene. The gene pair $tu\text{-}1^+/tu\text{-}1$ is located on the X chromosome while gene pair $tu\text{-}3^+/tu\text{-}3$ is located on chromosome 3. Tumor-head trait has 90 per cent penetrance and variable expressivity. The inheritance pattern of this trait is given in **Fig. 8.17**. The salient observations of this cross are: F_1 males and females have maximum penetrance when female parent is homozygous for *tu-1* and *tu-3* genes. Unequal results from reciprocal crosses suggest a cytoplasmic gene product in female egg cytoplasm that is controlled by *tu-1*.

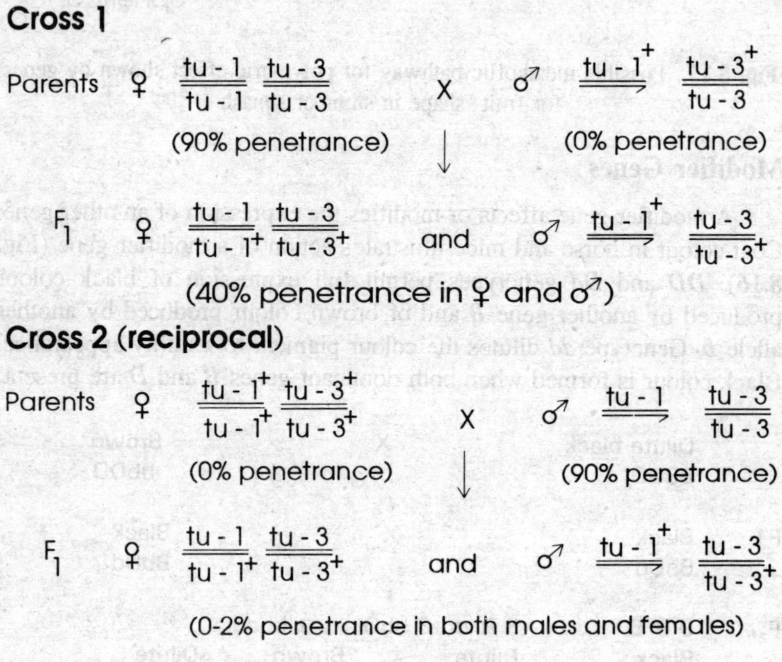

Cross 1

Parents $\female$ $\dfrac{tu-1}{tu-1}\dfrac{tu-3}{tu-3}$ × $\male$ $\dfrac{tu-1^+}{}\dfrac{tu-3^+}{tu-3}$

(90% penetrance) ↓ (0% penetrance)

F_1 $\female$ $\dfrac{tu-1}{tu-1^+}\dfrac{tu-3}{tu-3^+}$ and $\male$ $\dfrac{tu-1^+}{}\dfrac{tu-3}{tu-3^+}$

(40% penetrance in $\female$ and $\male$)

Cross 2 (reciprocal)

Parents $\female$ $\dfrac{tu-1^+}{tu-1^+}\dfrac{tu-3^+}{tu-3^+}$ × $\male$ $\dfrac{tu-1}{}\dfrac{tu-3}{tu-3}$

(0% penetrance) ↓ (90% penetrance)

F_1 $\female$ $\dfrac{tu-1}{tu-1^+}\dfrac{tu-3}{tu-3^+}$ and $\male$ $\dfrac{tu-1^+}{}\dfrac{tu-3}{tu-3^+}$

(0-2% penetrance in both males and females)

Fig. 8.17. Crosses showing inheritance pattern of tumor-head trait in *Drosophila melanogaster*

Explanatory Model. Molecular model explaining inheritance of tumor-head trait in *Drosophila* involving nuclear and cytoplasmic genes is given in **Fig. 8.18**. During early embryogenesis, the mutant *tu-3* gene produces a product that interferes with normal development in head of *Drosophila*. Even a little of this product can occasionally produce a *tu-h* phenotype

Gene-Gene Interactions 99

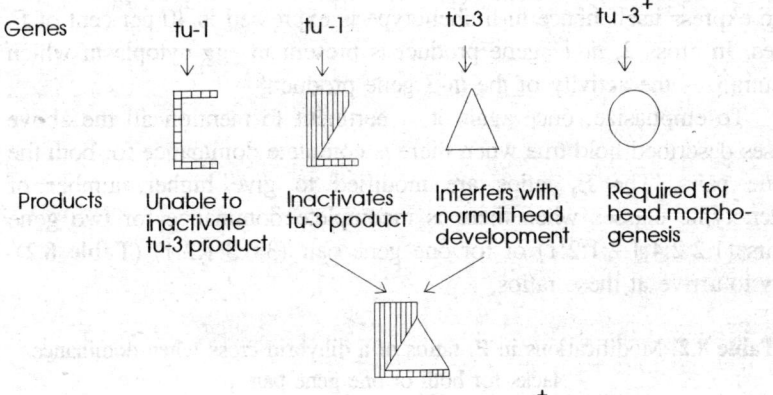

Genes	tu-1	tu-1	tu-3	tu-3$^+$
Products	Unable to inactivate tu-3 product	Inactivates tu-3 product	Interfers with normal head development	Required for head morpho-genesis

tu-3 product inactivated by tu-1$^+$ product

Cross 1 | **Cross 2**

Cytoplasm egg → Cytoplasm egg → Zygote: Head abnormalities

Cytoplasm egg → Cytoplasm sperm → Zygote: No head abnormalities

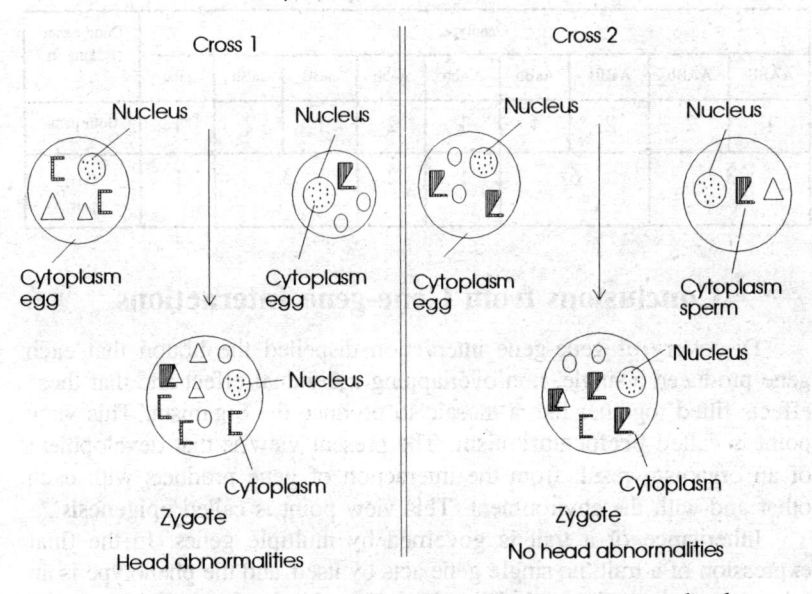

Fig. 8.18. Explanatory model explaining action of tumor-head growth genes in *Drosophila*

as in case of *tu-3/tu-3$^+$* heterozygote. The *tu-1* gene produces a product that blocks the action of the *tu-3* gene product but gene *tu-1$^+$* is active only during the formation of gametes. Eggs contain hundreds of times as much cytoplasm as the sperm, it can store more *tu-1$^+$* gene product. In cross 1, *tu-1$^+$* gene product is absent in egg cytoplasm. Here *tu-3* gene can express itself hence tu-h phenotype is expressed is 40 per cent of F_1 flies. In cross 2, $tu\text{-}1^+$ gene product is present in egg cytoplasm which neutralizes the activity of the $tu\text{-}3$ gene product.

To emphasize, once again it is pertinent to mention all the above cases described hold true when there is complete dominance for both the gene pairs. The F_2 ratios are modified to give higher number of phenotypic classes when there is incomplete dominance for two gene pairs (1:2:2:4:1:2:1:2:1) or for one gene pair (3:6:3:1:2:1) (**Table 8.2**). Try to arrive at these ratios.

Table 8.2. Modifications in F_2 ratios of a dihybrid cross when dominance lacks for both or one gene pair

| Genotype ||||||||| Domi-nance lacking in |
AABB	AABb	AaBB	AaBb	AAbb	Aabb	aaBB	aaBb	aabb	
1	2	2	4	1	2	1	2	1	Both pairs
3		6		1	2	3		1	One pair

Conclusions from Gene-gene Interactions

Discovery of gene-gene interaction dispelled the notion that each gene produced a single, non-overlapping individual effect and that these effects fitted together like a mosaic to produce the organism. This view point is called **preformationism**. The present view is that development of an organism result from the interaction of gene products with each other and with the environment. This view point is called **epigenesis**.

Inheritance of a trait is governed by multiple genes. In the final expression of a trait, no single gene acts by itself and the phenotype is an outcome of several integrated reactions. Gene interactions will change the expression of a phenotype beyond the limits of a gene.

Totipotency

Totipotency is the ability of a cell to proceed through all stages of development and thus produce a normal adult. Each cell does contain a complete set of DNA even after differentiation and hence each cell of a multicellular eukaryote is totipotent, i.e., it has the capability of developing into a complete individual. Thus all the genes of a cell have the potential to express themselves. There are cases of **multiple births** in which "identical" siblings are derived from a single fertilized human egg. The **Dionne quadruplets** were genetically identical. Totipotency of the cell is being used to maintain lines of trees and to produce many ornamental and cultivated plants In development, a particular gene expresses itself only when its product is required by the cell otherwise it remains turned off. The genes that are expressing may do so by performing some **primary cellular function**. The products of various genes interact to produce a particular phenotypic effect.

Pleiotropy

In general, one gene affects a single character, but many cases are known, where a gene influences more than one character. Such genes are known as **pleiotropic genes** and the condition is termed as **pleiotropy**. In Pea plant, the same gene that affects flower colour, also influences the colour of seed coat and colour of leaf-axil. In Cotton, the gene that affects the fibre (lint) character also affects plant height, boll size and fertility. In human beings, a recessive gene (H^S) in homozygous condition produces **sickle cell hemoglobin**. Persons suffer from **sickle cell anemia**. This gene in recessive form also produce offer effects mental retardation, poor physical development, paralysis, heart attack, pneumonia due to being damage, rheumatism, kidney damage and failure. In *Drosophila*, a recessive gene *vestigial* (*vg*) in homozygous condition produces vestigial wings. *vg* gene also affects haltares, bristles, spermatheca, number of egg strings in ovaries, length of life, fecundity, etc. In mice, a dominant gene (A^Y) for yellow coat colour in mice in heterozygous condition also affects survival of the animal. $A^Y A^Y$ codition is lethal.

Pleiotropy helps in understanding different developmental processes. Cystic fibrosis is a hereditary metabolic disorder in children that is controlled by a single autosomal recessive gene. This gene specifies an enzyme that produces a unique glycoprotein. This glycoprotein produces mucus with abnormally high viscosity. This causes interference with normal functioning of several exocrine glands including those in skin (sweat), lungs (mucous), liver and pancreas. Thus several different phenotypic effects result from the action of a single gene.

Most of the mutations detected as morphological abnormalities in vertebrates are clearly pleiotropic in their effect. One syndrome in mice with the members of *T* series of alleles such as taillessness, spina bifida, and absence of anus and urogenital openings is an example of pleiotropic effect.

In rats, homozygous condition for cartilage anomaly (caca) results in abnormal sternum, thickened ribs, and narrowed tracheal passages. This results into various symptoms, namely slow suffocation, poor lung circulation, general arrested development, displacement of chest organs, blocked nostrils and increased production of red blood corpuscles.

In humans, sickle cell hemoglobin results from replacement of glutamic acid in normal hemoglobin at position 6 of β chain with valine. This is due to a single autosomal recessive gene. The mutant red blood corpuscles are sickle-shaped whereas normal are round. This abnormal hemoglobin, in addition to sickling, is responsible for as many as 15 effects including paralysis, pneumonia, inability to fight infection, kidney failure, jaundice.

Homeostasis

The developmental pathways of organisms are genetically adjusted to produce the characteristics morphology of the species regardless of variations in internal and external conditions of development. This is termed as **genetic** or **developmental homeostasis**. Lerner (1954) feels that more heterozygous individuals should be characterized by increased **developmental stability**. This view has been supported by observations of Eanes (1978) in different organisms that variance of two **morphometric characters** among heterozygotes was lower than among homozygotes.

Genetic homeostasis is thus a mechanism of promoting the stability of phenotypic expression of a genotype when grown over a wide range of environments. The homeostatic genotypes are the ones which interact with variable environments in such a way as to maintain relative uniformity of an observable character. Stability of development in variable environment implies physiological adjustments and the capacity of such adjustment is a properly of the homeostatic genotypes.

Heterozygosity and less Asymmetry

Leary *et al.* (1983) determined genotypes of 40 loci for 50 rainbow trout (*Salmo gairdneri*). Number of heterozygous loci and asymmetric characters for these 50 individuals are given in **Table 8.3**. The number of loci that were heterozygous that number of characters that were asymmetric per fish both ranged from 0-5. The relationship between the number of heterozygous loci and the magnitude of asymmetry per individual was examined. The correlation between number of heterozygous loci and magnitude of asymmetry was significant. Next step was to test whether this relationship between heterozygosity and less **asymmetry** was associated with certain individual loci or was with a general heterozygous effect. For this, the amount of asymmetry was examined for homozygotes and heterozygotes at all the loci in which there were over 5 heterozygotes (**Table 8.4**). Heterozygotes have a smaller mean proportion of **asymmetric traits** at 7 of the 8 loci. The correlation between the proportion of the heterozygous loci per fish and the proportion of asymmetric characters remains significant.

Table 8.3. Number of heterozygous loci and asymmetric characters for 50 individuals in a population trout

No. of asymmetric characters	\multicolumn{6}{c}{No. of heterozygous loci}
	0	1	2	3	4	5
0	-	-	-	2	1	2
1	2	2	4	4	1	-
2	1	3	10	6	2	-
3	-	3	2	3	-	-
4	-	-	1	-	-	-
5	1	-	-	-	-	-
Mean	2.25	2.13	2.00	1.67	1.25	0.00

These results demonstrate that heterozygosity in general and heterozygosity at specific loci are associated with increased developmental stability. Heterozygosity may result in a more constant and efficient flux through **glycolysis** and the **citric acid cycle**, resulting in increased developmental stability.

Table 8.4. Mean percentage of asymmetric characters for homozygotes and heterozygotes at eight loci in rainbow trout

Locus	Homozygous	Heterozygous	Difference
Idh2	36.7	29.1	+7.6
Idh3,4	42.7	32.5	+10.2*
Ldh4	34.7	41.3	-6.6
Mdh3,4	36.7	33.3	+3.4
Pgm1-t	38.2	27.9	+10.3*
Pgm2	35.8	32.5	+3.3
Sdh	36.2	30.7	+5.5
Sod	35.7	34.2	+1.5

*$P<0.05$

Biochemical Basis of Homeostasis

Biemont (1983) observed that high developmental homeostasis and high degree of heterozygosity are associated with a low inbreeding depression in *Drosophila*. Trehan and Gill (1983) in *D. malerkotliana*, Shahi (1985) in *D. punjabiensis*, and Kumar (1985) in *D. malerkotliana* showed that acid phosphatase from heterozygotes exhibited better homeostatic ability than the enzyme from either of the two homozygotes. How can this be explained? Heterozygote is different from homozygotes by possessing two homodimeric allozymes alongwith a unique heterodimeric allozyme. Thus homozygotes and heterozygote differ from each other in three respects: (1) heterozygote carries three allozymes where each of the homozygotes carries only one allozyme, (2) heterozygote possesses the two parental allozymes at the same time, and (3) it also possesses a unique heterodimeric allozyme.

The role of each of the three **allozymes** present in heterozygote was assessed in terms of contribution of each component towards higher stability of the heterozygote. Stability pattern of 1:1 mixture of the flies of the two different homozygous genotype was compared with the heterozygote the heterozygote showed a much better stability. It shows that it is the unique heterodimeric enzyme which is responsible for increased stability in the heterozygotes. For supporting this contention, isolation and partial purification of the three allozymes revealed electrophoretically by the heterozygote was done. **Heterodimeric allozyme** was found to be superior in homeostatic ability to the two **homodimeric allozymes.** Thus it provided an evidence that increased stability of acid phosphatase in the heterozygote results from increased stability of the heterodimeric allozyme which constitute half of the total enzyme present in the heterozygotes.

Homeostasis as explained above may operate in cross-pollinated crops. Homeostasis in self-pollinated populations cannot be explained on the basis of heterozygous superiority. Perhaps homeostasis in such populations is brought about by selection in favour of genes which can perform optimally in a range of environments, rather than in a single environment.

Some Useful Reading Hints

Cuenot (1905) observed abnormal Mendelian ratios. Cuenot (1980) further discusses mechaisms responsible for these abnormalities. Dobzhansky (1927) Studied manifold effect of certain genes in *Drosophila melanogaster*. Biemont (1983) discusseed homeostasis, enzymatic heterozygosity and inbreeding depression in natural populations of *Drosophila melanogaster*. Trehan and Gill (1983) explianed phenomenon of homeostasis at the molecular level. Kumar (1985) reported his experiments on molecular basis of heterosis and homesotasis in *Drosophila malerkotliana*. Shahi (1985) reported results of his experiments on biochemical and physiological basis of heterosis and homeostasis in *Drosophila punjabiensis*. Leary *et al.* (1983) dealt with developmental stability and enzyme heterozygosity in rainbow trout.

Castle and Little (1910) observed abnormal Mendelian proportions among yellow mice. Eaton and Green (1962) dealt with implantation and lethality of the yellow mouse. Silvers (1979) dealt with gene inetraction in coat colors of mice. Jackson (1994) dealt with molecular and developmental genetics of mouse coat color. Lawrence and Sturgess (1957) studied genetics and chemistry of flower color in the garden forms, species, and hybrids. Lerner (1954) dealt with genetics of homeostasis. Gill (1981) discusseed developmental homeostasis in self- and cross- pollinated crops.

Questions

1. Differentiate between: Dominance and epistasis and Epistatic and hypostatic genes.
2. Describe one example illustrating interaction between two dominant genes. Give a metabolic pathway and a molecular model to explain results of the crosses.
3. Describe the following examples of recessive epistasis: Complementary gene interaction and Supplementary gene interaction. Give biochemical pathways and molecular models to explain results of the crosses in above examples.

106   Basic Genetics

4. What is dominant epistasis? Explain the following cases of dominant epistasis: Duplicate genes and Polymeric effect. Give biochemical pathways and molecular models to explain results of the crosses in above examples.
5. Explain the action of inhibitory genes exemplifying dominant recessive epistasis with the help of crosses, metabolic pathway and molecular model.
6. Represent all different types of gene interactions at a glance.
7. What are the modifier genes? Explain the action of modifier genes with the help of an example.
8. Gene interaction is not restricted among nuclear genes. The interaction has been reported to occur between the nuclear and cytoplasmic genes also. Illustrate this point with the help of an example giving crosses and molecular model.
9. How are the F_2 phenotypic ratios in a dihybrid cross modified when there is incomplete dominance for (a) both the gene pairs? (b) for one gene pair only?
10. What generalizations have been drawn from the study of examples of gene-gene interaction?
11. What do you understand by totipotency? How has totipotency of a cell been used in crop and animal improvement?
12. What is pleiotropy? In what way is this phenomenon important in genetic studies? Give some examples of pleiotropy?
13. Define homeostasis. What do you understand by genetic/developmental homeostasis? Give some examples of homeostasis.
14. What is relationship between heterozygosity and asymmetry?
15. What is the biochemical basis of homeostasis in cross-pollinated and self-pollinated crops?

Numericals

1. Rose-comb chickens mated with walnut comb chickens produce 15 walnut, 14 rose, 5 pea, and 6 single-comb chickens. Determine the probable genotypes of the parents.
2. A rose type crossed with a pea type produce six walnut and five rose offspring. Determine the genotypes of the parents.
3. Three walnut-combed chickens were crossed to single- combed individuals. In one case, the progeny was all walnut-combed. In another case, one of the progeny was single-combed. In the third case, the progeny were both walnut-combed and pea-combed. Give the genotype of all parents and offspring mentioned.
4. Matings between walnut-combed and rose-combed chickens gave 4 single, 5 pea, 13 rose and 12 walnut progeny in F_1. What are the most probable genotypes of the parents?
5. A mating of two walnut combed chickens produced the following F_1: 1 walnut, 1 rose, 1 single. Give the genotypes of the parents.
6. A hornless condition in cattle is due to a completely dominant gene *P*, normally horned cattle being *pp*. The gene for red colour (*R*) shows no dominance to that for white (*r*), the hybrid (Pr) being roan colour. Assuming independent segregation, give the genotypic and phenotypic expectations from the following matings: (a) *PpRR* X *ppRr*, (b) *PpRr* X *pprr*, (c) *PpRr* X *PpRr*, and (d) Hornless roan (whose father was horned) X horned white

7. A variety of pepper having brown fruit was crossed with a variety having yellow fruit. F_1 plants had red fruits. F_2 consisted of 182 plants with red fruits, 59 with brown fruit, 61 with yellow fruit and 20 with green fruits. What is the genetic basis of these fruit colours in peppers?

8. In weasel, two recessive genes, p and i, are known either or both of which in homozygous from results in platinum coat colour (*PPii, ppII, ppii*). The presence of both of the dominant alleles results in the brown colour (*P-I-*). Will platinum weasel necessarily breed true? Set up a cross between two platinum weasels which will result all brown offspring (*PPii* X *ppII*). What would be expected in the F_2 from this cross?

9. A cross was made between a red flowered plant and a white flowered plant. The F_1 plants were all red-flowered. In the F_2 the flower colours were : 92 red, 30 cream, and 42 white. Explain these results. On crossing the white-flowered plants of the F_2 among themselves, what proportion of the offspring would be red-flowered? Cream flowered? White?

10. In summer squashes white fruit colour is dependent upon a dominant gene W, coloured fruit upon its recessive allele, w. In the presence of ww, the colour may be yellow due to a dominant gene G, or green due to its recessive allele, g. How many different genotypes may be involved in the production of white fruits? If white fruited plant *WWGG* is crossed with green fruited, what will be the appearance of the F_1? of the F_2?

11. Two white flowered strains of sweet pea were crossed producing purple flowered F_1s. In F_2 53 plants had purple flowers and 43 had white flowers. What is the phenotypic ratio in F_2? What the of interaction is involved. What were the probable genotypes of the parental strains?

12. Red colour in wheat kernels is produced by genotype *R-B-*, white by double recessive (*rrbb*). The genotypes *R-bb* and *rrB-* produce brown kernels. A homozygous red variety is crossed to white. What phenotypic results are expected in F_1 and F_2?

13. When homozygous yellow rats are crossed to homozygous black rats, F_1 is all grey. F_2 progeny consisted of 10 yellow, 28 grey, 2 cream-coloured and 8 black rats. How are these colours inherited? Using appropriate genetic symbols, show genotypes of each colour? How many of 48 F_2 rats were expected to be cream-coloured?

14. White fruit colour in summer squash is governed by a dominant gene (W) and coloured fruit by its recessive allele (w). Yellow fruit is governed by an independently assorting hypostatic gene (G) and green by its recessive allele (g). When dihybrid plants are crossed the offspring appear in the form of 12 white : 3 yellow : 1 green. What fruit colour ratios are expected from the following crosses? (a) *Wwgg* x *WwGG*, (b) *WwGg* x Green, (c) *Wwgg* x *wwGg*, and (d) *WwGg* x *Wwgg*.

15. One chromosome 3 of corn there is a dominant gene (A_1) which together with the dominant gene (A_2) on chromosome 9 produces coloured aleurone. All other genetic combination produce colourless aleurone. Two pure colourless strains are crossed to produce all coloured F_1. What were the genotypes of the parental strains and the F_1s. What phenotypic proportions are expected in F_2? What genotypic ratios exist among the white F_2?

108 Basic Genetics

16. Two pairs of alleles govern the colour of onion bulls. A pure red strain crossed to a pure white strain produces all red F_1. The F_2 was found to consist of 47 white: 38 yellow: 109 red bulls. What epistatic ratio is shown by the data? What is the name of this type of gene interaction? If another F_2 is produced by the same kind of cross and 8 bulls of F_2 have double recessive genotype, how many bulls would be expected in each phenotypic class?
17. Three fruit shapes are recognized in summer squash, disc, elongated and spherical. A pure disc shaped variety was crossed to a pure elongated variety. The F_1 were all disc shaped. Among 80 F_2 there were 30 spherical, 5 elongated and 45 disc shaped. Reduce the F_2 numbers to the lowest ratio. What type of interactions is operative. What is the genotypic ratio of sphere-shaped F_2.

9

SYSTEMS USED IN GENETICS ANALYSIS

Experimental Method

Making Crosses

Most biological phenomenon must be investigated by experimental method. In the field of genetics this involves crossing individuals that differ in one or more characteristics and study of their descendants in an effort to determine the method by which these characteristics are inherited. We know that first important discoveries in genetics were made in studies on of crosses in plants (garden peas). Yet these discoveries have been found to hold true for most other forms of life, including man.

Inferences from Crosses

In multicellular eukaryotes, reproducing sexually and having biparental parentage, F_1 progeny of a cross between homozygous mutant female and homozygous wild type male gives, in most of the cases, a fairly good idea about genetic control of a difference (**Table 9.1**). The genetic explanation suggested by the F1 progeny can be tested in F_2, and if need be in F_3 generation.

Choice of Genetic System

Many discoveries in genetics today are being made through breeding experiments which various forms of plants and animals. These are four important factors which must be considered when one choose an organism for experiments in heredity: (i) short life cycle, (ii) large number of offspring, (iii) variation, and (iv) raising conveniently and cheaply.

Basic Genetics

Table 9.1. Genetic explanation inferred from some reciprocal crosses between wild type and mutant strains in animal and plant systems

Phenotype of F_1 progeny of Cross		Genetic explanation of mutant phenotype
Mutant ♀ X Wild type ♂	Wild type ♀ X Mutant ♂	
Animal and Plant Systems		
All wild type	All wild type	Nuclear recessive gene
All Mutants	All Mutants	Nuclear dominant gene
All mutants	All wild type	Cytoplasmic gene
Animals (having XY system)		
All ♀ and ♂ wild type	All ♀ and ♂ wild type	Autosomal recessive gene
All ♀ and ♂ mutant	All ♀ and ♂ mutant	Autosomal dominant gene
All ♀ wild type All ♂ mutant	All ♀ and ♂ wild type	Sex-linked recessive gene
All ♀ and ♂ mutant	All ♀ mutant All ♂ wild type	sex-linked dominant gene

Life Cycles

The significant stages in the process by which an organism gives rise to others of its kind make up the **life cycle** of that organism. Biologists often represent a particular life cycle by diagraming in a circle the salient events of a completed life history.

In this chapter we will discuss reproduction cycles of some of the organisms that are in use for genetic analysis. Here we will deal with only those organisms that reproduce sexually. Only in the sexually reproducing organisms that have **biparental parentage** do the principles of genetic hold valid. For sexual reproduction, there must be two different mating types. The organisms that reproduce asexually/vegetatively or through **apomixis** are not suitable for genetic study. While we discuss life cycles of various organisms we shall also refer to merits and demerits of each one with regard to their use in **genetic analysis**.

Man

Man as an experimental organism for genetic studies has more disadvantages than advantages. Schematic life cycle of man is given in **Fig. 9.1**. Man's life cycle very long. He is choosy for his mates. So crosses cannot be made at will. Size of progeny is too small to enable a geneticist to draw any inferences about genetic control of a trait. Hence man is not a suitable organism for genetic analysis.

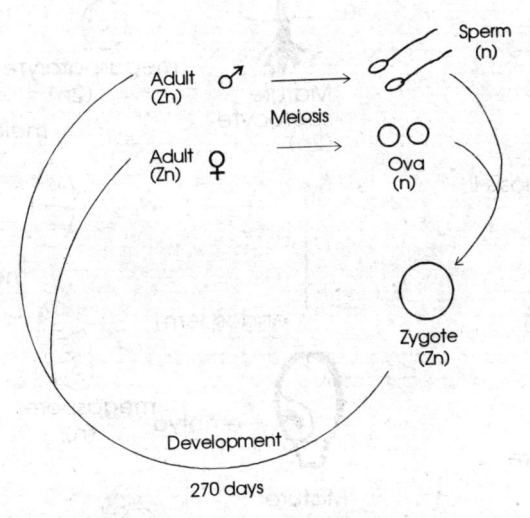

Fig. 9.1. Life cycle (schematic) of man

Now we will discuss life cycles of those organisms that have life histories that either are widely representative or represent some significant departure from generally families schemes of reproduction and those organisms that are important objects of genetic research either because they are widely studied or because they offer exceptional opportunities for investigation or both. Principles in genetics have been discovered on wide range of organisms, i.e., higher plants, animals, *Paramecium*, fungi, algae, bacteria, viruses. These principles have universal applicability.

Basic Genetics

Corn

Reproduction in corn in typical enough of what happens in most seed plants. Of all the plants its genetics has probably been investigated more thoroughly. Life cycle of corn is given in **Fig. 9.2**. Why geneticists have

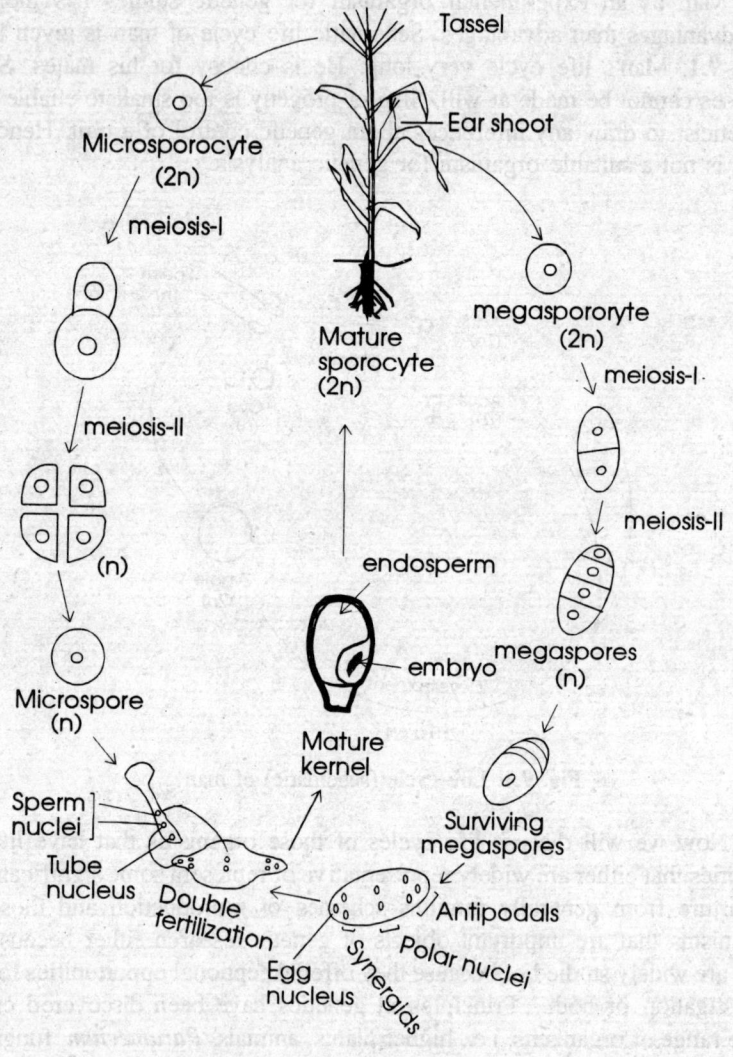

Fig. 9.2. Life cycle of *Zea mays*

been interested more in corn? It is because of several advantages this angiosperm offers to the geneticists. This crop is of agronomic importance. Single ear of corn represents a large progeny. It possesses numerous interesting characteristics. Being **monoecious**, there is relative ease of **controlled matings**. Its chromosomes are big enough for cytological study. This plant, like many others also has some disadvantages. it has comparatively long life cycle and it has high cost of production.

The genetic phenomenon can be conveniently studied in corn are: detection of **segregation, endosperm genetics**, and construction of **linkage maps**. We will return to these aspects in Chapter 10.

Drosophila melanogaster

Drosophila is genetically the best characterized multicellular eukaryote. Commonly known as fruit fly or vinegar fly, *Drosophila* is well suited for laboratory experiments because it offers several advantages. It breeds rapidly because of its short life cycle. **Development of adult from an egg takes 10 days at 25°C.** Because of its small size it occupies less laboratory space and at the same time it is large enough to note numerous morphological attributes. A large number of progeny of the insect is obtained from a single inseminated female and thus results in statistical accuracy. The insect breeds easily in the laboratory on a variety of simple culture media. It has many heritable characters, viz. eye colour, eye shape, body colour, wing size, wing shape, hairs on body. Large number of mutations are known. Mutant stocks are easily available and transportable all over the world. This insect is easy to handle by simple apparatus and technique. It has a small number of chromosomes: 2n = 8. The larvae of this insect possess large salivary-gland chromosomes in which a great deal of minute structural details can be seen cytologically. Almost any phenomenon in genetics can be conveniently and convincingly demonstrated in *Drosophila* through laboratory or controlled population experiments.

There are four distinct stages in the life cycle of the fruit fly: **egg, larva** (three instars), **pupa** and **adult** (Fig. 9.3). At 25°C, it completes its life cycle (egg to adult) in 10 days. The female begins to deposit eggs on about third day of adult life and continues until she dies. A female may be laying 50 to 70 eggs per day and perhaps a maximum of 400 to 500 in 10 days. The egg is visible as a small, white, oblong object with two filaments at the anterior end. After a day or two of egg deposition, small,

114   *Basic Genetics*

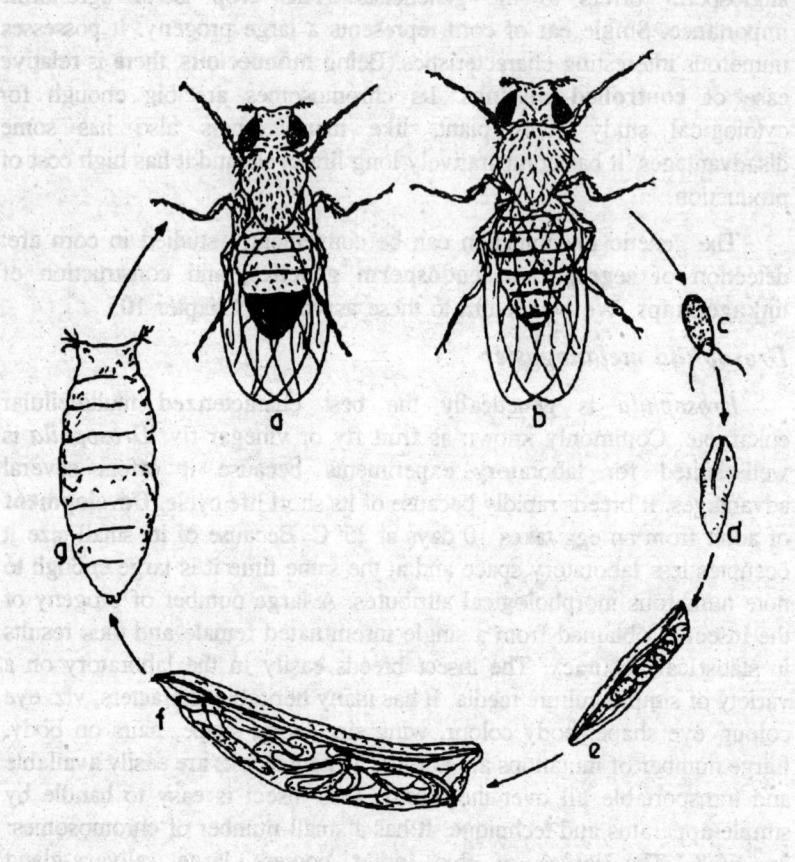

Fig. 9.3. Life cycle of *D. melanogaster* at 25°C: (a) adult male, (b) adult female, (c) egg, (d) larva, 1st instar, e) larva, 2nd instar, (f) larva, 3rd instar, and (g) pupa

white, worm-like larvae emerge from the eggs and begin to burrow into the medium. A larva has black mouth parts in the narrowed head region. For tracheal breathing it has a pair of **spiracles** both at anterior and posterior ends. The larvae are voracious eaters. Since insect skin does not stretch, a young larva must periodically shed its skin, mouth parts and spiracles (**moulting**). New expanded **integument** is laid down which larva proceeds to fill out by rapid feeding. There are three larval stages, called **instars**, separated from each other by moults. Both the size of the larva and the number of teeth on the dark-coloured jaw hooks are indications of the instar of the larva. During the third instar stage, the larva feeds until ready to pupate, then crawls out of the medium to a dry place where it ceases to move and averts its anterior breathing spiracles. The cuticle of the third instar larva shortens, hardens and darkens to become the puparium. **Metamorphosis** occurs within the puparium. The pupa starts darkening just prior to emergence, the folded wings appear as two dark elliptical bodies and the pigment in the eyes is visible through puparium. When metamorphosis is complete, the adult emerges by forcing its way through anterior end of the puparium. At first the fly is light in colour, the wings are unexpended, and the abdomen is long. In a few hours the wings expand, the abdomen gradually becomes round and the colour darkens. In 8 h adult fly matures sexually. Adult fly may live for several weeks. After maturity, fruit flies are fertile as long as they live.

Paramecium

Paramecium is a protozoan belonging to the group called ciliates. We will discuss *P. aurelia*. Its size is 150 X 50 microns. This is a one-celled eukaryote with complex **structural organization**. Its reproductive processes are not only complex but highly unusual. These unusual reproductive properties of *Paramecium* provide unusual possibilities of studying complex genetic phenomenon. Studies with *Paramecium* have led to the new concepts of relationships betweens nucleus and cytoplasm in inheritance.

Paramecium has two kinds of nuclei: **Macronucleus** - *P. aurelia* has only one very large macronucleus. This is the source of phenotypic control. Structure of macronuclei is incompletely understood. **Micronucleus** - *P. aurelia* has two nuclei in fertilization processes and serve to form new micronuclei and macronuclei.

Binnary Fission. Paramecium gives rise to more paramecia through **binnary fission** which is asexual reproduction and is essentially a division of one individual into two that are alike genetically **(Fig. 9.4)**. At fission each micronucleus divides mitotically and the macronucleus appears to elongate, constricts in the middle and then separates into two pieces. One of the two products of each of these nuclear divisions goes into the anterior half of the animals the others to the posterior portion. The animal constricts around the middle and finally separates into two daughter animals that contain nuclear constitution like that of the original animal.

116 Basic Genetics

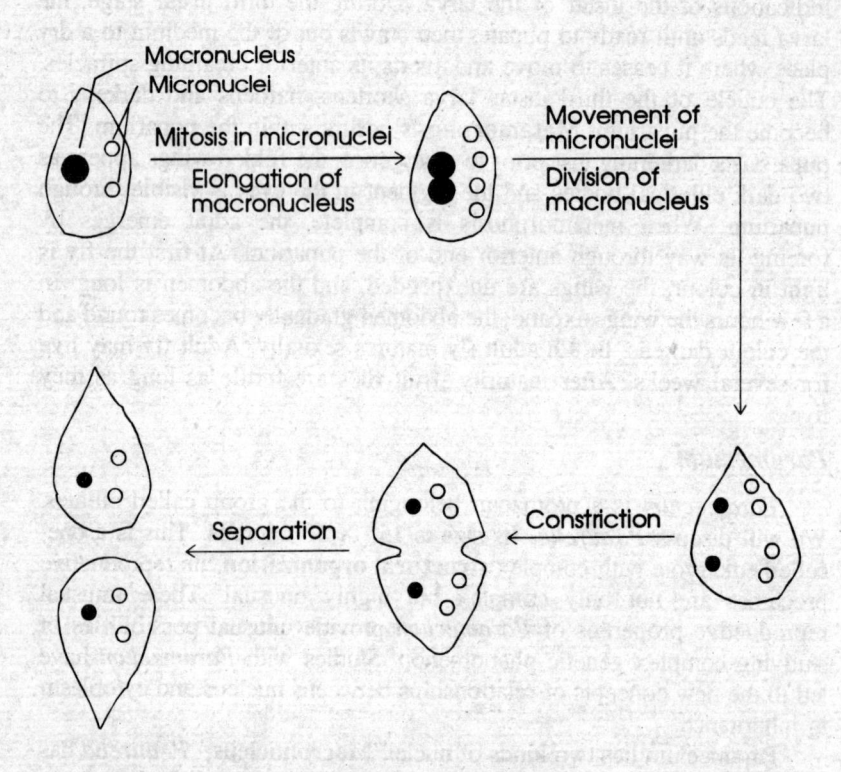

Fig. 9.4. Reproduction through binnary fission in *Paramecium aurelia* resulting into two genetically identical animals

Conjugation. Under appropriate conditions, paramecia of different mating types pair (conjugate) closely and carry out reciprocal fertilization. This type of mode is called sexual reproduction (**Fig. 9.5**). Within paired animals several important events occur before **nuclear fusions**. In each animal macronucleus breaks down into fragments that finally disintegrate, although disintegration may not be complete until after the conjugation cells have moved apart and undergone a number of fissions. In each animal, each of the two micronuclei undergoes meiosis. Each conjugant cells at some stage contains eight haploid nuclei. Seven of eight nuclei disintegrate. The remaining haploid nucleus divides mitotically. Each

Systems Used in Genetic Analysis 117

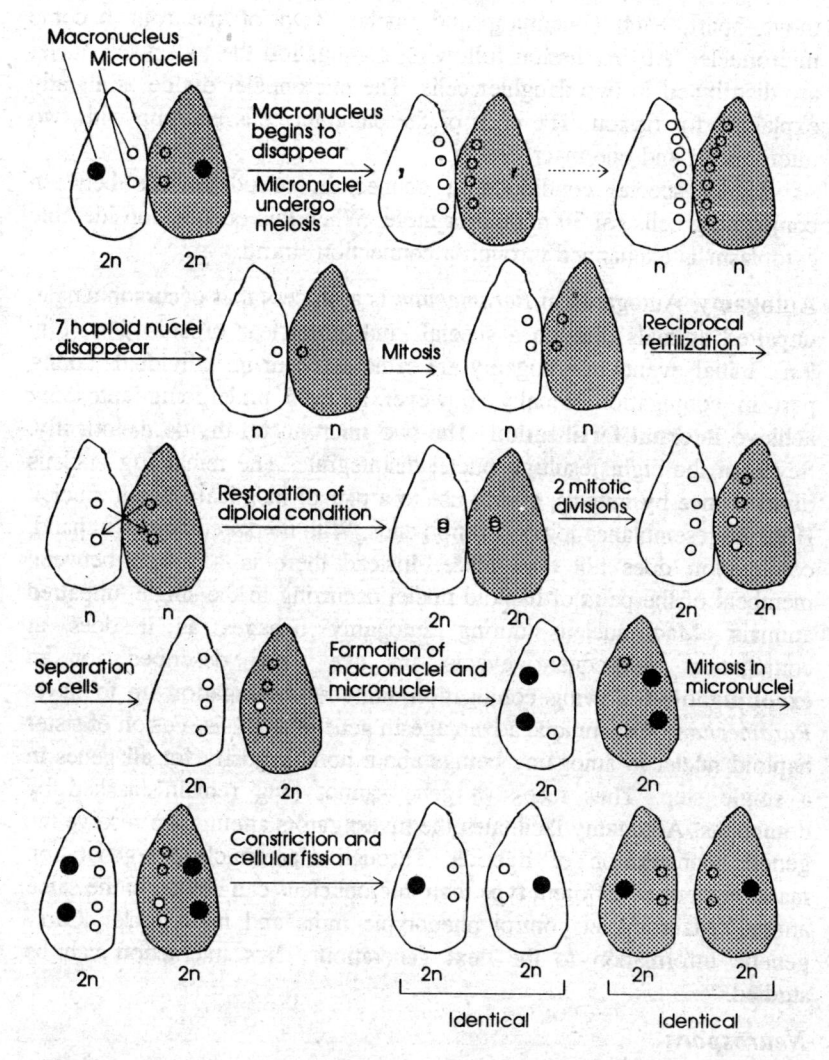

Fig. 9.5. Sexual conjugation in *Paramecium aurelia* producing 4 cells of genetically two types

conjugant cell thus contains two genetically identical **haploid nuclei**. This condition permits reciprocal fertilization. Thus diploid condition is restored. **Diploid nuclei** resulting from fertilization are alike. The fertilization nuclei then undergo two mitotic divisions. The paired animals move apart, each containing and nuclei. Two of the four become micronuclei. At first fission following conjugation the two macronuclei are distributed in two daughter cells. The micronuclei divide as already explained for fission. The each of the daughter cells ends up with two micronuclei and one macronuclei.

Under special conditions, a connection strand persists between conjugating cells for 30 minute or more. When this occurs, considerable cytoplasm is exchanged through a connection strand.

Autogamy. Autogamy in *Paramecium* is a process that occurs in single, unpaired animals through a special kind of nuclear organization (**Fig. 9.6**). Initial events in autogamy are same as occur in individuals taking part in conjugation. Finally, however animals undergoing autogamy achieve **internal fertilization**. The two micronuclei divide meiotically. Seven of the eight resulting nuclei disintegrate. The remaining nucleus divides once by mitosis, giving use to a pair of identical haploid nuclei. Here the resemblance to conjugation ends. With no paired animal at hand, conjugation does not take place. Instead there is a fusion between members of the pairs of haploid nuclei occurring in the single, unpaired animals. Macronucleus during autogamy behaves as it does in conjugation. Subsequent events are like those described for an **exconjugant**, following conjugation same as conjugation up to have. *Paramecium* offers unique advantagesin genetic analysis. Fusion of sister haploid nuclei in autogamy beings about homozygosity for all genes in a single step. Thus recessive gene cannot long remain masked by dominants. Autogamy facilitates the investigators attempts to resolve the genetic constitution of hybrids. Through macronuclear regeneration macronucleus of different type than micronucleus can be had in the same animal. Macronuclei control phenotypic traits and micronuclei. Carry genetic information to the next generation. Thus interaction can be studied.

Neurospora

Neurospora crassa is a bread mold belonging to group of fungi called **Ascomycetes**. It offers exceptional opportunities in genetic study as it provides many advantages like brief life cycle, well adapted for investigation on a mass scale, it may be readily propagated asexually, unlimited populations of a given genotype may be had easily, easily kept in pure culture on a chemically defined medium, nature of its life cycle

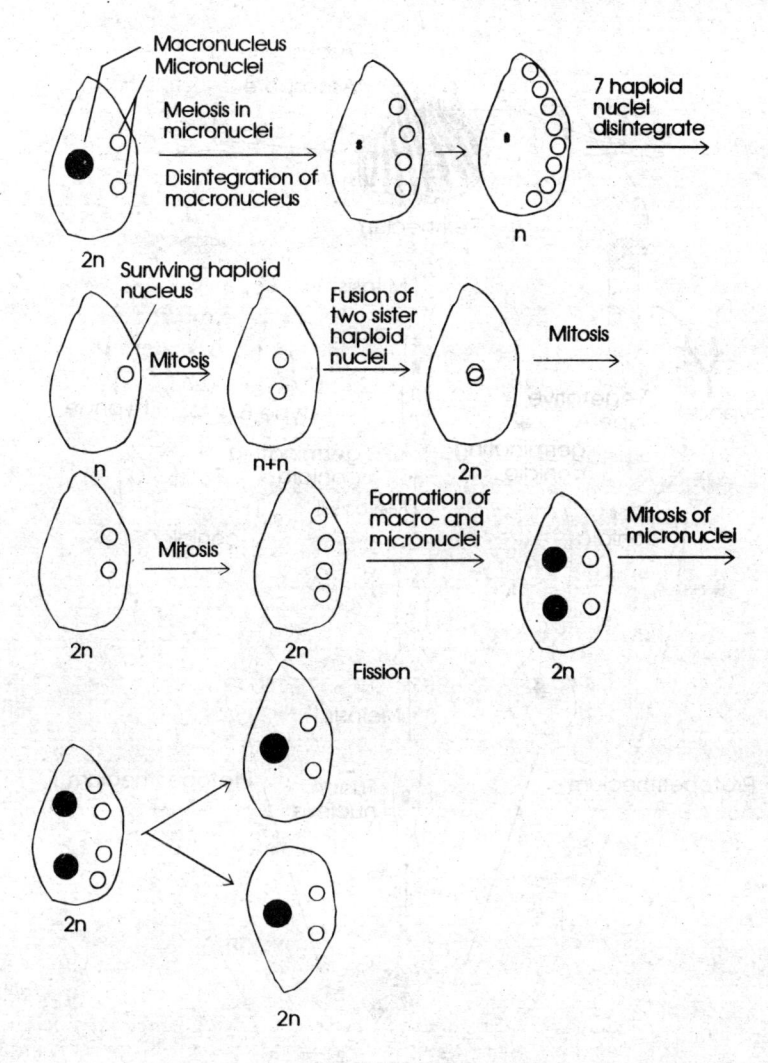

Fig. 9.6. Autogamy due to internal fertilization in *Paramecium aurelia* resulting into two homozygous animals at all the loci that were heterozygous prior to autogamy

permits unusually close genetic analysis, being haploid no dominance recessiveness relationship exists. Reproduction of this organism may be sexual or asexual.

120 *Basic Genetics*

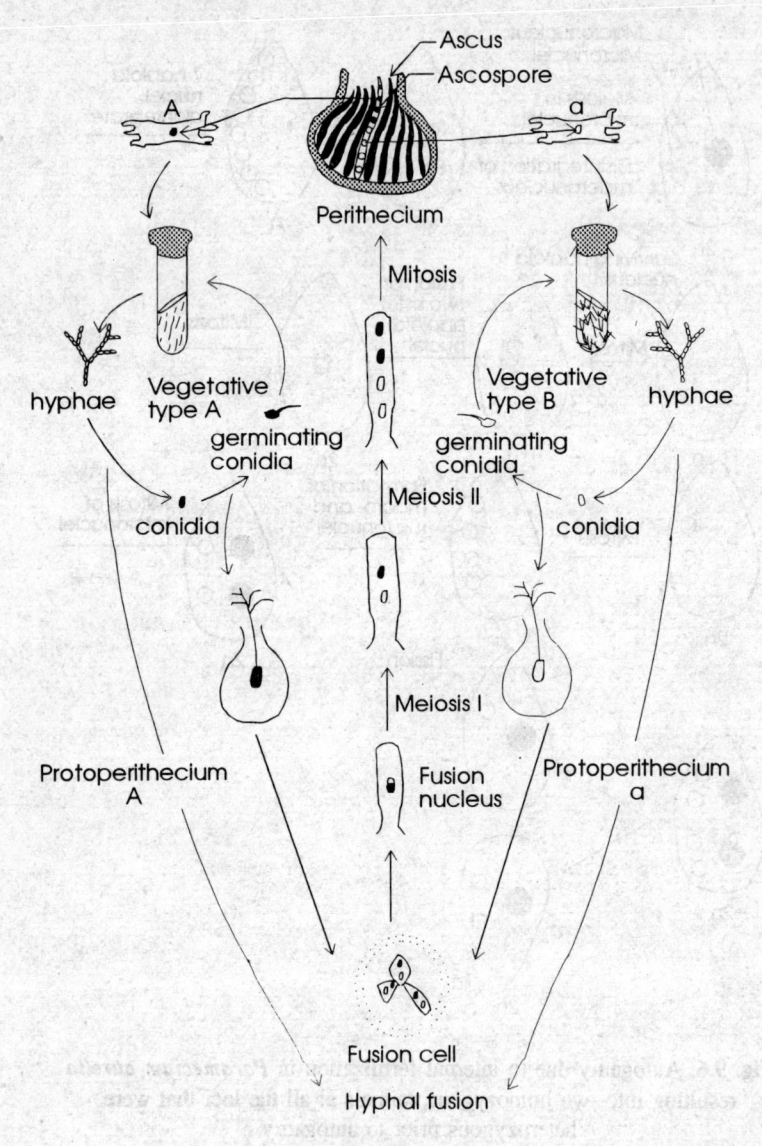

Fig. 9.7. Life cycle of bread mold *Neurospora crassa*

Asexual Reproduction. Asexual reproduction occurs by means of spores called **conidia** or simply by propagation through unspecialized fragments of **mycelium**. Asexual reproduction is based on mitotic divisions of haploid nuclei.

Sexual Reproduction. *Neurospora* has two **mating types,** *A* and *a*, which are determined by member of an allelic pair. Sexual reproduction occurs only when cells of opposite mating types unite. The fusion nucleus is the only diploid stage in this organism (**Fig. 9.7**). The zygote quickly undergoes meiosis in a sack like structure called **ascus**. The ascus is so narrow that divisions in tandem, and the resultant nucleic do not slip-past each other. Four haploid nuclei are produced at the end of meiosis. Each product of meiosis still in its place divides mitotically and give rise to eight nuclei in linear order in ascus. Heavy spore wall are formed around each nuclei and each nuclei thus becomes an **ascospore**. Each pair of ascospores from either end represents one of the products of meiosis. The members of a pair of ascospore are genetically alike. To know the genotype of a particular ascospore arising from crosses between strains of *Neurospora*, ascospores can be isolated and cultured separately.

Genetic Applications. *Neurospora* has some important genetic applications. Eight linearly arranged ascospores within an ascus provide an accurate diagrammatic record of what happened at meiosis. The eight ascospores thus represent an **ordered tetrad** of duplicate meiotic products. Eight ascospores are isolated and cultured separately. if eight such culture tubes are lined up in the same order as that of their corresponding ascospores in the ascus, one can observed directly which alleles went to which nuclei of meiotic division.

Yeast

Baker's or budding yeast, *Saccharomyces cerevisiae* (**Fig. 9.8**), is another fungus which belongs to Ascomycetes which is very favourable organism for genetic studies. Yeast produces **unordered tetrads** and is not a very appropriate organism for classical genetic analysis. However, yeast chromosomes have been a subject of detailed study to understand structure and synthesis of eukaryotic chromosome. Yeast exist in both, haploid and diploid forms. The haploid is usually formed under starvation. When better conditions return, haploid cells of two sexes, called **a** and α types, fuse to form the diploid. The haploid is again established through meiosis under starvation conditions. All the products of meiosis are contained in the ascus.

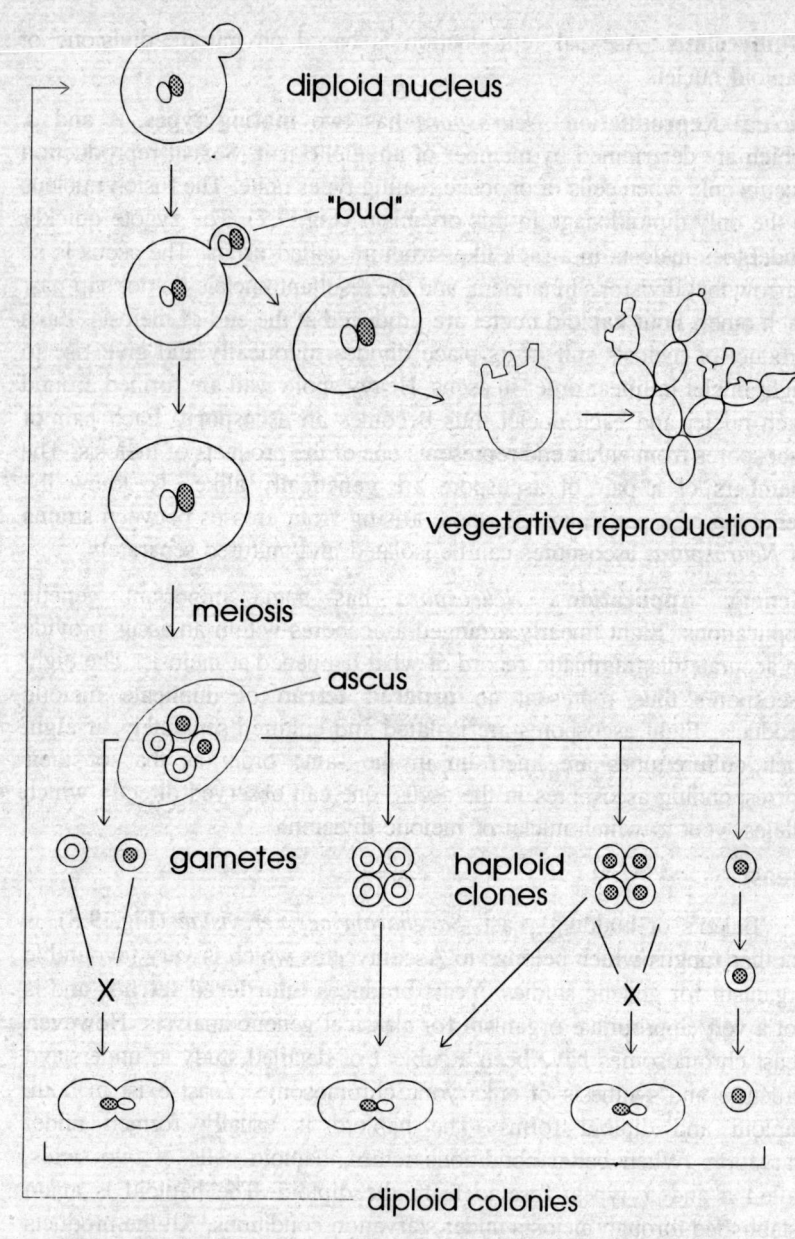

Fig. 9.8. Life cycle of yeast showing vegetative reproduction and formation of meiotic products

Chlamydomonas

It is a haploid **green algae**. Its life cycle is presented in **Fig. 9.9**. It has only one chloroplast and hence is very useful organism for study of **uniparental inheritance**. *Chlamydomonas* does not have sexes but does have mating types, mt^+ and mt^-. Only individuals of opposite type can mate. **Mating type** is inherited as a single locus with two alleles. When two haploid cells of opposite mating type fuse, they form a diploid zygote, which then undergoes meiosis to produce four haploid cells, two of mt^+ and two of mt^-. The mechanism of the extrachromosomal inheritance of *Chlamydomonas* is not known. It is especially mystifying in that both parent cells fuse and appear to contribute equally to the zygote.

Bacteria

Bacteria are single-celled prokaryotes. Three types of bacteria known are: rods (bacilli), spheres (cocci) and spirals. Bacterial cells have nucleus-like structure where DNA is located. In cytoplasm are present various components such as **ribosomes**, a cytoplasmic membrane, and a wall. Outside the wall one sometimes finds a protective sheath (capsule), whose composition is a genetic attribute, contain bacteria, those with flagella are motive. Bacterial nucleus does not have a membrane that separate it from cytoplasm unlike the nucleus of eukaryotes. Bacteria have internal structures. These is no spindle mechanism, nor a condensation cycle of DNA in bacteria.

Fission. A bacterial cell divides through fission to give rise to a pair of daughter cells that are genetically like each other and like the original. Nuclear material replicates. And since division of the nuclear material is not synchronous with cellular divisions, cells with more than one nucleus can occur. Bacteria can multiply with enormous speed. *Escherichia coli* doubles in number with each 20 minute interval. Such growth occurs under only under favourable conditions.

Favourable attributes. Rapid reproduction and small size favouable attributes for some one who wishes to conduct research on bacteria. There are certain phenotypes that are favourable for genetic research. These are not visually observable characteristics of size, shape, colour or the like as in higher organisms. Bacteria have biochemical and **physiological attributes** that provide genetic variables that can be determined with ease

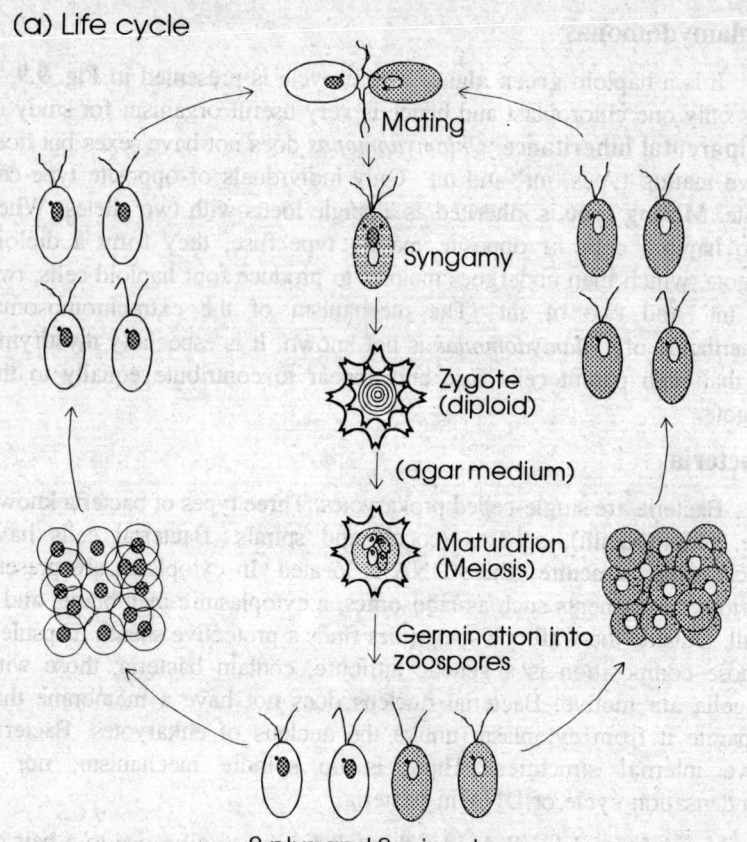

124 *Basic Genetics*

(a) Life cycle

- Mating
- Syngamy
- Zygote (diploid)
- (agar medium)
- Maturation (Meiosis)
- Germination into zoospores
- 2 plus and 2 minus types

(b) Representation based on electron micrograph

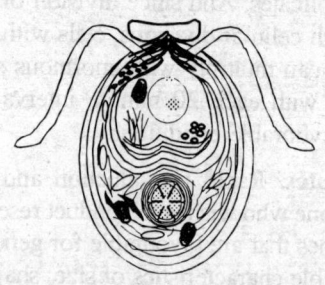

Fig. 9.9. (a) Life cycle and (b) representation based on electron micrograph of *Chalmydomonas reinhardtii*

Systems Used in Genetic Analysis

and precision. Such characteristics include sensitivity of **antibiotics**, ability or inability to grow in the absence of certain **vitamins**, the ability or inability to produce certain **antigens** or other chemical substances, and capacity or lack of it to be stained by particular dyes.

Quantitative Estimation. Quantitative estimation of bacteria is quite easy. Suppose there is a culture of *E. coli* growing in a liquid medium. Such a culture might have as many as 20 million live cells per ml of medium. A known volume, say 0.1 ml, is drawn from the culture of bacteria. This can be diluted 100 times by adding it to 9.9 ml of sterilized fluid. Further serial dilutions can be made. From the final dilution a known volume for inoculating onto solidified medium in Petri dishes. If solidified medium is appropriate for growth, each line bacterium inoculated will reproduce by fission and ultimately will give rise to a visible colony, composed of millions of cells. To make precise count of number of bacteria, the extent of dilution is standardized. Discrimination of phenotypes is usually done on colonies rather than on individual cells. This is a valid procedure genetically, since each colony is a vegetatively produced clone from a single cell.

Conjugation. Conjugation, a type of sexual reproduction, occurs only when matings occur between genetically different strains. When bacterial cells conjugate, continuity is established by bridges that mediate the injection of genetic material from one cell to another. With regard to the mating type, bacterial cells are of two types: F^- - This type of cell is recipient of the genetic material. **F factor** lacks in this type. This is a female type cell. F^+ - This type of cells is donor of the genetic material as F factor is present. This is a female type cell.

This particular factor which determines the mating type of bacteria is called **F-element** or F-factor. The possessors of F (donors) are actually of two types, designated as F^+ and **Hfr (high frequency of recombination)**. F-factor is of considerable interest in itself. F-factor has been formed to be a piece of genetic material that may (Hfr) or may not be (F^+) be directly attached with chromosome. Thus F-factor may be in free state in a chromosome or it may become an integral part of it. Such a entity is called an **episome**. When two mating types, F^+ and F^-, are mixed together, it produces two F^+ cells (**Fig. 9.10**). Conjugation between Hfr and F^- cells results into transfer of genetic material (**Fig. 9.11**). This may have different consequences, i. e., incomplete transfer or complete transfer of donor DNA to the recipient cell. In first case, recombination is possible between early entering donor genes and homologous genes of

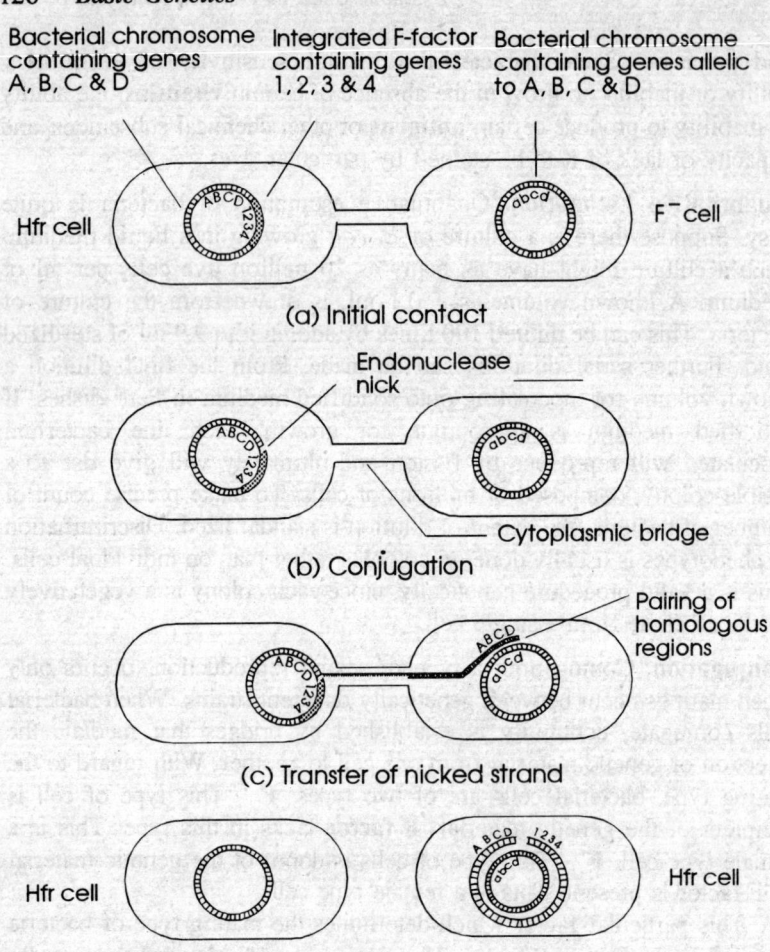

126 Basic Genetics

Fig. 9.10. Conjugation between Hfr and F- and cells of *E. coli*

the F recipient chromosome but the recipient cell remains F⁻ type. In second case, recombination is possible between all donor genes and F⁻ recipient chromosome. Here F⁻ recipient cell becomes Hfr type. Some electronmicrographs actually show bridge of cellular material between two different types through which the chromosome of the donor penetrates the recipient. Once the recipient cell receives either all or a part of donor chromosome (depending upon the time for which the conjugation takes place) the recipient may then incorporate some of the

Systems Used in Genetic Analysis 127

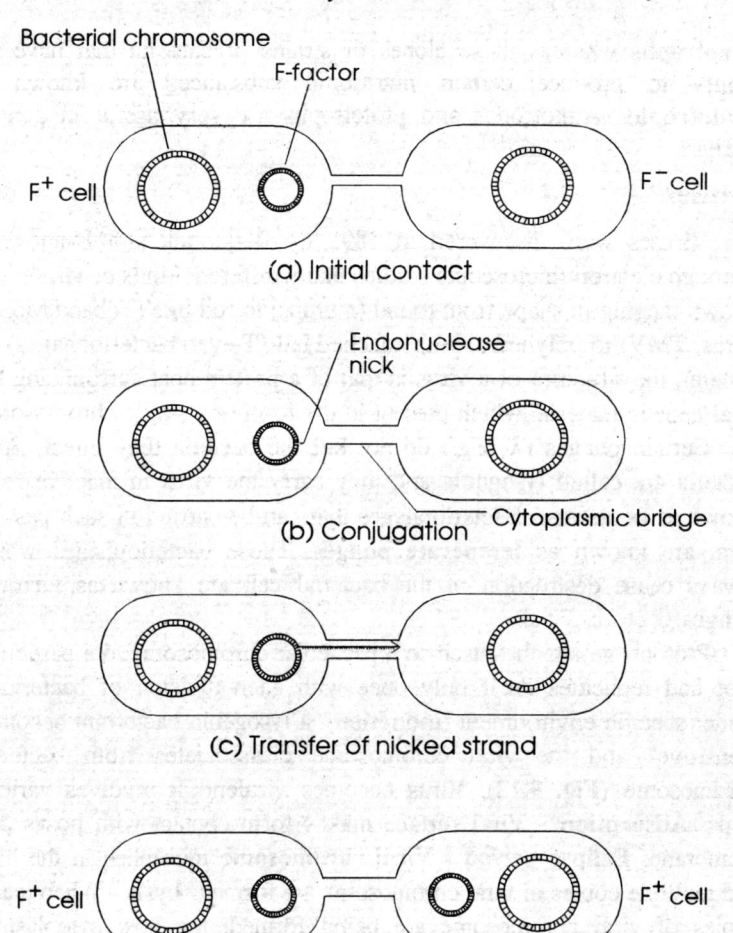

Fig. 9.11. Conjugation between F+ and F- cells of *E. coli*

inherit characteristics of the donor. Following conjugation, zygotes or partial zygotes remain in diploid state for a few generations of vegetative state of cells. Eventually segregation and reduction of haploidy occurs, but the process seems to be not as regular and rapid as in higher organisms.

Those clones or stains of bacteria that have lost the ability to produce certain nutritional substances (usually amino acids) are termed as

auxotrophs whereas those clones or strains of bacteria that have the ability to produce certain nutritional substances are known as **prototrophs**. Auxotrophs and prototrophs are very useful in genetic analysis.

Viruses

Viruses were discovered in 1892 by Beijerinck and Iwanowski. Through electron microscopic studies many different kinds of viruses are known ranging in shape from round (mumps) to rod like (Tobaco Mosaic Virus, TMV) to polyhedral with attached tail (T-even bacteriophages). In general, the structure of a virus is that of a protein coat surrounding the viral genetic materia, which present in the form of a single chromosome.

Certain phages (λ, e.g.) do not kill the bacteria they infect. Such bacteria are called **lysogenic** and they carry the virus in inactive form known as **prophage**. Bacteriophages that can be carried in such passive form are known as **temperate phages**. Those bacteriophages which always cause destruction of the bacterial cell are known as **virulent phages**.

Protophage attaches itself to a particular chromosome to a particular spot and replicates itself only once with each division of bacterium. Under specific environment (**induction**), a lysogenic bacterium becomes "sensitive" and the viral chromosome disassociates from bacterial chromosome (Fig. 9.12). Virus becomes virulent. It involves various steps: **Adsorption** - Viral surface makes form contact with host's cell membrane. **Eclipse period** - **Viral chromosome** multiplies in the host and multiple copies of viral chromosome are formed. **Lysis** - When many copies of viral chromosome are being formed, the host cytoplasmic machinery gets converted to synthesize **viral coat proteins**. Within about 30 minutes of entry of virus into the host, the host cell membrane is dissolved and infective viral particles are released.

A bacterial cell capable of **spontaneous lysis** due to uncoupling of a prophage from the bacterial chromosome is known as **lysogenic bacteria**. Viral chromosome when integrated into bacteria is known as prophage. Chromosome is in inactive form at this stage. A phage that can become a prophage is known as **temperate phage** whereas a phage that cannot become a prophage is called **virulent phage**. Infection by such a phage always leads to lysis of the host cell.

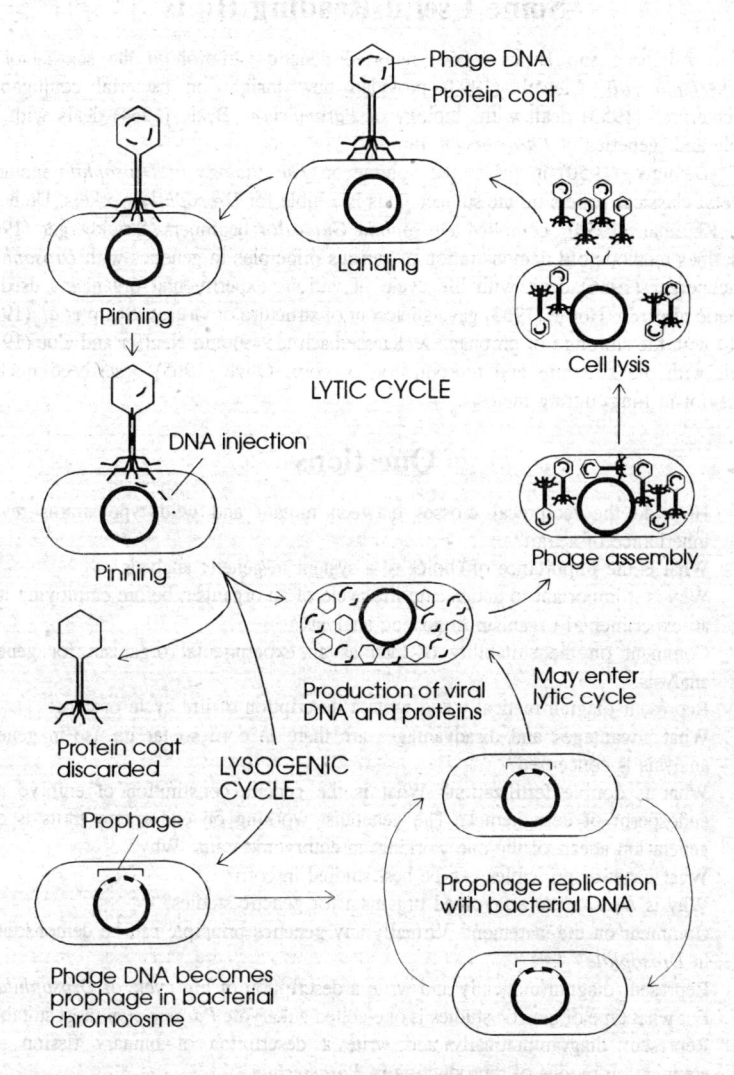

Fig. 9.12. Life cycle of bacteriophage - lytic and lysogenic cycle

Induction is the process in which under specific environment a lysogenic bacterium becomes sensitive and viral chromosome becomes dissociated from bacterial chromosome. Virus becomes virulent.

Some Useful Reading Hints

Adelberg and Burns (1960) reported genetic variation in the sex factor of *Escherichia coli*. Clewell (1993) provided new insight on bacterial conjugation. Wichterman (1953) dealt with biology of *Paramecium*. Beale (1954) deals with life cycle and genetics of *Paramecium aurelia*.

Demerec (1950) in his edited volume on *The Biology of Drosophila* included several classical papers on the subject. This is a bible for *Drosophila* workers. Demerec and Kaufman (1950) complied *Drosophila Guide* for beginners. Strickberger (1962) describes methods ofd demonstration of various principles in genetics with *Drosophila*. Strickberger (1986) dealt with life cycle of various experimental organisms used in genetic research. Horne (1963) gave an accont of structure of viruses. Sharp *et al.* (1972) dealt with the sturcture of prophage λ. Kiesselbach (1949) and Neuffer and Coe (1974) dealt with the structure and reproduction of corn. Olive (1965) describeed nuclear behavior in fungi during meiosis.

Questions

1. How do the reciprocal crosses between mutant and wild type strains reveal inheritance of a trait?
2. What is the importance of choice of a system in genetic analysis.
3. Why is it important to understand life cycle of an organism before employing it as an experimental organism in genetic analysis?
4. Comment on the suitability of man as an experimental organism for genetic analysis.
5. Represent diagrammatically and write a description of life cycle of corn.
6. What advantages and disadvantages are there in corn so far its use in genetic analysis is concerned?
7. What is double fertilization? What is the genetic constitution of embryo and endosperm of corn kernel? The geneticist working on endosperm traits is one generation ahead of the one working on embryonic traits. Why?
8. What genetics principles can be best studied in corn?
9. Why is *Drosophila* a favoured organism for genetic studies?
10. Comment on the statement: Virtually any genetics principle can be demonstrated in *Drosophila*.
11. Represent diagrammatically and write a description of life cycle of *Drosophila*.
12. For what type of genetic studies is one-celled eukaryote *Paramecium* most suitable?
13. Represent diagrammatically and write a description of binnary fission and conjugation modes of reproduction in *Paramecium*.
14. What is autogamy? Represent diagrammatically and write a description of this process. Explain how autogamy leads to homozygosity? What is the advantage of autogamy in genetic analysis?
15. What are the advantages of using *Neurospora* in genetic studies? What advantages does its haploid nature offer to a geneticist?
16. Represent diagrammatically and write a description of reproduction modes of *Neurospora*.

17. What is tetrad analysis? What is the advantage of ordered tetrads in genetic analysis?
18. Represent diagrammatically and write a description of life cycle of *Chlamydomonas*. For what type of genetic studies is this organism suited?
19. Describe results of conjugation between F^+ and F^- cells of *E. coli*.
20. Describe results of conjugation between Hfr and F^- cells of *E. coli* when transfer of genetic material from donor to recipient cell is (a) incomplete and (b) complete.
21. How is conjugation useful in mapping of genes in bacteria?
22. What are viruses? Represent diagrammatically and write a description of lytic cycle and lysogenic cycle in bacteriophages?

10

GENETIC ANALYSIS IN DIPLOID EUKARYOTES

Time of Segregation

Waxy and **non-waxy** pollen in corn are used to study time of segregation of members of the same allelic pair. Non-waxy pollen is stained blue with iodine. It is due to a dominant nuclear gene *Wx*. Waxy pollen is stained red with iodine. It is due to a homozygous recessive condition (*wx wx*). When a cross is conducted between non-waxy and waxy pollen producing plants and pollen produced by F_1 seeds (*Wx wx*) is stained with iodine, about half are red and half blue. Out of 6919 pollen studied by Demerc, 3,437 were stained blue and 3,482 were red. The result is indicative of the fact that segregation of allelic members is complete before the gametes are formed. To be precise we say that **segregation** occurs during **meiosis**.

Endosperm Genetics

Double fertilization results in formation of endosperm tissue which partakes of both paternal and maternal inheritance. In plant where the ovary wall and seed coat are thin and transparent, as in the kernel of corn, a direct effect of the male gamete on the character of the endosperm is evident. Thus, if an ear of maize from a type normally bearing white endosperm is pollinated by pollen from a yellow race, the endosperm of the seed produced will be yellow. This direct effect of male gamete on tissue other than embryonic ones is known as **xenia**. Xenia effect can be illustrated by the crosses between plants producing **sugary** and **starchy** pollen (**Fig. 10.1**).

In both the hybrids endosperm turns out to be starchy no matter which allele enters through the female line. One *Su* gene is dominant-

Genetic Analysis in Diploid Eukaryotes   133

Cross I

Starchy ♀	X	Sugary ♂
(Su Su)		(su su)

egg	pollen	→ Starchy embryo
(Su)	(su)	(Su su)

Polar nuclei	pollen	→ Starchy endosperm
(Su + Su)	(su)	(Su Su su)

Cross II

Sugary ♀	X	starchy ♂
(su su)		(Su Su)

egg	pollen	→ Starchy embryo
(su)	(Su)	(Su Su)

Polar nuclei	pollen	→ Starchy endosperm
(su + su)	(Su)	(Su su su)

Fig. 10.1. Results of the crosses between sugary and starchy pollen producing strains of corn for embryonic and endospermic trait

recessiveness over two *su* alleles. Other alleles however, may not show same kind of dominance relations in the endosperm. This is shown in crosses between corn plants producing **flinty** and **floury** seeds (**Fig. 10.2**). Flinty phenotype is due to a dominant gene *F* whereas floury is due to homozygous condition *ff*. In these reciprocal crosses, phenotype of endosperm is different. Well, this is **xenia effect**. Here two recessive alleles (*ff*) suppress the expression of a dominant allele (*F*). The geneticists studying the endosperm traits can keep one generation ahead in the game of genetic analysis as compared to some sporophytic character.

Linkage and Genetic Recombination

Genetic Recombination is the name given to the redistribution of information inherited from the parents in the progeny. Recombination involves the physical exchange of material between chromatids of homologous chromosomes. Non-parental combination of alleles, which are obtained in the segregating generations are called **recombinants**.

Cross I

| Flinty ♀ (FF) | X | Floury ♂ (ff) |

| Egg (F) | Pollen (f) | → | Flinty embryo (Ff) |

| Polar nuclei (F + F) | Pollen (f) | → | Flinty endosperm (FFf) |

Cross II

| Floury ♀ (ff) | X | Flinty ♂ (FF) |

| Egg (f) | Pollen (F) | → | Flinty embryo (Ff) |

| Polar nuclei (f + f) | Pollen (F) | → | Floury endosperm (Fff) |

Fig. 10.2. Results of the crosses between flinty and floury seed producing strains of corn for embryonic and endospermic trait

Genetic recombination is a fundamental property of all living systems starting from the RNA-containing viruses and ending with the higher plants and animals. Althoug this information applies to the recombination of both linked and non-linked genes, the term recombination of linked markers (between different genes or alleles). Whereas recombination of the unlinked genes is based on the machanism of free combination of chromosomes in through an exchange of segments of homologous chromosomes, process known as **"crossing-over"**. In crossing-over, there is a reciprocal exchange of information between homologous chromosomes, while in gene conversion, the exchange is predominantly non-reciprocal in character.

All the organisms have many more genes than the chromosomes. For example, *Drosophila* has only 4 pairs of chromosomes but over 5,000 genes. This implies that each chromosome must contain many genes. The genes located in the same chromosome would not be expected to assort independently. Such genes are said to show linkage.

Terminology

Linkage is the tendency of a parenta combination of characters to stay together in F_2. It is also defined as the tendency of a parental combination of gene tostay together in F_2. Association of genes as a result of their occurrence in the same chromosome. Tendency of two dominant characters, inherited by F_1 from one parent to stay together in F_2 is termed as **coupling linkage** (Fig. 10.3). Tendency of two dominant characters, in an F_1, inherited one from one and the other from the other parent, to stay apart in the F_2 generation is known as **repulsion linkage**. **Crossing-over** is the exchange of corresponding segments between chromatids of homologous chromosomes, by breakage and reunion, following pairing. This process is inferred genetically from the

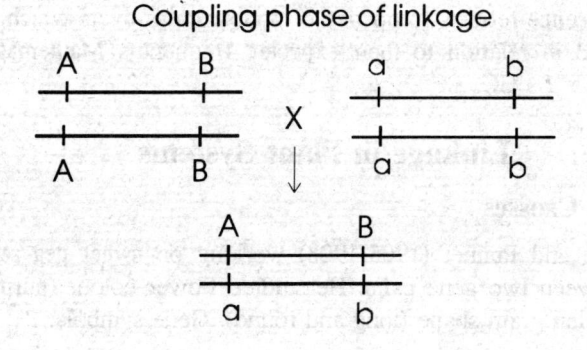

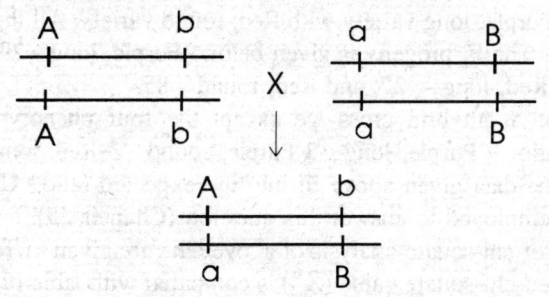

Fig. 10.3. Coupling and repulsion linkage

recombination of linked factors in the progeny of heterozygotes and cytologically from the formation of chiasmata between homologous chromosomes. **Chiasma** (pl. **Chiasmata**) is X-shaped configuration of the chromosomes in a bivalent in prophase of meiosis I, usually the visible result of prior cytological crossing-over. Distane between genes is expreesed as map unit (centimorgan). One **map unit** equals 1 per cent crossing-over. A group of genes showing linkage with one another forms one linkage group, i.e., a group of genes located in the same chromosome. **Linkage map** is the arrangements of genes in a linkage group in such a way that the distance between any two of them reflects per cent crossing-over between them. **Coefficient of coincidence** (CC) refers to the phenomenon which expresses ratio of frequencies observed and expected double crossing overs. It expresses strength of interference, and is expressed as a ratio of observed frequency of double cross overs divided by expected frequency of double cross overs.

Interference reflects frequency of double cross overs which fail to be recovered in relation to their expected frequency. Mathematically, interference = 1 - CC.

Linkage in Plant Systems

Two-point Crosses

Bateson and Punnet (1905-1908) working on sweet pea reported linkage between two gene pairs. He studied flower colour (purple and red) and pollen grain shape (long and round). Gene symbols:

R = Purple flowers; R_o = long pollen grain

r = Red flowers; and r_o = round pollen grain.

He crossed Purple, long variety with Red, round variety. All the F_1s were purple, long. The F_2 progeny is given below: Purple, long - 296, Purple, round - 19, Red, long - 27, and Red, round - 85.

In F_2 of a dihybrid cross we except the four phenotypes in the following ratio: 9 Purple, long : 3 Purple, round : 3 Red, long : 1 Red, round. Do the data given above fit into the expected ratio? **Chi-square (X^2) test** is employed to answer this question (Chapter 25).

Results of chi-square analysis of above data are given in **Table 10.1**. The calculated chi-square value (X^2_c) is compared with table (theoretical) chi-square value (X^2_t). Accepted **probability value (p)** is 0.05. **Degree of**

freedom (d.f.) for this test will be = number of classes (4) - 1 = 3. X^2_t 0.05,3 = 7.815. Since $X^2_c < X^2_t$, the observed numbers in different phenotypic classes do not fit into the expected 9:3:3:1 ratio. It means the two gene pairs do not assort independently. Deviation from Mendelian ratio is not sufficient justification for stating that the two gene pairs in question are linked. Linkage between two gene pairs is not the only cause that can be expected to affect the frequencies of certain classes. Abnormal

Table 10.1. Chi-square analysis of two-point cross data on sweet peas

Phenotype	Observed	Expected	(Obs. - Exp.)² / Exp.
Purple, long	296	240	$(56)^2/240$ = 13.066
Purple, round	19	80	$(61)^2/80$ = 90.262
Red, long	27	80	$(53)^2/80$ = 35.112
Red, round	85	27	$(58)^2/27$ = 124.592
			X^2_c = 263.032

segregation for alleles at a locus can be an other reason for departure from expected ratio. Chi-square must first test whether each gene pair segregate as expected in Mendellian fashion, and then whether two gene pairs segregate independently of each other again as expected in Mendellian fashion.

A mating between two heterozygotes $RrR_o r_o$ X $RrR_o r_o$ should produce four classes of offspring: purple long ($RrR_o r_o$), purple round ($Rrr_o r_o$), red long ($rrR_o r_o$) and red round ($rrr_o r_o$) in ratio 9:3:3:1, respectively, If assortment between R, r and R_o, r_o is independent. If linkage is complete, only two parental classes of offspring are produced. If linkage is incomplete and some recombination occurs, then recombinant classes are also produced but these will not be as frequent as the parental classes. Consequence of **incomplete linkage** is departure from the frequencies of phenotypes expected according to independent assortment.

Detection of Linkage Without Partitioning Chi-square

In cases where segregation is non-Mendelian, a cross Rr X Rr dose not produce purple and red in expected 3:1 ratio. Similarly, cross $R_o r_o$ x

$R_o r_o$ dose not produce long and round in expected 3:1 ratio. For detection of linkage, three chi-square values must be calculated; one each for segregation of the two individual gene pairs and a third for their joint segregation. If the gene pairs singly show **Mendelian segregation** and jointly they fail to show independent assortment it can be concluded with certainty that the genes in question are linked. This is called **detection of linkage**. Some principle is applied for detection of linkage in a test cross progeny. Here segregation at individual loci is tested using 1:1 ratio and independent assortment is tested using 1:1:1:1 ratio.

Estimation of Linkage

Estimation of linkage means measuring the distance between linked genes. For estimation of linkage, per cent recombination is calculated as follows: In the F_2 progeny, two classes purple, long and red, round are recombinant type. How the recombinant types appear in the F_2 progeny? The recombinant types in the F_2 progeny most likely originated from new combinations of the linked genes. The mechanism responsible for origin of new phenotypic combinations is **crossing-over** which is defined as the exchange of corresponding segments between chromatids of homologous chromosomes, by **breakage and reunion**, following pairing.

Whereas independent assortment generates new combinations of genes in unlinked genes, crossing-over is important mechanisms for new combinations of linked genes. Thus crossing over is **exception to linkage**. Crossing-over occurs during **pachytene** of meiosis. Visible consequences of crossing-over is seen in form of X-shaped configurations, called **chiasma**, during diplotene of prophase I of meiosis. Gene hypothesis of above cross is given in **Fig. 10.4**. Recombination frequency is a function of distance between the genes in the chromosomes. More is the distance between the two genes, greater is the recombination frequency observed. Let us estimate recombination frequency in the case already discussed. The two parental classes, purple, long and red, round, appear more frequently than expected and the two recombinant classes, purple, round and red, long, appear less frequently than expected because of linkage. The following steps procedure is followed for estimation of linkage.

Mark the parental pair of **complementary classes** as "P_1" and "P_2". The pair showing maximum number of flies belongs to the parental types. Mark the pair of complementary classes which arises due to crossing-over as "R_1" and "R_2". The pair showing the least number of flies belongs to this category.

Genetic Analysis in Diploid Eukaryotes 139

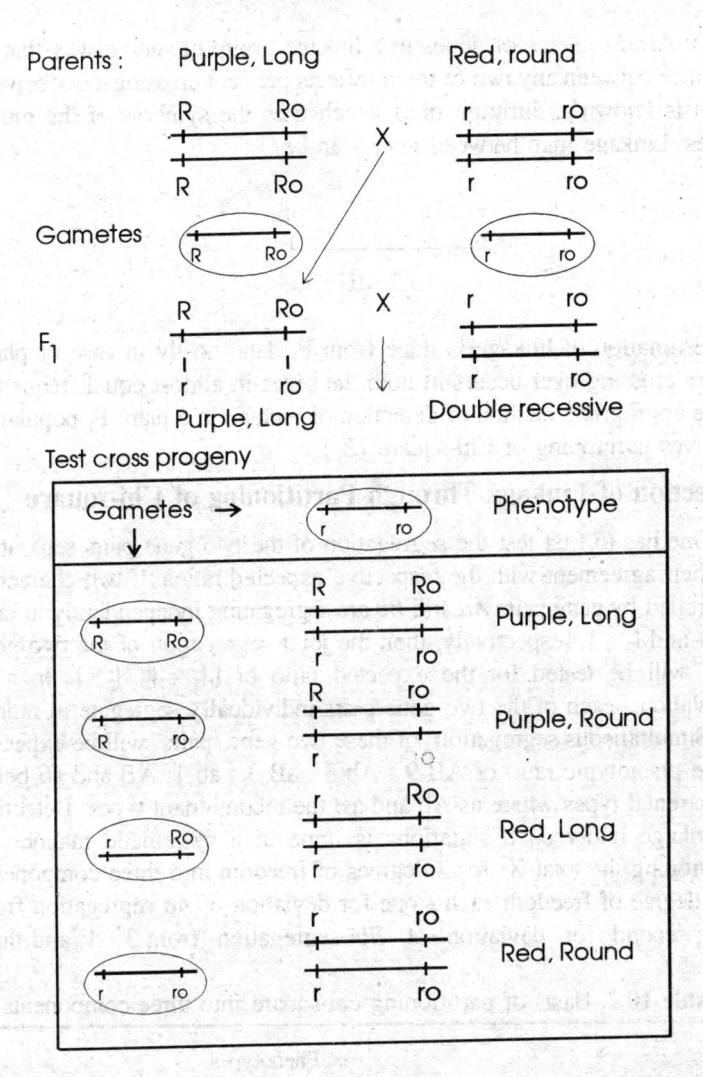

Fig. 10.4. Gene hypothesis showing a two-point test cross in sweeet peas

To calculate per cent recombination use the following formula: Number of recombinants/Total progeny X 100. The per cent recombination in the above case comes out to be 46/427 X 100, i.e., 10.7.

Distance between linked genes is measured in **map units**. One map unit equals 1 per cent crossing-over which is equal to one **centimorgan**

140   Basic Genetics

(cM). Arrangements of genes in a linkage group in such a way that the distance between any two of them reflects per cent crossing-over between them is known as **linkage map** which uses the symbols of the mutant genes. Linkage map between gene r and r_o is:

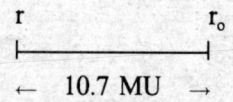

$$\underset{\leftarrow \ 10.7 \text{ MU} \ \rightarrow}{\overset{r \qquad\qquad\qquad r_o}{\vdash\!\!\!-\!\!\!-\!\!\!-\!\!\!-\!\!\!-\!\!\!-\!\!\!-\!\!\!-\!\!\!-\!\!\!\dashv}}$$

Estimation of linkage is done from F_2 data mostly in case of plants where crossing-over occurs in both the sexes in almost equal frequency. More appropriate method of detection of linkage in a plant F_2 population involves paritioning of Chi-square (X^2).

Detection of Linkage Through Partitioning of Chi-square

One has to first test the segregation of the two gene pairs separately for their agreement with the respective expected ratios. If two characters controlled by gene pairs Aa and Bb are segregating independently in ratio $l_1 : 1$ and $l_2 : 1$, respectively, then the joint segregation of the two gene pairs will be tested for the expected ratio of $l_1l_2 : l_1 : l_2 : 1$. In a F_2 population, each of the two gene pairs individually segregate in ratio 3 : 1. Simultaneous segregation of these two gene pairs will be expected in the phenotypic ratio of AB 9 : Ab 3 : aB 3 : ab 1; AB and ab being the parental types where as Ab and aB the recombinant types. Detection of linkage under such situations is done in a systematic manner by partitioning the total X^2 for 3 degrees of freedom into three components of 1 degree of freedom each - one for deviation of Aa segregation from 3 : 1, second for deviation of Bb segregation from 3 : 1, and third

Table 10.2. Basis of partitioning chi-square into three components

Segregation	Phenotypes			
	AB	Ab	aB	ab
A/a	+1	+1	-3	-3
B/b	+1	-3	+1	-3
A/B	+1	-3	-3	+9

detecting association of the two gene pairs by testing for deviation for joint segregation from 9 : 3 : 3 : 1. The information provided in Table 10.2 forms the basis for X^2 formula to partition the 3 degrees of freedom into three components.

If the observed frequencies in four F_2 phenotypic classes AB, Ab, aB and ab are a_1, a_2, a_3 and a_4, respectively, for the practical working of the method the following X^2 formulae can be deduced and used as indicated in **Table 10.3**.

Table 10.3. Formulae for partitioning of Chi-square

Formula	Used for testing
$X^2_A = \dfrac{(a_1 + a_2 - 3a_3 - 3a_4)^2}{3n}$	Segregation for A,a
$X^2_B = \dfrac{(a_1 - 3a_2 + a_3 - 3a_4)^2}{3n}$	Segregation for B,b
$X^2_{A/B} = \dfrac{(a_1 - 3a_2 - 3a_3 + 9a_4)^2}{9n}$	Independent assortment for A/B

Where, $n = (a_1 + a_2 + a_3 + a_4)$

If X^2_A and X^2_B show goodness of fit but $X^2_{A/B}$ value is above the theoretical value this provides a very strong evidence of linkage. Similar formulae can be deduced for detection of linkage in test cross data of a plant population. Seek guidance from your teacher to take up this exercise.

Estimation of Linkage in F_2 Plant Population

When the existence of linkage has been conclusively proved, then **intensity of linkage** can be measured. Hypothesis for estimation of linkage is provided by modern genetic theory which regards these deviations from independence as resulting from the genes controlling the characters being situated on the same chromosome. The intensity of linkage is measured in inverse sense as the fraction of total number of chromosome pairs in which crossing-over takes place at gametogenesis.

This is known as **recombination fraction**. The smaller this fraction, the more intense the linkage.

In a double heterozygote, two relations in arrangement may exist between genes. Either A - B and a - b may be on the same chromosomes or A - b and a - B may be the arrangement found. In former case A - b and a - B are the recombinant types. In the latter case recombinants are A - B and a - b. These arrangements are differentiated under the terms **coupling** and **repulsion**. Thus coupling is the AB/ab type and repulsion is the Ab/aB type in F_1 individuals.

Several methods of estimating linkage have been proposed from time to time. A good method should fulfil two criteria: the estimation obtained should tend to the theoretical value as the sample is enlarged (Consistency) and the estimate should have the lowest possible variance (efficiency). The recombination fraction is denoted by p. It is a measure of frequency of crossing-over between the two chromosomes in the region delimited by the two genes under consideration.

Expected frequencies of four phenotypes AB, Ab, aB and ab can be easily determined. If p and p' are the recombination fractions in males and females, respectively, both assumed to be in the coupling phase, the frequencies of four types of gametes would be as given in **Table 10.4**.

Table 10.4. Expected frequencies of gametes F_2 generation

Parent	AB	Ab	aB	ab
Male	1/2(1 - p)	1/2p	1/2p	1/2(1 - p)
Female	1/2(1 - p)	1/2p	1/2p	1/2(1 - p)

From the above gametic frequencies, genotypic frequencies can be obtained which can be grouped into following four phenotypic classes as given in **Table 10.5**.

Table 10.5. Gametic and genotypic frequencies

AB	Ab	aB	ab
1/4θ	1/4(1 - θ)	1/4(1 - θ)	1/4θ

Where $\theta = (1 - p)(1 - p)$.

Since frequencies depend upon θ only, this is the quantity that can be estimated from the observed frequencies for this type of data. Three methods are available for estimation of linkage.

Emerson's method

$$\theta = \frac{E - M}{n}$$

Where E = sum of observed frequencies of AB and ab classes, M = sum of observed frequencies of Ab ans aB classes, and n = E+M. This will provide an unbiased estimate of θ. The standard error of the estimate of θ can be obtained from the formula:

$$V_\theta = \frac{1 - \theta^2}{n}$$

If p = p, then $\theta = (1 - p)^2$

$p = 1 - \sqrt{\theta}$

$$V_p = \frac{v\theta}{4\theta}$$

S.E. (p) = $\sqrt{V_p}$

Maximum-liklihood method

In case of F_2, it can shown that

$$\frac{a1}{2+\theta} - \frac{a2}{1-\theta} - \frac{a3}{1-\theta} - \frac{a4}{\theta} = 0$$

On simplification, this gives the following quadratic equation for θ: $2a_4 + \{ a_1 - 2 (a_2 + a_3) - a_4 \} \theta - n \theta^2 = 0$. Substitute the values of a1, a2, a3, and a4 and solve the quadratic equation for θ using the formula

$$\theta = \frac{-b \pm \sqrt{(b^2 - 4ac)}}{2a}$$

Where,

$a = -n$, $b = (a_1 - 2a_2 - 2a_3 - a_4)$, and $c = 2a_4$

We know that $\theta = (1-p)(1-p)$. If $p = p$, then $\theta = (1-p)^2$, and $p = 1 - \sqrt{\theta}$. In order to calculate standard error for estimate of θ, first calculate V_θ

$$V_\theta = \frac{2\theta(1-\theta)(2+\theta)}{n(1+2\theta)}$$

$$V_p = \frac{V_\theta}{4\theta}$$

S.E. $(p) = \sqrt{V_p}$

Product-ratio method

In this method we equate the ratio of the product of extreme classes to the product of middle classes, with its theoretical value. This gives us the required equation for estimation of θ.

$$z = \frac{a_1 a_4}{a_2 a_3} = \frac{(2+\theta)\theta}{1 - 2\theta + \theta^2}$$

This is quadratic in θ. Solving this equation we get

$$\theta = \frac{1 + z - \sqrt{(1 + 3z)}}{z - 1}$$

Using value of z calculated above, per cent recombination values can be found out from **Table 10.6** which would directly give you estimate of recombination fraction p in coupling or repulsion phase of linkage. The standard error of estimate of p from V_θ.

$$V_\theta = \frac{1 - \theta^2}{n}$$

Then, $Vp = \dfrac{V_\theta}{4\theta}$

where Vp is the variance of p. Standard error of p is calculated by using the formula S. E. (p) = $\sqrt{V}$.

Linkage in Animal Systems

Two-point Crosses

Crossing-over does not occur in both the sexes in some animal species. For example, crossing-over is known to be absent in *Drosophila* males. This is a case of **complete linkage**. Such situation is also known to exist in females of silkworm *Bombyx mori*. In such cases F_2 progeny of such organisms cannot be used for constructing linkage maps. In such cases, F_1 female is test crossed and the test cross progney is used for detection and estimation of linkage. Let us consider two traits in *Drosophila melanogaster*: (a) Wing size: Wildtype (+), *dumpy* (*dp*) (b) Body colour: Wildtype (+), *black* (*b*). A cross between *dumpy black* females and wildtype males yields F_1 flies with both traits wildtype. When F_1 wildtype females are crossed with X *dumpy black* males, the following TC_1 progeny was recovered: wildtype - 335, *dumpy* - 185, *black* - 170, and *dumpy black* - 310. Gene hypothesis of the above cross is given in **Fig. 10.5**. Detection of linkage in case of a test cross progeny is done by testing segregation for +/*dp* and +/*b* using 1:1 ratio. The observed numbers in the four phenotypic classes are tested for independent assortment using for 1:1:1:1 ratio.

Estimation of recombination in a test cross progeny is done the same way as for F_2 progeny. Detect and estimate linkage for the above test cross progeny in *Drosophila*.

Three-point Crosses

Detection of linkage for autosomal genes. A mating between a heterozygote *AaBbCc* and a triple homozygote *aabbcc* should produce eight classes of offspring: ABC (*AaBbCc*), Abc (*Aabbcc*), aBC (*aaBbCc*),

Table 10.6. Per cent recombination in crosses between F_1 individuals heterozygous for two gene pairs A,a and B,b in coupling or repulsion

z	Percent recombination (p)	
	Coupling	Repulsion
0.0010	2.7	2.2
0.0015	3.3	2.7
0.0020	3.7	3.2
0.0025	4.2	3.5
0.0030	4.6	3.9
0.0035	4.9	4.2
0.0040	5.2	4.5
0.0045	5.5	4.7
0.005	5.6	4.9
0.006	6.4	5.5
0.007	6.8	5.9
0.008	7.3	6.3
0.009	7.7	6.7
0.010	8.1	7.0
0.012	8.8	7.7
0.014	9.4	8.3
0.016	10.0	8.9
0.018	10.6	9.4
0.020	11.1	9.9
0.022	11.6	10.3
0.024	12.1	10.8
0.026	12.5	11.2
0.028	12.9	11.6
0.030	13.3	12.0
0.035	14.3	13.0
0.040	15.1	13.8
0.045	15.9	14.6
0.050	16.7	15.3

Table 10.6 contd...

z	Per cent recombination (p)	
	Coupling	Repulsion
0.055	17.4	16.0
0.060	18.0	16.7
0.065	18.7	17.4
0.070	19.2	18.0
0.075	19.8	18.5
0.080	20.3	19.1
0.085	20.9	19.6
0.090	21.4	20.1
0.095	21.8	20.6
0.100	22.3	21.1
0.12	24.0	22.9
0.13	24.8	23.7
0.14	25.5	24.5
0.15	26.2	25.2
0.16	26.9	25.9
0.17	27.5	26.6
0.18	28.1	27.2
0.19	28.7	27.9
0.20	29.3	28.5
0.22	30.4	29.6
0.24	31.3	30.7
0.26	32.3	31.7
0.28	33.2	32.6
0.30	34.0	33.5
0.32	34.8	34.3
0.34	35.5	35.1
0.36	36.2	35.8
0.38	36.9	36.6
0.40	37.6	37.2

Table 10.6 contd...

z	Per cent recombination (p)	
	Coupling	Repulsion
0.42	38.2	37.9
0.44	38.8	38.5
0.46	39.4	39.1
0.48	39.9	39.7
0.50	40.5	40.3
0.55	41.8	41.6
0.60	42.9	42.8
0.65	44.9	43.9
0.70	45.0	45.0
0.75	46.0	45.9
0.80	46.9	46.9
0.85	47.7	47.7
0.90	48.5	48.5
0.95	49.3	49.3
1.00	50.0	50.0

ABc (*AaBbcc*), abC (*aabbCc*), AbC (*AabbCc*), aBc (*aaBbcc*) and abc (*aabbcc*) (**Fig. 10.6**). If assortment between *Aa*, *Bb* and *Cc* is independent, each of the above eight classes should be in equal ratios of 1/8. If linkage is complete, only two parental classes of offspring are produced. If linkage is incomplete and some recombination occurs, then recombinant classes are also produced but these will not be as frequent as the parental classes. Consequence of incomplete linkage is departure from the frequencies of phenotypes expected according to indepenent assortment.

In case segregation is non-Mendelian, a cross *Aa* x *aa* does not produce expected 1/2 *Aa* : 1/2 *aa*, cross *Bb* x *bb* does not produce expected 1/2 *Bb* : 1/2 *bb*, and cross *Cc* x *cc* does not produce expected 1/2 *Cc* : 1/2 *cc* offspring. Due to non-Mendelian segregation at one or more loci the eight phenotypes in the three-point test cross progeny will not be observed in equal ratios of 1/8.

Genetic Analysis in Diploid Eukaryotes 149

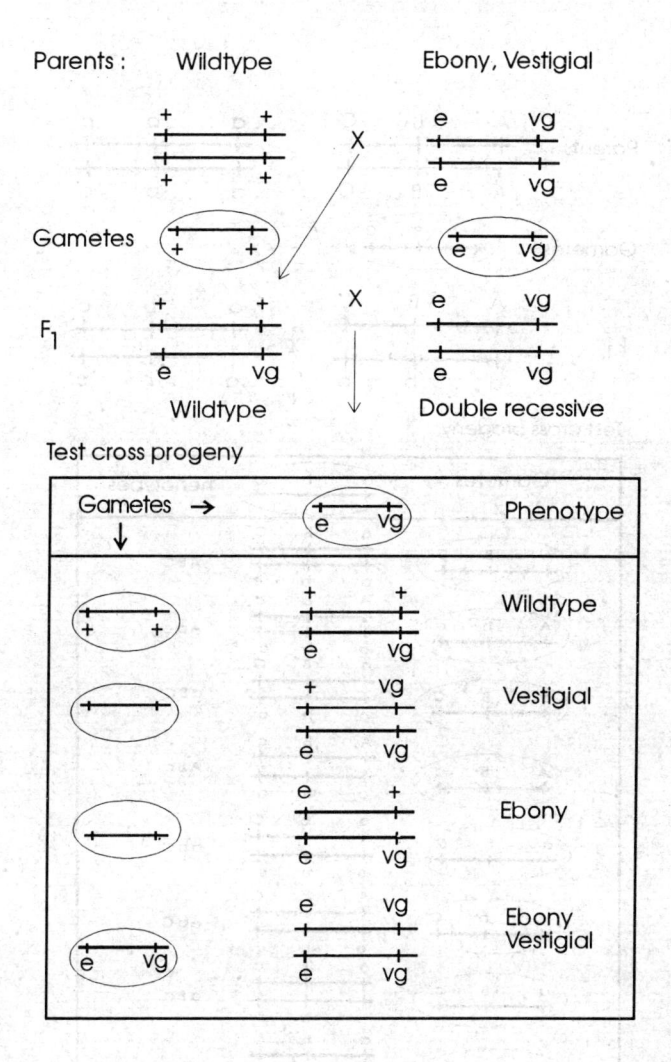

Fig. 10.5. Gene hypothesis showing a two-point test cross in *Drosophila*

Thus for detection of linkage in a three-point test cross, four X^2 values must be calculated: one each for segregation of the three individual gene pairs and the fourth for their joint segregation.

Detection of Linkage for sex-linked Genes. A mating between a heterozygous female *LlMmNn* and hemizygous recessive male *lmn*/Y

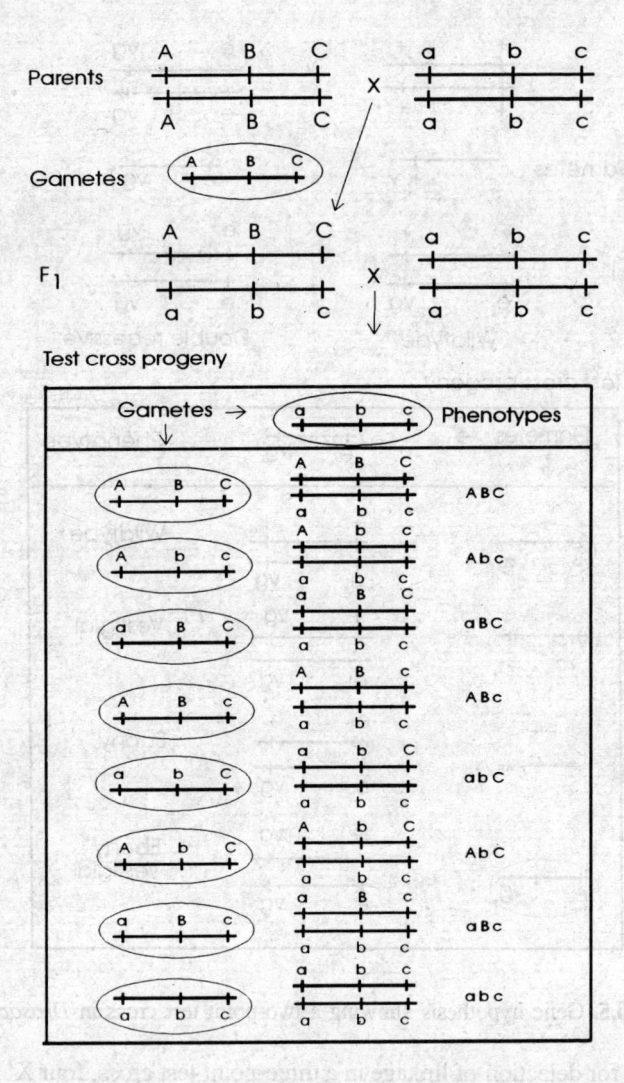

Fig. 10.6. Gene hypothesis of a three-point test cross involving autosomal genes

should produce eight classes of female offspring: LMN (*LlMmNn*), Lmn (*Llmmnn*), lMn (*llMmNn*), LMn (*LlMmnn*), lmN (*llmmNn*), LmN (*LlmmNn*), lMn (*llMmnn*) and lmn (*llmmnn*) (**Fig. 10.7**). Same eight

Genetic Analysis in Diploid Eukaryotes 151

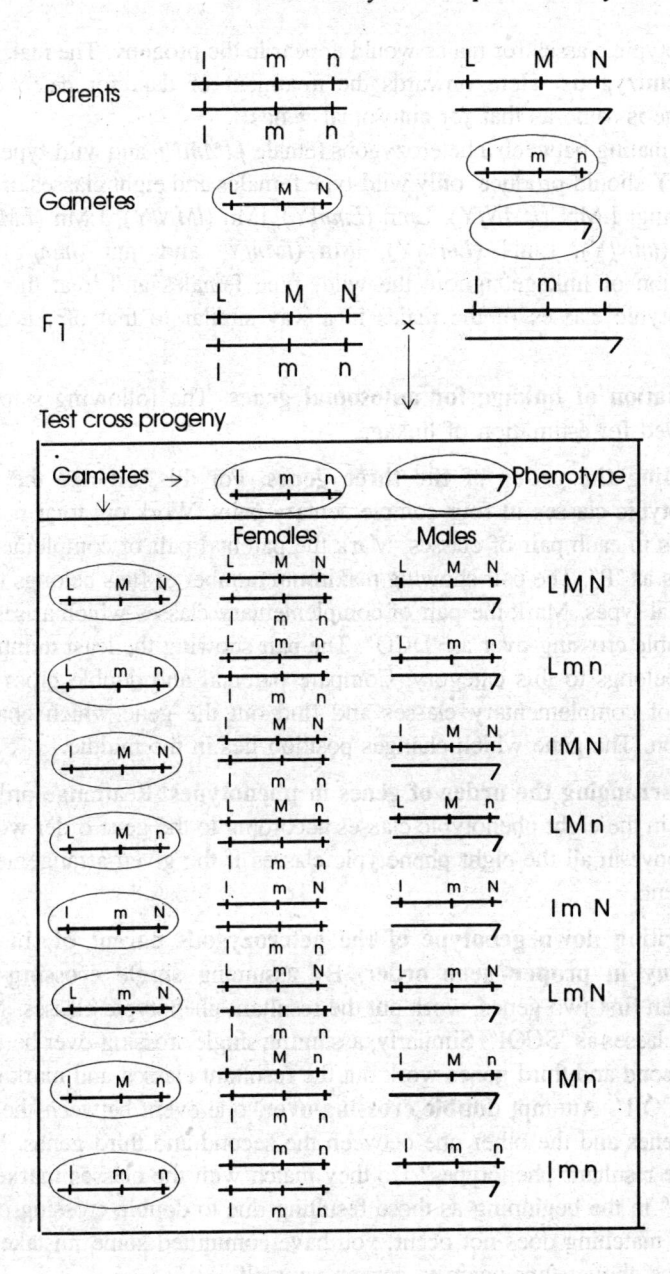

Fig. 10.7. Gene hypothesis of a three-point test cross involving sex-linked genes

phenotypic classes for males would appear in the progeny. The males will be **hemizygous**. Here onwards the treatment of data for detection of linkage is same as that for autosomal genes.

A mating between a heterozygous female *LlMmNn* and wild-type male *LMN*/Y should produce only wild-type females and eight classes of male offspring: LMN (*LMN*/Y), Lmn (*Lmn*/Y), lMn (*lMN*/Y), LMn (*LMn*/Y), lmN (*lmN*/Y), LmN (*LmN*/Y), lMn (*lMn*/Y) and lmn (*lmn*/Y). For detection of linkage, ignore the wild- type females and treat the eight phenotypic classes of the males in a way similar to that of autosomal genes.

Estimation of linkage for autosomal genes. The following steps are followed for estimation of linkage:

Finding the order of the three genes. For this, arrange the eight phenotypic classes in four complementary pairs. Work out total number of flies in each pair of classes. Mark the parental pair of complementary classes as "P". The pair showing maximum number of flies belongs to the parental types. Mark the pair of complementary classes which arises due to double crossing-over as "DCO". The pair showing the least number of flies belongs to this category. Compare parental and double cross over pairs of complementary classes and find out the gene which changes position. The gene which changes position lies in the middle.

2. **Rearranging the order of genes in phenotypes.** Rearrange order of genes in the eight phenotypic classes according to the gene order worked out above in all the eight phenotypic classes if the given arrangement is different.

3. **Writing down genotype of the heterozygous parent of the TC$_1$ progeny in proper gene order.** By assuming **single crossing-over** between first two genes, work out the resultant phenotypic classes. Mark these classes as "SCOI". Similarly, assuming single crossing-over between the second and third genes work out the resultant classes and mark them as "SCOII". Attempt **double crossing-over**, one event between the first two genes and the other one between the second and third genes. What are the resultant phenotypes? Do they match with the classes marked as "DCO" in the beginning as those resulting due to double crossing-over? If this matching does not occur, you have committed some mistake. Go over the above steps again to correct yourself.

Genetic Analysis in Diploid Eukaryotes 153

4. Calculating per cent recombination. Calculate per cent recombination between the first two genes and between the second and third genes. Calculate per cent recombination between first two genes by using the formula:

$$\% \text{ Recombination} = \frac{SCO1 + DCO}{N} \times 100$$

Calculate per cent recombination between second and third genes by using the formula.

$$\% \text{ Recombination} = \frac{SCO2 + DCO}{N} \times 100$$

5. Constructing a linkage map. Show the symbols of the genes indicating distance between them in map units along a straight line.

6. Calculating coefficient of coincidence (CC)

$$CC = \frac{\text{Observed frequency of double cross overs } (O_f)}{\text{Expected frequency of double cross overs } (E_f)}$$

Where, O_f = DCO/N and E_f = [(SCO1 + DCO)/N] x [(SCO2 + DCO)/N]. See whether the value of CC is less or more than one? What does this value signify?

7. Calculating interference. Mathematically, interference = 1-CC.

Estimation of linkage for sex-linked genes. In the test cross progeny resulting from mating between a heterozygous female *LlMmNn* and hemizygous recessive male *lmn*/Y eight classes [LMN (*LlMmNn*), Lmn (*Llmmnn*), lMn (*llMmNn*), LMn (*LlMmnn*), lmN (*llmmNn*), LmN (*LlmmNn*), lMn (*llMmnn*) and lmn (*llmmnn*)] of female and male offspring are expected. Back cross progeny data of this test cross are to be treated the same way as that of an autosomal test cross. A mating between a heterozygous female *LlMmNn* and wild-type male *LMN*/Y should produce only wild-type females and eight classes of male offspring: LMN (*LMN*/Y), Lmn (*Lmn*/Y), lMn (*lMN*/Y), LMn (*LMn*/Y), lmN (*lmN*/Y), LmN (*LmN*/Y), lMn (*lMn*/Y) and lmn (*lmn*/Y). For detection of linkage, ignore the wild- type females and treat the eight phenotypic classes of the males in a way similar to that of autosomal genes.

Complete Linkage

Crossing-over does not occur to same extent in males and females of different organisms. For example, Crossing-over is known to be absent in males of fruit fly *Drosophila melanogaster* and females of silk worm *Bombyx mori*. In *Drosophila* absence of crossing-over can be observed through a simple experiment. A cross between a trihybrid male (ABC) *AaBbCc* and triple recessive (abc) *aabbcc* female of *Drosophila melanogaster* will in the TC_1 progeny produce only two parental classes, (ABC) *AaBbCc* and (abc) *aabbcc*. In absence of crossing-over in *Drosophila melanogaster* males, no recombinant classes are expected.

Cytological Basis of Crossing-over

There is consistent view that cross-shaped structures (chiasma) observed during diplotene of prophase I of meiosis are directly related with crossing-over. Direct evidence that homologous chromosome exchange their parts during crossing-over was obtained by Stern (1931) in *Drosophila melanogaster* (**Fig. 10.8**). Stern studied whether the progeny with recombinant genotypes carried X-chromosomes with recombination of morphological markers. Recombinant male progeny with Bar- shaped red eyes were found to carry the short X chromosome but now with translocated piece of the Y-chromosome at one end. Males with normal-shaped carnation eyes contained long X chromoosme but without attached piece of the chromosome Y.

Creighton and McClintock (1931) also showed that genetic crossing-over is accompanied by exchange of chromatin and demonstrated the relationship of the chiasma to crossing-over, in maize. They made use of a knob on the end of a certain chromoosme and a visible irregularity on the homologue as cytological markers.

Crossing-over Occurs at Four-strand Stage

Evidence that crossing-over takes place at four-strand stage comes from *Drosophila* and *Neurospora*. These experiments are described below briefly:

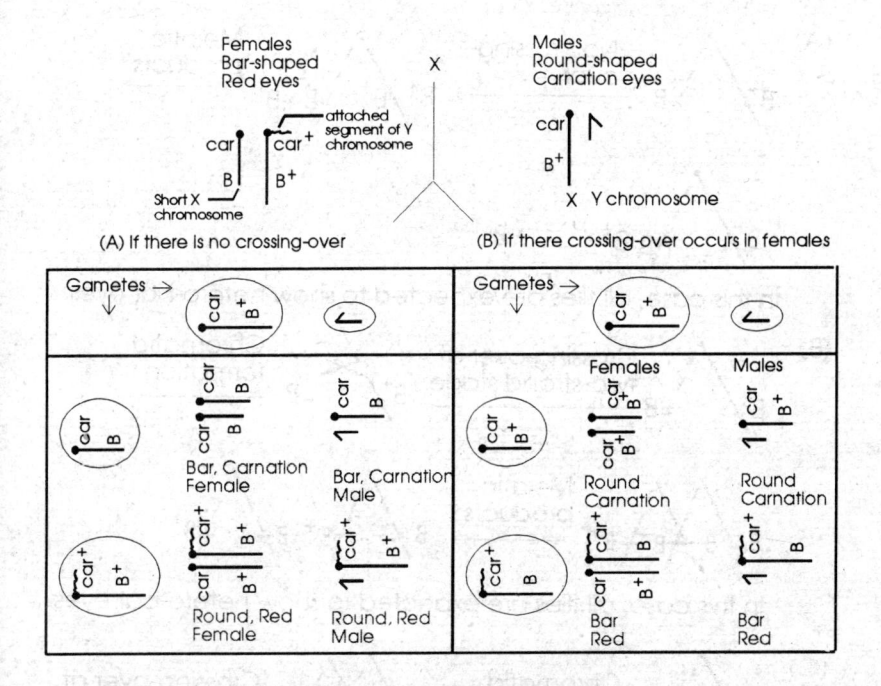

Fig. 10.8. Stern's experiment on *Drosophila* providing cytological evidence of crossing-over

Attached X chromosomes (X^X) in *Drosophila*

The experiment involving a cross between an attached XX female which is heterozygous at *Bar* locus (B^+/B) and a normal male is expected to produce all flies with hetero-bar eyes when it is assumed that there is no crossing-over (**Fig. 10.9A**) or when there is to crossing-over is assumed to occur at two-strand stage (**Fig. 10.9B**) between the centromere and the *Bar* locus, and Bar and wildtype flies in 1:1 ratio when crossing-over is assumed to take place at four-strand stage (**Fig. 10.9C**). The above cross when conducted gives the results of the second assumption thus proving that crossing-over takes place at four stage stage.

Tetrad Analysis in *Neurospora*

A cross between a mutant strain carrying linked gene markers and a wildtype strain is expected to produce all progeny of recombinant types

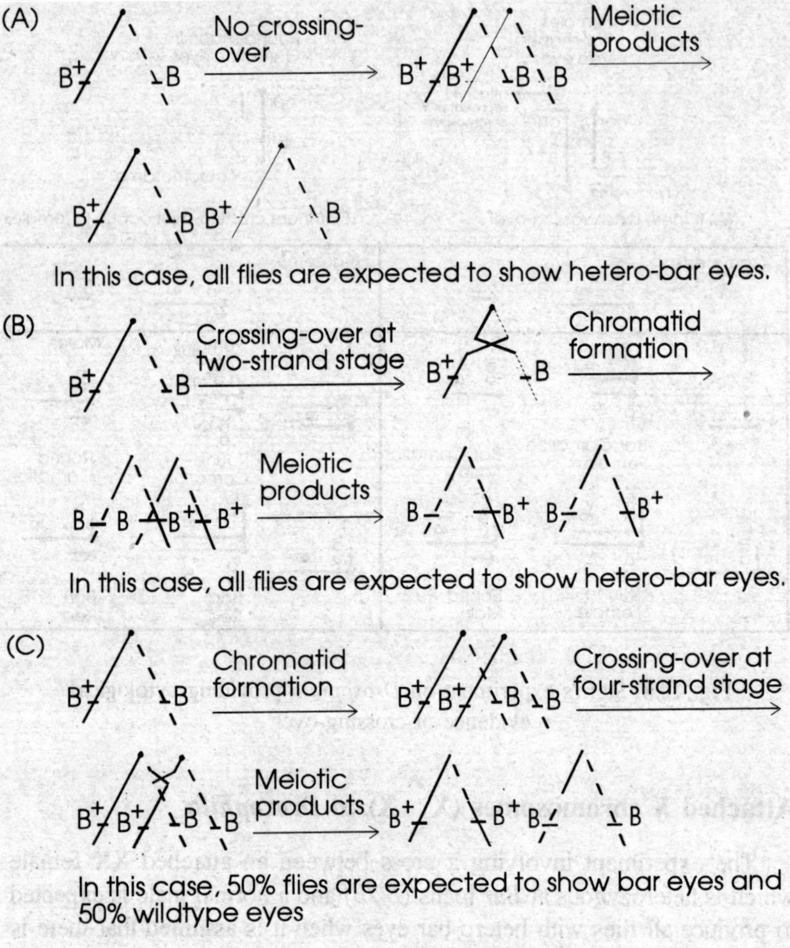

156 Basic Genetics

(A) No crossing-over → Meiotic products

In this case, all flies are expected to show hetero-bar eyes.

(B) Crossing-over at two-strand stage → Chromatid formation → Meiotic products

In this case, all flies are expected to show hetero-bar eyes.

(C) Chromatid formation → Crossing-over at four-strand stage → Meiotic products

In this case, 50% flies are expected to show bar eyes and 50% wildtype eyes

Fig. 10.9 Attached X chromosome experiment in *Drosophila*

when crossing-over is assumed to take place at two-strand stage (**Fig. 10.10A**). However, if crossing-over is assumed to take place at four-strand stage, the parental and recombinant type progeny will be expected to appear in 1:1 ratio (**Fig. 10.10B**). Actual cross yields the progeny as expected in the second assumption, thus proving proof that crossing-over takes place at four-strand stage.

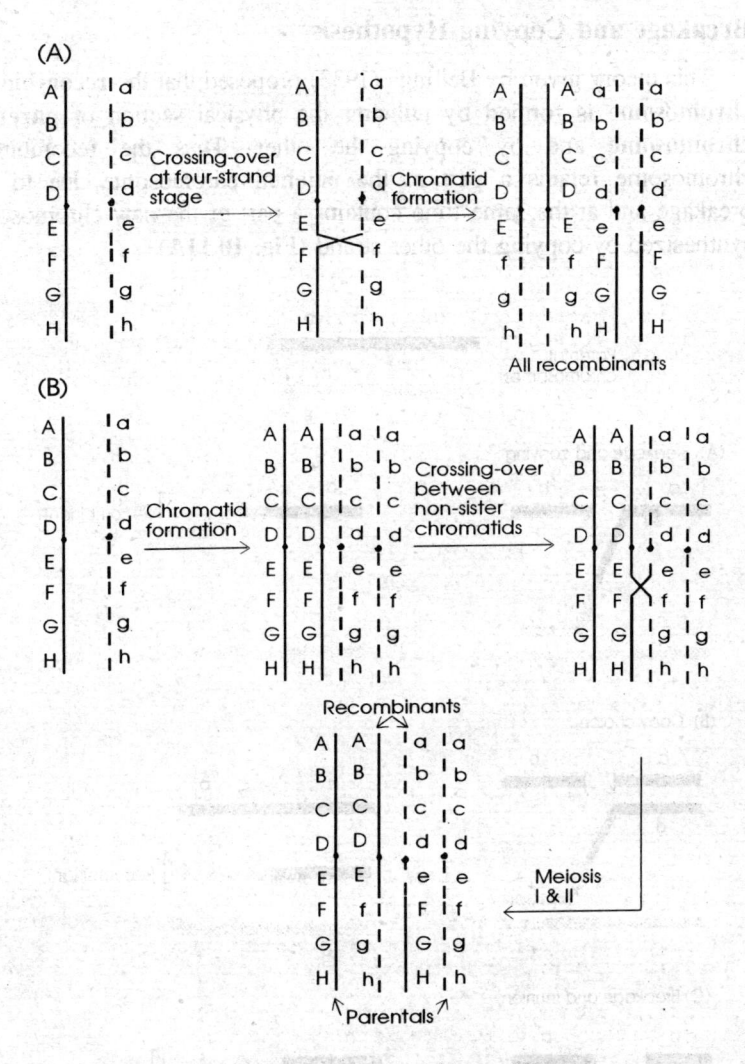

Fig. 10.10. Tetrad analysis *Neurospora*

Machanism of Recombination

Theories explaning mechanism of recombination may be divided into three general groups: breakage and copying, complete copy choice, and breakage and reunion.

Breakage and Copying Hypothesis

This theory given by Bellings (1933) proposed that the **recombinant chromosome** is formed by utilizing the physical section of **parental chromosome** and by copying the other. Thus the recombinant chromosome retains a part of the parental chromosome due to the breakage and at the same time contains a part of the new chromosome synthesized by copying the other strand (**Fig. 10.11A**).

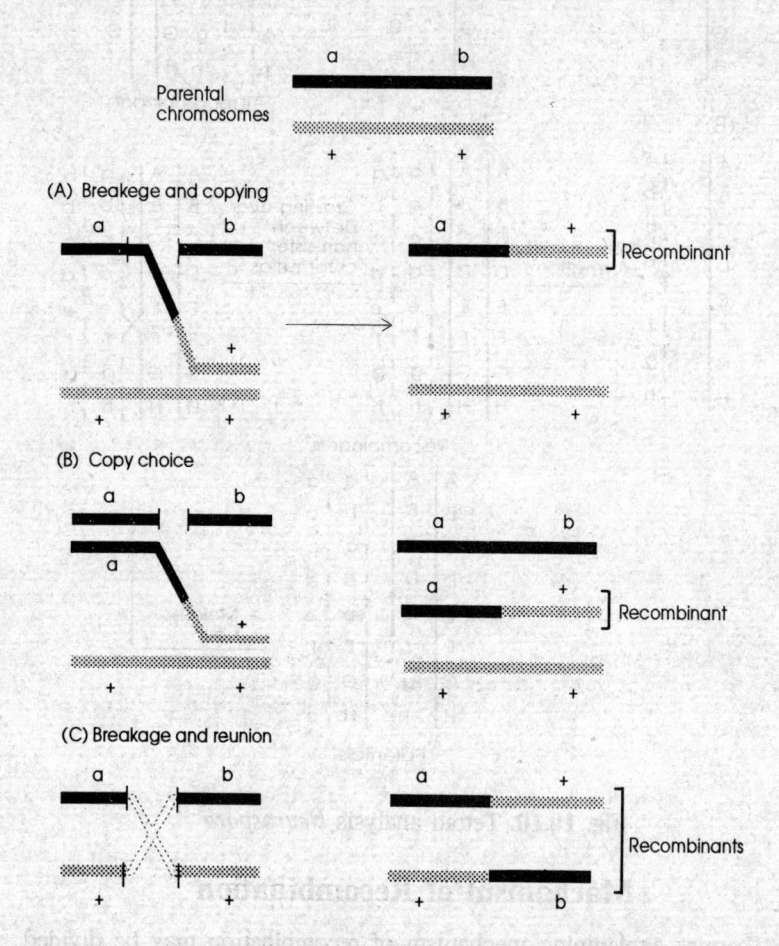

Fig. 10.11. Possible explanations for recombination events: (A) Breakage and copying, (B) Copy choice and (C) Brekage and runion

Complete Copy Choice Hypothesis

A modified version of Belling's theory was given by Lederberg (1955). The entire recombinant chromosomes arises from newly synthesized sections which have copied a part of their sequence from a section of one parental and other part of their sequences from the section of other parental chromosome (**Fig. 10.11B**). This hypothesis relates recombination to replication of the hereditary material. This hypothesis requires that homologous chromosome shall be paired atleast at certain points along their length at the time of replication and that the new strands may be replicas of one parental strand is one region and of the other in another region. This theory signifies that the replication is at the chromatid·level. It also provides a neat explanation of how within a tetrad on site may show a 2:2 ratio and a neighbouring one a 3:1 ratio (gene conversion).

Breakage and Reunion Hypothesis

According to this theory proposed by Darlington (1937), the two pieces of parental chromosomal information that are combined in the recombinant chromosome arise from the physical breaks in the parental chromosomes with subsequent physical exchange (**Fig. 10.11C**). The paired parental chromosome break and rejoin at meiosis resulting in the formation of "**chiasmata**" case, recombination does not depend upon the synthesis of new DNA molecules since the information entering the recombinants is merely transferred from the parentals.

Molecular Models of Recombination

Whitehouse-Hastings Model

In homologous chromatids, strands of different polarity are broken. The complementary nucleotide chains, one from each DNA molecule, break nearly but not exactly at homologous points. The parts of the two brokenn chains between the two breakage points uncoil from their complementary chains and then coil around each other so that complementary base pairing occurs between their terminal portions to form a cross over molecule (Whitehouse, 1963, 1966, 1967; Hastings and Whitehouse, 1964, 1965). New nucleotide chains are synthesied alongside the old chains that have not crossed over. The new chains uncoil from their complemetns, peel back and unite by complementary base pairing to form the other cross over molecule. A fundamental feature of this theory is thus the occurrence of **hybrid DNA** in the region of cross over.

Holliday's Model

According to Holliday (1962, 1964, 1968), the first event in recombination is recognition and alignment of homologous nucleotide sequence in two **duplex chromosomes**. Breaks are next made at comparable position in two of the polynucleotide chains having the same polarity. Each broken chain then invades the opposite helix, establishing base pairs with complementary nucleotides. Ligase enzyme then seals the discontinuities to produce a molecule with the shape of the Greek letter **Chi (X)** which is referred to as a "**Holliday intermediate**". Once created connection is free to move ot the right or to the left. Such shifts are called "**bridge migrations**". The hybrid DNA thus formed may carry a mismatched base pair. It is, therefore, said to contain **heterologous base sequence** and is called a **heteroduplex**. Holliday model is the most satisfactory to explain recombination at molecular molecule.

In a **reciprocal exchange** all four polynucleotide chains participating in the exchange events are conserved. Reciprocal exchange are thought to accompany most general recombinational events, but there are important cases like bacterial transformation and transduction where exchange appear to be non-reciprocal event, a piece of donor DNA is inserted into a recipient chromosome and may "extra" or "**dangling**" **DNA** is degraded by endonuclease.

Recombination involving reaction between homologous sequences of DNA is called **generalized recombination**. Recombination between chromosomes involves a physical exchange of parts. The structure created by this exchange is visible at meiosis in the form of a chiasma. The chiasma represents the results of **breakage and reunion**, in which two **nonsister chromatids** (each containing a duplex of DNA) have been broken and then linked each with the other. Recombination occurs between precisely corresponding sequences, so that not a single base pair is added to or lost from the recombinant chromosomes.

Synapsis. Two homologous duplex molecules of DNA must be brought into close contact so that the corresponding sequences can be exchanged. How do the DNA molecules some into juxtaposition? Contact between a pair of parental chromosomes occurs early in meiosis. The process is called **synapsis** or **chromosome pairing**. Homologous chromosomes (each actually consisting of the two sister chromatids produced by the

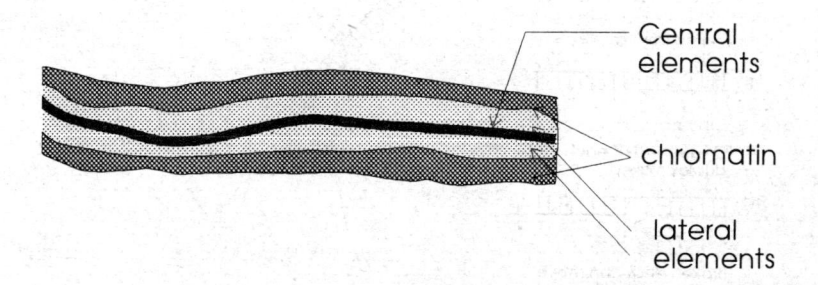

Fig. 10.12. Synaptonemal complex of *Neotellia*

prior replication) approach one another. They become laterally associated in the form of a **synaptonemal complex (Fig. 10.12)**. Each chromosome at this stage appears as a mass of chromatin bounded by a **lateral element** (which in this case has a striated structure). The two lateral elements are separated from each other by a **central element**. The triplet of parallel danse strands lies in a single plane that curves and twists along its axis. The distance between the homologous chromosomes is cosniderable in molecular terms, more than 200 nm. The generation of the synaptonemal complex coincides with the presumed time of crossing-over, although there is no direct evidence that recombination occurs at the stage of synapsis.

Recombination joint. The process of crossing-over, as shown in **Fig. 10.13,** starts with breakage at the corresponding points of the homologous strands of two paired DNA duplexes. The breakage allows movement of the free ends created by the nicks. each strand leaves its partner and crosses over to pair with its complement in the other duplex. The reciprocal exchange creates a connection between the two DNA duplexes. Initially, this is sustained only by hydrogen bonding; at some point, it is made covalent by sealing the nicks at the sites of exchange. The connected pair of duplexes is called a **joint molecule**. The point at which an individual strand of DNA crosses from one duplex to the other is called the **recombination joint**. At the site of recombination, each duplex has a region consisting of one strand fro meach of the parental DNA molecules. This region is called **hybrid DNA** or **heteroduplex DNA**.
Branch migration. Once the strand exchange has been initiated, it can move along the duplex. Such mobility is called branch migration. **Fig.**

162 Basic Genetics

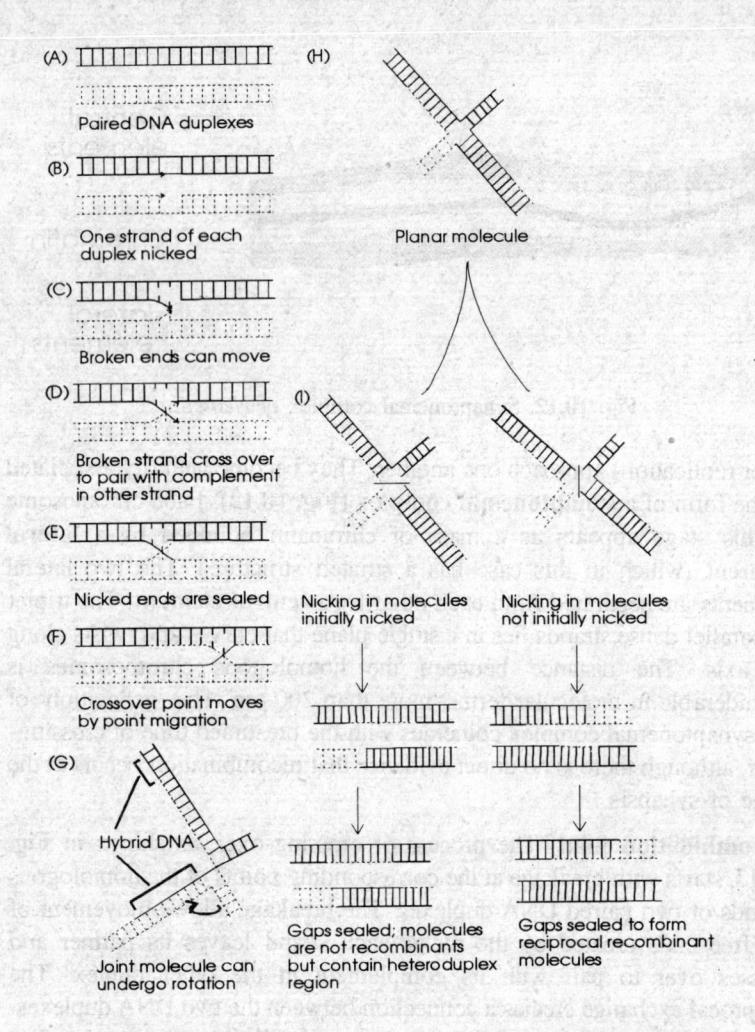

Fig. 10.13. The Holliday model of genetic recombination

(A) Paired DNA duplexes
(B) One strand of each duplex nicked
(C) Broken ends can move
(D) Broken strand cross over to pair with complement in other strand
(E) Nicked ends are sealed
(F) Crossover point moves by point migration
(G) Hybrid DNA — Joint molecule can undergo rotation
(H) Planar molecule
(I) Nicking in molecules initially nicked / Nicking in molecules not initially nicked → Gaps sealed; molecules are not recombinants but contain heteroduplex region / Gaps sealed to form reciprocal recombinant molecules

10.14 illustrates the branch can migrate in either diection as one strand is displaced by the other. The rate of branch migration is uncertain, but probably ∼30 bp/sec. Although the branch point is not fixed, the rate of migration is probably inadequate to support the formation of extensive regions of heteroduplex DNA in natural conditions. Any extensive branch migration *in vivo* must therefore be catalyzed by a **recombination**

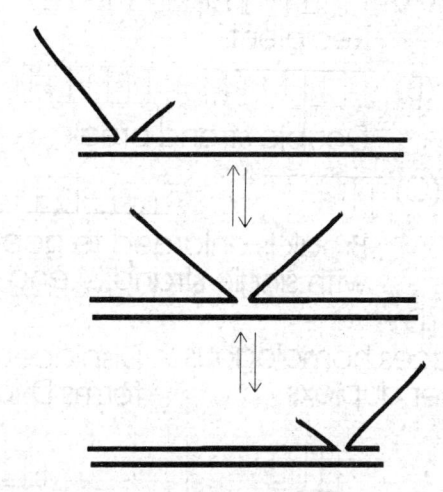

Fig. 10.14. Branch migration can occur in either directly by by displacement of unpaired single strand with a paired strand

enzyme. The joint molecule formed by strnad exchange must be **resolved** into two separate duplex molecules. This requires are visualized on the right side of Fig. 10.13 in terms of the **planar molecule** generated by rotating one of the duplexes of the recombination intermediate. The consequences depends on which pair of strands in nicked. A strand exchange between duplex DNAs always leaves behind a region of heteroduplex DNA, but the exchange may or may not be accompanined by recombination of the flanking regions. Formation of an intermediate involving heteroduplex DNA that can be extended by branch migration.

Initiation of genetic exchange. Genetic exchange is initiated by a double-strand break. Then, one of the single strands migrates to the other duplex, so that the duplex molecules become connected by a stretch of heteroduplex DNA. The model is illustrated in **Fig. 10.15**. Recombination is initiated when an endonuclease makes a double-strand break in one chromatid, the "recipient". The cut is enlarged to a gap, probably by **exonuclease** action. The exonuclease(s) nibble away one strand on either side of the break, generating 3' single-stranded termini. One of the free 3' ends then invades a homologous region in the other, "donor" duplex.

Formation and extension of D-loop. The formation of heteroduplex DNA generates a D-loop, in which one strand of the donor duplex is

(A) Recipient

(B) Double strand break

(C) Break is enlarged to gap with single strand 3' end

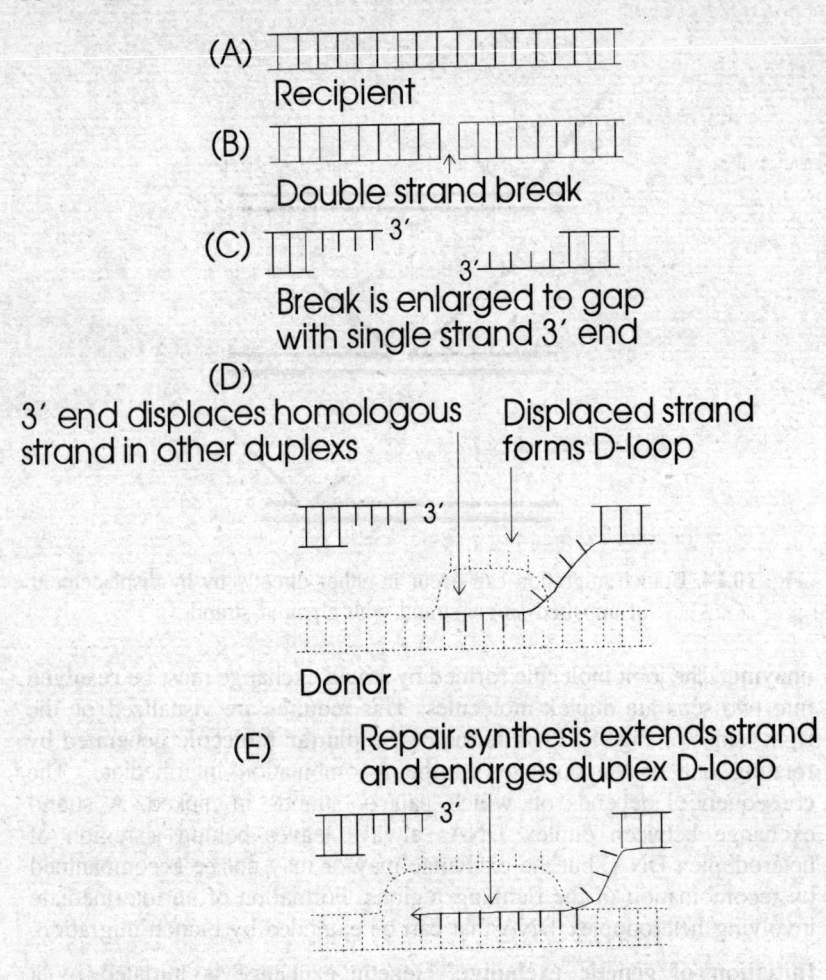

(D) 3' end displaces homologous strand in other duplexs   Displaced strand forms D-loop

Donor

(E) Repair synthesis extends strand and enlarges duplex D-loop

Fig. 10.15. Recombination by Double-strand break followed by formation of single stranded 3' end, one of which migrates to a homologous duplex DNA

displaced. The D-loop, in which one strand of the donor duplex is displaced. The D-loop is extended by **repair synthesis**, using the free 3' end as a primer. Eventually, the D-loop becomes large enough to correspond to the entire length of the gap on the recipient chromatid. When the extruded single strand reaches the far side of the gap, the complementary single-stranded sequences anneal. Now there is heteroduplex DNA on either side of the gap. and the gap itself is

(F) Donor D-loop pairs with recipient 3' end

(G) Repair synthesis converts donor D-loop to duplex

(H) Heteroduplex DNA   Recipient gap has been replaced by donor sequence

Branch migration →

Resolution by cutting strands

Non crossovers

or

Crossovers

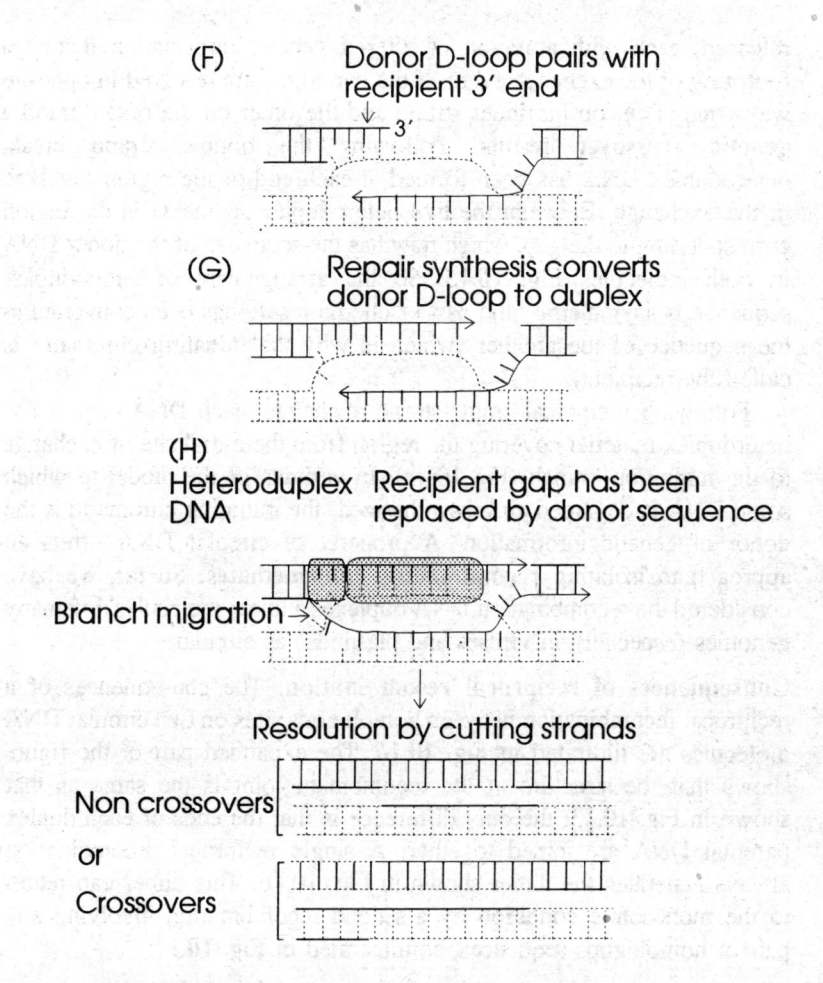

Fig. 10.15. contd...

represented by the single-stranded D-loop. The duplex integrity of the gapped region can be restored by repair synthesis using the 3' end on the left side of the gap as a primer. Overall, the gap has been repaired by two individual rounds of **single strand DNA synthesis**. Branch migration converts this structure into a molecules with two recombinant joints. The joints must be resolved by cutting.

If both jonts are resolved in the same way, for example, the inner strands are cut at each joint, the original noncrossover molecules will be released, each with a region of altered genetic information that is a footpring of the exchange event. If the two joints are resolved in opposite ways one is cut on the inner strand and the other on the outer strand a **genetic crossover** results. Following the double strand break, heteroduplex DNA has been formed at each end of the region involved in the exchange. Between the two heteroduplex segments is the region corresponding to the gap, which now has the sequence of the donor DNA in both molecules **Fig. 10.15**. So the arrangement of heteroduplex sequence is asymmetric, and part of one molecule has been converted to the sequence of the another (which is why the initiating chromatid is called the recipient).

Following reciprocal single-strand exchange, each DNA duplex has heterduplex material covering the region from the initial site of exchange to the migration branch (Fig 10.13). In variants of the model in which some DNA is degraded and resynthesized, the initiating chromatid is the donor of genetic information. A property of circular DNA offers an approach to isolating **recombination intermediates**. So far, we have considered the recombination DNA duplexes a linear molecules, but many genomes (especially in viruses and plasmids) as circular.

Consequences of reciprocal recombination. The consequences of a reciprocal recombination between homologous sites on two circular DNA molecules are illutrated in **Fig. 10.16**. The expanded part of the figure shows that the structure at the recombinant joint is the same as that shown in Fig. 10.13; the only difference is that the ends of each duplex parental DNA are joined together. A single reciprocal recombination always generates the dimer shown in Fig. 10.16. This dimer can return to the monosomic condition by a second recombination involving any pair of homologous sequences, as illustrated in **Fig. 10.17**.

Figure-eight and Chi-structure

A practical consequence of recombination between circular DNAs is that the recombination intermediate should take the form of a figure-eight, as depicted in the expanded structure of Fig. 10.16. The nature of the figure-eight can be distinguished by cleaving the isolated molecule with a restriction enzyme that cuts each monomeric circle only once. The two parts of the figure-eight then will fall apart unless they are covalently connected. If the figure-eight is a **recombination intermediate**, cleavage will generate a structure in which the parental duplex molecules are held together by a region of heteroduplex DNA at the point of fusion. The

Genetic Analysis in Diploid Eukaryotes 167

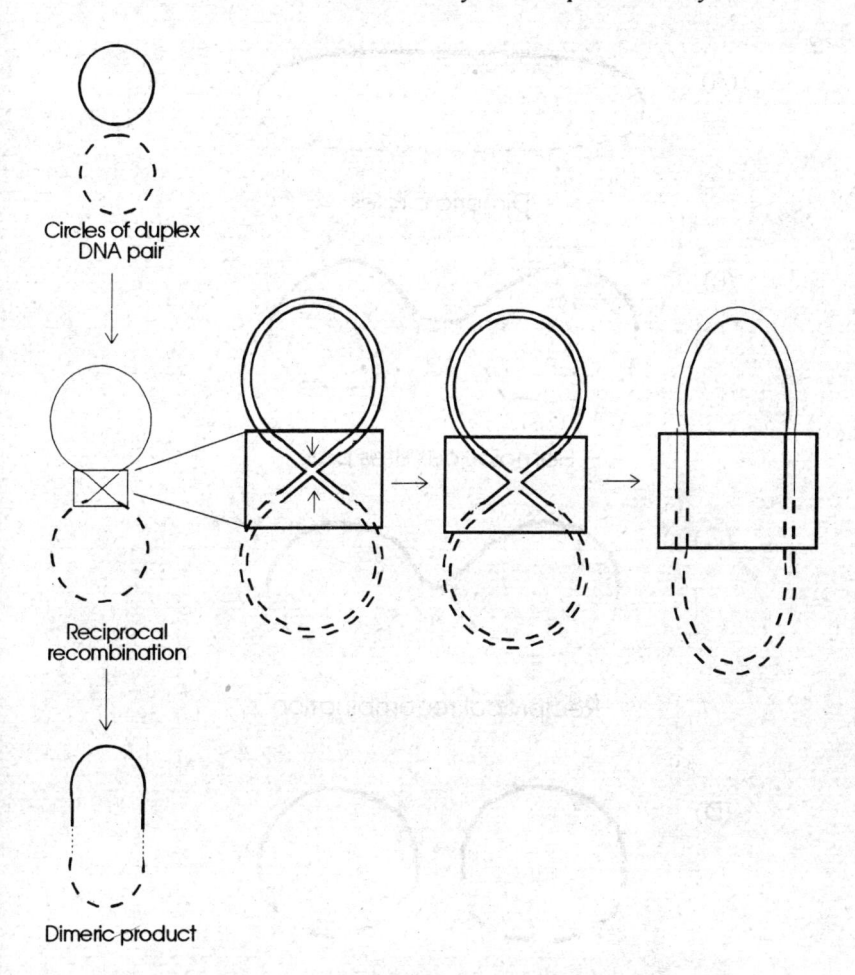

Fig. 10.16. Generation of a dimeric circle through a figure-eight intermediate structure resulting in reciprocal recombination between homologous duplex circles

cleaved molecule has four arms, each pair of the arms corresponding to each of the original monomeric circles. In the cleaved form, this molecules is called a **chi structure**. An example is shown in **Fig. 10.18**. The four arms of duplex DNA are connected by a region in which the heteroduplex segments have been pulled apart into their consituent single strands (probably by the conditions of preparation).

(A)

Dimeric circles

(B)

Homologous sites pair

(C)

Reciprocal recombination

(D)

Monomeric circles

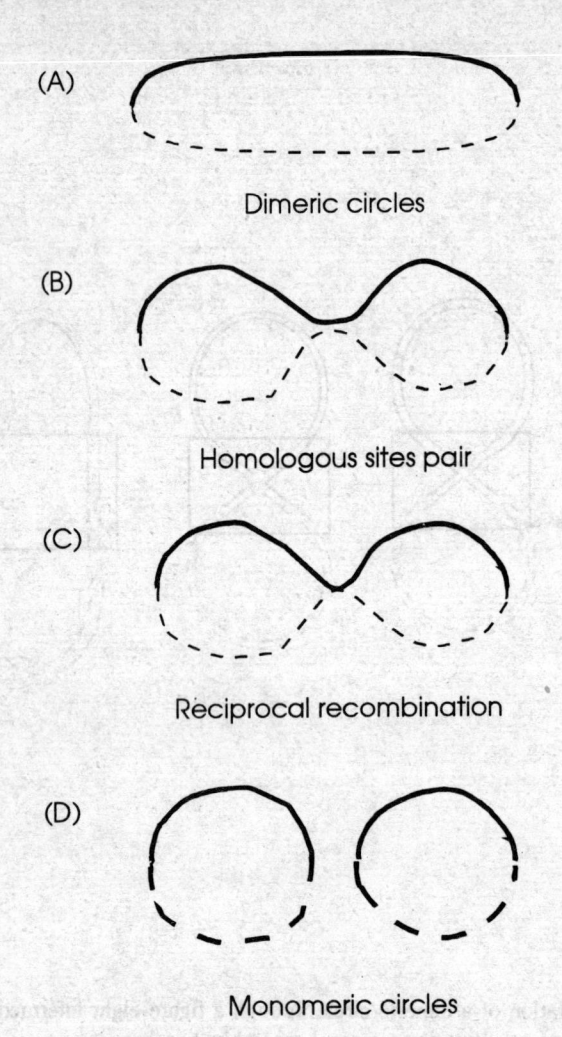

Fig. 10.17. Generation of two monomeric circles by a reciprocal recombination between homologous sequences

Gene Conversion

The involvement of heteroduplex DNA explains the characteristics of recombination between alleles; indeed, allelic recombination provided the impetus for the development model. When recombination between alleles was discovered, the natural assumption was that it makes place by the

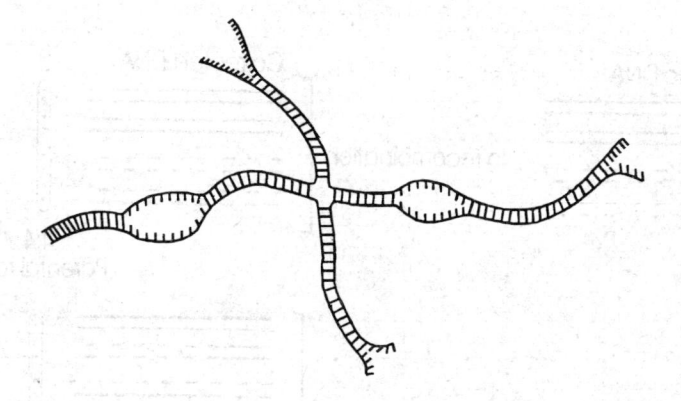

Fig. 10.18. Chi-structure with four duplex arms generated from figure eight

same mechanism of reciprocal recombination that applies to more distant loci. That is to say that an invidual breakage and recunionevent occurs within the locus to generate a reciprocal pair of recombinant chromosomes. However, in the close quarters of a single gene, the formation of heteroduplex DNA itslef is usually responsible for the recombination event.

Individual recombination events can be studied in the **Ascomycetes** fungi, because the products of a single meiosis are held together in a large cell, the ascus. Even better, the four haploid nuclei produced by meiosis are arranged in a linear order. Actually, a mitosis occurs after the production of these four nuclei, giving a linear series of eight haploid nuclei. **Fig. 10.19** shows that each of these nuclei effective represents the genetic character of one of the eight strands of the four chromosomes produed by the meiosis. Meiosis in a heterozygote should generate four copies of each allele. This is seen in the majority of the spores. But there are some spores with abnormal ratios. They are explained by the formation and correction of heteroduplex DNA in the region in which the alleles differ.

Some asci display ratios of 3:5 or 2:6, in which one or two spores, respectively, that should have been of one allelic type actually are of the other type. These ratios are a consequence of the process of recombination. An uneven ratio (3:5 or 5:3) can result only from segregation of two **mismatched strands** in one DNA duplex; only the other heteroduplex must have been corrected. Even ratios (2:6 or 6:2)

(A)
Meiotic DNA — No recombination — Corrected DNA

4:4 Parental ratio

(B)
Meiotic crossing-over — Correction — 3:5 Postmeiotic segregation

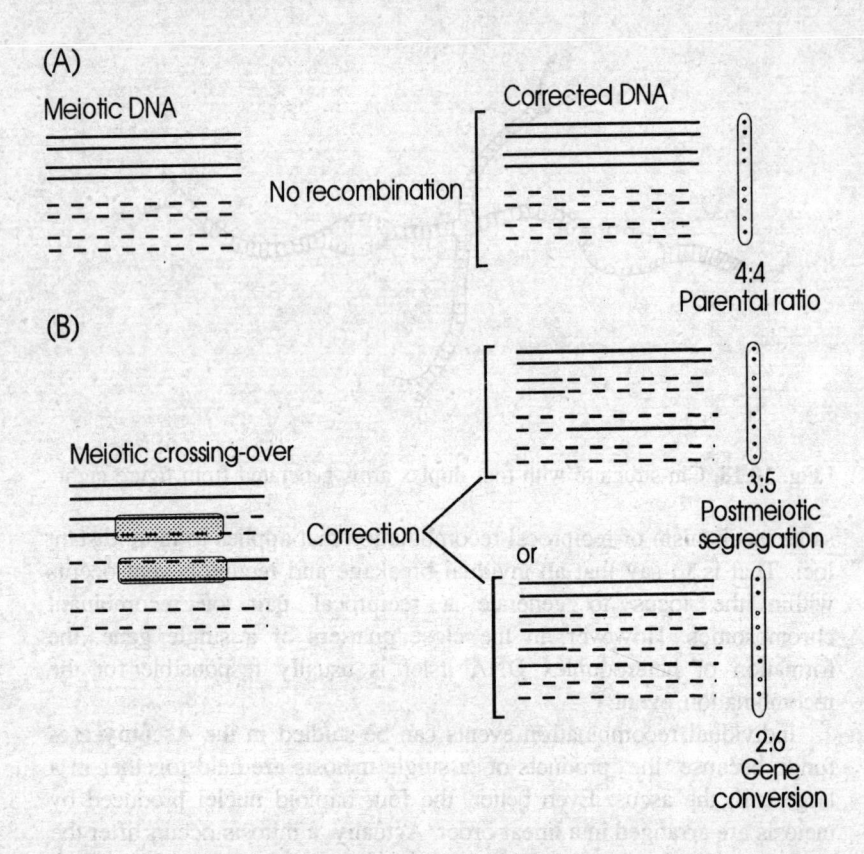

or

2:6 Gene conversion

Fig. 10.19. Gene conversion in *Ascomycetes*

could in principal result from independent correction of two heteroduplexes, but are more likely to result from a repair mechanism such as the copying event resulting from a duble-strand break, as illustrated in Fig. 10.15. Gene conversion does not depend on crossing-over, but is correlated with it. A large proportion of the aberrant asci show genetic recombination between two markers on either side of a site of interallelic gene conversion.

Significance of Crossing-over

Crossing-over is a means of introducing new combinatins of genes and hence traits. It thus increase **genotypic** and **phenotypic variability** which is useful for natural selection and for adjustment under the changed environment. Frequency of crossing-over depends upon the distance between the two genes. This serves as a basis for perparing linkage **chromosome maps**. Linkage maps which use phenomenon of crossing-over as basis, proves that genes lie in linear fashion in the chromosome. Useful recombinants produced by crossing-over are picked up by breeders to produce new varieties of crop plants and animals.

Some Useful Reading Hints

Gowen and Gowen (1922) dealt with complete linkage in *Drosophila melanogaster*. Belling (1933) gave an over view of corssing-over and gene rearrangement in flowering plants. Randolph (1936) deals with developmental morphology of the caryopsis in maize. Slizynski (1964) dealt with chiasmata in spermatocytes of *Drosophila melanogaster*. Immer (1930) gave formulae and tables for calculating linkage intensities. Creighton and McClintock (1931) were first to report a correlation of cytological and genetical crossing over in *Zea mays*. Whittinghill (1950) described consequences of crossing over in oogonial cells. Muller (1916) suggested the mechanism of crossing-over. Owen (1950) dealt with the theory of genetical recombination. Levine (1955) discussed chromosome structure in relation to the mechanism of crossing over. Weinstein (1958) dealt with the geometry and mechanics of crossing-over. Jenkins (1975) and Strickberger (1986) dealt in length on construction of genetic maps in diploid eukaryotes.

Mather (1951) gave us a must use book *The Measurement of Linkage in Heredity* without which proper understanding of the subject is not possible. Green (1963) gave methods for testing linkage. Emerson (1969) discussed linkage and recombination at the chromosome level. Peacock (1971) discussed cytogenetic aspects of the mechanism of recombination in higher organisms. Howard-Flenders (1973) discusseed role of DNA repair and recombination. Whitehouse (1963) and Whitehouse and Hastings (1965) gave a theory of crossing over by means of hybrid DNA. Meselson (1964) gave an excellent historical review of the mechanism of genetic recombination between DNA molecules. Meselson and Radding (1975) giae a general model for genetic recombination. Dressler and Potter (1982) described molecular mechanisms in genetic recombination. Hasting and Whitehouse (1982) reviewed mechanisms of genetic recombination. Flanders *et al.* (1984) described role of RecA protein spiral filaments in genetic recombination. Howard-Flenders *et al.* (1984) dealt with role of RecA protein spiral filaments in genetic recombination. West (1992) explained the enzymes involved in genetic recombination.

Lindsley and Zimm (1992) compiled all gene chromosomal and molecular markers known in the genome of *Drosophila melanogaster*. McPeek and Speed (1995) gave model explaining interference in genetic recombination. Stahl and Lander (1995) referred to estimating interference and linkage map distance from two-factors tetrad data. Risch (1992) dealt with interpreting of LOD scores in genetic linkage. Morton (1995) dealt with LODs past and present.

Questions

1. How can time of segregation be studied in corn?
2. What is xenia effect? Is xenia effect shown for all the endospermic traits? Answer this question with the help of appropriate reciprocal crosses.
3. What is linkage? What is its relatinship with independent assortment and crossing-over?
4. What is a linkage group?
5. What do you understand by genetic recombination?
6. Differntiate with the help of a diagram between coupling and repulsion phase of linkage.
7. What type of gamets will be produced from an individual of genotypes Ab/aB, with and without crossing-over between two genes?
8. What is the procedure for detection and estimation of linkage in a two-point test cross and F_2 progeny in a plant system?
9. How does the procedure of detection and estimation of linkage in animal systems differ from that in plant systems?
10. What is linkag map? What is its basis?
11. Outline the procedure for detection and estimation of linkage in a three-point test cross involving autosomal genes?
12. Outline the procedure for detection and estimation of linkage in a three-point test cross involving sex (X)-linked genes?
13. What do you understand by coefficient of coincidence and interference/ How are these concepts related to each other and what do they signify?
14. What is the theoretical basis of detection of linkage through partitioning of chi-square? Describe the procedure along with the formulae used for detection of linkage in a F_2 plant population. Also describe similar procedure for a test cross plant population.
15. Describe the following methods for estimation of linkage in F_2 plant population: Emerson's method, maximum-liklihood method, and product ratio method. Which one of these methods is more efficient and why?
16. What is complete linkage? Illustrate this phenomenon with the help of a diagram.
17. Give an experiment for cytological proof of crossing-over.
18. Give experiment evidence using *Drosophia* that crossing-over occurs at four-strand stage.
19. Through tetrad analysis in *Neurospora* show that crossing-over occurs at four-strand stage.
20. Explain with the help of suitable diagrams the following hypotheses of machanism of recombination: beakage and copying hypothesis, complete copy choice hypothesis, and breakage and reunion hypothesis.
21. Explain with the help of appropriate diagrams the following molecular events of recombination: (a) Formation of synaptonemal complex, (b) Recirocal single strand exchange, its extension by branch migration and resolution by nicking, (c) Occurrence of branch migration in either direction, (d) Initiation of recombination by a double-strand break, formation of single-stranded 3' ends annd migration of one of them to a homologous duplex, (e) Reciprocal recombination between

homologous duplex circles produced through figure-eight intermediate, (f) Generation of two monomeric circles from a dimeric circle by reciprocal recombination between homologous sequences, and (g) Generation of a chi structure by cleavage of a figure-eight.
22. How does gene conversion account for interallelic recombination?
23. What is the significance of crossing-over? In what way has it been used during evolution and how has this phenomenon been used for plant and animal improvement?

Numericals

1. In tomatoes, round fruit shape (O) is dominant over oblate (o) and smooth fruit skin (P) is dominant over peach (p). Plants having round smooth fruit were crossed with those having elongate peach fruit. The F_1 plants were test crossed and the following progeny were obtained: Round, Smooth - 360, Round, Peach - 32, Elongate, Smooth - 38, and Elongate, Peach - 370. Test if the genes o and p linked? If so, in which phase? Estimate the frquency of recombination between the genes and prepare their linkage map.

2. In tomato, compound inflorescence (s) is recessive to solitary (S), peach fruit shape (p) is recessive to smooth (P) and oblate fruit shape (o) is recessive to round (O). When plants heterozygous for genes o, p and s were test crossed, the following progeny were obtained: Round, Smooth, Solitary - 73, Oblate, Peach, compound - 63, Round, Smooth, compound - 348, Oblate, Peach, Solitary - 306, Round, Peach, Compound - 3, Oblate, Smooth, Solitary - 4, Round, Peach, Solitary - 96, and Oblate, Peach, Compound - 110. Find out the genotypes and phenotypes of the two parents and of F_1 individuals. What different types of gametes will be produced by the F_1 plants? Calculate frequency of recombination between genes o, p and s and prepare their linkage map. Calculate coefficient of coincidence and interference and comment on their nature.

3. A cross was made between yellow bodied (y), echinus (ec), white eyed (w) female ($y\ ec\ w/y\ ec\ w$) flies and wild males. The F_1 females were mated with $y\ ec\ w$ males. The following genotypes were present in a sample of 1000 progeny flies.

Females	+	Males	=	Total
+ + +/y ec w		+ + +/Y		475
y ec w /y ec w		y ec w/Y		469
y + +/y ec w		y + +/Y		8
+ ec w/y ec w		+ ec w/Y		7
y + w/y ec w		y + w/Y		18
+ ec +/y ec w		+ ec +/Y		23
+ + w/y ec w		+ + w/Y		0
Y ec +/y ec w		y ec +/Y		0

Determine the order in which three loci y, ec, and w occur on the chromosome and prepare a linkage map. Calculate coefficient of coincidence and interference and comment on their nature.

4. A cross involving X-linked genes was made between *yellow, Bar, verminion* female flies and wild males, and the F_1 females were crossed with $y + v$ males. The following phenotypes were obtained when 1000 progeny were examined: $y B v$ and $+ + + = 546$, $y + + $ and $ + B v = 244$, $y + v$ and $+ B + = 160$, and $y B +$ and $+ + v = 50$. Determine the order in which the three loci occur on the chromosome and prepare a linkage map. Calculate coefficient of coincidence and interference and comment on their nature.

5. Female *Drosophila* heterozygous for *ebony* (e^+/e), *scarlet* (st^+/st), and *spineless* (ss^+/ss) were test crossed, and the followin(g progeny were obtained: Wild type - 67, *ebony* - 8, *ebony scarlet* - 68, *ebony spineless* - 347, *ebony scarlet spineless* - 78, *scarlet* - 368, *scarlet spineless* - 10, and *spineless* - 54. (a) Are these genes linked? Justify your answer. (b) Write genes given on a chromosome symbol with the genes in correct order. (c) Write the genotypes of flies involved in the parental cross and test cross. (d) What are the map distances between the three loci? (e) Calculate the cofficient of coincidence and interference and comment on their nature.

6. Assume that in *Drosophila* there are three pairs of alleles, $+/x$, $+/y$ and $+/z$. As shown by the symlols, each mutant gene is recessive to its wild type allele. A cross between females heterozygous at these three loci and wild type males yielded the following progeny: Females: $+ + + = 1010$; Males: $+ + + = 39$, $+ + z = 430$, $+ y z = 32$, $x + + = 27$, $x y + = 441$, and $x y z = 31$. Draw the appropriate linkage map for the above data showing the order of the three markers and the map distances for each marked interval. Calculate the co-efficient of coincidence and interference for these data and comment on their nature.

7. Singed bristles (*sn*), crossveinless wins (*cv*), and vermilion eye colour (*v*) are due to recessive mutants alleles of three sex-linked genes in *Drosophila melanogaster*. When a female heterozygous for each of the three genes was test the following progeny were obtained: *singed, crossveinless, vermillion* - 3, *crossveinless, vermillion* - 392, *vermillion* - 34, *crossveinless* - 61, *singed, crossveinless* - 32, *singed, vermition* - 65, *singed* - 410, and wild type - 3.
(a) What is the correct order of the three genes on the chromosome? (b) What are genetic map distance between *sn* and *cv*, *sn* and *v*, and *cv* and *v*? (c) What are the co-efficient of coindence and interference values? What do these values signify?

8. Genes *a* and *b* are linked with 20 per cent crossing-over. An $a^+ b^+/a^+ b^+$ individual was mated with an *a b/a b* individual. (a) Represent the cross on the chromosomes and illustrate the gametes produced the F_1 individual. (b) If F_1 was crossed with double recessive, what offspring would be expected and in what proportion? (c) Is this a case of coupling or repulsion linkage?

9. Genes a and b are linked with 20 per cent crossing-over. An F_1 individual of genotype $a^+ b/a b^+$ was mated with an *a b/a b* individual. (a) Represent the cross between the original parents on the chromosomes. (b) Illustrate the gametes produced the F_1 individual. (c) If F_1 was crossed with double recessive, what offspring would be expected and in what proportion? (d) Is this a case of coupling or repulsion linkage?

10. What phenotypic classes would be expected in F_2 of a dihybrid cross and in what proportion, if crossing over is 20%.

11. Assume that genes *a* and *b* are linked and show 30% recombination. If a wild type individual is crossed with double recessive individual, what will be the genotype of F_1? What gametes with the F_1 produce and in what proportion? If F_1 is crossed with a double recessive, what will be the appearance and genotypes of the offspring.
12. Plants heterozygous for genes *a* and *b* were selfed and gave the following number of plants in four phenotypic classes: A_B_ = 260, A_bb = 40, aaB_ = 30, and aabb = 70. Detect linkage by partitioning chi-square. Estimate linkage by Emerson's, Maximum-liklihood and product ratio methods. Comment on the efficiency of these methods.
13. A fully heterozygous F_1 corn plant was red with normal seed. This plant was crossed with green plant (*b*) that had tassel seed (*ts*). The following progeny was obtained: Red, Normal = 129, Red, Tassel = 121, Green, Normal = 127, and Green, Tassel = 123. Do the genes *b* and *ts* show linkage? If so what is the percentage crossing-over? If these genes do not show linkage show that recombination frequency is 50%. Diagram the cross showing the arrangement of genetic markers on the chromoosmes.

11

GENETIC ANALYSIS IN HAPLOID EUKARYOTES

Gene Mapping in Neurospora

First- and Second-division Segregation

In order to know genotype of a particular ascospore arising from crosses between strains of *Neurospora crassa*, ascospores must be isolated and cultured separately. This is called **tetrad analysis**. Because of ordered tetrads, in *Neurospora* the distance between *a* gene locus and its centromere can be determined. When two members of an allelic pair separate during meiosis I, they are said to show **first-division segregation (Fig. 11.1A)** and when two members of an allelic pair do not separate during meiosis I, rather they do so at meiosis II, they are said to show **second-division segregation (Fig. 11.1B)**. *Neurospora* has ordered tetrads. The tetrads resulting from first- and second division segregation can be identified.

Different Types of Tetrads and Their Origin

In order to understand types of tetrads, let us consider a cross + + X *a a* in *Neurospora crassa*. In tetrad analysis, if the ascospores have genotype of one or the other parent, (e.g. *a b, a b,* + +, + +), the tetrad is classified as **parental ditype** (PD). If the ascospores have genotype of neither of the parents, i.e., all ascospores have recombinant types (e.g. *a* +, *a* +, + *b,* + *b*), the tetrad is classified as **nonparental ditype** (NPD), If the ascospores of a tetrad include two parental types and two nonparental types (e. g. *a* +, + *b, a b,* + +), the tetrad is classified as **tetratype** (TT). Origin of different type of tetrads is explained in **Table 11.1**.

Gentic Analysis in Haploid Eukaryotes 177

(A) First-division segregation with no crossover between gene a and centromere

(B) Second-division segregation with crossover between gene a and centromere

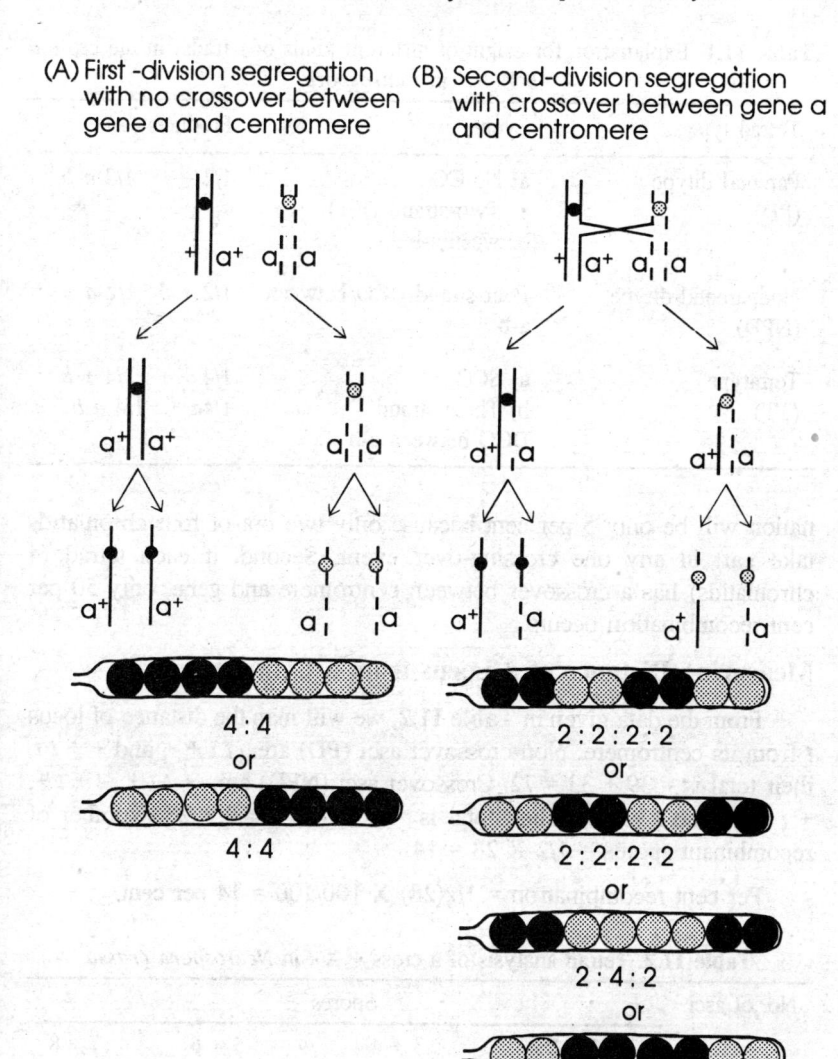

Fig. 11.1 (A) First-and (B) second-division segregation and their products

Two things must be remembered while mapping genes in *Neurospora crassa*. First, if 10 per cent of the asci in an experiment show crossing-over (second division segregation), the percentage of recombi-

178 Basic Genetics

Table 11.1. Explanation for origin of different kinds of tetrads in the cross *a b* X + + in *Neurospora*

Tetrad type	Event	Products
Parental ditype (PD)	a) No CO b) Two-strand DCO between a-b	1/2 + + : 1/2 *a b*.
Nonparental ditype (NPD)	Four-strand DCO between a-b	1/2 + *b* : 1/2 *a* +
Tetratype (TT)	a) SCO b) Three-strand DCO between a-b	1/4 + + : 1/4 + *b* : 1/4*a* + : 1/4 *a b*

nation will be only 5 per cent because only two out of four chromatids take part in any one crossing-over event. Second, if each tetrad (4 chromatids) has a crossover between centromere and gene, only 50 per cent recombination occurs.

Measuring Distance of a Locus from a Centromere

From the data given in **Table 11.2**, we will map the distance of locus *t* from its centromere. Non-crossover asci (PD) are: *t t* + + and + + *t t*, their total is = 39 + 33 = 72. Crossover asci (NPD) are: + *t t* +, *t* + *t* +, + *t* + *t*, and *t* + + *t*). Their total is = 5 + 6 + 9 + 8 = 28. Number of recombinant spores = 1/2 X 28 = 14.

Per cent recombination = 1/2(28) X 100/100 = 14 per cent.

Table 11.2. Tetrad analysis of a cross + X *t* in *Neurospora crassa*

No. of asci	Spores			
	1 + 2	3 + 4	5 + 6	7 + 8
5	+	*t*	*t*	+
6	*t*	+	*t*	+
9	+	*t*	+	*t*
8	*t*	+	+	*t*
39	*t*	*t*	+	+
33	+	+	*t*	*t*

Two-point Cross

Distance of genes in relation to centromere. Let us assume two genes *a* and *b* in *Neurospora* are segregating. The data recorded are given in **Table 11.3**. Classify the tetrads as PD, NPD and TT. We see that PD = (A) + (C) = 58 + 59 = 117 and NPD = (B) + (D) = 57 + 58 = 115. There are no TT **tetrads**. As a first step, one needs to detect linkage. When there are only two types of tetrads (PD and NPD), check whether PD and NPD are 1:1 ratio or not. This should be done statistically, using chi-square. If PD and NPD are in 1:1 ratio, linkage is absent but if PD and NPD are not 1:1 ratio linkage exists. In the above example, it can be seen that PD and NPD are in 1:1 ratio and hence linkage is absent which means that the genes *a* and *b* are located on different chromosomes. Further, the data show no **second division segregation** which means that there is no crossing-over between genes and their centromere, i.e., the genes *a* and *b* are very close to their respective centromeres.

Table 11.3. Results of tetrad analysis of cross *a b* X + + in Neurospora crassa

PD	NPD	PD	NPD
a b	+ b	+ +	a +
a b	+ b	+ +	a +
+ +	a +	a b	+ b
+ +	a +	a b	+ b
FDS 58 FDS	FDS 57 FDS	FDS 59 FDS	FDS 58 FDS

Detection and estimation of linkage. Let us consider results of a cross between two strains *pdx* + X + *pan* of *Neurospora* given in **Table 11.4**. For detection of linkage of linkage, classify all the asci [(A) to (G)] as PD, NPD and TT. In each class, mark the genes showing **first division segregation** (FDS) and **second division segregation** (SDS) as done in Table 11.4. Compare PD (showing FDS for both loci) with NPD (showing FDS for both loci) for 1:1 ratio, statistically. Compare all NPDs with all TTs. See whether ratio NPD/TT is less or more than 0.25. If PD = NPD and NPD/TT is more than 0.25, the conditions reveal **independent assortment** between *pdx* and *pan*. If either one or both the conditions are not met with, it shows existence of linkage between the genes in question. In the given example, PD (showing FDS for both loci)

180 Basic Genetics

Table 11.4. Results of tetrad analysis of cross *pdx* + X + *pan* in *Neurospora crassa*

(A)	(B)	(C)	(D)
pdx +	*pdx pan*	*pdx* +	*pdx* +
pdx +	*pdx pan*	*pdx pan*	+ +
+ *pan*	+ +	+ +	*pdx pan*
+ *pan*	+ +	+ *pan*	+ *pan*
FDS 15 FDS PD	FDS 1 FDS NPD	FDS 17 SDS TT	SDS 1 FDS TT

(E)		(F)	(G)
pdx +		*pdx pan*	*pdx* +
+ *pan*		+ +	+ *pan*
pdx +		*pdx pan*	*pdx pan*
+ *pan*		+ +	+ +
SDS 13 SDS PD		SDS 0 SDS NPD	SDS 2 SDS TT

15 is statistically not equal to 1. Ratio of NPD (all) 1 to TT (all) 20 is 0.05, i.e., less than 0.25. In the given example, according to both the criteria, linkage is present.

For **estimation of linkage**, one first needs to know the origin of seven types of tetrads given in the above example The events leading to these tetrads are shown in **Table 11.5**. One needs to understand this step thoroughly.

Are the genes located on the opposite sides or same side of the centromere (c)? The two possibilities that exist are diagrammatically depicted below:

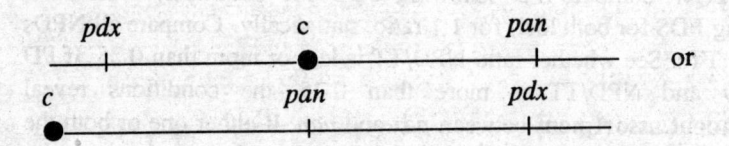

```
    pdx         c              pan
    |───────────●───────────────|          or
    c           pan             pdx
    ●───────────|───────────────|
```

Table 11.5. Origin of different types of tetrads in a two-point cross
$pdx + \times + pan$, in *Neurospora* assuming that the two genes lie
in the same arm of a chromosome

Class	Event	Tetrad	Segregation pdx	Segregation pan	Tetrad type
(A)	No CO	pdx + pdx + + pan + pan	F	F	PD
(B)	Four-strand DCO a) pdx-pan 1 & 3 b) pdx-pan 2 & 4	pdx pan pdx pan + + + +	F	F	NPD
(C)	SCO pdx-pan 2 & 3	pdx + pdx pan + + + pan	F	S	TT
(D)	Two-strand DCO a) cent-pdx 2 & 3 b) pdx-pan 2 & 3	pdx + + + pdx pan + pan	S	F	TT
(E)	SCO cent-pdx 2 & 3	pdx + + pan pdx + + pan	S	S	PD
(F)*	Four strand DCO a) pdx-cent 1 & 4 b) cent-pan 2 & 3	pdx pan + + pdx pan + +	S	S	NPD
(G)	Three strand DCO a) cent-pdx 2 & 3 b) pdx-pan 1 & 3	pdx + + pan pdx pan + +	S	S	TT

*Type (F) tetrads are seen only when the two genes in question lie on opposite sides of the centromere

To decide this, compare PD (SDS for both loci) and NPD (SDS for both loci). If one of the classes has relatively large number of tetrads, the two genes must be located on the same chromosome arm. In the present example, PD = 13 and NPD = 0. This means that the two genes *pdx* and *pan* are located on the same chromosome arm. If PD and NPD would have been approximately equal, the two genes would have been on opposite side of centromere. In other examples, one may come across with such a situation.

Linkage between genes. Now we need to find out the gene which is closer to the centromere. For this, compare tetrads that show SDS for *pdx* with those that show SDS for *pan*. In the present case, classes (D), (E) and (G) show SDS for *pdx* and their total is 16 whereas classes (C), (E) and (G) show SDS for *pan* and their total is 32. Greater frequency of SDS can be observed for a gene if it lies farther from the centromere than the one which shows lesser frequency. This comparison thus shows that *pdx* is closer to centromere; *pan* is farther.

Now we come to calculating per cent recombination between centromere and gene *pdx* and centromere and gene *pan*. The following formula is used for calculating distance of a gene from centromere.

$$\text{Per cent recombination} = \frac{\frac{1}{2}(\text{S.C.O.}) + \text{D.C.O.}}{\text{Total number of tetrads}} \times 100$$

Let us first calculate distance between centromere (c) - *pan*. Classes (C) and (E) are due to S.C.O. and (B), (D) and (G) are due to D.C.O. (see this from Table 11.5). Per cent recombination between centromere - *pan* is = [{½(17+13)} + {1+1+2}]/49 X 100 = 38.7 map units. While calculating distance between centromere (c) - *pdx* we must keep in mind that double cross overs can not be detected between two markers unless a known third marker (or a centromere) lies in between. So double cross overs in this example will be zero; only single cross overs will be found. Classes (D), (E) and (G) are due to single crossing-over. Per cent recombination is = ½(1 + 13 + 2)/49 X 100 = 16.3 map units. Genetic map for *pdx* and *pan* in relation to centromere will be:

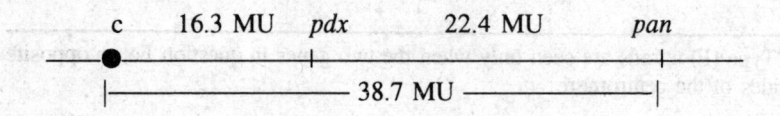

```
     c      16.3 MU    pdx          22.4 MU         pan
    ●─────────────────┼──────────────────────────────┤
    ├─────────────────── 38.7 MU ────────────────────┤
```

Once distance between gene *pdx* and centromere and *pan* and centromere is known, distance between genes *pdx* and *pan* can be calculated. Thus the distance between *pdx* and *pan* comes out to be 22.4 map units.

Three-point cross

We classify each tetrad as to whether it is PD TT or NPD with regards to each of the three pairs of segregating genes. We can also differentiate among the two, three and four stranded **double cross overs** (**Table 11.6**). The gene order is most likely *a-b-c* if class ③ (double crossover) is less frequent than classes ② and ④

Table 11.6. Determination of gene order in a three-point cross by observation of the relative frequencies of different tetrad classes, three of which ②, ③ and ④ represent single-crossover tetrads for two of the three gene orders and a two-strand double-crossover tetrad for the third*

Parental strains *a b c* X + + +
Tetrad classes

	①	②	③	④
	a b c	a b c	a b c	a b c
	a b c	a b +	a + c	a + +
	+ + +	+ + c	+ b +	+ b c
	+ + +	+ + +	+ + +	+ + +

Gene order				
a-b-c	no exchange	single crossover (b-c)	double crossover (a-b, b-c)	single crossover (a-b)
a-c-b	no exchange	double crossover (a-c, c-b)	single crossover (c-b)	single crossover (a-c)
b-a-c	no exchange	single crossover (a-c)	single crossover (b-a)	double crossover (b-a, a-c)

Consider the following example. The data of a cross involving strains *me a +* and *+ ad A* are given in **Table 11.7**. On the basis of explanation regarding origin of different types of tetrads given in **Table 11.8**, one can

Table 11.7. Results of tetrad analysis of cross between strains *me + a* and *+ ad A* of *Neurospora crassa*

(1)	(2)	(3)	(4)	(5)
me + a	me + a	me + a	me + a	me + a
me + a	+ + a	+ ad A	+ ad a	+ + A
+ ad A	me ad A	+ ad a	me a A	me ad a
+ ad A	+ ad A	+ ad A	a ad A	+ ad A
46	6	4	2	2

(6)	(7)	(8)	(9)	(10)
me + a	me + a	me + a	me + a	+ ad a
+ + A	+ + A	+ ad A	+ ad a	+ + a
me ad A	me ad a	me + a	me ad A	me ad A
+ ad a	+ ad A	+ ad A	+ + A	me + A
1	0	0	0	0

Table 11.8. Origin of different types of tetrads in a three-point cross *me + a* X *+ ad A*, in *Neurospora* assuming that the two genes *me* and *ad* lie in one arm and gene *a* lies in the other of a chromosome

①	②	③	④	⑤
none	single in I	single in III	single in II	two-strand double in I and II

⑥	⑦	⑧	⑨	⑩
three-strand double in I (2nd and 3rd) and III (2nd and 4th)	two-strand double in I and II	two-strand double in I and II	three-strand double in I (2nd and 3rd) and II (2nd and 4th)	four-strand double in I and II

Region I = *me-ad*; Region II = *ad*-cent; Region III = cent-*A*

see that class (1) represents PD type tetrads as they arise form non-cross over chromosomes. Class (2) is the next most prevalent which arises due to single **crossing-over** between *me* and *ad*. Next most frequent is class

(3) which results due to single crossing-over between *ad* and *A*. Class (4) also arises due to single crossing-over between *ad* and *A* but has in addition second division segregants (SDS) for *me* and *ad*. This means class (4) represents crossing-over between *me* and *ad* and their centromere which has also produced recombination between *ad* and *A*. This now gives us the gene order:

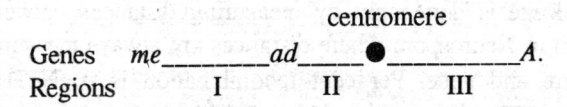

```
                       centromere
Genes    me_____ad_____●_____A.
Regions       I          II       III
```

Now recombination frequency can be calculated. As seen above, there are three cross over regions (I, II and III). The zero classes, (7) to (10), in the data as a sample of additional tetrads that may occasionally occur. There are 61 tetrads in all. Classes (2), (5) and (6) show exchange in region I. Per cent recombination between *me* and *ad* is ½(6+2+1)/61 = 7.35. **Recombination frequency** in regions II and III can be found out to be 1.5 and 5.7 per cent, respectively. Thus the gene map constructed from the above data is:

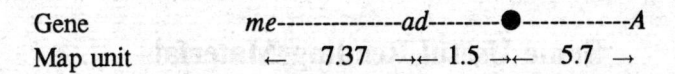

```
Gene          me--------------ad--------●------------A
Map unit      ←    7.37    →←  1.5  →←   5.7   →
```

Interference. In the three-point linkage data, two kinds of information can be obtained from these data. First, what nonsister chromatids are involved in crossing over in the adjacent region? Normally, we expect two-strand double exchange, three-strand double exchanges and four-strand double exchanges in ratio 1:2:1, if there were no effect of chromatids in one exchange on the chromatids of an other exchange. A departure from 1:2:1 ratio would indicate some sort of **chromatid interference**. This interference could be "positive" or "negative" (see Chapter 10).

Second form of interference also occurs in double exchanges between adjacent regions which arises due to **chiasma interference**. Accordingly, **a cross over (chiasma)** in one region interferes in some way with crossing over (chiasma) in an adjacent region no matter what chromatid are involved. In *Neurospora*, negative interference is generally observed.

Gene Mapping in Yeast and *Chlamydomonas*

Yeast and *Chlamydomonas* have unordered tetrads. it is not possible in such cases to distinguish between tetrads resulting from first- and second-division segregation. For detection and estimation of linkage, the tetrads are classified as PD, NPD, and TT type. **Detection of linkage** in yeast and *Chlamydomonas* is done the same way as in *Neurospora*. **Estimation of linkage** is done only by measuring distances between genes, compared to in Neurospora where distances are always measured between centromere and gene. Per cent recombination is = (NPD + ½TT)/(PD + NPD + TT). Let us assume data of Table 11.3 giving results of tetrad analysis of cross *pdx* + X + *pan* in *Neurospora crassa* as of unordered tetrads. Analysis has already established linkage between *pdx* and *pan* genes. For estimation of linkage, compute total number of PD, NPD and TT tetrads. Classes (A) and (E) represent PD, class (B) represents NPD and classes (C), (D) and (G) represent TT tetrads. Thus PD, NPD and TT come out to be 1, The distance between *pdx* and *pan* comes out to be 28, 1 and 20, respectively. Distance between genes *pdx* and *pan* = {1 + ½(20)} /49 X 100 = 22.4 per cent which is same as observed above.

Some Useful Reading Material

Lindergren (1933) while dealing with genetics of *Neurospora*, gave his experimental results on pure bred stocks and crossing-over in *N. crassa*. Barratt *et al.* (1954) dealt with the method of map construction in *Neurospora crassa*. Emerson (1963) dealt with meiotic recombination in fungi with special reference to tetrad analysis. Fincham and Day (1963) described at length life cycle and genetic analysis methods in fungi. Holliday (1964) and Fogel and Mortimer (1969) described informational transfer by meiotic gene conversion. Holliday (1974) dealt with molecular aspects of genetic exchange and gene conversion. Hall and Linder (1993) described the early days of yeast genetics. Roman and Ruzinski (1990) dealt with mechanisms of gene conversion in *Saccharomyces cerevisiae*.

Questions

1. Explain differences between first- and second-division segregation with the help of a diagram.
2. Describe different types of tetrads and their origin.
3. What important things are kept in mind while mapping genes in haploid eukaryotes?

4. Describe the procedure of measuring distance of a locus from a centromere.
5. In a two point cross, how distances of genes are calculated in relation to the centromere?
6. What is the procedure of detection and estimation of linkage in a two-point cross in *Neurospora*.
7. What is the procedure of detection and estimation of linkage in a three-point cross?
8. What is interference? What different types of interferences can be observed in data showing double exchanges?
9. What is the basic difference between the type of tetrads in *Neurospora* on one hand and yeast and *Chlamydomonas* on the other? In what way do methods of estimating linkage in these two systems differ?

Numericals

1. For each of the following traits in *Neurospora* in which first- and second-division segregation was observed, state the gene-centromere distance (data of Lindergren).

Trait	Number of Asci	
	First-division segregation	Second-division segregation
Mating type (A vs. a)	331	51
Pale conidial color vs. orange	73	36
Fluffy growth form vs. normal	42	67

2. The following are the results of ordered tetrad analysis from a cross between a *Neurospora* strain carrying albino (al) that was also unable to synthesize inositol (inos) and a wild-type strain (+ +) (adapted from data of Houlahan, Beadle, and Calhoun, 1949):

①	②	③	④
al inos	al +	al +	al +
al inos	al +	al inos	+ +
+ +	+ inos	+ +	al inos
+ +	+ inos	+ inos	+ inos
4	3	23	36

⑤	⑥	⑦
al inos	al +	al +
+ +	+ inos	+ inos
al inos	al +	+ +
+ +	+ inos	al inos
15	16	22

(a) Determine whether these two genes are linked and, if so, the linkage distance between them. (b) which gene has the longest gene-centromere distance?

3. A cross between a *Neurospora* stock bearing the mutant gene snowflake (sn), which was otherwise wild type, to a colonial temperature-sensitive stock (cot), which was otherwise wild type, produced a number of regular tetrads showing first- and second-division segregation as follows (from data of M.B. Mitchell). Determine whether these genes are linked, and their gene-centromere distances.

(A)	(B)	(C)	(D)	(E)	(F)
sn cot	sn +	sn cot	sn cot	+ +	sn +
sn cot	sn +	sn +	sn +	+ cot	sn cot
+ +	+ cot	+ cot	+ +	sn cot	+ +
+ +	+ cot	+ +	+ cot	sn +	+ cot
25	16	11	12	8	8

4. A *Neurospora* stock which was adenine-requiring (*ad*) and tryptophan-requiring (*tryp*) was crossed to a wild type stock (+ +) and produced the following tetrads:

①	②	③	④	⑤	⑥	⑦
ad tryp	ad +	ad tryp	ad tryp	ad tryp	ad +	ad tryp
ad tryp	ad +	ad +	+ tryp	+ +	+ tyrp	+ +
+ +	+ tryp	+ tryp	ad +	ad tryp	ad +	+ tryp
+ +	+ tryp	+ +	+ +	+ +	+ tyrp	ad +
49	7	31	2	8	1	2

Determine whether these two genes are linked and, if so, draw a linkage map including the centromere.

5. In addition to the *ad* and *tryp* loci in the above problem, there are other such loci in the *Neurospora* genome which also affect the requirements for adenine and tryptophan. In a cross similar to the one in Problem 18-4 but using different *ad* and *tryp* gene, the same classes of tetrads were observed but the numbers of tetrads in each class were different as follows:

① = 36, ② = 1, ③ = 39, ④ = 21, ⑤ = 1, ⑥ = 1, ⑦ = 1.

Determine whether these *ad* and *tryp* genes are linked, if so, draw a linkage map that includes the centromere.

6. In a cross between a *Neurospora* strain requiring histidine (*hist*-2) and a strain requiring alanine (*al*-2), Giles and co-analyzed 646 unordered tetrads with the following compositions and numbers.

Number	Tetrad Composition
115	2 hist al⁺ : 2 hist⁺ al
45	2 hist al : 2 hist⁺ al⁺
484	1 hist al : 1 hist al⁺ : 1 hist⁺ al : 1 hist⁺ al⁺

Determine whether the two genes are linked and, if so, the linkage distance between them.

7. In an experiment with yeast (unordered tetrads) a cross was made between two strains each different in respect to three different genes, a, b, c, not necessarily linked on the same chromosome. The cross was $+ b c \times a + +$ and produced the following tetrads:

Numbers	Tetrad Composition
407	2 $+ + c$: 2 $a b +$
396	2 $+ b c$: 2 $a + +$
104	1 $+ b c$: 1 $+ b +$: 1 $a + c$: 1 $a + +$
92	1 $+ + c$: 1 $+ + +$: 1 $a b c$: 1 $a b +$
1	2 $+ b +$: 2 $a + c$
1	2 $+ + +$: 2 $a b c$

From these data determine which genes are linked, if any, and the linkage distances.

8. A *Neurospora* cross was made between an albino strain of A mating type that was wild type for leucine requirements ($al + A$) to a leucine-requiring strain of a mating type ($+ leu\ a$). Derive the linkage order and linkage distances for the three genes from the following 1000 ordered tetrads that were obtained from this cross:

(1)	(2)	(3)	(4)	(5)
$al + A$	$al + A$	$al + A$	$al + A$	$al + A$
$al + A$	$+ + A$	$al\ leu\ A$	$al\ leu\ A$	$+ leu\ a$
$+ leu\ a$	$al\ leu\ a$	$+ + a$	$+ + a$	$al + A$
$+ leu\ a$	$+ leu\ a$	$+ leu\ a$	$+ leu\ a$	$+ leu\ a$
591	142	134	123	1

(6)	(7)	(8)	(9)	(10)
$al\ leu\ a$	$al + A$	$+ + A$	$al + A$	$al\ leu\ A$
$+ + A$	$+ leu\ a$	$al\ leu\ a$	$+ leu\ A$	$+ + A$
$al + A$	$+ + A$	$+ + A$	$al + a$	$al + a$
$+ leu\ a$	$al\ leu\ a$	$al\ leu\ a$	$+ leu\ a$	$+ leu\ a$
1	1	1	1	1

(11)	(12)	(13)	(14)
$al + A$	$al + A$	$al + A$	$+ leu\ A$
$+ leu\ A$	$al\ leu\ a$	$al + a$	$al + a$
$al\ leu\ a$	$+ leu\ A$	$+ leu\ A$	$+ + A$
$+ + a$	$+ + a$	$+ leu\ a$	$al\ leu\ a$
1	1	1	1

12

GENETIC ANALYSIS IN BACTERIA AND VIRUSES

Recombination in Bacteria

Recombination in bacteria may take place through transformation, conjugation or transduction.

Transformation

The first evidence of genetic recombination or exchange of hereditary material in bacteria was noted by Griffith in the transformation of harmless pneumococci into virulent ones. Avery, MacLeod and McCarty (1944) demonstrated the transforming agent as pure DNA. **Transformation** seems to arise from some form of recombination mechanism which produces gene exchange similar to that produced by sexual reproduction. Some **complementary base pairing** between single stranded DNA sections appears necessary in transformation. A transforming molecule carrying gene A may also carry gene B. If these genes are closely linked, there is a good likelihood that transformation at the locus A produced transformation at the B locus (double transformation). If genes A and B are not linked within one transforming molecule, double transformation is caused by two independent events.

Two genes affect synthesis of tryptophan (trp_2^-) and tyrosine (tyr_1^-) in *Bacillus subtilis* so that double mutants do not grow on unsupplemented media. The double mutant, however, may be transformed by DNA of other strains so that single and double transformants can be selected and scored.

Let us consider the following example where two strains $trp_2^+ \, tyr_1^-$ and $trp_2^- \, tyr_1^+$ are given to a third strain $trp_2^- \, tyr_1^-$. Three transformant classes arose as a result of this treatment, of which two were **single transformants** and one was double type. Results are given in **Table 12.1**.

Double transformant class is present in very low frequency in cross (1) when DNA is used from trp_2^+ tyr_1^+ donor (cross (2), the frequency of double transformants rises significantly. This shows that those two genes are linked. The **extent of linkage** can be calculated on the basis of recombination between two markers of the donor DNA as a proportion of all the cells that were transformed. Here also frequency of recombination depends on the distance between the two markers. Distance between *trp* and *tyr* thus is (196 + 328)/(328 + 196 + 367) = 58.8 per cent.

Table 12.1. Results of a transformation experiment where two strains trp_2^+ tyr_1^- and trp_2^- tyr_1^+ are given to a third strain trp_2^- tyr_1^-

Donor DNA	Recipient Cells	Transformant classes	No. of colonies
trp_2^+ tyr_1^- Plus trp_2^- tyr_1^+	trp_2^- tyr_1^-	trp^+ tyr^- trp^- tyr^+ trp^+ tyr^+	190 256 2
trp_2^+ tyr_1^+	trp_2^- tyr_1^-	trp^+ tyr^- trp^- tyr^+ trp^+ tyr^+	196 328 367

When differences at three or more linked loci are simultaneously involved in a transformation experiments, linkage distances as well as gene order can be calculated. This is explained in the following example. The prototrophic strain trp_2^+ his_2^+ tyr_1^+ is the **donor strain** and the auxotrophic strain trp_2^- his_2^- tyr_1^- is the **recipient strain**. Results are the following seven types of transformant classes (**Table 12.2**). Three individuals linkage

Table 12.2. Seven types of transformant classes obtained when a prototrophic strain trp_2^+ his_2^+ tyr_1^+ is given to an auxotrophic strain trp_2^- his_2^- tyr_1^-

(1)	(2)	(3)	(4)	(5)	(6)	(7)
trp^- his^- tyr^+	trp^- his^+ tyr^-	trp^- his^+ tyr^+	trp^+ his^- tyr^-	trp^+ his^- tyr^+	trp^+ his^+ tyr^-	trp^+ his^+ tyr^+
685	418	3660	2600	107	1180	11940

Table 12.3. Computation of recombination distances from the transformation data given in Table 12.2

Genes	Recombinants	Parentals	Frequency
trp2 - his2	2600 + 107 + 418 + 3600 = 6785	1180 + 11940 = 13120	6785/19905 = 0.34
trp2 - tyr1	2600 + 1180 + 685 + 3660 = 8125	107 + 11940 = 12047	8125/20172 = 0.40
his2 - tyr1	418 + 1180 + 685 + 107 = 2390	3660 + 11940 = 15600	2390/17990 = 0.13

distances can be calculated by considering two genes at a time. The frequencies of recombination obtained are shown in **Table 12.3**. Thus the genetic map between *trp2, his2* and *tyr1* genes is as under:

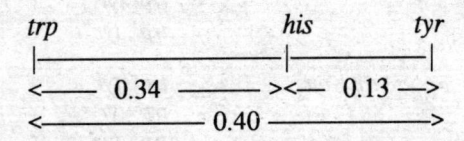

```
        trp                his        tyr
        |------------------|----------|
        <----- 0.34 ------><— 0.13 —>
        <----------- 0.40 ----------->
```

Gene order can also be confirmed by noting phenotype of the rarest transformant class trp_2^+ his_2^- tyr_1^+.

The F Factor. The transmission of F factor in bacteria seemed to be independent of the transmission of **chromosomal genes**. **Hfr strain** has proven to be very useful to mapping genes of **bacterial chromosome**. When recombinants of a cross Hfr ($A^+B^+C^+D^+$....) x ($A^-B^-C^-D^-$....) different + type genes appeared in different frequency. A^+, for example appeared most frequently, D^+ less frequently, J^+ still less frequently and Z^+ showed itself least frequent of all. If we can consider that genes to A to Z are in one **linkage group**, it appeared as though the Hfr donor gene A consistently entered into the recipient first and therefore had the greatest opportunity for recombination. Genes B to Z sequentially entered in that order and their recombination were determined in order of entry.

In some, F strain became Hfr strains because of entry of Hfr in the recipient cell. The Hfr factor appeared to be directly connected to the terminal end of the chromosome. We don't have similar situation in all the cases. Hfr can break the bacterial chromosome and orient its transfer at different points. This means in one cross gene A^+ can show the highest while in orders gene D^+, H^+ or K^+ can show the highest frequency. In all these cases if both ends of the chromosome are connected to form a circle, the sequence is identical for each strain. This shows that bacterial chromosome in circular and Hfr can break this chromosome at different points.

F factor is composed of double stranded DNA having about 105 nucleotide pairs. The **bacterial chromosome** is 40 times longer than the F factor. In F^+ stain, F factor is circular and one copy is present in each bacterial cells. This factor replicates midway through each bacterial chromosome **replication cycle**, thereby maintaining its frequency. Effective contact of different type of cells is followed by **cytoplasmic connection**. F factor may remain outside the bacterial chromosome as an independent cytoplasmic inclusion, or it recombines with a section of bacterial chromosome so that chromosome now has a **directional orientation**. In first case, only F factor will enter the recipient cell and F^- becomes F^+ which may now act as donors. In the second case, the chromosome itself enters the recipient cell. When F^+ is integrated into the bacterial chromosome it becomes Hfr strain. Occasionally Hfr becomes F^+.

F factor acts as a **replicating unit** that begins replication of DNA upon **sexual contact** and transfer only one of the replication donor DNA strands to the recipient which enters the recipient at 5' end. This strand the acquires a complementary strand through **DNA synthesis** in the recipient before under going recombination with the recipient chromosome.

The length of the donor chromosome transferred to the bacterial **recipient cell** depending upon the time for which conjugating cells keep in contact with each other.

Conjugation mapping

"Interrupted mating technique" is used as conjugation can be interrupted by violent agitation in a Waring Blendor. Thus, length of donor chromosome that entered the recipient cell is controlled. An 5-8 minute interval after conjugation is necessary for chromosome transfer to begin. The relationship between genes and their position on chromosome could be mapped in terms of time unit, in which one time unit is equal to 1 minute. **Bacterial chromosome** can be mapped as a complete length of 90 time units (**Fig 12.1; Table 12.4**). This method is suitable for genes that are located three or more time units apart.

194   *Basic Genetics*

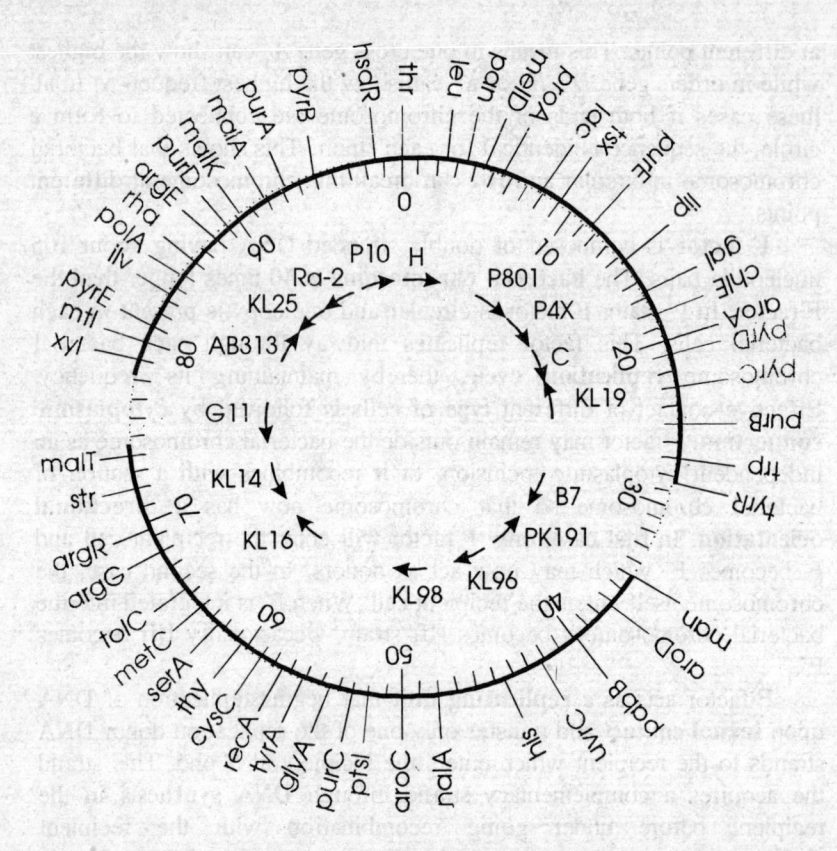

Fig. 12.1. Selected loci on circular map of *E. coli*. Definitions of loci can be found in Table 12.4. Units on the map are in minutes. Arrows within the circle refer to Hfr strain transfer starting points, with directions. The two thin regions are the only areas not covered by P1 transduction phages. (Bachmann *et al.*, 1976).

Recombination Mapping. To detect linkage order between genes separated by distances smaller than 3 time units, recombination mapping can be used. It has been found that 20 per cent recombination = 1 time unit. This method is suitable for mapping only those genes that are separated by less than 3 time units. Why?

Transduction

Generalized Transduction. This is a most widely used **bacterial mapping** approach. It is an effective way to map closely linked bacterial

Table 12.4. Genetic symbols, mutant character and ezzyme or reaction affected in the gene map of *E. coli*

Genetic symbols	Mutant character	Enzyme or reaction affected
araD	Cannot use the sugar arabinose as a carbon source	L-Ribulose-5-phosphate-4-epimerase
araA		L-Arabinose isomerase
araB		L-Ribulokinase
araC		
argB	Requires the amino acid arginine for growth	N-Acctylglutamate synthetase
argC		N-Acetyl-γ-glutamokinase
argH		N-Acetylglutamic-γ-semialdehyde dehydrogenase
argG		Acetylornithine-d-transaminase
argA		Acetylornithinase
argD		Ornithine transcarbamylase
argE		Argininosuccinic acid snthetase
argF		Argininosuccinase
argR	Ariginine operon regulator	
aroA, B,C	Requires several aromatic amino acids and vitamins for growth	Shikimic aid to 3-enolpyruvyl-shikimate-5-phosphate
aroD		Biosynthesis of shikimic acid
azi	Resistant to sodium azide	
bio	Requires the vitamin biotin for growth	
carA	Requires uracil and arginine	Carbamate kinase
carB		

Table 12.4 contd...

Genetic symbols	Mutant character	Enzyme or reaction affected
cblA-E	Cannot reduce chlorate	Nitrate-chlorate reductase and hydrogen lysase
cysA	Requires the amino acid cysteine for growth	3-Phosphoadenosine-5-phosphosulfate to sulfide
cysB		Sulfate to sulfide; four known enzymes
cysC		
dapA	Requires the cell-wall component diaminopimelic acid	Dihydrodipicolinic acid synthetase
dapB		N-Succinyl-diaminopimelic acid deacylase
dap + bom	Requires the amino acid precursoe homoserine and the cell-wall component diaminopilmelic acid for growth	Aspartic semialdehyde dehydrogenase
dnaA-Z	Mutation, DNA replication	DNA biosynthesis
Dsd	Cannot use the amino acid D-serine as a nitrogen source	D-Serine deaminase
fla	Flagella are absent	
galA	Cannot use the sugar galactose as a carbon source	Galactokinase
galB		Galactose-1-phosphate uridyl transferase
galD		Uridine-diphosphogalactose-4-epimerase
glyA	Requires glycine	Serine hydroxymethyl transferase
gua	Requires guanine for growth	

Table 12.4 contd...

Genetic symbols	Mutant character	Enzyme or protein affected
H	The H antigen is present	
his	Requires the amino acid histidine for growth	Ten known enzymes*
hsdR	Host restriction	Endonuclease R
ile	Requires the amino acid isoleucine for growth	Threonine deaminase
ilvA	Requires the amino acids isoleucine and valine for growth	α-Hydroxy-β-Keto acid rectoisomerase α, β-dihydroxyisovaleric dehydrase*
ilvB		
ilvC		Transaminase B
ind	Cannot grow on tryptophan as a carbon source	Tryptophanase
λ (attλ)	CHromosomal location where prophage λ is normallly inserted	
lacI	Lac operon regulator	
lacY	Unable to concentrate β-glactosides	Galactoside permease
lacZ	Cannot use the sugar lactose as a carbon source	β-Galactosidase
lacO	Constitutive synthesis of lactose operon proteins	Defective operator
leu	Requires leucine for growth	Three known enzymes*
lip	Requires lipoate	
lon	Filament formation and radiation sensitivity are affected	

Table 12.4 contd...

Genetic symbols	Mutant character	Enzyme or protein affected
lys	Requires the amino acids lysine for growth	Diaminopimelic acid decarboxylase
lys + met	Requires the amino acids lysine and methionine for growth	
λ rec, malT	Resistant to phage λ and cannot use the sugar maltose	Regulator for two operons
malK	Cannot use the sugar maltose as a carbon source	Maltose permease
man	Cannot use mannose sugar	Phosphomannose isomerase
melA	Cannot use melibiose sugar	Alpha-galactosidase
metA-M	Requires the amino acid methionine for growth	Ten or more genes
mtl	Cannot use the sugar mannitol as a carbon source	Two enzymes
muc	Forms mucoid colonies	Regulation of capsular polysaccharide synthesis
nalA	Resistant to nalidixic acid	
O	The O antigen is present	
pan	Requires the vitamin pantothenic acid for growth	
pabB	Requires p-aminobenzoate	
phe A, B	Requires the amino acid phenylalanine for growth	
pho	Cannot use phosphate esters	Alkaline phosphatase
pil	Has filaments (pili) attached to the cell wall	
plsB	Deficient phospholipid synthesis	Glycerol 3-phosphate acyltransferase

Table 12.4 contd...

Gene symbols	Mutant character	Protein or enzyme affected
polA	Repairs deficiencies	DNA polymerase 1
proA	Requires the amino acids proline for growth	
proB		
proC		
ptsI	Defective phosphotransferase system	Pts-system enzyme 1
purA	Requires certain purines for growth	Adenylosuccinate synthetase
purB		Adenylosuccinase
purC, E		5-Aminoimidazole ribotide (AIR) to 5-aminoimidazole-4-(N-succino carboximide) ribotide
purD		Biosynthesis of AIR
pyrB	Requires the pyrimidine uracil for growth	Asparate transcarbamylase
pyrC		Dihydroorotic acid dehydrogenase
pyrD		Orotidylic acid pyrophosphorylase
pyrE		Orotidylic acid decarboxylase
pyrF		Repressor for enzymes involved in galactose production
R gal	Constitutive production of galactose	Alkaline phosphatase repressor

Table 12.4 contd...

Gene symbols	Mutant character	Protein or enzyme affected
R1 pho, R2 pho	Constitutive synthesis of phosphatase	Repressor for enzymes invovled in tryptophan synthesis
R try	Constitutive synthesis of tryptophan	
RC	Uncontrolled synthesis of RNA	
recA	Cannot repair DNA radiation damage or recombine	
rhaA-D	Cannot use the sugar rhamnose as a carbon source	Isomerase, kinase, aldolase and regulator
rpoA-D	Problems of transcription	Subunits of RNA polymerase
serA	Requires the amino acid serine for growth	3-Phosphoglycerate dehydrogenase
serB		Phosphoserine phosphatase
str	Resistant to or dependent on streptomycin	
suc	Requires succinc acid	
supB	Suppresses ochre mutations	t-RNA
tonA	Resistant to phages T1 and T5 (mutants called B/1, 5)	T1, T5 receptor sites absent
tonB	Resistant to phage T1 (mutant called B1)	T1 receptor site absent
T6, colK rec	Resistant to phage T6 and colicine K	T6 and colicine receptor sites absent
T4 rec	Resistant to phage T4 (mutant called B/4)	T4 receptor site absent
tsx	T6 resistance	
tbi	Requires the vitamin thamine for growth	

Table 12.4 contd...

Gene symbols	Mutant character	Protein or enzyme affected
tolC	Tolerance to colicine E1	
thr	Requires the amino acid threonine for growth	
thy	Requires the pyrimidiine thymine for growth	Thymidylate synthetase
trpA	Requires the amino acid trytophan for growth	Tryptophan synthetase, A protein
trpB		Tryptophan synthetase, B protein
trpC		Indole-3-glycerolphosphate synthetase
trpD		Phosphoribosyl anthranilate transferase
trpE		Anthranilate synthetase
tyrA	Requires the amino acid tyrosine for growth	Chorismate mutase AT-prephenate dehydrogenase
tyrR		Regulates three genes
uvrA-E	Resistant to ultraviolet radiation	Ultraviolet-induced lesions in DNA are reactivated
valS	Cannot charge Valyl-tRNA	Valyl-tRNA synthetase
xyl	Cannot use the sugar Xylose as a carbon source	

*Denotes enzymes control by the homologous genes loci of *Salmonella typhimurium*. (From Bachmann and Low, 1990).

gene as well as those that are more widely separated. **Generalized transduction** occurs when a **bacteriophage** picks up a random piece of host DNA during the **lytic phase** in one bacterium and carries this DNA into a second bacterium. Entrant and host DNAs have **homology**. The **genetic exchange** (recombination occurs). Two, three or more markers can be followed and data can be examined for co-transduction of the markers.

In a transduction experiment in *E. coli* the donor and recipient strains were *supC⁺ pyrF⁺ trpA⁺* and *supC pyrF trpA*, respectively.

Table 12.5. Transductants obtained when transduction was brought about by P_1 bacteriophage between donor $supC^+$ $pyrF^+$ $trpA^+$ and recipient supC pyrF trpA bacterial strains

Transductant as given in the experimental data	Transductants according according to the gene order	Number
$supC^+$ $pyrF$ $trpA^-$	$supC^+$ $trpA^-$ $pyrF$	453
$supC^+$ $pyrF^+$ $trpA^-$	$supC^+$ $trpA^-$ $pyrF^+$	0
$supC^+$ $pyrF$ $trpA^+$	$supC^+$ $trpA^+$ $pyrF$	114
$supC^+$ $pyrF^+$ $trpA^+$	$supC^+$ $trpA^+$ $pyrF^+$	36

Transduction was accomplished with P_1 **bacteriophage**. The results are given in **Table 12.5**. First step in construction of a **linkage map** is to find out the gene order. For this, the rarest class is compared with the donor and recipient strains. Accordingly, transductant class $supC^+$ $pyrF^+$ $trpA^-$ has 0 individual. The **gene order** thus is $supC$ - $trpA$ - $pyrF$. Next step is of rewriting the data in proper gene order. This has been done in Table 12.5. Now transduction frequency is calculated. Cotransduction frequency is equal to number of times a particular transduction occurred divided by total number of transductants. One needs to carefully note the co-transductants for different gene pairs. There are a total of 603 transductants in the given example. For genes $supC$ and $trpA$, two cotranductant classes are $supC^+$ $trpA^+$ $pyrF$ (114) and $supC^+$ $trpA^+$ $pyrF^+$ (36). Similarly, for genes $trpA$ and $pyrF$, the only co-tranductant class is $supC^+$ $trpA^+$ $pyrF^+$ (36). Cotransduction frequency for genes $supC$ and $trpA$ is $(36 + 114)/603 = 0.25$ and for genes $trpA$ and $pyrF$ the frequency is $36/603 = 0.06$. One must bear in mind that **closer are the genes higher is the cotransduction frequency**. Genetic map obtained from the data of Table 12.5 on the basis of cotransduction frequency is:

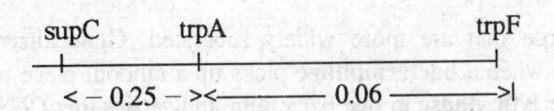

```
  supC        trpA                    trpF
   |-----------|-----------------------|
   <— 0.25 —><———— 0.06 ————>|
```

Sexduction/F-duction/F-mediated sexduction. F-element of *E. coli* is capable of independent existence and it can integrate into the **bacterial chromosome**. Therefore, F factor qualifies as an **episome**. When F-element his in bacterial cell independent of chromosome, the cell is termed as is F^+ and when F-element is integrated into the bacterial chromosome, the cell is termed as Hfr. Following is the diagrammatic representation of F-element as a plasmid.

$$F^+ \xleftarrow[\text{Excision of F element}]{\text{Integration of F element}} Hfr$$

Error may occur during excision and genes from bacterial chromosome become included in the F-element which is after excision called F'-element. Thus, F-element when carries a few bacterial chromosomal genes becomes **F'-element**.

When F' element is infectiously transmitted to an F- cell of differing genetic constitution, the recipient cell and its descendants become **partially diploid,** also known as **merozygote,** for the bacterial genes introduced by F' element. Bacterial genes in F' element may frequently take part in exchange with the recipient's chromosomal genes to produce **true** (reciprocal) **recombinants.** This phenomenon is known as **sexduction** or **F-duction.**

Various F' elements that arrive independently within a single Hfr strain can be characterized and frequencies with which different genes are transmitted together can be calculated. There frequencies are similar to co-transduction frequencies and are used in constructing genetic maps in the same way that co-transduction frequencies are used.

Recombination in Viruses

Genetic recombination was found to occur between different strain of bacteriophages by Delbruck and Bailey (1946) and Hershey (1946). In bacteriophages, **phenotypic differences** are most easily recognized by different effects on the bacterial cell hosts, type of **plaques** formed after **bacterial lysis** and **host range.**

Phenotypic Mixing

If two kinds of viruses T_2 and T_4, infect a cell, two types of protein envelopes and two types of **viral chromosomes** are formed. Some T_2 viral chromosomes may find their way into T_4 protein envelopes, and vice versa. Such "**mixed**" **particles** will grow only for one generation on the same culture of bacteria. For example, T_2 chromosome covered with T_4 envelop could not grown on B/4 bacteria alone. However by infecting the B bacteria in a B and B/4 mixture these particles produced a new generation of T_2 phages which infected the B/4 bacteria. This arises due to replication mechanism of bacteriophages. This is not synonymous with genetic recombination.

Viral Chromosome Replication

Virus must exchange genetic material in a host cytoplasm. Since many copies of viral chromoqomes are present in any one cell, genetic exchange may occur between members of numerous populations of chromosomes. **Viral chromosome multiplication** is geometric or clonal, i.e., one viral chromosomes replicates to produce two progeny, each of which replicates to produce two, and so on.

Recombination and Mapping

Hershey and Rotman (1942) conducted crosses utilizing h and three r mutants, $r1$, $r7$ and $r13$ that were induced independently. Data are given in **Table 12.6**, where h = **host range** and r = **rapid lysis** mutants. A perusal of the data reveals that the two recombinant classes of each cross

Table 12.6. Data of Hershey and Rotman (1942) involving crosses utilizing host range (h) and rapid lysis (r) mutants

Cross	Progeny (percent)			
	h^+r^+	hr^+	h^+r	hr
$hr1^+ \times h^+r1$	12	42	34	12
$hr1 \times h^+r1^+$	44	14	13	29
$hr7^+ \times h^+r7$	5.9	56	32	6.4
$hr7 \times h^+r^+7$	42	7.8	7.1	4.3
$hr13^+ \times h^+r13$	0.74	59	39	0.94
$hr13 \times h^+r13^+$	50	0.83	0.76	48

appear in approximately equal frequency. This shows that in phages also genetic recombination is a reciprocal event involving a physical exchange between two chromosomes, both of which are recovered. Recombination frequency varies from 1 to 2 (h-$r13$) to 25 to 30 (h-$r1$) per cent. This indicates linkage relationships between the three independently induced r mutations. The above data are not sufficient for construction of a linkage map. So additional crosses were made between the r mutants themselves and between another gene called *minute* (*m*) which produced very small plaques. It was noted that although there was a variety of linkage distances, no distance was ever greater than 30 percent. There were numerous genes linked together by small recombination frequencies which, as groups, always showed the same 30 percent linkage distance with gene *minute*.

Hershey and Rotman (1949) hypothesized three **linkage "groups"** in T_2 and that **"independent assortment"** between these linkage groups was characterized by 30 per cent recombination rather than by 50 per cent recombination expected in higher organisms (**Fig. 12.2**).

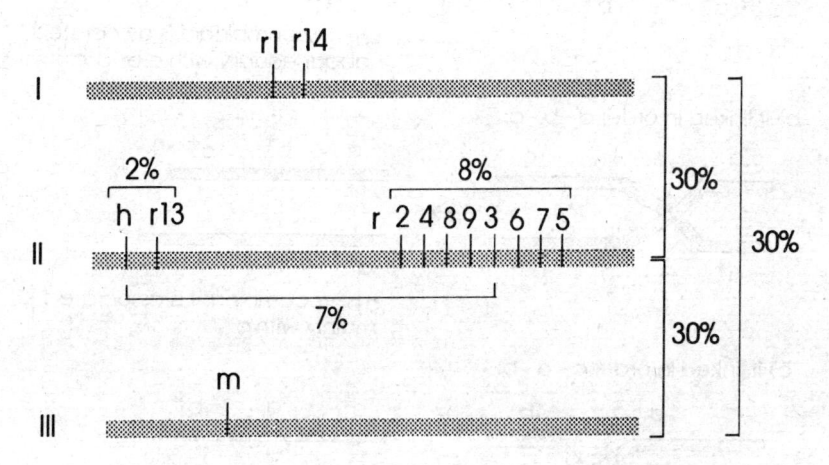

Fig. 12.2. Linkage relations found by Hershey and Rotman (1949)

The smaller frequency of recombination by independent assortment in phages arises because different kinds of **phage genomes** introduced into a cell can "mate" homozygously as well as heterozygously. A bacterial cell infected with *a b* phage as well as *a⁺ b⁺* phage will from many chromosomes of each kind. Matings then may be of varieties *a b/a b*, *a b/a⁺ b⁺*, *a⁺ b⁺/a⁺ b⁺*. If such matings are random *a b/a⁺ b⁺* have 50 per cent chance of occurring. However, if a and b are widely separated, or are not linked at all, these two loci will assort independently and **recombinational products** (*a⁺ b, a b⁺*) of such heterozygous matings will be as frequent as **non-recombinational products**. Thus, only 1/2 of 1/2, i.e. 1/4, of mating products will be recombinants after first round of mating. To some extent recombination values may be

a) If a - b unlinked to c:

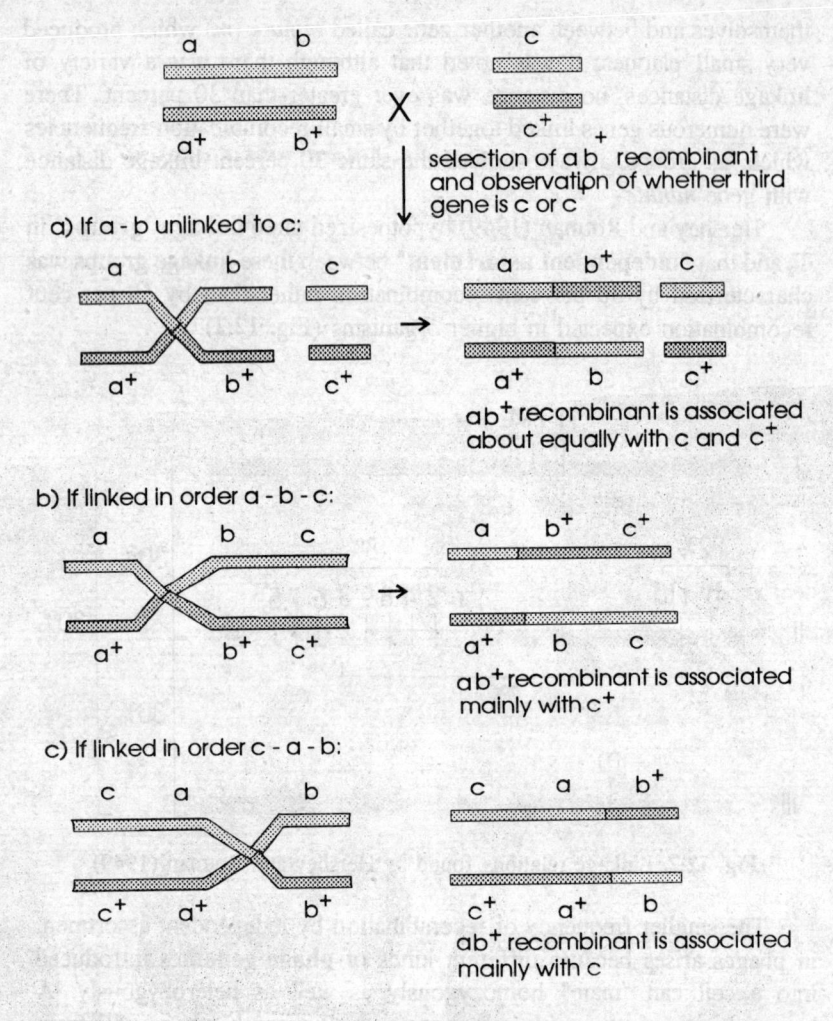

ab^+ recombinant is associated about equally with c and c^+

b) If linked in order a - b - c:

ab^+ recombinant is associated mainly with c^+

c) If linked in order c - a - b:

ab^+ recombinant is associated mainly with c

Fig. 12.3. Presence or absence of linkage between a gene c and two known linked genes (a and b) when a cross is $abc \times a^+b^+c^+$ and the selection is for recombinant $a\ b^+$. The frequency with which c and c^+ associates with these recombinants determines its relative position

altered by changing the number of rounds of mating. This is obtained by controlling the time of lysis. Thus, the comparisons between recombination frequency in phage crosses must be done under uniform conditions of phage maturation.

Viral recombination is not confined to simple pairing between two homologous chromosomes and to only one set of recombinational events in the "zygotes". Same viral chromosomes may pair and exchange genetic material more than once and may possibly mate with more than one chromosome at a time. Thus **viral recombination** is a population phenomenon whose events would have to be described statistically.

Circular Map of T Even Phages

Streisinger and Bruce demonstrated that three presumed linkage groups within each species actually form single linkage maps. The method used by these workers utilizes selection of particular class of mutants and is shown in **Fig. 12.3**. By use of this method the maps of T_2 and T_4 bacterio phages were shown to be connected end-to-end in the form of a circle.

Some Useful Reading Hints

Lederberg (1947) reported gene recombination and linked segregation in Escherichia coli. Hotchkiss and Weiss (1956) dealt with transformation in bacteria. Wollman and Jacob (1956) Wollman et al. (1956) discussed sexuality, conjugation and genetic recombination in *Escherichia coli* K-12. Jacob and Adelberg (1959) and Jacob and Wollman (1961) described transfer of genetic characters by incorporation in the sex factor of *Escherichia coli*. Cairns (1963) dealt with the chromosome structure of *Escherichia coli*. Taylor and Thoman (1964) constructed genetic map of *Escherichia coli* K-12. Bachmann and Low (1980) gave a linkage map of *Escherchia coli* K-12. Siddiqui and Fox (1973) described the mechanism of integration of donor DNA in bacterial conjugation. Scaife et al. (1985) dealt with genetics of Bacteria. Ippen-Ihler and Minkley (1986) explained the conjugation system of the fertility factor of *Escherichia coli*. Lederberg (1987) gave a discovery account of genetic recombination in bacteria. Mayloy (1989) described experimental techniques in bacterial genetics. Brock (1990) summarised the emergence of bacterial genetics. Drlica and Riley (1990) discussed the structure of bacterial chromosome. Miller (1992) offered a short course in bacterial genetics. Lloyd and Buckman (1995) reported conjugational recombination in *Escherichia coli* by conducting genetic analysis of recombinant formation in Hfr X F- crosses.

Luria and Delbruck (1943) conducted basic researches on Mutations of bacteria form virus sensitivity to virus resistance. Hershey and Rotman (1949) reported genetic recombination between host-range and plaque-type mutants of bacteriophage in single bacterial cells. Doermann and Hill (1953) explained genetic structure of bacteriophage T_4 as described by recombination studies of factors influencing plaque morphology. Visconti and Delbruck (1953) explained the mechanism of genetic recombination in phage. Lennox (1955) reported experiments on transduction of linked genetic characters of the host by bacteriophage P_1. Levinthal (1959) dealt with basics of bacteriophage genetics. Streisinger (1956) and Streisinger and Bruce (1960) reported early studies on phenotypic mixing of host range and serological specificities in bacteriophages T_2 and T_4 and linkage of genetic markers in these phages. Streisinger et al. (1964) studied vhromosomes structure in phage T$_4$ and determined circularity of the linkage map. Meselson and Weigle (1961) observed that chromosome breakage accompanied genetic recombination in bacteriophage. Stent (1963) and Stent and Calendar (1978) gave textbooks on molecular biology of bacterial viruses. Tomizawa and Anraku (1964) explained that molecular mechanisms of genetic recombination in bacteriophage involves joining of parental DNA molecules of phage T$_4$. Baylor et al. (1965) gave a circular linkage map of bacteriophage T$_2$. Rothman (1965) conducted one of early transduction studies on the relation between prophage and host chromosome. Cairns et al. (1966) revealed role of phage and the origin of molecular biology. Echols and Murialdo (1978) worked out genetic map of bacteriophage lambda. Wood and Revel (1976) gave an over view of the genome of bacteriophage T$_4$. Stahl (1979) and Stahl (1985) reviewed phenomenon of genetic recombination in phage and fungi. Berlyn and Letovsky (1992) explained COTRANS, a program for cotransduction analysis in phages. Hershey (1958) reported production of recombinants in phage croses. Hayes (1968) and Birge (1981) dealt with genetics of bacteria and bacteriophages. Wood (1980) used bacteriophage T4 morphogenesis as a model for explaining assembly of subcellular structure.

Questions

1. Define the following terms: transformation, conjugation mapping, recombination mapping, generalized transduction, Sexduction, and F-duction, F-mediated sexduction phenotypic mixing.
2. How are linkage maps prepared in bacteria through transformation?
3. How are linkage maps prepared in bacteria through conjugation mapping?
4. How are linkage maps prepared in bacteria through recombination mapping?
5. How are linkage maps prepared in bacteriophages through Sexduction/F-duction/F-mediated sexduction?
6. What are special features of bcateriophage chromosome replication?
7. Write a short note on recombination and mapping in bacteriophages.
8. How was it shown by Streisinger and Bruce that T even phages have a circular map?

Numericals

1. DNA was extracted from a wild-type strain of bacteria and used to transform a mutant strain unable to synthesize the amino acids alanine (ala), proline (pro), and arginine (arg). The number of colonies produced in the different transformant classes were as follows:

Number	Transformant class	Name	Transformant class
7200	ala$^+$ pro$^+$ arg$^+$	100	ala$^-$ pro$^+$ arg$^+$
200	ala$^+$ pro$^-$ arg$^-$	400	ala$^-$ pro$^+$ arg$^-$
500	ala$^+$ pro$^-$ arg$^+$	800	ala$^-$ pro$^-$ arg$^+$
800	ala$^+$ pro$^+$ arg$^-$		

What are the linkage distances between these genes? What is the linkage order?

Genetic Analysis in Bacteria and Viruses 209

2. If a donor-transforming strain of bacteria is wild type for three genes a^+, b^+ and c^+, and the recipient strain carries the mutations a^-, b^- and c^- which transformant classes would you expect to be least frequent. (a) If a and b are linked together but c is unlinked? (b) If a and c are linked, but b is unlinked? (c) If all three genes are linked in the order a-c-b? (d) If they are linked in the order c-a-b?

3. Using the technique of interrupted mating, five Hfr strains (1, 2, 3, 4, 5) were tested for the sequences in which they transmitted nine different gene markers, (F, G, O, P, Q, R, S, W, X, Y) to an F- strain. Each Hfr strain was found to transmit its gene markers in a unique sequence, as follows (only the first six gene marker transmitted were scored for each strain).

Order of Transmission		Hfr Strain				
		1	2	3	4	5
(First) ↑		Q	Y	R	O	Q
(Second)		S	G	S	P	W
Order of (Third)		R	F	Q	R	X
Transmission		P	O	W	S	Y
		O	P	X	Q	G
		F	R	Y	W	F

(a) What is the gene sequence in the original strain from which these Hfr strains derived? (b) For each of these Hfr strains, state which donor-gene marker should be selected in the recipients after conjugation to obtain the highest proportion of recombinants that will be Hfr.

4. In a P_1 transcution experiment, the donor bacteria were $synP^+$ $supM^-$ $trpZ^1$ the recipients were $synp^-$ $supM^-$ $trpZ^-$. Initial selection was for SupM transductants. 48 of these were M^+ P^+ Z^+, 120 of these were M^+ P^+ Z^-, 500 of these were M^+ P^- Z^-, 0 of these were M^+ P^- Z^+
   (a) What is the marker order? (b) What is the co-transduction frequency between M and P? M and Z?

5. In a generalized transduction experiment, donor *E. coli* cells have the genotype $trpC^+$ $pyrF^-$ $trpA^-$ and recipient cells have the genotype $trpC^-$ $pyrF^+$ $trpA^+$. P_1-mediated transductants for $trpC^+$ were selected and their total genotypes are determined with the results given below:

Genotypes	Number of progeny	Genotypes	Number of progeny
$trpC^+$ $pyrF^+$ $trpA^+$	274	$trpC^+$ $pyrF^+$ $trpA^-$	2
$trpC^+$ $pyrF^-$ $trpA^+$	279	$trpC^+$ $PyrF^-$ $trpA^-$	46

(a) Determine the order of the three markers. (b) Determine the cotransduction frequencies for *trpC* and *pyrF* and for *trpC* and *trpA*. (Note All cases in which two markers are co-transducted must be scored. (c) calculate physical map distances between these markers assuming the P_1 chromosome to be 10μ in length.

6. A three-factor cross between 2 strains of a virus (*a b c* and + + +) was performed. The results are given as follows:

Genotype of progeny	Number of plaques	Genotype of progeny	Number of plaques
+ + +	1200	a b c	1100
a + +	280	+ b c	300
a b +	180	+ + c	160
a + c	85	+ b +	75

Determine the linkage order of these three genes and linkage distance between *a-b*, *b-c* and *a-c*. Comment on your result.

7. A number of mixed infections were carried out between strains of T_4 phage that differed at three loci. Let us designate the gene simply as *a*, *b* and *c*. The following data indicate the crosses made the results obtained.

Genotype of progeny		Number of plaques		
$a^+ b^- \times a^- b^-$	$a^+ c^+$	$a^- b^-$	$a^+ b^-$	$a^- b^+$
	18	22	160	200
$b^+ c^+ \times b^- c^+$	$b^+ c^+$	$b^- c^-$	$b^+ c^-$	$b^- c^+$
	42	48	115	95
$a^+ c^- \times a^- c^+$	$a^+ c^+$	$a^- c^-$	$a^+ c^-$	$a^- c^+$
	11	9	35	45

Calculate the recombination frequencies and determine the order of these genes and the distance between them.

13

GENETIC MATERIAL

Nature of Genetic Material

"Genes lie in the chromosomes" is the thesis of the **chromosomal theory of inheritance**. Obviously some component of the chromosome must be the genetic material. Chromosome is made up of proteins and nucleic acids. Proteins are composed of polypeptides which are made up of amino acids linked together. Twenty kinds of essential amino acids are known to be present in proteins. Thus proteins are quite complex in shape and composition and they appear to qualify as possible genetic substances. Nuclear acids are of two kinds - deoxyribonucleic acid (DNA) and ribonucleic acid (RNA). Nucleic acids are long molecules (approximately 200,000 nucleotides made up four tpes of nucleotides. A nucleotide is in turn composed of three compounds linked together (a) a phosphoric acid, (b) a pentose sugar, and (c) a nitrogen-containing aromatic base. In nucleic acids, nucleotides are hooked together in form of sugar-phosphate-sugar-phosphate... linkage to form a polynucleotide chain. Nitrogen-containing aromatic basis are attached to sugar molecules as side chains. DNA molecules are very complex and, therefore, DNA is also candidate as a genetic material. These are evidences to show that DNA is the genetic material in all the eukaryotes and most of the prokaryotes. Some viruses have RNA as the genetic material. A few evidences are described below:

Transformation

Transformation is the process in which genetic material is transferred from one bacteria to another without involving any intermediate organism. A bacterial cell may incorporate in itself genetic material from medium also. This is also termed as **transformation**.

Griffith Effect. Griffith (1928) worked on bacterium *Diplococcus pneumonia* which is associated with certain types of pneumonia. This bacteria occurs in two forms: The first type have smooth (S) cells which secrete a covering capsule of polysaccharide materials causing the colonies on agar to be smooth and shiny. This type is virulent. It produces pneumonia in mice. The second type has rough (R) cells which lack a polysaccharide capsule. Colonies appear rough and dull. This type is non-virulent. It does not cause pneumonia in mice.

Smooth (S) and rough (R) traits are genetically determined. Griffith performed the experiment shown in **Fig. 13.1**. When laboratory mice were injected with S strain of bacteria, mice died. When he injected mice with R strain of bacteria, mice lived. When mice were injected with heat killed S strain of bacteria, mice lived. When mice were injected with R strain plus heat killed S strain of bacteria, mice died. Heat killed S strain somehow transformed R strain to virulence. This process was called "transformation". The agent that was responsible for transformation was called **transforming principle**.

Transforming principle. Avery, MacLeod and McCarty (1944) were able to identify nature of transforming principle. They separated extract of S

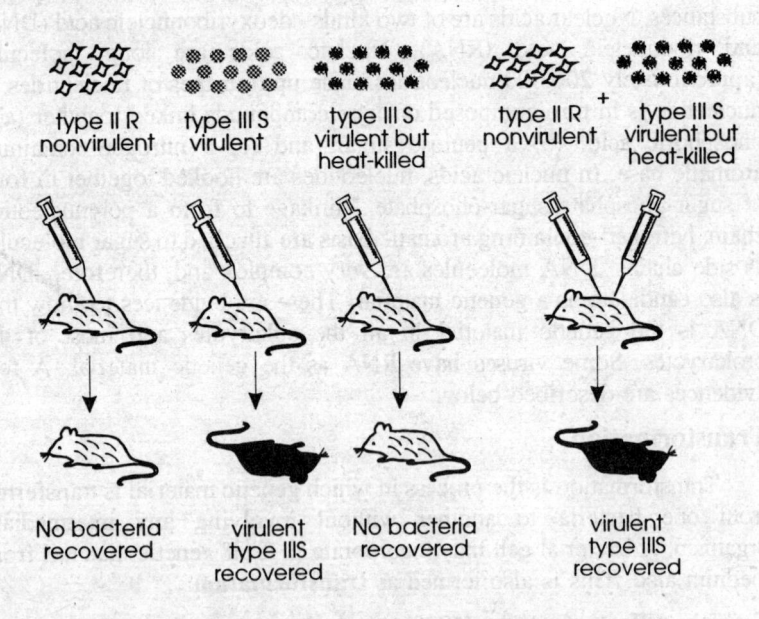

Fig. 13.1. Transformation experiment by Griffith

strain into different fractions, viz. polysaccharides, lipids, proteins, RNA and DNA. Only extract containing DNA were effective. This was shown by an experiment given in **Fig. 13.2**. This experiment demonstrated that DNA is the transforming principle. It was thus shown that genes, the units of heredity, are made up of DNA.

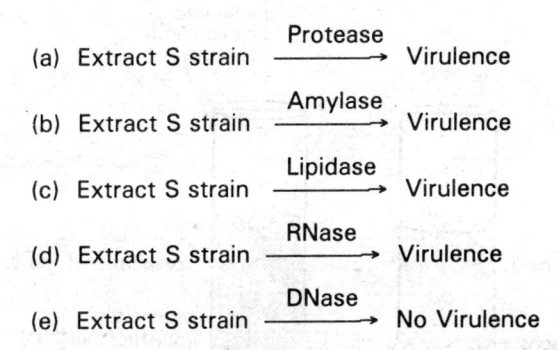

(a) Extract S strain $\xrightarrow{\text{Protease}}$ Virulence

(b) Extract S strain $\xrightarrow{\text{Amylase}}$ Virulence

(c) Extract S strain $\xrightarrow{\text{Lipidase}}$ Virulence

(d) Extract S strain $\xrightarrow{\text{RNase}}$ Virulence

(e) Extract S strain $\xrightarrow{\text{DNase}}$ No Virulence

Fig. 13.2. Results of experiment by Avery, MacLeod and McCarty (1944)

Transduction

Clear implication of DNA as the genetic material came from transformation experiments by Zinder and Lederberg (1952) on mouse typhoid bacterium *Salmonella typhimurium*. Their experiment involved process of transduction. In this process, a bacterium-infecting virus (bacteriophage) serves as the vector transferring DNA from one bacterial cell to another. He used two strains of bacteria, LA2 (phe^+ trp^+ met^- his^-) and LA22 (phe^- trp^- met^+ his^+). Zinder and Lederberg placed each strain in one of the two arms of a **Davis-U-tube** separated by a sinerted glass filter (**Fig. 13.3**). This filter prevents the transfer of bacterial cells but permits the free passage of growth medium between the wo arms through the application of suction and pressure. Phage P22 is capable of **generalized transduction**. As a result a "filterable agent" (FA) arose in connection with LA2 that could produce phototrophs in LA22. This FA was phage P22. It was formed that "phe^+ trp^+ factor" could pass through the filters that hold bacteria but allow viruses to pass through. "phe^+ trp^+ factor" was located inside the virus. Production of recombinant viral DNA was result of transduction. **Transduction** is defined as the process of the transfer of genetic material from one genotype to another through agency of bacteriophages.

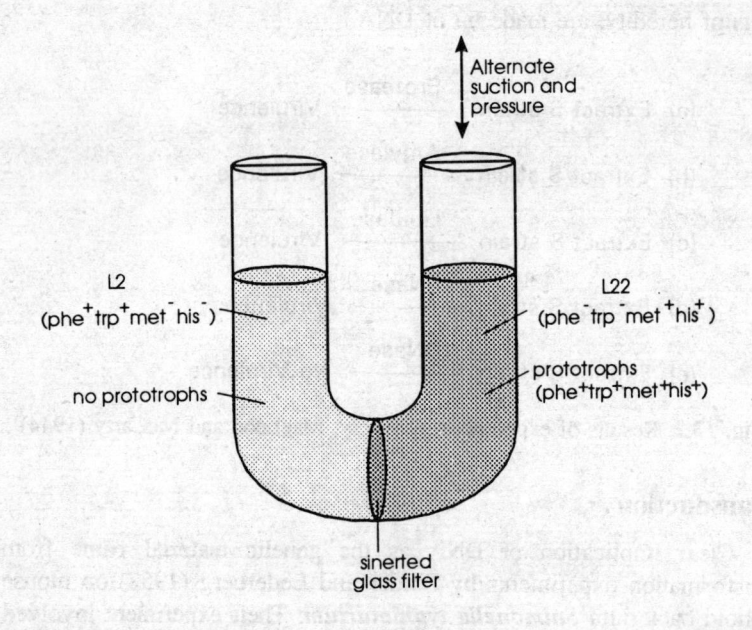

Fig. 13.3. Transduction experiment of Zinder and Lederberg (1952) in *Salmonella typhimurium* providing one line of evidence for DNA as genetic material

Transformation and transduction showed that DNA is the genetic material in bacteria.

Viral Infection Experiments

Hershey and Chase (1952) conducted experiments on phage T_2. This phage is relatively simple in molecular constitution. Most of its structure is protein with DNA contained inside the protein sheath of its head. Phosphorus is not found in proteins but is an integral part of DNA. Sulphur is not found in DNA but is present in proteins. They labelled phage DNA with radioisotope of phosphorous ^{32}P and phage proteins with that of Sulphur ^{35}S in different cultures (**Fig. 13.4**). Hershey and chase observed that after infection labelled ^{32}P went inside the bacteria

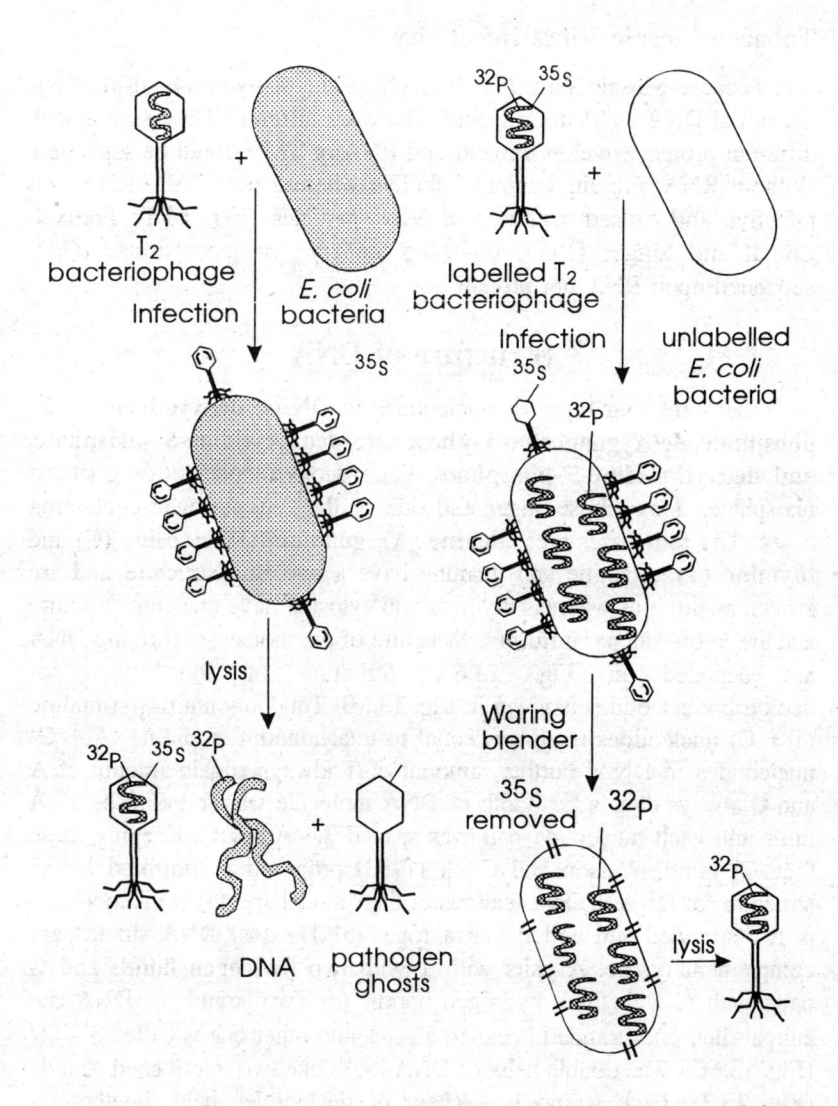

Fig. 13.4. Experiments of Hershey and Chase (1952) which showed that labelled DNA was transmitted whereas no labelled protein was transmitted to T$_2$ during infection of bacteria

whereas label 35S was left outside. They thus demonstrated that genetic material of phage T$_2$ is DNA, not protein.

Tobacco Mosaic Virus Infectivity

Tobacco Mosaic Virus (TMV) is one of the many viruses using RNA instead of DNA as its nucleic acid. There are different TMV strains with different protein envelop. Protein and RNA of TMV could be separated. Without RNA, protein was not infective whereas pure TMV RNA was infective and caused synthesis of Viral particles (**Fig. 13.5**). Frankel-Conrat and Singer (1957) observed the type of reconstituted TMV depended upon RNA not protein.

Structure of DNA

There are four types of nucleotides in DNA : **deoxyadenosine-5'-phosphate, deoxyguanosine-5'-phosphate, deoxycytidine-5'-phosphate, and deoxythmidine-5'-phosphate**. Each deoxyribonucleotide contains phosphate, deoxyribose sugar and one of the four nitrogen-containing bases. The four bases are: **adenine** (A), **guanine** (G), **cytosine** (C) and **thymine** (T). Adenine and guanine have a two ring structure and are known as **purines** whereas thymine and cytosine have one-ring structure and are known as **pyrimidines**. Structure of the molecules forming DNA are depicted in **Fig. 13.6A**. Structure of four types of deoxyribonucleotides is given in **Fig. 13.6B**. Total amount of pyrimidine (T + C) nucleotides is always equal to total amount of purine (A + G) nucleotides in DNA. Further, amount of T always equals amount of A and C always equals G. Width of DNA molecule was found to be 20 Å units and each nucleotide pair was spaced 3.4 Å apart. Keeping these facts in mind, Watson and Crick (1953) proposed a "**doubled helix**" structure for DNA. Salient features of their model are: (a) DNA molecule is two-stranded and coiled like a rope. (b) The two DNA strands are complementary, i.e. A pairs with T with two **hydrogen bonds** and G pairs with C with three hydrogen bonds. (c) Two strands of DNA run antiparallel, one stranded is called $5' \rightarrow 3'$ and other one is called $3' \leftarrow 5'$ (**Fig. 13.6C**). The double helix of DNA looks like two interlocked strands (**Fig. 13.7**). Each strand is a chain of nucleotides held together by **phosphodiester bonds**. The two polynucleotide strands are held together by hydrogen bonds between bases. The purines and pyrimidines are joined with dexyribose by β-**glycosydic** bonds. The intertwining of two strands of DNA is just like a staircase in which railings are made up of sugar-phosphate backbones and stirs are made up of bases. The bases

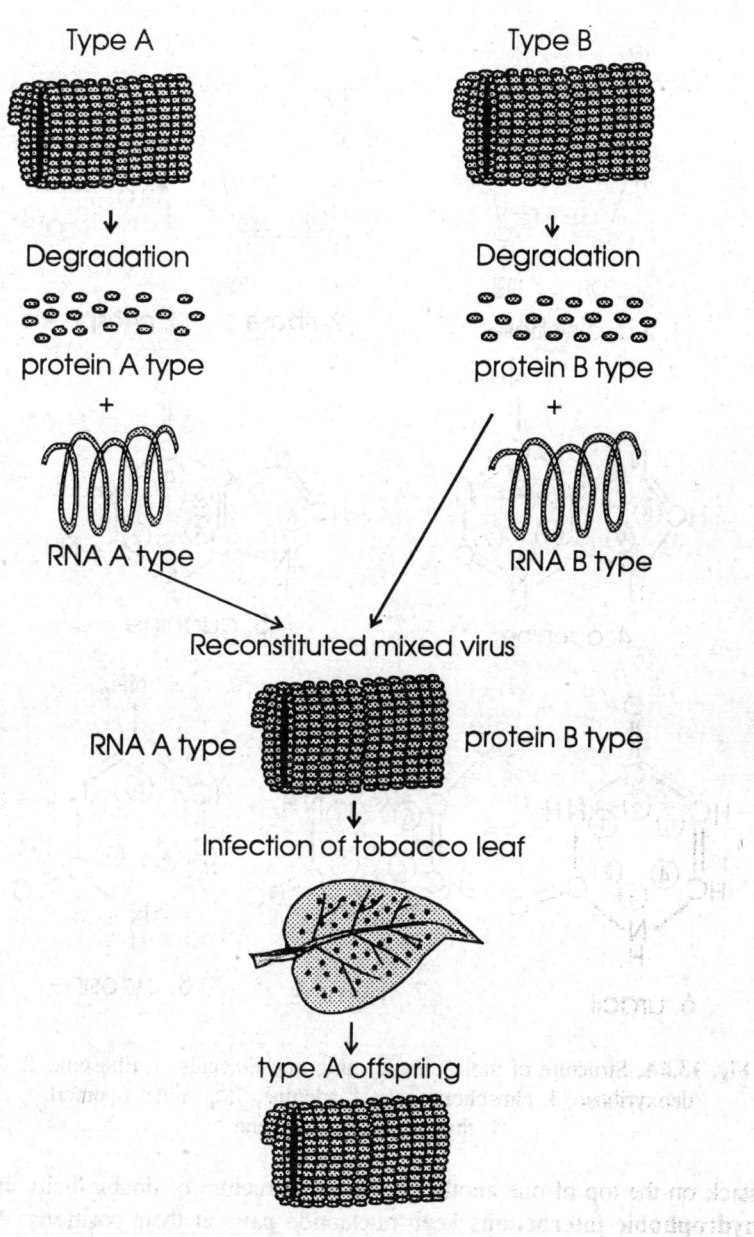

Fig. 13.5. Viral reconstitution experiment of Frankel-Conrat and Singer (1957) showing that RNA is the genetic material in TMV

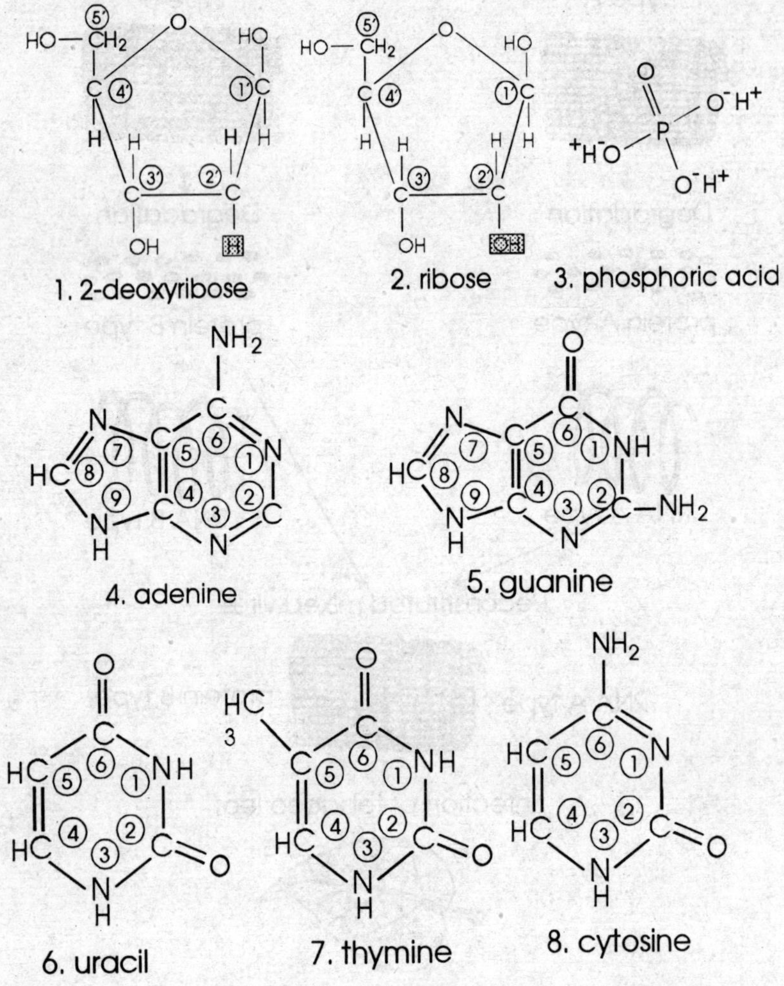

Fig. 13.6A. Structure of molecules forming nucleic acids: 1. ribosome, 2. 2-deoxyribose, 3. phosphoric acis, 4. adenine, 5. quanine, 6. uracil, 7. thymine and 8. cytosine

stack on the top of one another in twisted structure of double helix and **hydrophobic interactions** keep nucleotide pairs at their positions. All different types of bonds that are required in the formation of DNA are

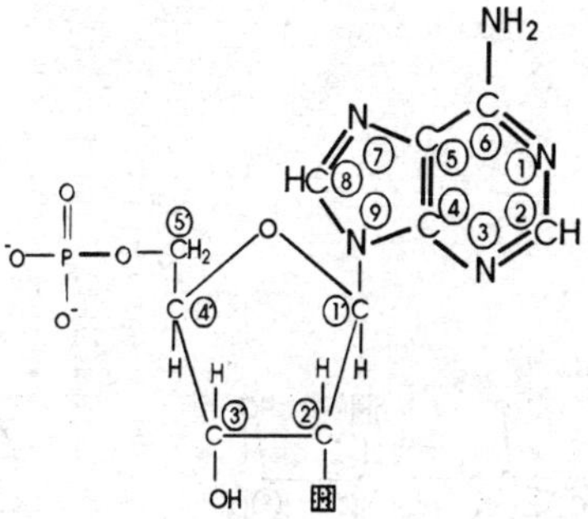

1. deoxyadenylic acid
(deoxyadenosine-5'-phosphate)

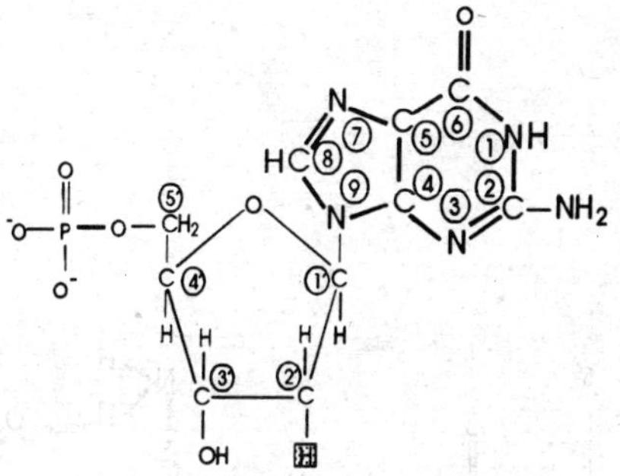

2. deoxyduanylic acid
(deoxyguanosine-5'-phosphate)

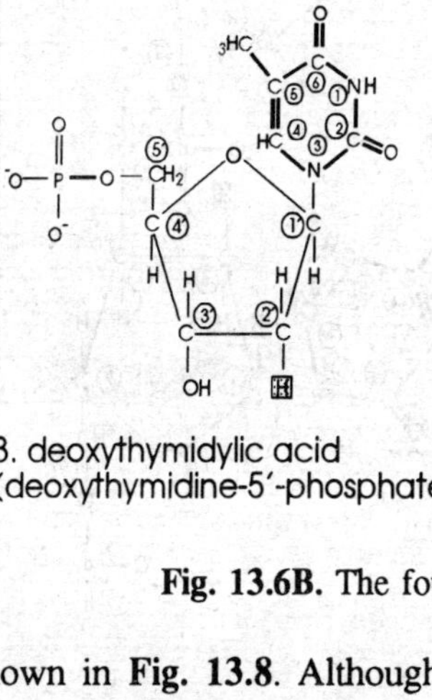

3. deoxythymidylic acid
(deoxythymidine-5'-phosphate)

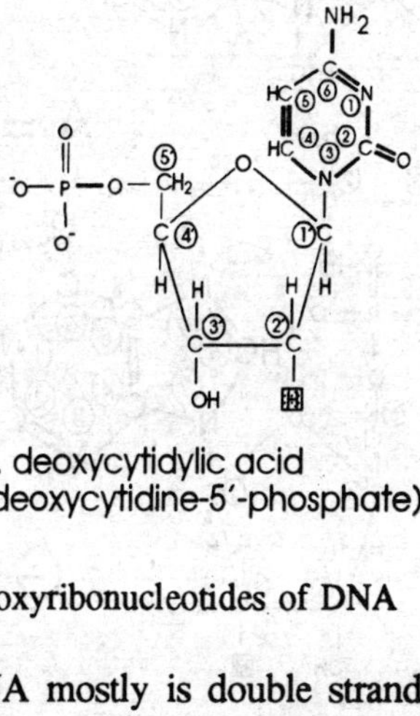

4. deoxycytidylic acid
(deoxycytidine-5'-phosphate)

Fig. 13.6B. The four deoxyribonucleotides of DNA

shown in **Fig. 13.8**. Although DNA mostly is double stranded, single-stranded DNA has also been found to occur as in φ X 174 virus).

Advantages of Watson-Crick Model. Double helical model of DNA proposed by Watson and Crick (1953) explained a few important biological phenomenon. How is biological information stored in DNA? How does DNA replicate? How does the genetic recombination take place? How do the mutations occur? Biological information is contained in DNA in form of sequence of deoxyribonucleotides. One gene differs from another in nucleotide sequence. It was suggested that DNA replicates in semi-conservative mode; the most accepted hypothesis of

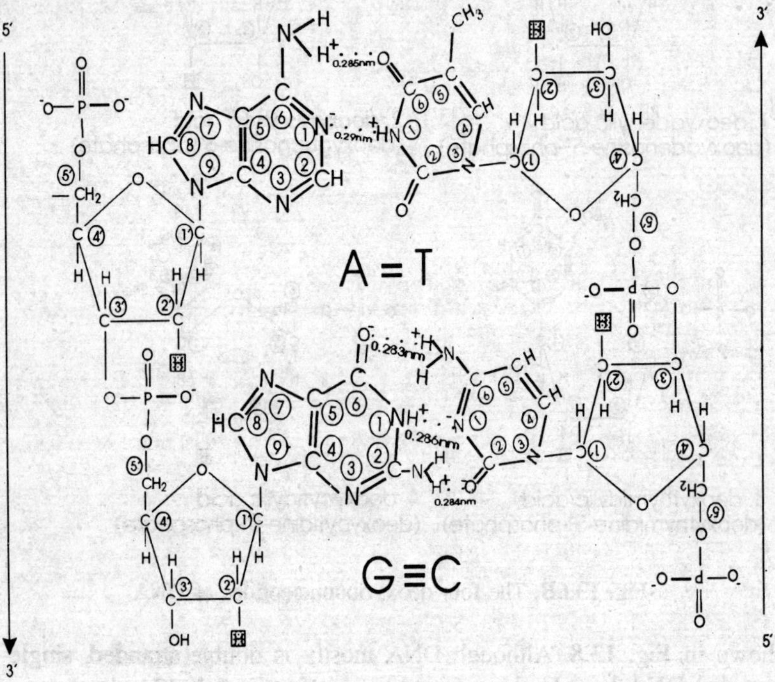

220 *Basic Genetics*

Fig. 13.6C. DNA structure comprising of two complementar base showing that two DNA strands are (1) complementary to each other (i.e., adenine paris with thymine with two hydrogen bonds and guanine paris with cytosine) with three hydrogen bond, (2) antiparallel in terms of direction of 5′ and 3′ ends of deoxyribose

recombination is breakage and reunion hypothesis. Molecular basis of mutation lies in change in nucleotide sequence.

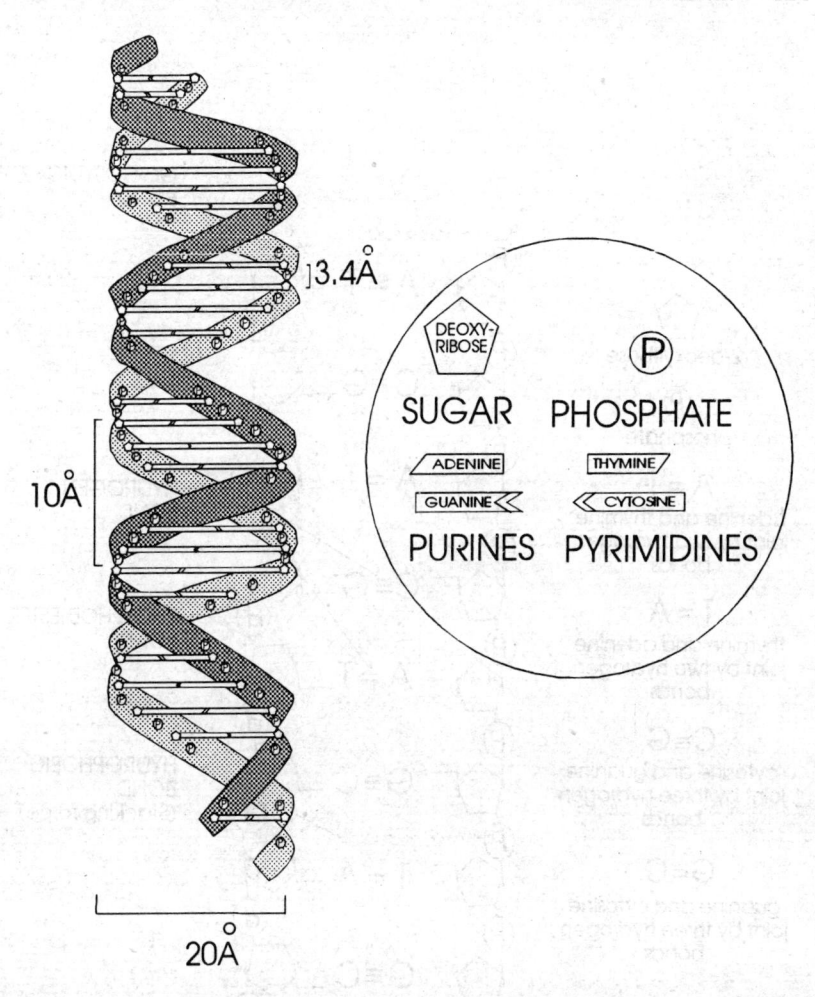

Fig. 13.7. Watson-Crick model of DNA structure

Structure of RNA

Molecules of RNA are generally single-stranded, linear polymers of ribonucleotides. **Double stranded RNA** is also known to exist. Some RNA molecules assume, in part, a doulbe helix configuration through complementary base pairing. Sugar of sugar-phosphate back bone is **ribose**. Successive ribonucleotides are joined by phosphodiester bonds. Uracil replaces thymidine of DNA (**Fig. 13.9**).

222 Basic Genetics

2-deoxyribose

(P) phosphate

A = T
adenine and thymine joint by two hydrogen bonds

T = A
thymine and adenine joint by two hydrogen bonds

C≡G
cytosine and guanine joint by three hydrogen bonds

G≡C
guanine and cytosine joint by three hydrogen bonds

— GLYCOSYDIC BOND
— HYDROGEN BOND
— PHOSPHODIESTER BOND
— HYDROPHOBIC BOND (Stacking force)

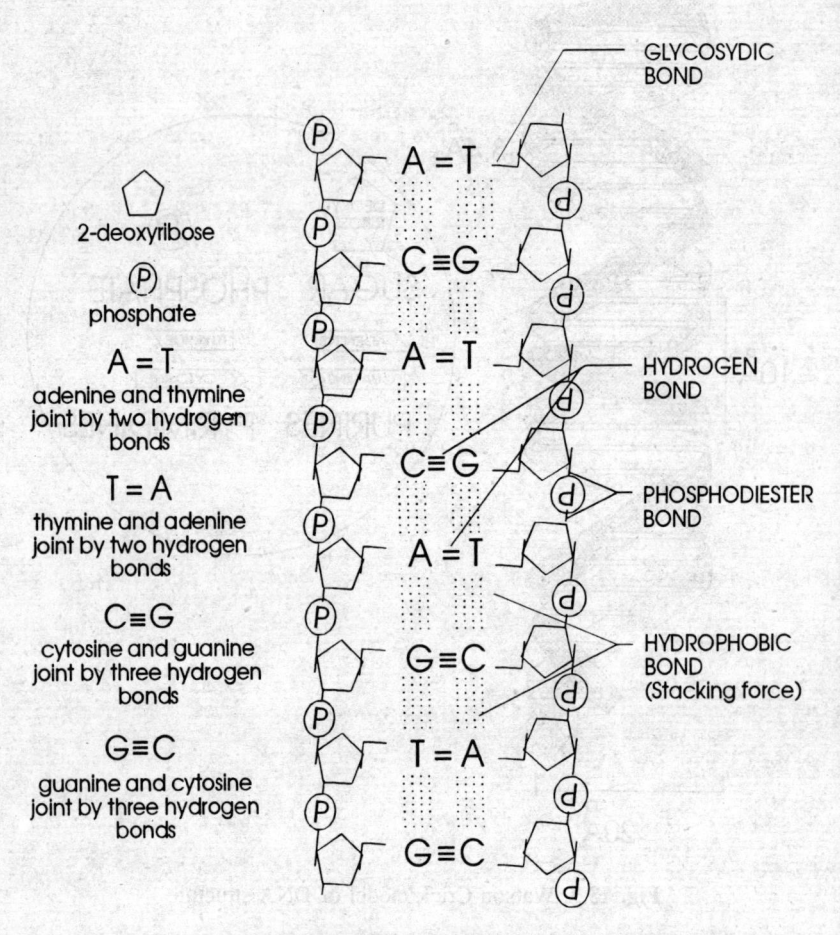

Fig. 13.8. Glycosydic, phosphodiester, hydrogen and hydrophobic bonds hold DNA together

Properties of Genetic Material

Storage of Information

Genetic material carries from one generation to the next the information that specifies the characteristic of the plant or animal. This

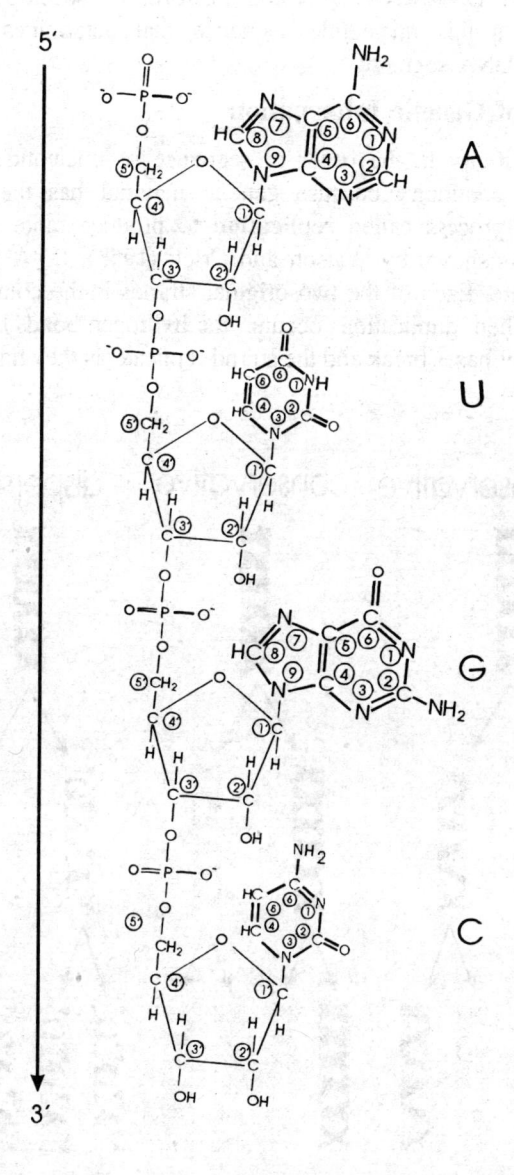

Fig. 13.9. Structure of ribonucleotides joined together forming RNA. Usually RNA is single stranded, doubel stranded RNA is also known

information in DNA/RNA is stored in form of specific nucleotide sequence. It is this nucleotide sequence that determines biological function of a DNA segment.

Continuity of Genetic Information

The specificity in the order of sequence of nucleotides in DNA maintains its continuity because genetic material has the ability of undergoing a process called **replication** to produce more copies like themselves. As shown by Watson and Crick (1953), DNA is a double helical structure. Each of the two original strands is the complement of the other. When duplication occurs, the hydrogen bonds between the complementary bases break and the strand replicate as they unwind. Each

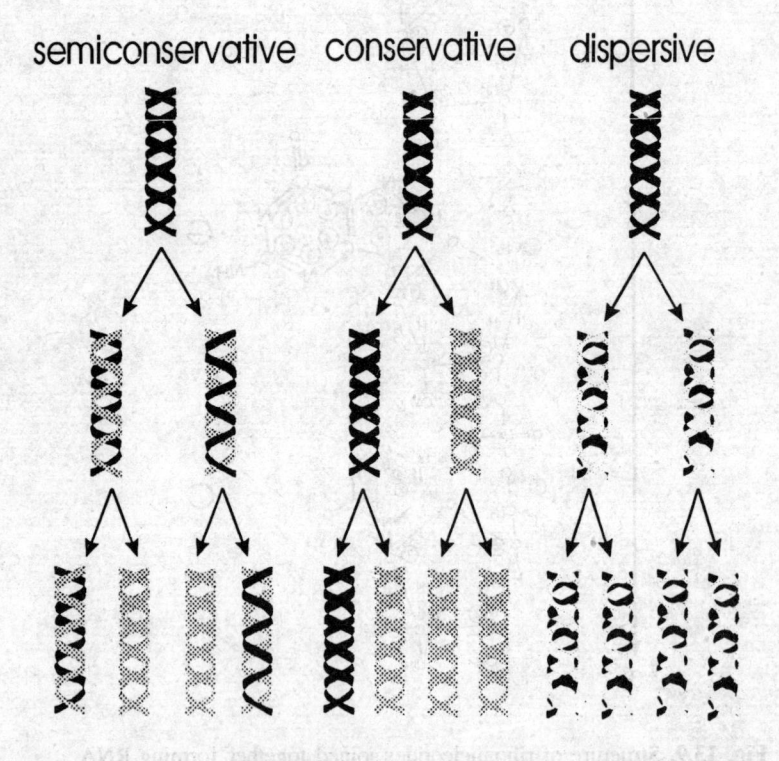

Fig. 13.10. Three modes of DNA replication: expectations in two cycles of replication. Dark colour indicates label

strand acts as a template for the formation of a new complementary chain. Two pairs of chain thus appear where before only one pair existed. The mechanism of synthesis of DNA proposed by Watson and Crick (1953) offered the important advantage of explaining how now DNA molecules could form exact replicates of the old. According to Watson and Crick, each single strand is a template or mold for its complement, and a new helix has one old strand and one that is newly synthesized. There are three possible **modes of DNA replication** (Fig. 13.10) (a) **Semi-conservative** - the new helix has a one old strand and one strand that is newly synthesized. (b) **Conservative** - the two new strands are synthesized in form of a double helix while the old double helix remains unchanged. (c) **Dispersive** - Double helix does not unwind but double helical strands break along their length into smaller pieces. After each individual piece replicates, it would then be randomly reconnected with newly synthesized piece to form a patch work single string of "dispersed" old and new pieces.

Mutation

The genetic material is subject to low rate of alteration, the changes being transmissible to next generation. A change in the genetic material at a particular locus in an organism is termed as mutation. How change in the sequence of nucleotide(s) leads to a mutation is shown in **Fig. 13.11**. The term mutation includes point/gene mutations involving a single and chromosomal changes. Mutations will be dealt with at length in Chapter 21.

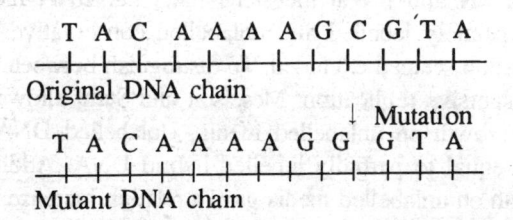

```
        T A C A A A A G C G T A
        -+-+-+-+-+-+-+-+-+-+-+-
        Original DNA chain
                              ↓ Mutation
        T A C A A A A G G G T A
        -+-+-+-+-+-+-+-+-+-+-+-
        Mutant DNA chain
```

Fig. 13.11. Change in sequence of nucleotide(s) leads to mutation

Recombination

DNA breakage and reunion is responsible for genetic recombination in prokaryotes as well as eukaryotes. **Endonuclease** produces **nicks** in single strands of DNA and a **ligase** rejoins broken DNA strands.

Eukaryotes mostly show reciprocal recombination products. Various aspects of recombination have already been considered in Chapters 10, 11 and 12.

Replication of DNA in Prokaryotes

In order to decide the actual mode of replication of DNA, chemical labelling of DNA was done. In later duplications the label was detected. Our expectations in the three modes of replication are shown earlier in Fig. 13.10. In case of conservative replication, both strands of **newly synthesized DNA** double helix should be labelled while the old double helixes remain unlabelled. In case of semi-conservative replication, the replicated double helixes carry one old and one new stand, and all are, therefore, labelled. In dispersive mode of replication also, label in both the strands of both the helixes will be expected. Semi-conservative and dispersive modes can be differentiated in next cycle of DNA replication.

Meselson-Stahl Experiment

Meselson and Stahl (1958) labelled the DNA of *Escherichia coli* bacteria with "heavy" nitrogen, ^{15}N, by growing them on an ^{15}N containing medium that was sole source of nitrogen for many generations. The DNA extracts of such cells gave a characteristic UV-light pattern showing a dark band near one and of a tube that had been spun at high speed in a ultracentrifuge. When these labelled ^{15}N cells were grown on nonlabelled media, the DNA extracted was shown to consist of a **hybrid DNA** carrying both ^{14}N and ^{15}N at the same time; i.e. DNA had not replicated in the separately labelled and unlabelled conservative form. Conservative replication seemed excluded. To distinguish between semi-conservative and dispersive replication. Meselson and Stahl allowed the next generation of growth on unlabelled media. Unlabelled DNA was formed in amounts equal to partially labelled hybrid DNA. Additional generations of growth on unlabelled media gave a relation increase in the amount of unlabelled DNA.

Since unlabelled DNA was always formed despite the presence of labelled hybrid DNA, duplication evidently did not involve random dispersive labeling of all newly formed DNA. (**Fig. 13.12**). The dual $^{15}N/^{14}N$ composition of the "hybrid DNA" was also demonstrated by heating this band of DNA to $100°C$ and then reanalyzing it by ultracentrifuging. Denaturation with heat broke the hydrogen bonds

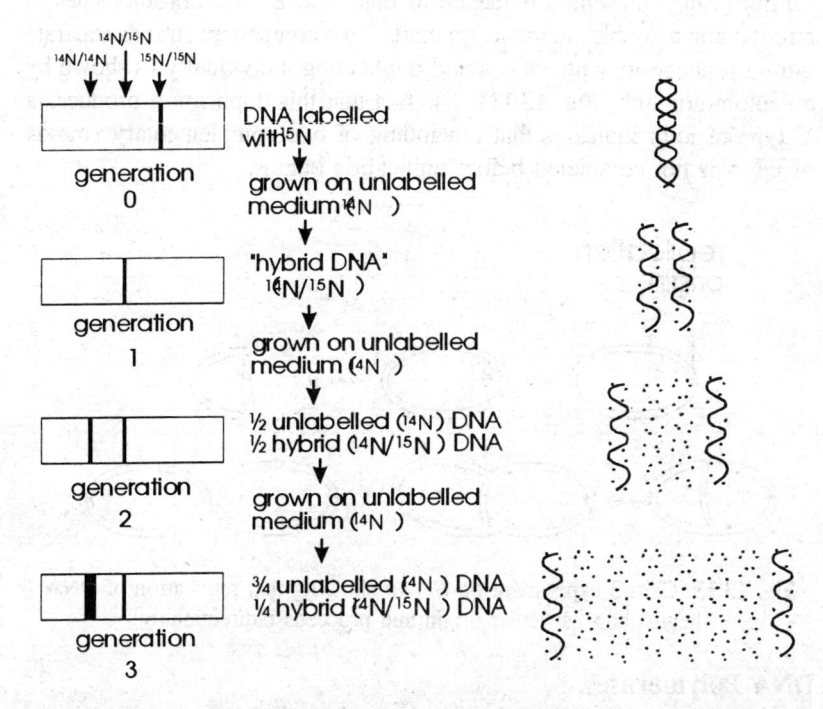

Fig. 13.12. Meselson-Stahl experiment and its interpretation providing proof that mode of DNA replication is semi-conservative

between two completementary DNA strands and two single strands of DNA were produced. Two separate bands were formed in ultracentrifuge density gradient, one heavy (^{15}N) and one light (^{14}N). So conclusion drawn was that DNA replication is semi-conservative.

Objection Against Semi-conservative Replication

Objection was raised with respect to the amount of time it would take for such replication to occur in a very long molecule. T_2 phage DNA is 200,000 nucleotide pairs. Since these are 10 base pairs to each rotation, such a molecules would have 20,000 rotations. The molecules will be 20,000 X 34 Å = 68 microns in length. Such a molecule would have to be folded a number of times to fit into cells or nuclei. This problem of unwinding of DNA and replication appears profound. Answer to this is that replication proceeds before the parental DNA is completely unwound.

Cairns (1963) presented evidence to this. The *E. coli* chromosomes is circular and a double stranded structure. Two component strands separate during replication, with each strand duplicating individually as shown by an autoradiograph (**Fig. 13.13**). The fact that this duplication produces a Y type of joint indicates that unwinding of two complementary strands of DNA is not completed before replication begins.

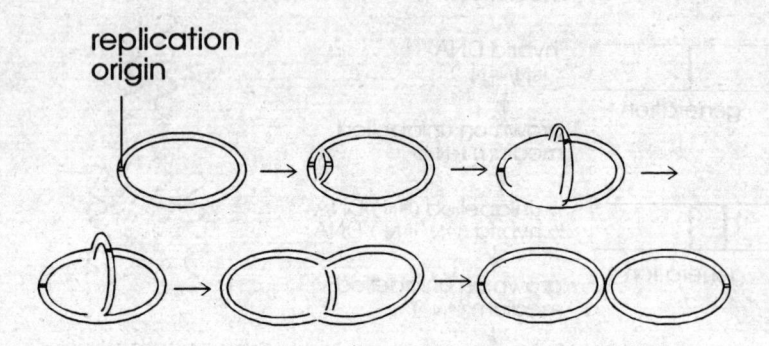

Fig. 13.13. Cairn's experiment on *E. coli* showing that replication of DNA begins at a particular origin and proceeds bidirectionaly

DNA Polymerases

Three types of DNA polymerases are known: DNA polymerase I may be used to help termination of replication. DNA polymerase II is the major enzyme involved in DNA replication. It can synthesize DNA only in 5'→3' direction. New nucleotides are added at 3' position. The 5'→3' strand of DNA is thus continuously synthesized. DNA polymerase III is necessary for *in vivo* replication.

If the major enzyme (DNA polymerase II) can function only in one direction, this means that only one of the parental DNA strands is being continuously replicated while DNA helix unwinds. How is the second strand being replicated? The second strand, having opposite polarity can best be replicated in a discontinuous fashion. The 3'←5' strand replicates more or less continuously whereas for the other strand only short fragments, known as **Okazaki fragments** will be produced during replication. In this way replication is **bidirectional**. The short fragments of DNA are then united enzymatically. Major protein and enzyme activities during DNA replication are shown in **Fig. 13.14**.

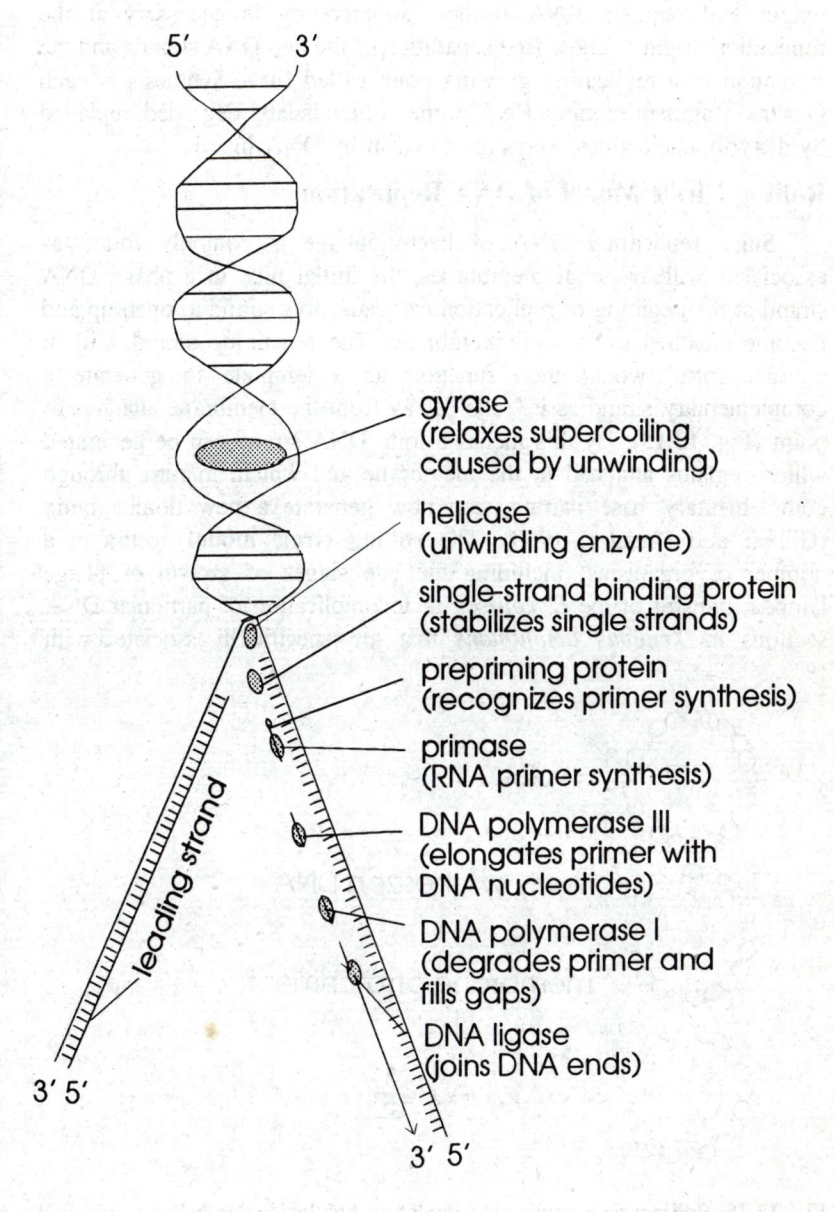

Fig. 13.14. Some of the enzyme and protein activities in DNA replication

DNA replication begins at a unique site on the chromosome, called **origin** and requires RNA primer. Supercoiling is necessary at the replication origin to allow first separation of the two DNA strands and the formation of a replicating growing point called **fork**. Synthesis is each Okazaki fragment requires RNA primer which is later degraded, replaced by dexyribonucleotides. Gaps are filled in by DNA ligase.

Rolling Circle Model of DNA Replication

Since replicating DNA of bacteriophage is comonly found as associated with bacterial membranes, the initial nick in a phage DNA strand at the begining of replication can cause this strand to open up and become attached to the cell membrane. The remaining strand, still in circular form, would then function as a template to generate a complementary strand as it "rolls" away from the membrane attachment point (**Fig. 13.15**). By this menas a long DNA strand can be generated which remains attached to the membrane and which, in turn, through complementary base pairing, can now generate a new doulbe helix (Gilbert and Dressler, 1968). The **rolling circle model** occurs in a number of organisms, including the late stages of growth of phage lambda, transfer of the *E. coli* sex-factor, replication of particular DNA sections in *Xenopus amphibians* that are specificalli associated with

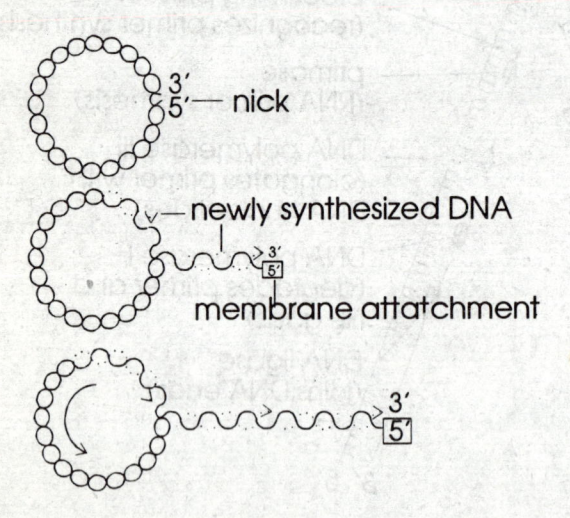

Fig. 13.15. Rolling cycle replication model of circular double helix

production of ribosomal RNA, and replication of some single-stranded DNA phages such as φX174.

DNA Replication in Eukaryotes

DNA replication takes place during **S phase**. The trigger for DNA synthesis takes place seems to occur during G_1 **period**. Taylor *et al.* (1957) labelled dividing root-tip cells of the broad bean *Vicia faba* with ^{3}**H-thymidine** which is specifically incorporated into newly synthesized DNA (Fig. 13.16). It was found that a new chromosome was replicating semi-conservatively because it consisted of half radioactive newly formed chromosome and half parental chromosome. When the new chromosomes (hybrid) were allowed to grow and divide on unlabeled media, half of the new chromosomes carried labelled and other half were unlabeled. Thus random labelling of chromosomes did not occur, and the mode of replication was consequently not dispersive. Semi-conservative replication of the eukaryotic chromosome is based essentially on an underlying structure that is a continuous length of DNA double helix.

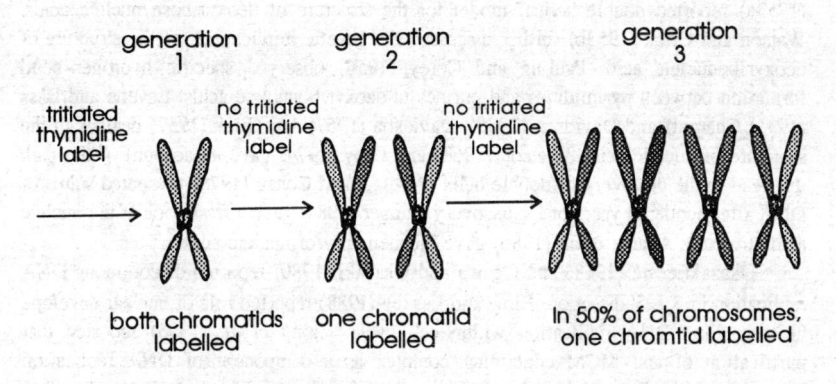

Fig. 13.16. Experiment of Taylor and his associates on root tips of *Vicia faba* showing semi-conservative replication of eukaryotic chromosomes

Each round of **eukaryotic replication** involves the separation and unwinding of DNA from its histone structure, then enzymatic synthesis of new DNA. Histone components are doubled and rewinding of DNA occurs. Both DNA and histones replication occurs during S phase of cell cycle. DNA replication occurs at many points along the eukaryotic chromosome, each replicating region known as, **replicon**, is generated into a bubble structure. Number of replicons is in thousands in eukaryotic chromosome.

Some Useful Reading Hints

Miescher (1871) was first to report presence of nucleic acids in pus cells. Griffith (1928) gave understanding about the significance of pneumococcal types by injecting them into rat. Avery *et al.* (1944) through their studies discover the chemical nature of substance inducing transformation of pneumococcal types. They found out that induction of transformation was brought about by a deoxyribonucleic acid fraction isolated from pneumococcus type III. Hershey and Chase (1952) worked on understanding independent functions of viral protein and nucleic acid in growth of bacteriophage. Fraenkel-Conrat and Singer (1957) conducted virus reconstitution experiments by combination of protein and nucleic acid from different strains and demonstrated RNA as the genetic material in TMV. Zinder and Lederberg (1952) reported genetic exchange in *Salmonella*. Zinder (1958a) and Zinder (1958b) reported transduction in bacteria giving an evidence for DNA as the genetic material. Zinder (1992) commented on the discovery of bacterial transduction. Sinsheimer (1959) reported a single-stranded deoxyribonucleic acid from bacteriophage φX174. Wilkins *et al.* (1953) and Wilkins (1963) reported their experiments on molecular structure of deoxypentose nucleic acids. Watson and Crick (1953a) reported double helical model for the structure of deoxyribose nucleic acids. Watson and Crick (1953b) further discussed the genetic implications of the structure of deoxyribonucleic acid. Pauling and Corey (1956) observed specific hydrogen-bond formation between pyrimidines and purines in deoxyribonucleic acids. Levene and Bass (1931) Chargaff and Davidson (1955), Davidson (1957) and Crick (1957) dealt with the structure of nucleic acids. Watson (1968) and Olby (1974) gave an account of the path that lead to the discovery of double helix. Portugal and Cohen (1978) presented status of DNA after hundered years of its discovery. Sanger (1984) gave a fresh look at the nucleic acid structure. Adams *et al.* (1986) gave biochemistry of nucleic acids.

Okazaki *et al.* (1968) and Ogawa and Okazaki (1980) reported discontinous DNA replication in $5' \rightarrow 3'$ direction. Blow and Laskey (1988) reported role of nuclear envelope in controlling DNA replication within cell cycle. Clong *et al.* (1995) reported that purification of and MCM containing complex as a component of DNA replication licensing system. Kamada *et al.* (1996) provided insight into structure of a replication-termination protein complexed with DNA.

Gilbert and Dressler (1968) gave the rolling circle model of DNA replication. Shekman (1974) discussed multienzyme systems of DNA replication. Kornberg (1974) and Kornberg (1980), Kornberg and Baker (1992) isolated and characterised enzyme of DNA replication. Morrison and Wang (1991) gave the properties and action of eukaryotic DNA polymerase.

Questions

1. Are the proteins or nucleic acids the genetic material? What different types of evidences are available to demonstrate nature of the genetic material?
2. What is transformation, Griffith effect and transforming principle? Who discovered these processes? Explain these processes with the help of experiments which showed that DNA is the genetic material in *Diplococcus pneumonia*.
3. What is transduction? Who used transduction as a means to demonstrate that DNA is the genetic material in *Salmonella typhimurium*? Explain the transduction experiment with the help of a suitable diagram.
4. Who conducted viral infection experiments to demonstrate that DNA is the genetic material in phage T_2? Explain the the experiment with the help of a suitable diagram.
5. What was the objective of tobacco mosaic virus infectivity experiment? Who performed this experiment and how was this objective achieved?
6. Who were the key persons involved in deciphering the structure of DNA? Explain various attributes of DNA structure.
7. In what ways structure of RNA differs from that of DNA?
8. Describe various properties of genetic material (DNA) which are were explained as built in part of its structure.
9. Write brief account of the following properties of DNA: storage of information, mutation, and recombination
10. What are our expectations in labelling experiment under three modes of replication?
11. Describe Meselson and Stahl (1958) experiment showing that DNA replication is semi-conservtive in *E. coli*.
12. What were the objections against semi-conservative mode of DNA replication and how were these objections answered?
13. How was it experimentally shown that DNA replication proceeds before the parental strands are completely unwound?
14 How was it experimentally shown that DNA replication is bidirecional?
15. What are different proteins and enzymes involved in DNA replication in prokaryotes?
16. Describe experiment of Taylor *et al.* (1957) showing that DNA replication is semi-conservtive in *Vicia faba*.

14

GENE STRUCTURE

Concept of Gene

Structure of gene can be understood in terms of its two functions - recombination and mutation. The concept of gene structure has been changing with the developments in the subject. This changing scenario can be discussed under three phases - classical, transitional and modern.

Classical Phase

In the beginning, gene was regarded as a hypothetical particle, performing a specific function and recombining with other genes. A gene was considered a unit of function and the unit of recombination was regarded no larger than a gene. After the independent discovery of Sutton (1903) and Boveri (1903), gene was considered a concrete particle - a small part of the chromosome. There are several genes carried by a chromosome. Genes located in the same chromosome do not show independent assortment. Genes showing their location on the same chromosome can also show recombination by crossing-over. Studies of Morgan showed that multiple allelism arises from alternative modification of the same gene (e.g., *white* locus in *Drosophila melanogaster*). Multiple alleles (w and w^e) do not show complementation in F_1 (F_1 is not wild-type) and do not show recombination in F_2. Alleles determined on the bases of complementation are termed **functional alleles**. While those determined on the bases of recombination was regarded no smaller than a gene. A gene thus was consider a unit of recombination.

Morgan and sturtevent (1923) studied *Bar* locus in *D. melanogaster*. A *Bar* mutation consists of insertion of a *Bar* gene (duplication) at that locus (16A). When this repeated segment was lost and the mutant reverted back to wildtype. **Ultrabar** had two repeats. Duplication of 16A reduced number of facets in the eye: B^+/B^+, 779; B^+/B, 358; B/B, 68,

chromosome constitution	phenotype	number of facets
X, X	wildtype	779
X, X	heterobar	358
X, X	bar	68
X, X	ultrabar	45

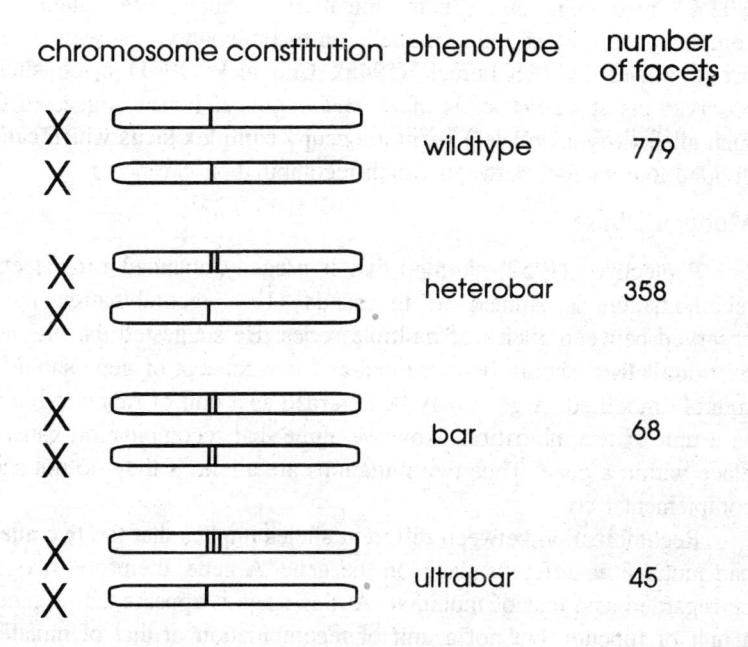

Fig. 14.1. Number of facets in different phenotypes of *bar* locus of *Drosophila melanogaster*

Ultrabar, 45 (Fig. 14.1). Ultrabar arose due to **unequal pairing** and crossing-over. These observations lead to the conclusion that a gene does not necessarily represent a constant amount of genetic material and its alleles may differ from each other in this regard. When such recombined sections appear in *cis*-**position**, the phenotype produces may be different than when they are in *trans*-**position**. Muller (1927) observed that mutation could be produced after being hit by a particles which clearly indicated particulate nature of the gene. At this stage, gene was regarded as a unit of mutation. Muller (1940) defined gene as simultaneously a unit of function and a unit of recombination. This concept also implied that gene is differentiated internally.

Transitional Phase

This phase deals with pseudoalleles. Two mutations *Star* (*S*) and *asteriod* (*ast*) showed no complementation and *S* and *ast* mutations were concluded as allelic. Lewis (1941), however, showed that *S* and *ast* show

0.02% recombination. Such mutations which are allelic, in complementation test and non-allelic on recombination test were termed **pseudoalleles** by McClintock (1945). Chovnick (1961) made similar observations at *garnet* locus of *D. melanogaster*. It was suggested that such alleles (pseudoalleles) seem to occupy **complex locus** which can be divided into subloci between which recombination can occur.

Modern Phase

Pontecorvo (1952) claimed that if progeny obtained for detecting recombination is studied in thousands (10^5), recombination can b observed between alleles of multiple series. He suggested that the term **pseudoallelism** should be discarded and our concept of gene should be suitably modified. A gene may be regarded as a unit of function but not as a unit of recombination. Now we agree that recombination can take place within a gene. Thus two mutations are allelic if they do not show complementation.

Recombination between different alleles implies that the two alleles had mutated at different places in the gene. A gene, therefore, may not be regarded as a unit of mutation. At this stage it appears that a gene is a unit of function but not a unit of recombination or that of mutation. Gene can no longer be considered as a structureless molecule but one with a well-defined internal structure.

Complementation Mapping. *Lozenge* (*lz*) in *D. melanogaster* is a complex locus (Green and Green, 1949) which consists of 4 groups, each containing one or more different *lozenge* alleles (**Fig. 14.2**). Alleles of different groups recombine. All these mutants are recessive. When two mutations at two subloci are present in ***cis*-position**, they have their wild type alleles on the other homologue. However, when two mutations are placed in ***trans*-position** many heterozygous combinations of different *lozenge* alleles do show mutant effects. From this, Lewis hypothesized that this effect arose from sequential nature of gene product synthesis.

Discovery of *cis-trans* effect enabled simple functional test for allelism with similar effect. Two mutations is ***trans*-heterozygote** are allelic if they produce a mutant phenotype and non-allelic if they produce a wild type phenotype. The ability of two recessive mutations to restore wild type phenotype (partially or completely) has been termed as **complementation (Fig. 14.3)**.

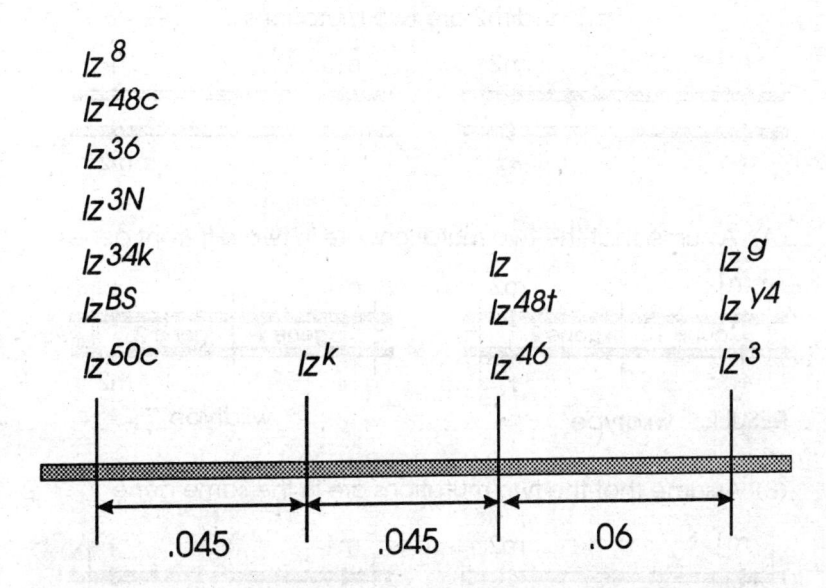

```
lz⁸
lz⁴⁸ᶜ
lz³⁶
lz³ᴺ
lz³⁴ᵏ              lz                          lz ᵍ
lzᴮˢ              lz⁴⁸ᵗ                       lz ʸ⁴
lz⁵⁰ᶜ      lzᵏ    lz⁴⁶                        lz³
```

Fig. 14.2. Some alleles of *lozenge* occupying different subloci within the *lozenge* locus of *Drosophila melanogaster*

The *rII* Locus. Complementation test determines if two mutations are in the same gene. Work on *rII* locus of bacteriophage T_4 has brought interesting results. Certain mutations in rII region influence length of life cycle of T_4 in bacterial cell (r = rapid lysis). The *rII* mutants also make larger plaques than wild type phages. The *rII* mutants and wild type T_4 phage grow equally well on *E. coli* strain B but on *E. coli* K (λ) strain only wild type can multiply. Benzer's selection procedure is described in **Fig. 14.4**). Region rII of T_4 phage consists of two genes, *rIIA* and *rIIB*.

In a typical cross, *E. coli* strain B bacteria were infected with two T_4 phage particles were bearing independently isolated *rIIA* (or *rIIB*) mutation. As viruses multiplied, recombination occurred. The progeny were then grown on *E. coli* (λ) strain (where only wild type can grow) to test for wild-type. Normal particles were found in may large fraction of crosses, indicating that recombination had occurred within a gene (**Fig. 14.5**).

cis-heterozygote / trans-heterozygote

m1 and m2 are two mutations

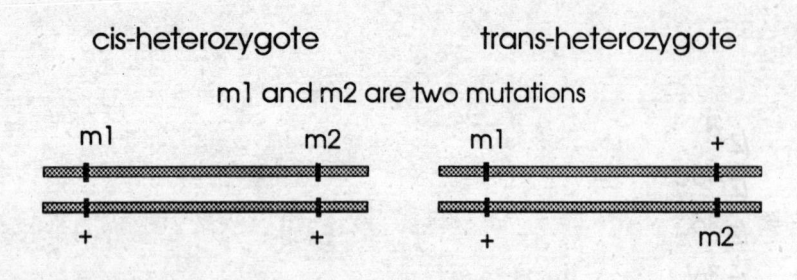

(A) Assume that the two mutations are in two different genes

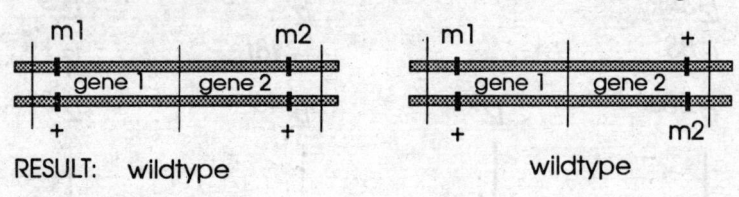

RESULT: wildtype / wildtype

(B) Assume that the two mutations are in the same gene

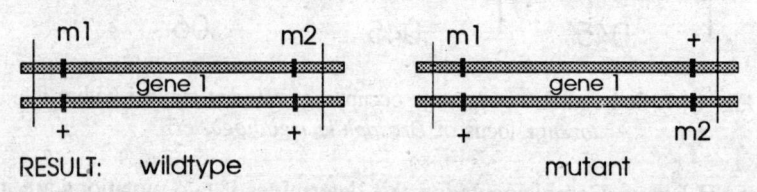

RESULT: wildtype / mutant

Fig. 14.3. *Cis-trans* complementation test for knowing whether two mutations are located in two different genes or in the same gene

Deletion Mapping Method. Benzer employed deletion mapping method to determine whether or not deletion includes site of a mutation (**Fig. 14.6**). Benzer found that through recombination he could distinguish between "**point**" **mutations** affecting only one mutations site and longer "**multisite**" **mutations**. The distinction was made by observing whether r^+ recombinants were produced from a cross between as unknown r mutant and r mutants previously known to recombine with each other to producer r^+ recombinants. Were such r^+ recombinants formed, the mutation could be considered a point mutation. If r^+ recombinants were not formed, the mutation was of multisite nature. For example, r mutants 155 and 274, known to recombine with each other to produce wild-type

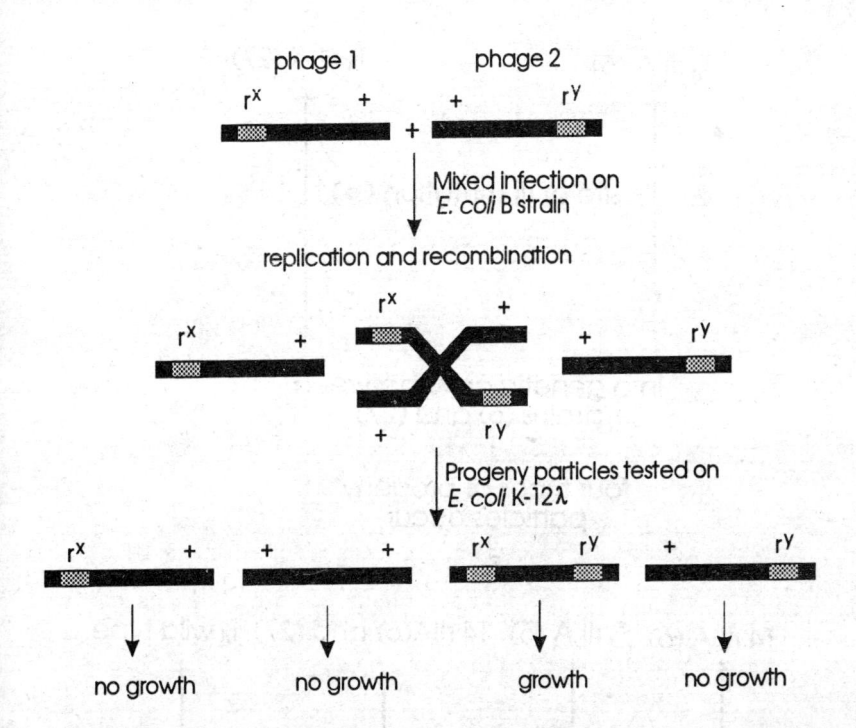

Fig. 14.4. Benzer's selection procedure for recovery of wildtype recombinants

phages, do not produce r^+ recombinants when either of them is crossed to r mutant 164. Mutant *r164* is there fore considered to be a multisite mutation that covers both the 155 and 274 regions. The fact that the presence of such multisite mutations can be shown to decrease recombination frequencies for mutations on either side of them (e.g., 288 and 196) indicates that they are probably deletions. Also, as expected of **deletions**, such mutations never show spontaneous **reversion** to wild type, whereas the point mutations occasionally do revert.

Benzer used 47 large and small overlapping deletions to locate *rII* mutations within cistrons A and B. Procedure used for locating position of unknown mutations is described in **Fig. 14.7.**

Complementation Matrix. The abundance of nutritional mutations in microorganisms and the ease with which they can be detected have made them especially suitable for complementation studies. In *Neurospora.* for example, many mutations produce strains that lack the ability to

T₄ rII A (6)   **T₄ rII A (27)**

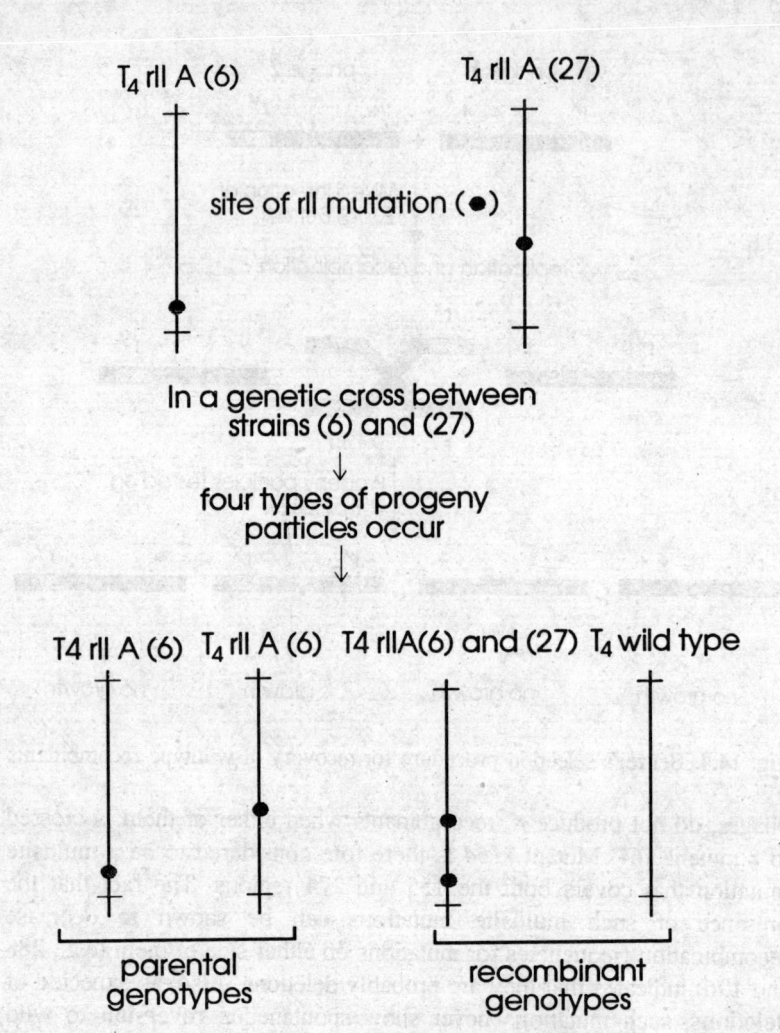

site of rII mutation (●)

In a genetic cross between
strains (6) and (27)
↓
four types of progeny
particles occur
↓

T4 rII A (6)   T₄ rII A (6)   T4 rIIA(6) and (27)   T₄ wild type

parental genotypes   recombinant genotypes

Fig. 14.5. The use of T4 rII mutations in demonstration of crossing-over within the gene

synthesize the amino acid histidine, and such strains will consequently be unable to grow on minimal medium. However, by introducing nuclei from each of two such different strains into a **heterokaryon**, complementation may be observed through the appearance of wild-type

(A) Deletion includes the site of mutant r

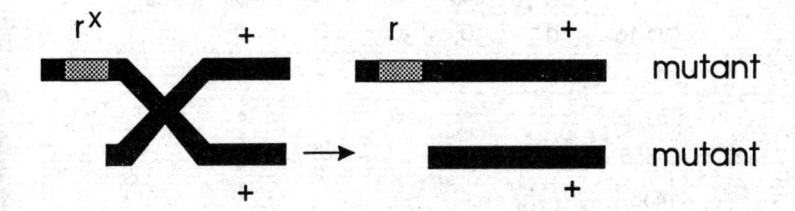

![diagram showing crossover between r^x and r chromosomes producing mutant outcomes]

(B) Deletion does not include the site of mutant r^y.

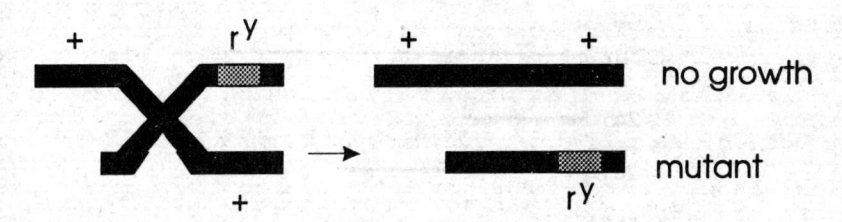

![diagram showing crossover between r^y chromosomes producing no growth and mutant outcomes]

Fig. 14.6. Principle of deletion mapping

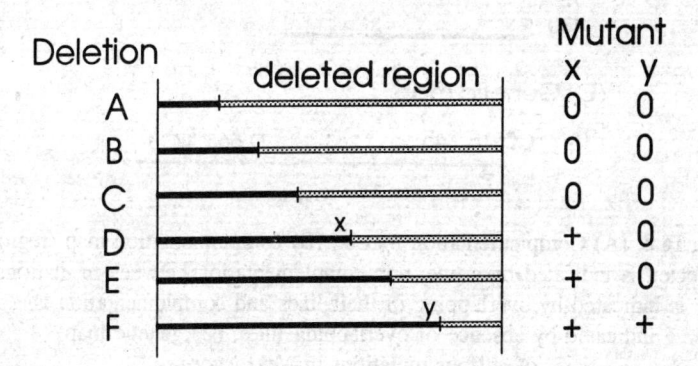

Deletion	deleted region	Mutant x	y
A		0	0
B		0	0
C		0	0
D	x	+	0
E		+	0
F	y	+	+

Fig. 14.7. Procedure for locating the position of unknown mutations r^x and r^y

growth. Let us consider only five such histidine mutations that map in the same *his-3* locus on the *Neurospora* first chromosome, *CD-16, 245, 261,*

(A) Complementation matrix

	CD-16	245	261	D-566	1438
CD-16	0	0	0	0	0
245		0	+	+	+
261			0	+	+
D-566				0	0
1438					0

(B) Complementation map

Complementation regions

```
              I        II       III
CD-16  ·─────────────────────────────·
 245   ·───────────·    :        :
 261   :      ·─────────·        :
D-566  :           :    ·────────·
1438   :           :    ·────────·
 430   ·──────────────────────────·
```

(C) Genetic map

```
CD-16 245        261      D-566 1438
──┴────┴──────────┴─────────┴─────┴──
```

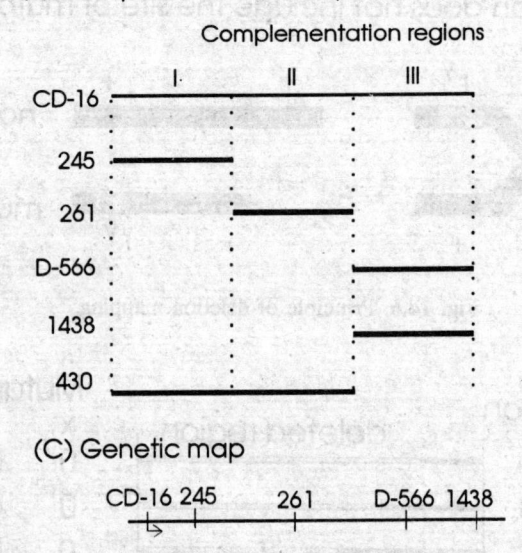

Fig. 14.8. (A) Complementation matrix, (B) complementation map (region affected is indicated by a line; non-complementation between m utations is indicated by overlapping of their lines and complementation is indicated by absence of overlapping lines, (C) genetic map of vairous mutations in *Neurospora*

D-566, and *1438*. If we draw a complementation matrix as shown in **Fig. 14.8**, we are not surprised to see that no complementation occurs in the heterokaryon between each of these mutations and itself. For *CD-16*, the absence of complementation extends also to all other strains. Mutations

245 and *261*, however, complement with all other strains except *CD-16*, while *D-566* and *1438* complement with *245* and *261* but not with each other. Using these data we can then draw a complementation map in which the functional region affected by a mutation is indicated by a line; noncomplementation between mutations is indicated by overlapping of their lines; and complementation between mutations is indicated by the absence of overlapping between their lines. Thus, as illustrated in Fig. 14.8, *CD-16* is represented as a straight line overlapping (noncomplementation) all others, while *245* and *261* are represented as small lines as they overlap only *CD-16*. *D-566* and *1438* overlap each other and *CD-16*, but do not overlap *245* and *261*. On this basis three functional regions can be distinguished within which all mutations except *CD-16* affect only one region, while *CD-16* affects all three. The order in which the mutations should be placed within these regions cannot of course be determined from these data, but as more data are accumulated the complementation map becomes clearer. For example, another mutation, *430*, overlap *245* and *261* but does not overlap *D-566* and *1438*. By similar reasoning, further complementation analysis with additional mutations indicate that the functional order is *245, 261-D, 566, 1438*. Since *CD-16* affects all three functional regions, its complementation position cannot be determined from these data alone, but other data indicate that it extends to the left of *245*.

A comparison between the complementation map and the genetic map determined by recombinational analysis is of great interest. As shown in Fig. 14.8, the genetic map for these closely linked mutations preserves the same order as the complementation map; that is, they are colinear. However, colinearity between the genetic and complementation map is not always true, and allele that maps in one position genetically may affect another functionally. The *CD-16* allele on the left of the genetic map, for example, affects distant functional regions at the extreme right.

Mapping by overlapping deletions is illustrated through a matrix given in **Fig. 14.9**.

Fine Structure of *rII* Locus. Benzer (1955, 1956) used extensively *cis-trans* **complementation test** in his study of *rII* locus in T4 phage. He observed a group of non-complementary mutants falling within a short map segment, corresponding to a unitary function. Since *cis-trans* complementation test was used by Benzer, he proposed the term **cistron**. The term cistron and gene are used synonymously. Benzer located 2400 independent mutants at over 300 ites within *rII* locus (**Fig. 14.10**). The

Complementation matrix

	1	2	3
1	0	0	+
2	0	0	0
3	+	0	0

Complementation map

```
        ┌─── 2 ───┐
┌── 1 ──┤         ├── 1 ──┐
│       │    0    │       │
│       └─────────┘       │
│            +            │
│                         │
│            0            │
```

Fig. 14.9. Complementation matrix and complementation map of three mutations

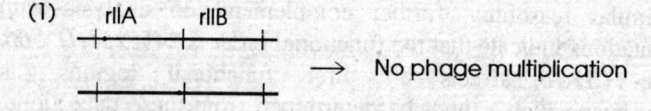

(1) rIIA   rIIB
    ├──┼──────┼──┤
                        → No phage multiplication
    ├──┼──────┼──┤

Simultaneous infection of *E. coli* K 12 (λ) with two phage particles each containing a separate rII A mutation.

(2) rIIA   rIIB
    ├──┼──────┼──┤
                        → No phage multiplication
    ├──┼──────┼──┤

Simultaneous infection of *E. coli* K12 (λ) with two phage particles each containing a separate rII B mutation.

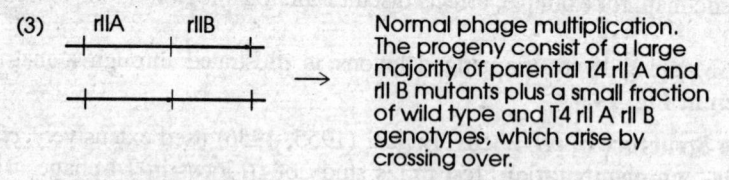

(3) rIIA   rIIB
    ├──┼──────┼──┤   Normal phage multiplication.
                     →   The progeny consist of a large
    ├──┼──────┼──┤   majority of parental T4 rII A and
                          rII B mutants plus a small fraction
                          of wild type and T4 rII A rII B
                          genotypes, which arise by
                          crossing over.

Simultaneous infection of *E. coli* K12 (λ) with two phage particles one an rII A and the other an rII B mutation.

Fig. 14.10. Demonstration that rII region consists of two distinct genes which can complement each other during simultaneous infection

existence of many mutational sites in the *rII* locus implies that this locus possesses an intricate fine structure. The unit of mutational sites show recombination. Gene is, therefore, not a unit of recombination; the unit of recombination is much is much smaller unit, termed by Benzer as **recon**.

Operationally, a cistron is defined by complementation tests. Two mutations belong to the same cistron if they do not show complementation in cis-trans test or in a simple trans complementation test. Both **muton** and **recon** are defined by **recombination** test. Two mutations belong to the same recon or muton if they do not show recombination. In light of the molecular structure of DNA (as given by Watson and Crick, 1953) gene can be considered as a part of the DNA molecule possessing a specific base-pair sequence and a mutation as a change in the base-pair sequence.

Comparing Benzer's observations with general estimates of size of phage chromosome, T_4 chromosome has a linear length of 200,000 nucleotides, 1,500 map units (MU). Thus, 1 MU = 200,000/1,500 = 133 nucleotides. Cistron rIIA = 6 MU and cistron rIIB = 4 MU. Number of nucleotides in rII region = 133 (6+4) = 1330. Within this length, recombination values range from 10 to 0.02 per cent. In other words, there are 10/0.02 = 500 recombinational units within *rII* locus. In all, Benzer estimated 400-500 mutational sites in rII region. Number of nucleotides per mutational sites = 1330/500 =3. Same was suggested to be the size of the recombinational unit. These estimates were not considered as precise. With the advances in detection and estimation of recombination frequency and mutational activity the present view is that the smallest recombinational unit probably is two neighbouring nucleotides and probably the smallest mutational unit is a single nucleotide. In light of insights into the structure of the gene, from a geneticist's point of view a gene is a discrete chromosomal region which is responsible for synthesis of a specific **cellular product** (mRNA, tRNA, rRNA), and it consists of a linear collection of potentially mutable sites each of which can exist in several alternative forms and between which crossing-over can occur.

What is a gene?

A continuous sequence in a nucleic and that specified a particular polypeptide or **functional RNA**. This condition exists in the simple RNA viruses. The functional portion of the gene obviously is the sequence of nucleotides that encodes a particular macromolecule which plays a role in the structure or metabolic processes of the cell. Coding sequence is transcribed into RNA. Because first product of gene is RNA whose sequence is complementary to the **coding strand** of DNA, it is customary to represent the gene by complementary strand sequence of **duplex DNA**. A gene in a double stranded DNA molecule is located in **antisense strand** but in literature it is usually represented by the **sense strand** because its sequence is same as that of its primary transcript except that where there is thymine in DNA there is uracil in RNA. A **transcriptional unit** is a sequence of DNA transcribed into a single RNA, starting at the promoter and ending at the terminator. **Intergenic spacers** are the genomic regions that are not directly involved in the final product and are found between actual coding sectors of two adjacent genes. These spacers vary in length from a single base pair to many thousands. These are frequently referred to as **flanking regions**. Certain parts of these sequences exhibit specialization of function. The sector of the noncoding gene proper (**mature gene**) is called **leader** while that following the 3' end is the **train**.

Leader

Leader is actually located before the 3' end of the functional region of the coding strand and the train follows the 5' end. One given strand of DNA does not necessarily bear the coding sequences of all the genes located in any single sector of the genome; sometimes some genes are located on one strand while others are located on the opposite one. Since cistrons are always read (transcribed) in 3' to 5' direction on the coding strand, those on opposite strands also differ in polarity. Thus train of one gene may be located under a mature gene of opposite polarity or even below its leader.

Promoter. To assist in locating, reading and processing the gene and its product, a number of signalling devices appear to exist. Promoters are specific sites of a given gene that serve in recognition of the gene and subsequent attachment with the **transcribing enzyme**. Promoter is part of the leader region. Promoter contains **Pribnow box** (term used for prokaryotic genes) or **Goldberg-Hogness box** (term used for eukaryotic genes). In certain genes coding for small species of RNA, e.g. tRNAs, 5S rRNA, promoter sequences are located within the mature gene itself.

The sequence of DNA needed for **initiation reaction** for transcription defines the promoter. Transcription enzyme, **RNA**

polymerase binds at this site. The site and which the first nucleotide is added during transcription is called the **startise** or **startpoint**.

RNA polymerase finds target promoter rapidly on DNA by random diffusion. Promoters belong to two general classes. Some promoters can be recognized by RNA polymerase **holoenzyme** alone. Other promoters are not by themselves adequate to support transcription. **Ancillary protein factors** are needed for initiation to occur. These proteins recognize the sequence on DNA close to, or overlap with, the sequence bound by RNA polymerase itself.

Promoter is recognized as a particular sequences to DNA with which RNA polymerase binds. Minimum size of promoter is 12 bp. **Consensus sequences** is defined by aligning all known examples so as to maximize their homology. Some generalizations in this regard are known in *E. coli*. The start point in greater than 90 per cent cases a purine. Just upstream of the start point, a 6 bp region is recognized in almost all the promoters. The centre of the hexamer generally is close to 10 bp upstream of the start point. This hexamer is, therefore, often called the -10 sequence. This -10 sequence is TATAAT conservation of these bases varies from 45 to 96 per cent. Capital letters T_{80} A_{95} t_{45} A_{60} a_{50} T_{96} are used to indicate bases conserved 75.4 per cent. At -35 sequence of promoter, consensus sequence is T_{82} T_{84} G_{78} A_{65} C_{54} A_{45}. Distance between -10 sequence and -35 sequence in 90 per cent of cases is between 16 and 18 bp. Actually sequence of these 16-18 bp may not be important, the distance between two sites may be important in appropriate separation for the geometry of RNA polymerase.

Thus an **optimal promoter** is defined as a sequence consisting of the -35 hexamer, separated by 17 bp from the -10 hexamer, lying 7 bp **upstream** of the start point. Typical promoter can use the -35 and -10 sequences to be recognized by RNA polymerase.

Down mutations in the promoter decrease the transcription rate whereas up mutations in the promoter increase the transcription rate compared to transcription. Study of down and up mutations of the promoter suggest that the function of the **-35 sequence** is to provide the signal for recognition by RNA polymerase while the **-10 sequence** allows the complex to convert from closed to open form.

RNA polymerase may initially "touch down" at the -35 sequence and then extend its contact over the -10 region. The enzyme covers $\sim$ 60 bp of DNA. The initial "closed" binary complex is converted by melting of a sequence of $\sim$ 12 bp that extends from the -10 region to the start point.

The A-T rich base pair composition of the -10 sequence may be important for melting reaction.

Ancillary Sites. These sequences are located between -30 and -40 position. Also referred to as -35 sequence. In bacterial genes, it is known as **CAP site (cAMP-activated protein)**. The term ancillary site applies to both prokaryotic and eukaryotic genes.

Enhancers. First enhancer found was 72 bp tandemly repeated sequence located 100 nucleotides upstream of the ancillary site in DNA of SV40. A common feature of the series is GGTGTGGAAAG. Enhancers have been detected in eukaryotic genes also. Enhancers mostly are *cis*-acting but *trans*-acting ones are also known. They are effective whether lying upstream or downstream from the promoter. They are active whether they lie in same or opposite polarity as the mature gene. They are equally effective regardless of the organism from which the gene is derived when attached to foreign DNA.

Train

Train contains signals concerned with termination of transcription. Typically there exists a series of T-A bp, in prokaryotes usually six or more in number while in metazoans four times seems to suffice. A **stem-and-loop structure** could conceivably form in the transcript.

Terminators. RNA polymerase continues transcription until it meets a **terminator sequence**. At this point, the enzyme stops adding nucleotides to the growing RNA chain, releases the transcript and dissociates from the DNA template. Identification of terminators is provided by systems in which RNA polymerase terminates *in vitro*. Many prokaryotic and eukaryotic terminators require a **hairpin** to form in the secondary structure of the RNA being transcribed. This indicates that termination depends on the RNA product and is not determined simply by scrutiny of the DNA sequence during transcription. At some terminators, termination event can be prevented by specific **ancillary sequences** (factors) that interact with RNA polymerase.

Antitermination causes the enzyme to continue transcription past the terminator sequence, an event called **read through**. Antitermination is used as a control mechanism in both bacterial operons and phage regulatory circuits. Antitermination is used as a control mechanism in operon controlled by **attenuation** (to be discussed later) to provide a link between translation and transcription. In this case termination of transcription when the ribosome is unable to move along a leader segment of the mRNA. During phage infection different **ancillary proteins (antitermination factors)** allow RNA polymerase to bypass specific terminator sequences. Termination (as well as initiation) of transcription requires: breaking of hydrogen bonds and additional proteins to interact with core enzyme.

Responsibility for termination lies with the sequences already transcribed by RNA polymerase. With regard to requirement of **rho (ρ) factors** in termination, terminators are of two types - **rho-independent and rho-dependent terminators**. Rho-independent or simple terminator core enzyme can terminate *in vitro* at certain sites in absence of any other factor. These terminators have two structural features, a hairpin in secondary structure and a run of $\sim$ 6 U's at the very end of the unit. Both features are needed for termination. The hairpins contain a G-C rich region near base of the stem. Probably all hairpins that form in the RNA product cause the polymerase to slow or pause in RNA synthesis. Pausing creates an opportunity for termination to occur. Rho-independent terminators include palindromic regions that form hairpins varying in length from 7-20 bp. The stem-loop structure is followed by a run of U residues. The string of U residues probably provides the signal that allows RNA polymerase are to dissociate from the template when it pauses at the hairpin.

In case of rho-dependent terminators, there is a need for addition of rho (ρ) factor for termination *in vitro*. Rho factor is required for termination *in vivo* also. Most of the known rho-dependent terminators are found in phage genomes.

Attenuation. Attenuation is a mechanism that links the supply of an aminoacyl-tRNA to the ability of RNA polymerase of read through a termination site. The terminator is located at the beginning of the cluster of structural genes coding for the enzymes that synthesize amino acid carried by tRNA. In this way synthesis of amino acid responds to the level of aminoacyl-tRNA; if aminoacyl-tRNA is available, synthesis is inhibited, but if aminoacyl tRNA runs out, more amino acid is synthesized.

Attenuation was discovered in the tryptophan (*trp*) operon whose five structural genes are arranged in a contiguous series. Transcription starts at a promoter at the left and of the cluster. Transcription of the structural genes is partially terminated at a rho-independent site, *trpt*, 36 bp beyond the end of the last coding region. About 250 bp later there is a rho-dependent terminator *trpt'*.

In addition to the promoter-operator complex, a sequence between the operator and the *trp* E coding region (denoted as L) which is 162 nucleotides long leader that precedes the initiation codon for the *trp* E gene. This regulator site is known as attenuator which serves as a barrier to transcription. The termination event at this site responds to the level of transcription. RNA polymerase terminates at attenuator and leader RNA and polymerase is released in presence of tryptophan whereas in absence of tryptophan transcription continues past attenuator into structural genes.

Internal control signals. In certain genes coding for small species of RNA - tRNAs, 5S rRNA, etc. - promoter sequences are located within the mature gene itself.

A simple gene may be as long as 100,000 bp, including leader and train regions.

Split Gene

In split genes, coding sequences (exons) are separated by non-coding sequences (introns). Based on the number of introns in a split gene, the genes are of two types. First are the **monintron genes** as they have only one intron. Such gene have been reported in case of $tRNA^{tyr}$ gene in *Xenopus* and $tRNA^{leu}$ gene in *Drosophila*. The transcripts of such genes are not used for translation and they as such perform a biological function. The importance of this intron is indicated by the fact that removal of this intron leads to lack of processing and loss of function of the transcript. The second are the **multintron genes** as they have several introns. Transcripts of multintron genes are translated into proteins, e.g., rat muscle α-acting gene has six exons. Size of introns is usually much larger than the exons. Introns are spliced out during processing of the pre-mRNA. Collagen gene has 52 introns.

Intervening Sequences. Let us consider the following questions: Why the intervening sequences(IVS) that are not expressed in proteins but are nevertheless quite long are maintained in the eukaryotic genome? What is an IVS sequence for one mRNA may be included at another time in another mRNA? All IVS may not exist for the same reason. For instance, those at the 5' end in particular are involved in regulating transcription or translation. But certainly there are IVS in certain genes that may never be used as parts of an mRNA. Perhaps many eukaryotic cells simply do not have facility for deleting "excess" sequences so that non-functional "intervening" sequences would be maintained for a longer time in eukaryotes. A genome that build by recruitment of ligatable pieces might have been successful only if its system for deleting unnecessary DNA was inefficient.

Sharing of Exons in Gene Evolution. One likely mechanism for sharing of exons coding for different proteins is through the duplication and migration of exons, during evolution.

Complex Gene

Three types of genes - immunoglobulin genes, dimorphic genes and cryptomorphic genes are classified as complex genes. They are so called because either there is rearrangement of DNA at different levels before $3' \rightarrow 5'$ strand of DNA is available for transcription (as is case of immunoglobulin gene) or pre-mRNA is cleaved through a series of steps before mRNA in finished form is available (as is the case of dimorphic genes, as exemplified by the kallikrein gene) or these genes have a cryptic structure in that the ultimate active product or products are carried within the precursorial protein (as is the case of cryptomorphic genes, exemplified by yeast sex pheromone gene ($MF\alpha 1$), mammalian pancreatic glucagon genes) which are released after enzymatic breakdown of the precursor and become functional following further processing.

The gene is a sequence in a nucleic acid (DNA, except in RNA viruses) that provides a code for a protein, polypeptide, or RNA of direct value to cellular metabolic processes, plus those parts, adjacent or internal, important to its being located, identified transcribed, often translated, and processed. Internal transcribed regions, such as introns, are also a part of the gene.

Colinearity

A prokaryotic gene and its protein product has a point-for-point correspondence. In other words, a prokaryotic gene and its protein product were colinear. Sequence of nucleotides in a prokaryotic gene determines sequence of amino acids in a protein/polypeptide. This conclusion was drawn by a study carried out by Yanofsky *et al.* (1987) on many mutants of tryptophan synthetase A polypeptide. Different mutants were crossed together and a genetic map was obtained, showing the order of the mutated sites along the DNA of the gene. Tryptophan synthetase A protein was purified from each mutant strain, and the

252 Basic Genetics

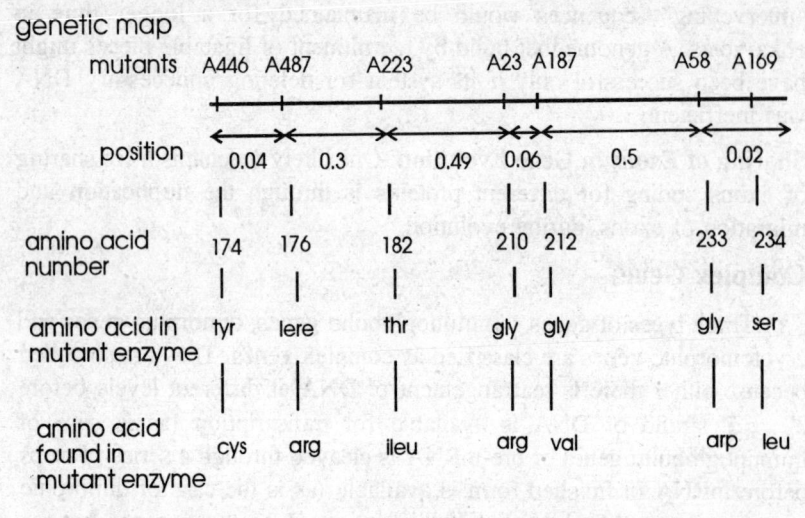

genetic map mutants	A446	A487	A223	A23	A187	A58	A169
position	0.04	0.3	0.49	0.06	0.5	0.02	
amino acid number	174	176	182	210	212	233	234
amino acid in mutant enzyme	tyr	lere	thr	gly	gly	gly	ser
amino acid found in mutant enzyme	cys	arg	ileu	arg	val	arp	leu

Fig. 14.11. Colinearity shown by tryptophan synthetase gene in *E. coli* and its product

sequence of amino acids compared to that of wildtype (**Fig. 14.11**). The order of the changes in the amino acid sequence correspond exactly to the order of the genetic changes in DNA.

Overlapping Genes

In strictest senese overlapping genes are defined as a single nucleotide sequence coding for more than one polypeptide. Such an arrangement of genes was first discovered by Barrel *et al.* (1976) in phage φX174. Later these genes in the genomes of bacteriophages, animal viruses, bacteria, and mitochondria (Normark *et al.*, 1983).

Some Useful Reading Hints

Morgan (1926) gave a theory of the Gene. Green and Green (1949) reported crossing over between alleles at the *lozenge* locus in *Drosophila melanogaster*. Lewis (1951) discussed role of pseudoallelism and gene evolution Green (1961) reported **phenogenetics** of the lozenge locus in *Drosophila melanogaster*. Chovnick (1961) reported pseudoallelism while working at the garnet locus in *Drosophila melanogaster*. Benzer (1961) and Benzer (1962) worked fine structure of the gene through *cis-trans* complementation test. Dillon (1987) reviewed structural, functional and evolutionary aspects of gene. Fincham (1966) dealt with the concept and use of genetic complementation.

Crick (1979) discussed split genes and RNA splicing. Chambon (1981) gave insight into the split genes. Yanofsky *et al.* (1967) determined complete amino-acid sequence of the tryptophan synthetase A protin (α subunit) and its colinear relationship with the genetic map of the A gene. Yanofsky (1984) conducted comparison of regulatory and structural regions of genes of tryptophan metabolism. Lawn and Vehar (1986) discussed the molecular genetics of hemophilia.

Barrell *et al.* (1976) reported overlapping genes in bacteriophage ϕX174 and Normark *et al.* (1983) discussed various aspects of the overlapping genes.

Questions

1. What was concept of gene structure during classical phase?
2. What is the significance of transitional phase in changing the scenario of gene structure?
3. What are pseudoalleles? What is the difference between structural and functional alleles? In what way has the concept of pseudoallelism helped in understanding gene structure?
4. What event marks the beginning of modern phase in the concept of gene structure?
5. What is complementation mapping?
6. Describe *cis-trans* complementation test.
7. How has complementation matrix been used to understand the structure of *rII* locus of T4 bacteriophage?
8. Describe deletion mapping method used by Benzer to understand fine structure of *rII* locus.
9. Classify genes on the basis of their structural complexity.
10. What are the simple genes? What flanking regions are present on both sides of a mature gene?
11. What is leader and where is it located with respect to the gene? Explain structure of leader with special reference to important features and functions of the promoter, ancillary site, and enhncers.
12. What is train and where is this structure located with respect to the gene? Explain structural and functional attributes of terminators.
13. What is attenuation? Describe structural properties and functions of attenuators. Give the answer with reference to attenuator of the *trp* gene.
14. Differentiate between rho-independent and rho-dependent attenuators.
15. Describe structural features of rho-independent attenuator with the help of a suitable diagram. What is role of run of U residues in attenuation?
16. How does rho-dependent trp attenuator perform its function in presence or absence of tryptophan?
17. Describe diagrammatically structural features of bacterial *trp* operon. Where are terminators of this operon located?
18. Show diagrammatically complementary sequences present in bacterial trp rho-dependent attenuator. What alternate base pair conformations of this attenuator are formed?
19. What are internal control signals and how they differ from external control signals in structure and function?

254 Basic Genetics

20. What are compound genes? Why are they also known as split genes?
21. What are unintron genes? Are their transcripts used for translation. If not, then what the function of such transcripts?
22. What are multintron genes? Are the transcripts produced by these genes used for translation?
23. What are intervening sequences. What do you know about their evolution and functions?
24. Duplication and migration of exons has been considered to be responsible for evolution of different genes. Illustrate this point with the help of the following examples: (a) human lipoprotein receptors, (b) immunoglobin heavy chain, (c) haemoglobin, (d) glyceraldehyde-3-phosphate dehydrogenase in chicken, and (e) globin genes
25. What are complex genes. How have these genes been classified?
26. What is meant by the term colinearity between genes and its product?
27. Yanofsky et al. (1967) demonstrated colinearity between gene *trpA* and its polypeptide product tryptophan synthetase A of *E. coli*. Describe the method and results which helped in understanding this phenomenon.
28. On the basis of all the discussion in this chapter, answer the following simple question: What is a gene?
29. What are overlapping genes and how were they detected?

Numericals

1. In the synthesis of the amino acid histidine in *Salmonella* bacteria, a number of different mutations (A to D) affect one step in this process by interfering with the production of a particular enzyme. These mutant stocks may be "mated" together (through transduction) to produce a wild-type recombinant; as follows (+ means production of wild-type recombinant; 0 means lack of appearance of wild type):

Mated to Mutant Stock

Mutant Stock	A	B	C	D
A	0	+	0	+
B		0	0	0
C			0	+
D				0

Considering that some or all of these mutation may be deletions, draw a possible topological map of this area. If a point mutation produced wild-type recombinants with all the above mutations except C, at which position on this map would it most likely be located?

2. Using complementation tests, Catcheside has classified numerous *Neurospora* mutations that affect the synthesis of tryptophan (*trp*-1 locus) into nine groups, A to I, whose complementation patterns are as follows:

Gene Structure   255

	A	B	C	D	E	F	G	H	I
A		0	0	0	0	0	0	0	0
B	0		+	+	+	+	+	+	+
C	0	+		0	0	0	+	+	+
D	0	+	0		0	0	0	+	+
E	0	+	0	0		0	0	0	+
F	0	+	0	0	0		0	0	0
G	0	+	+	0	0	0		0	0
H	0	+	+	+	0	0	0		0
I	0	+	+	+	+			0	

Draw a complementation map of this area.

3. Seven arginine-requiring mutants of *E. coli* were independent by isolated. All pairwise matings were done (by transduction) to determine the number of loci (complementation groups) involved. If a (+) indicates growth and a (-) no growth on minimal medium, how many complementation groups are involved here? Why is only "half" a table given? Must the upper left to lower right diagonal be all (-)?

	1	2	3	4	5	6	7
1	−	+	+	+	+	−	−
2		−	+	+	−	+	+
3			−	−	+	+	+
4				−	+	+	+
5					−	+	+
6						−	−
7							−

4. Several *rII* mutations (M to S) have been localized to the A cistron because of their failure to complement with a known deletion of the A cistron. The phages carrying these mutations are then mated pairwise with the following series of subregion

Mutant	Deletion			
	1	2	3	4
M	+	+	−	+
N	+	−	−	+
O	+	+	+	−
P	+	+	−	−
Q	+	−	+	+
R	−	−	+	+
S	−	−	−	+

deletions. The mating is done on *E. coli* B and plated out on *E. coli* K12. A (+) shows the presence of plaques on K12, whereas a (-) shows an absence of growth. Provide a deletion map of the area and localize each of the rII A mutations on this map.

5. Eight independent mutants of *E. coli*, requiring tryptophan (*trp*), are isolated. Complementation tests are performed in all pairwise combinations. Based on the following results, determine how many genes you have identified and which mutants are in which genes (+ = complementation, - = no complementation).

	1	2	3	4	5	6	7	8
1	-	+	-	-	+	+	+	-
2		-	+	+	-	+	+	+
3			-	-	+	-	-	-
4				-	+	+	+	-
5					-	+	+	+
6						-	-	+
7							-	+
8								-

6. A series of overlapping deletions in phage T4 are isolated. All pairwise crosses are performed, and the progeny scored for wild-type recombinants. In the following table, + = wild-type progeny recovered; - = no wild-type progeny recovered.

	1	2	3	4	5
1	-	+	-	-	-
2		-	+	+	-
3			-	+	+
4				-	-
5					-

15

GENE FUNCTION

Relationship Between Gene and Enzyme

First clue to the nature of the primary gene function came from the studies on humans. Garrod, a physician, in 1909, noted that several **hereditary human defects** are produced by recessive mutations. Various human disorders of **tyrosine metabolism**, studied by Garrod are given in **Fig. 15.1**. In case of **phenylketonuria**, large amounts of phenylalanine is accumulated and the affected person has lighter pigment but is not a complete albino since tyrosine is available in diet. In case of **alkaptonuria**, large amounts of homogentisic acid are released in urine. Urine turns black in presence of light. Various other metabolic disorders were attributed to an absence or a defect in an enzyme. Thus, a relationship between genes and enzymes (which are responsible for conducting various biochemical reactions) was suggested.

Genetic Control of Biochemical Reactions

Lawrence and co-workers, began the biochemical analysis of flower pigments and eventually showed a precise correlation between genetic and biochemical changes. For example, in the **anthocyanin pigments** of the cape primrose (*Streptocarpus*), three individual biochemical effects can be traced to separate genes. The three genes, called *R, O, D*, appear to produce their effects by adding or subtracting hydroxyl (OH) units, methoxyl (O-CH$_3$) units, or sugar molecules. The *rroodd* triple recessive is salmon-coloured, containing mainly the pigment whose structure is shown in **Fig. 15.2(A)**. Presence of the dominant allele R (R-oodd) results in hydroxylation or methoxylation at the 3' position and causes a change to rose color (**Fig. 15.2(B)**). The addition of O causes hydroxylation or methoxylation at both the 3' and 5' positions, and the color changes the

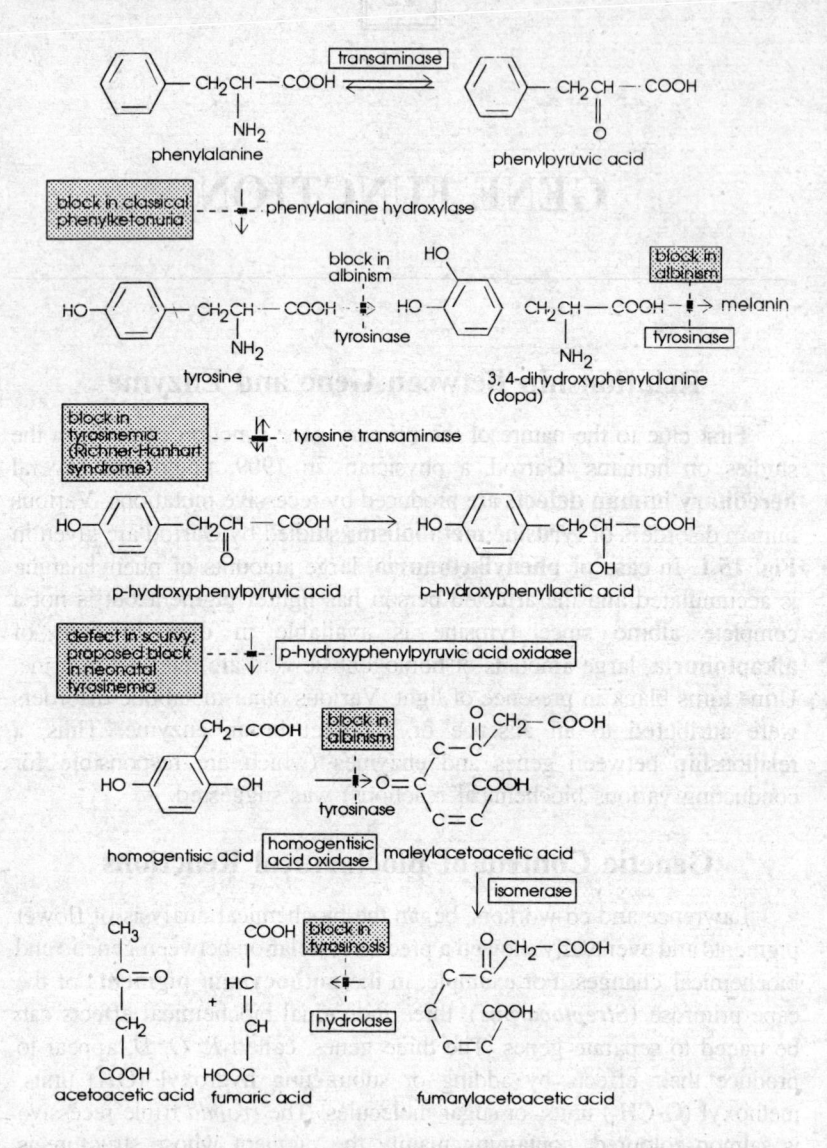

Fig. 15.1. Some biochemical reactions in humans beings with metabolism of phenylalanine. Various metabollic blocks are shown in filled boxes. Defective enzymes responsible for a metallic disorder are mentioned in empty boxes

(a) rroodd (b) R-oodd (c) R-O-dd

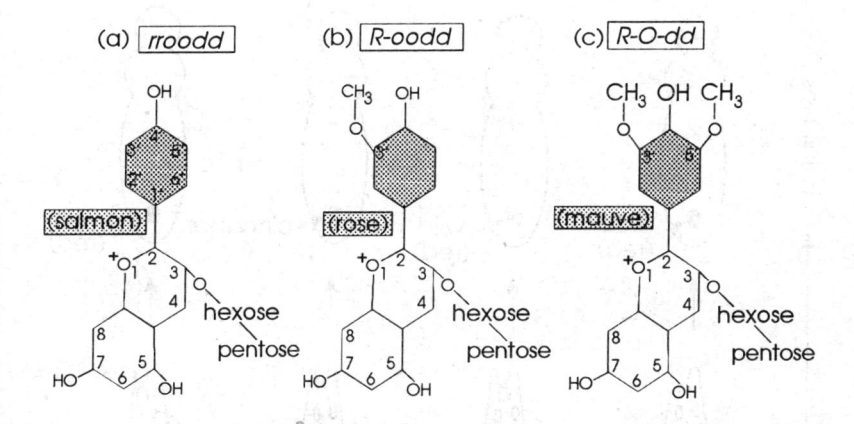

Fig. 15.2. Anthocyanin pigments produced by some genotypes among garden forms of *Streptocarpus*

mauve (**Fig. 15.2(C)**). In plants dominant for gene D, heose sugars are found at the 3' and 5' positions instead of the hexose-pentose combination at the 3' position, and the anthocyanin appears bluer.

Investigations of numerous other plants have shown similar findings; the **anthocyanin pigments** are usually bluer with the addition of sugar and hydroxyl groups, and somewhat redder when the hydroxyl groups are methylated. Acidity of the cell, another hereditary factor in some plants, is also of considerable importance, since dyes, such as cyanin, change from blue to red with increased acidity. Pigment inheritance, therefore, helped to show the effect of genes in controlling specific biochemical steps.

Relationship Between Genotype and Phenotype

Beadle and Ephrussi (1937) conducted **eye disc transplantation experiments** in *Drosophila*. The experimental approach used by them is illustrated in **Fig. 15.3**. The results of the experiment are explained in **Table 15.1**.

Experiments 1-5 show nonautonomous development. In this case v, or cn transplanted discs developed into normal eyes in wildtype host. Wild type host in this case provided a diffusible substances that **the discs used to bypass the genetic block and to make brown pigment**. Experiment 6 is a case of **autonomous development**. Here wildtype host was unable

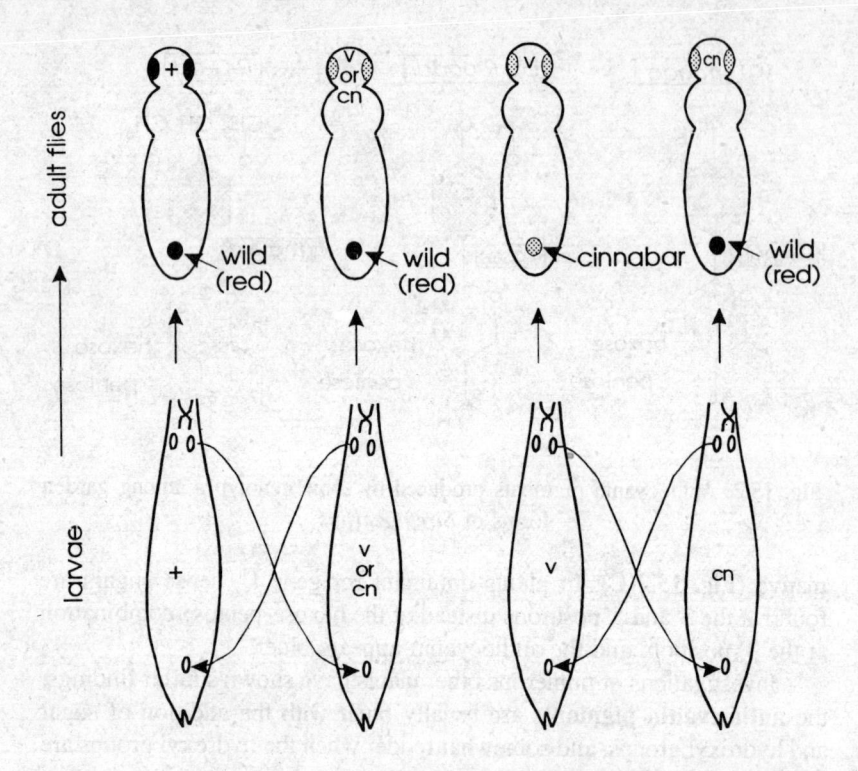

Fig. 15.3. Eye transplantation experiment in *Drosophila* by Beadle and Ephrussi showing autonomous and nonautonomous development of implanted larval imaginal discs into various phenotypes

to provide some substance to the disc to enable the brown pigment to be produced. Experiments 7 and 8 showed that production of brown pigment involved a biochemical sequence with at least two precursors, v^+ and cn^+. wildtype having both v^+ and cn^+ substances, *cinnabar* having one (v^+), and v having neither. The *vermillion* host is deficient in v^+ substance so no brown pigment can be produced. *Cinnabar* host makes up for the deficiency of *vermillion* disc by supplying it with diffusible substance. Sequence of metabolic events suggested by the experiments of Beadle and Ephrussi (1937) are given in **Fig. 15.4**. Two main types of pigments are known to be present in *Drosophila* eyes, the **pterins** (red pigments) and **ommochromes** (brown pigments), each effected by different set of genes. Both the pigments function by becoming attached to protein granules in

Table 15.1. Results of eye disc transplantation experiment

Experiment No.	Source of eye disc	Host fly	Colour of transplanted eye in adult
1	+	v	+
2	v	+	+
3	+	cn	+
4	cn	+	+
5	+	st	+
6	st	+	st
7	cn	v	cn
8	v	cn	+

(v = vermillion; cn = cinnabar; st = scarlet; + = wildtype)

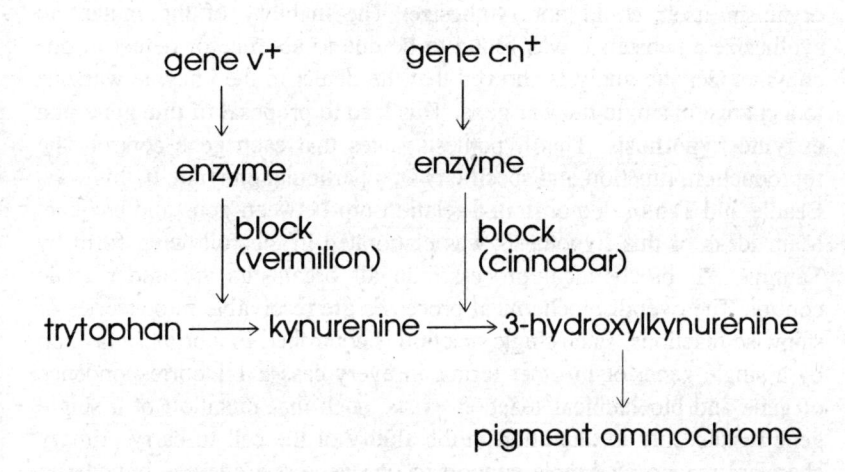

Fig. 15.4. Some biochemical reactions for biosynthesis of eye pigments in *Drosophila*

which form they are deposited in **ommatidia**. A fly deprived of both types of pigments will have *white* eyes. A fly not capable of forming ommochromes will show only bright pterine pigments, e.g., *vermillion, scarlet,* or *cinnabar*. A fly in which pterins are absent will have dull eye colour, such as brown, raspberry, or garnet. Flies deprived of both the ommochromes and the pterins, such as double mutants *cinnabar brown* and *scarlet brown* will have no eye colour and hence will appear white. This work indicated a strong link between phenotype (eye colour) and genotype (*v* and *cn*). Beadle found *Drosophila* too complex an organism to address this question of gene function further. He chose for his further work a simple, unicellular eukaryote, *Neurospora*.

One Gene - one Enzyme

Nutritional Mutants of *Neurospora*

Work of Beadle and Tatum (1941) provided further clarification on the function of gene. For their work they were awarded a Nobel Prize in Medicine and Physiology in 1958. They induced nutritional mutants with X-rays in bread mold *Neurospora crassa* The procedure used in their experiments is shown in **Fig. 15.5**). A nutritional mutant would grow only when **minimal medium** was supplemented with a substance that the organism itself could not synthesize. The inability of the mutant to synthesize a substance was found to be due to absence or defect in one enzyme. Genetic analysis showed that the defect in the enzyme was due to a change in single nuclear gene. This lead to proposal of **one gene-one enzyme hypothesis**. This hypothesis states that each gene controls the reproduction, function and specificity of a particular enzyme. In this way, Beadle and Tatum demonstrated relationship between gene and enzyme. Main ideas of this hypothesis was elaborated in the following form by Tatum: "All biochemical processes in all organisms are under genic control. The overall biochemical processes are resolvable into a series of stepwise reactions. Each single reaction is controlled in a primary fashion by a single gene, or in other terms, in every case a 1:1 correspondence of gene and biochemical reaction exists, such that mutation of a single gene results only in alteration in the ability of the cell to carry primary chemical reaction." Ample support to one gene-one enzyme hypothesis came from studies on microorganisms.

Histidine Metabolism in *Salmonella*

More than 500 mutations for histidine metabolism in *Salmonella* were mapped. These were linked in nine closely linked loci, *A* to *H*, each affecting a different enzyme in the synthesis of histidine. The gene sequence was found to follow largely the metabolic sequence shown in **Fig. 15.6**.

Gene Function 263

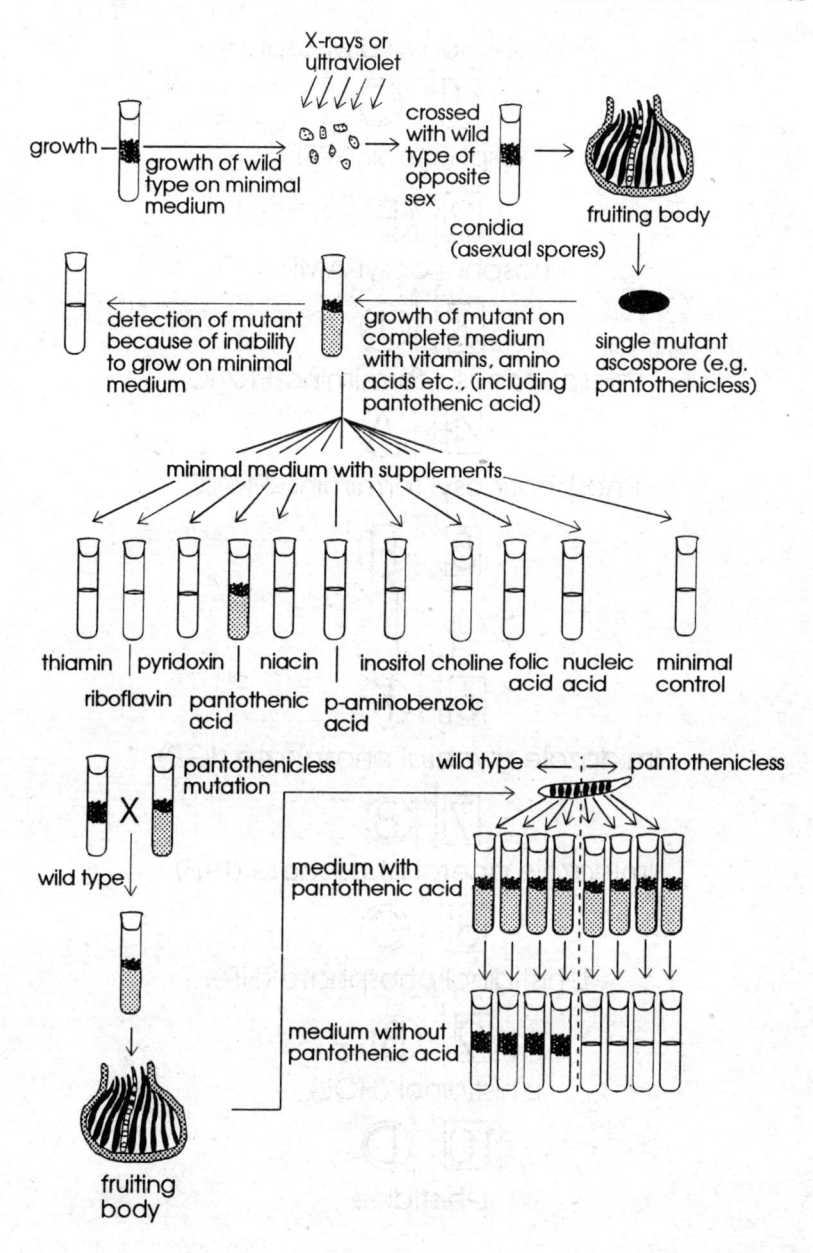

Fig. 15.5. Procedure for induction and detection of nutritional mutations in *Neurospora crassa*

264 Basic Genetics

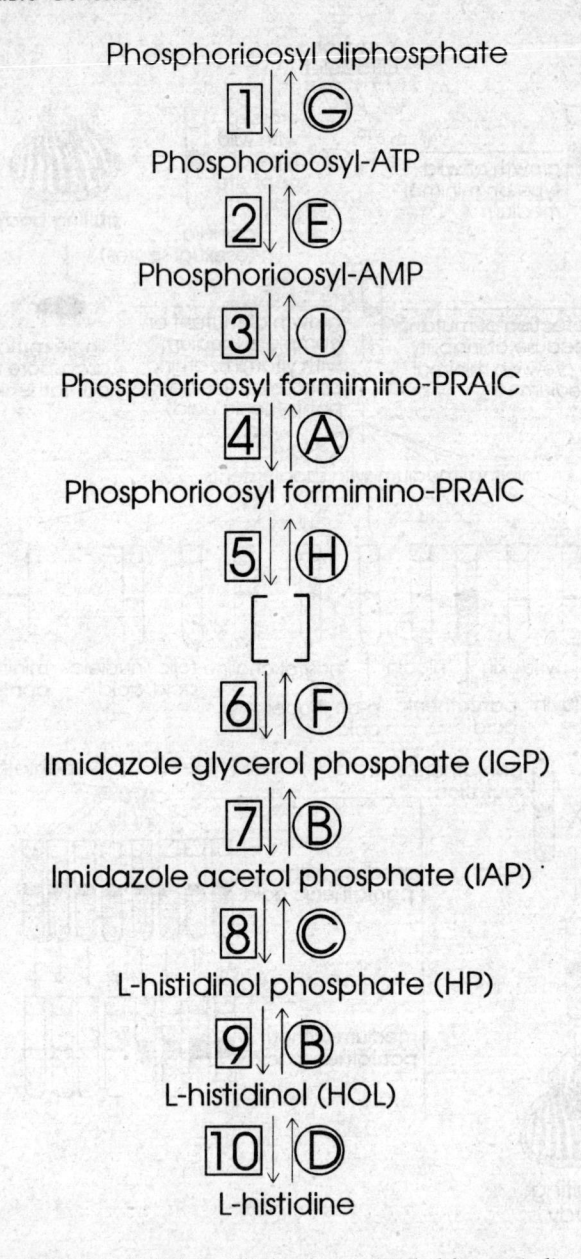

Phosphorioosyl diphosphate
[1]↓ ↑Ⓖ
Phosphorioosyl-ATP
[2]↓ ↑Ⓔ
Phosphorioosyl-AMP
[3]↓ ↑Ⓘ
Phosphorioosyl formimino-PRAIC
[4]↓ ↑Ⓐ
Phosphorioosyl formimino-PRAIC
[5]↓ ↑Ⓗ
[ ]
[6]↓ ↑Ⓕ
Imidazole glycerol phosphate (IGP)
[7]↓ ↑Ⓑ
Imidazole acetol phosphate (IAP)
[8]↓ ↑Ⓒ
L-histidinol phosphate (HP)
[9]↓ ↑Ⓑ
L-histidinol (HOL)
[10]↓ ↑Ⓓ
L-histidine

Fig. 15.6. Pathway of histidine biosynthesis. Circled letters indicate genes and numbers in rectangles symbolize enzymes

Arginine Synthesis in *Neurospora*

In *Neurospora*, following three kinds of mutations for arginine synthesis were isolated. (1) Mutations, which grow only when arginine is supplied and do not grow with the help of either ornithine or citrulline alone. (2) Mutations, which grow only when either citrulline or arginine is supplied, but do not grow when only ornithine is supplied. (3) Mutations which can grow only, when either ornithine or citrulline or arginine is supplied. These mutations indicate that, in the first case: (i) mutants are incapable of utilizing ornithine or citrulline; in the second case, (ii) mutant are incapable of utilizing ornithine; but in the last case, (iii) only the precursor cannot be utilized. The biosynthetic pathway, suggested on the basis of these mutation studies, in shown is **Fig 15.7**.

```
    mutation      mutation      mutation
      (c)           (b)           (a)
       |             |             |
       ↓             ↓             ↓
precursor ⟶ ornithine ⟶ citrulline ⟶ arginine
```

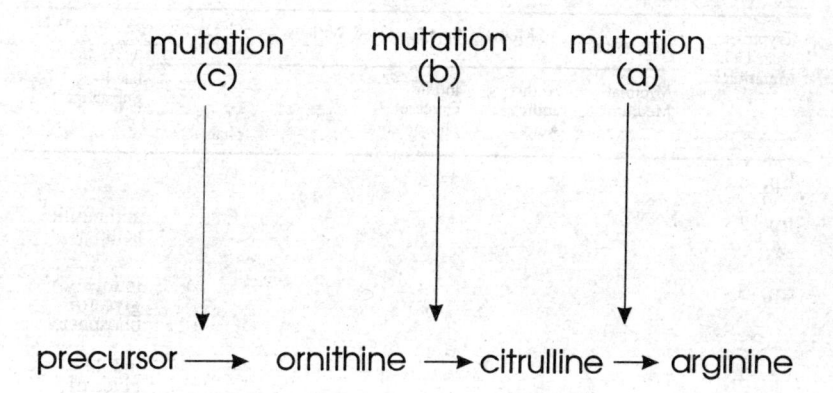

Fig. 15.7. Different steps in synthesis of arginine and three different mutations blocking three steps

One Gene-one Polypeptide

One gene-one enzyme hypothesis could not be universally applied. This was shown by studies on haemoglobin, lactate dehydrogenase, and tryptophan synthetase.

Haemoglobin

Smith and Torbert (1958) and Itano and Robinson (1960) studied abnormal haemoglobins and discovered that haemoglobin was made up of two different polypeptides, each controlled by a separate gene. It suggested that a protein may be made up of more than one polypeptide chain, each controlled by a different locus. Thus one gene-one enzyme relationship does not hold good for all the structural (protein-ecoding) genes.

Lactate Dehydrogenase

Lactate dehydrogenase (LDH) in horse is an tetrameric enzyme which comprises of two polypeptides A and B. These polypeptides can combine in five possible ways - A4, A3B1, A2B2, A1B3, B4 thus revealing maximum number of five electrophoretically distinguishable isozymes.

Table 15.2. Effect of different media on the growth response[*] of tryptophan mutations in *Salmonella typhimurium*

Trypto-phan Gene Mutation	Medium Supplemented With					Accumu-lated Substance
	Minimal Medium	Anthr-anilic Acid	Indole Glycerol Phosphate	Indole	Trypto-phan	
trp, -8	-	+	+	+	+	
trp, -2, -4	-	-	+	+	+	anthranilic acid
trp, -3	-	-	-	+	+	indole glycerol phosphate
trp, -1, -6, -7. -9, -10, -11	-	-	-	-	+	indole glycerol phosphate indole

([*]+ = growth; - = no growth); Data of Brenner

Tryptophan Synthetase

Tryptophan synthetase in, *Neurospora, Salmonella typhimurium* and *E. coli* (**Table 15.2**; **Fig. 15.8**) is made up of two polypeptides A and B, each synthesized by a gene. All these examples challanged one gene-oneenzyme hypothesis and lead to **one gene-one polypeptide** relationship. One gene-one polypeptide hypothesis states that one gene controls the synthesis of one polypeptide.

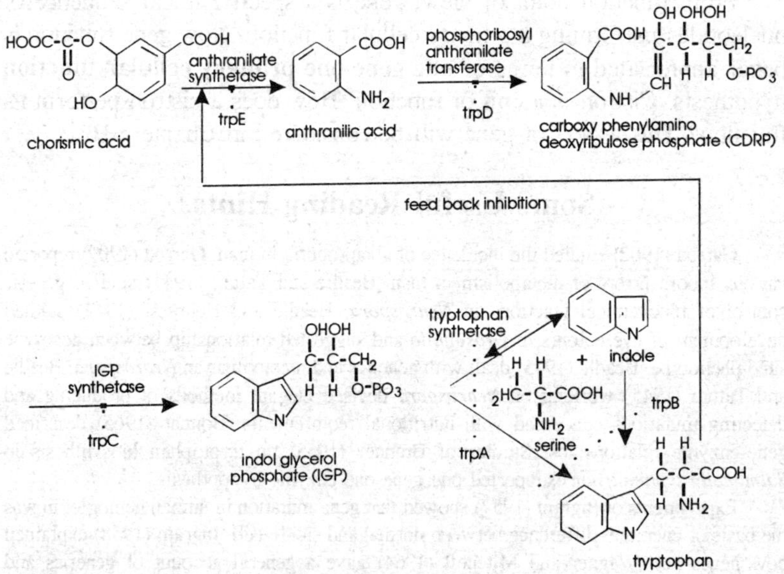

Fig. 15.8. Biosynthetic pathway of tryptophan from chrismic acid showing the enzymes and genes at each step

One Cistron - one Polypeptide

Since cistron is recognized as a unit of function it has been suggested that one gene-one polypeptide hypothesis should be more appropriately known as **one cistron-one polypeptide** hypothesis.

One Gene - one Primary Cellular Function

Above mentioned hypotheses put forward to explain gene function were based on the studies on structural genes. In addition to the **structural genes**, which encode for polypeptides, there are other genes (tRNA genes and rRNA genes) which produce RNAs required in protein biosynthesis. The three types of RNAs (mRNAs, tRNAs and rRNAs) are considered **primary cellular products** of genes. Thus a gene performs one primary cellular function. This gave birth to one gene-one primary cellular function hypothesis.

Gene as Functional Entity

From function point of view, gene is a specific linear sequence of nucleotides performing a primary cellular function. Now gene function is better appreciated in terms of **one gene-one primary cellular function** hypothesis. Cistron is a unit of function. How does a cistron perform its function? This aspect of gene will be considered in Chapters 16.

Some Useful Reading Hints

Garrod (1902) studied the incidence of alkaptonuria in man. Garrod (1909) reported various inborn errors of metabolism in man. Beadle and Tatum (1941) studied genetic control of biochemical reactions in *Neurospora*. Beadle and Ephrussi (1943) studied development of eye colours in *Drosophila* and suggested relationship between genotype aand phenotype. Beadle (1945) dealt with genetics and metabolism in *Neurospora*. Beadle and Tatum (1945) working on *Neurospora* devised elegant methods of producing and detecting mutations concerned with nutritional requirements. Bonner (1965) described gene-enzyme relationships. Studies of Brenner (1955) on tryptophan biosynthesis in *Salmonella typhimurium* supported one gene-one enzyme hypopthesis.

Experiments of Ingram (1957) showed that gene mutation in human hemoglobin was the basis of chemical difference between normal and sickle-cell. Ingram (1958) explained how genes act. Wagner and Mitchell (1964) gave a general account of genetics and metabolism. Woods (1973) dealt with biochemical aspect of gene. Hood (1972) discussed whether "two genes, one polypeptide chain" is fact or fiction.

Questions

1. Describe the work of Garrod. How did these experiments help in understanding gene function? What relationship did they suggest between gene and enzyme?
2. How did the studies of Lawrence and coworkers on cape primrose (*Streptocarpus*) establish that genes control biochemical reactions?
3. Describe the eye disc transplantation method used by Beadle and Ephrussi.
4. What do you understand by nonautonomous and autonomous development of a transplant.
5. Describe salient observations of eye disc transplantation experiments of Beadle and Ephrussi conducted on *Drosophila*. In what way were these experiments significant in understanding gene function?
6. What question did Beadle and Tatum address themselves when they used *Neurospora crassa* as an experimental organism? Why did they prefer *Neurospora* over *Drosophila* in their experiment?
7. Describe the method used to induce and detect nutritional mutants in *Neurospora crassa*?
8. Define (a) complete medium, (b) minimal medium, and (c) supplemented medium.
9. What observations on *Neurospora* made Beadle and Tatum propose one gene-one enzyme hypothesis?

10. What studies on microorganisms supported one gene-one enzyme hypothesis?
11. Was one gene-one enzyme hypothesis applicable to all the proteins and enzymes? What studies posed a challenge to one gene-one enzyme hypothesis?
12. Define one gene-one polypeptide hypothesis. What observations led to this hypothesis. What studies supported this hypothesis?
13. What do you known about one cistron-one polypeptide hypothesis?
14. What do you understand by one gene-one primary cellular function hypothesis? Is this hypothesis more acceptable than one gene-one polypeptide hypothesis and why?

16

TRANSCRIPTION

Central Dogma and Its Reversal

Genetic information from DNA is transcribed to RNA and that information from RNA is translated to polypeptide chain. This concept about flow of biological information from DNA to RNA and then to protein was given by Crick and is known as central dogma. This concept diagrammatically presented in **Fig. 16.1** holds true for those organisms where DNA is the genetic material.

Crick's concept of central dogma was modified with increasing knowledge in the subject. In some plant viruses, for example, tobacco mosaic virus (TMV), where RNA is the genetic material, genetic

```
         Self-replication
            ⤺
         DNA  ──Transcription──▶  RNA  ──Translation──▶  Protein
```

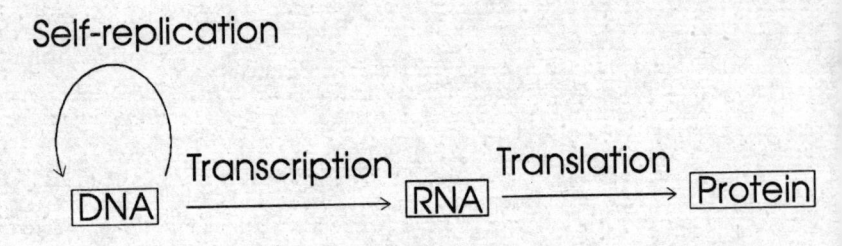

Fig. 16.1. The original central dogma of Crick depicting the flow of genetic information

information flows from RNA to DNA with the help of an enzyme reverse transcriptase. This concept was termed as **reverse transcription** by Tamin and Mizutani (1970). Under laboratory conditions proteins can be synthesized directly using DNA as a template. RNA can undergo self-replication. The modified central dogma is shown in **Fig. 16.2**.

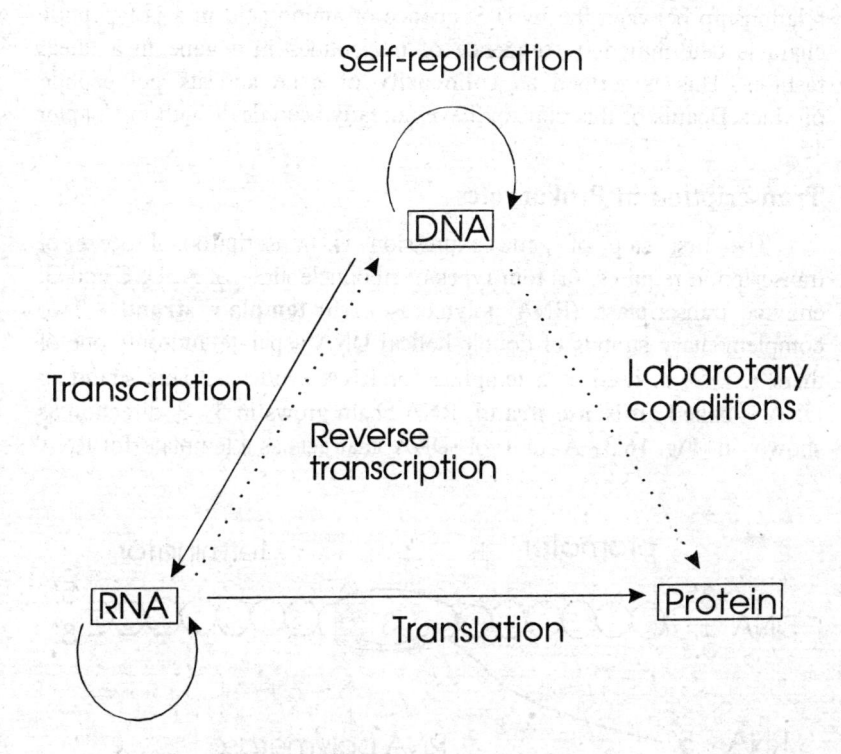

Fig. 16.2. Updated central dogma of molecular biology, showing all known parthways of genetic ifnormation transfer

Gene Expression

How do the genes perform their phenotypic functions, i.e. how do genes exert their effects on the phenotype of a virus, a cell, or an organism? Different genes exert their effects in different ways, genes may express themselves in several ways. On the basis of gene expression genes may be divided into several classes: **tRNA genes** - synthesize pre-tRNA, **rRNA genes** - synthesize pre-rRNA, and **structural genes** - synthesize mRNA (in prokaryotes) or pre-mRNA (in eukaryotes); these transcripts after processing become functional molecules tRNA, rRNA and mRNA, respectively. Whereas tRNA and rRNA do not encode for polypeptides, mRNA contains codons and acts as a template for transcription. As we have seen earlier, gene is a linear sequence of nucleotides. Polypeptide is a linear sequence of amino acids. What is the relationship between the two? Sequence of amino acid in a polypeptide chain is determined by sequence of nucleotides in a gene in a linear fashion. This is termed as **colinearity** of gene and its polypeptide product. Deatils of this concept have already been dealt with in Chapter 15.

Transcription in Prokaryotes

The first step of gene expression is transcription. Process of transcription requires: (a) four types of ribonucleotides of A, U, C and G, enzyme transcriptase (RNA polymerase), (b) **template strand** - Two complementary strands of double helical DNA separate and only one of them ($3'\leftarrow 5'$) is used as a template for RNA synthesis. This strand of DNA is called **antisense strand**. RNA chain grows in $5'\rightarrow 3'$ direction as shown in **Fig. 16.3**. A unit of DNA that acts as a template for RNA

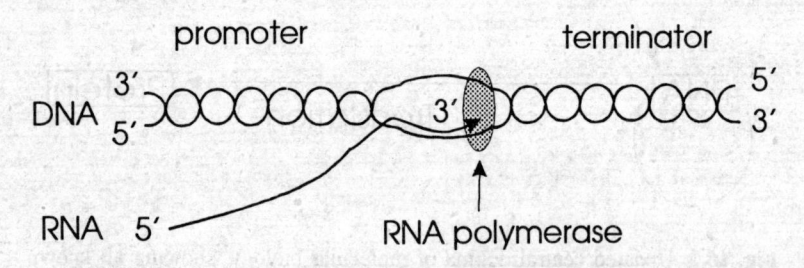

Fig. 16.3. A transcribed piece of RNA and its template DNA showing position of promoter, terminator and direction of transcription

transcription is called an **operon** in prolaryotes. A operon contains a **promotor**, an **operator**, and **structural gene**(s). (c) **DNA-dependent RNA polymerase** - In bacteria, there is only one species of this enzyme which is composed of six polypeptides ($\alpha_2\ \beta\ \beta'\ w\ \sigma$) of five different kinds. This means five different genes ($\alpha,\ \beta,\ \beta',\ w,\ \sigma$) are needed to make *E. coli* polymerase. Active form of enzyme is called **holoenzyme**. Holoenzyme is made up of a core enzyme and a σ **(sigma) factor**. **Core enzyme** has 5 subunits ($\alpha_2\ \beta\ \beta'\ w$). No covalent bond runs between the various chains. Aggregation results from formation of secondary bonds. Core enzyme catalyzes formation of internucleotide $3'$-$5'$ phosphodiester bonds equally well in absence or presence of σ factor. Only holoenzyme can initiate transcription; but then σ factor is released, leaving the core

Holoenzyme
   ↓ DNA binding
Holoenzyme.DNAclosed binary complex
   ↓ DNA melting
Holoenzyme.DNAopen binary complex
   ↓ Phosphodiester bond formation
Ternary complex.Core enzyme.DNA.nascent RNA

Fig. 16.4. Initial reactions involved in conversion of RNA polymerase holoenzyme to core enzyme or transcription start

enzyme to undertake elongation (**Fig. 16.4**). Sigma factor and core enzyme recycle at different points in transcription. **Rho (ρ) factor** plays role in **termination of transcription**. This is shown in **Fig. 16.5**. The termination of transcription occurs at specific terminator sequences in DNA. Sequence 3'-AAAAAAT- 5' in sense strand of DNA is terminator sequence. Probably Rho factor recognizes this sequences. Double helix of DNA opens up a short reactions thereby allowing free bases on one of the DNA strands to base pair with ribonucleoside- (P)∼(P)∼(P) precursors. As RNA polymerase moves along the DNA template, the growing RNA strand is peeled off allowing hydrogen bonds to reform between the two strands of DNA. Immediately after RNA synthesis commences, the front end becomes available to bind to a ribosome.

Recognition of Start Signals. Initiation of transcription requires **startise**, a point at which first nucleotide is incorporated during transcription. σ factor recognizes start signals along the DNA molecules and also the correct strand of DNA to be used as template. Presence of σ factor leads to very tight binding of holoenzyme to promotor regions of DNA which contain start signals. Then very localized unwinding occurs, allowing synthesis RNA polymerase are not themselves transcribed. Soon after this initial binding of RNA polymerase recognition site, RNA polymerase diffuses to RNA polymerase binding site which is A=T rich region. This binding region is called TATA box.

Termination of Transcription. Chain termination in certain cases requires a rho (ρ) factor stop signals seem to exist on DNA template. They are of two kinds (1) those read by RNA polymerase . It is a poly (A) site, and (2) those read only by ρ factor- recognizes symmetrical DNA sequences. Attenuators (discussed in Chapter 15) are a very important mechanism of termination of transcription.

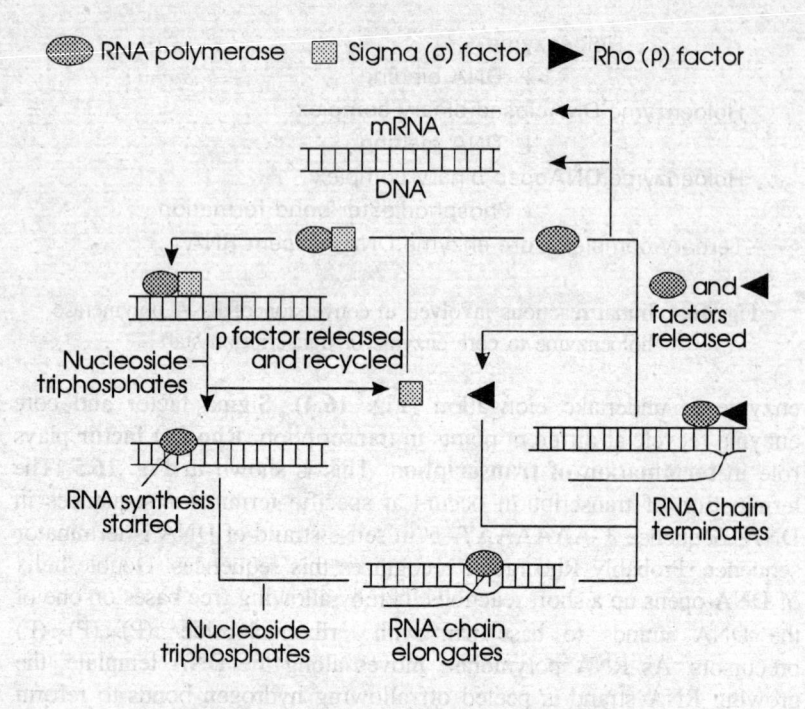

Fig. 16.5. Start and termination of transcription

DNA Topoisomerases. The enzymes that introduce or remove turns from the double helix by transient breakage of one or both polynucleotides are knwn as DNA topoisomerases. These enzymes are important during transcription. RNA polymerase generates **positive supercoiling** ahead and leaves **negative supercoiling behind**. **Type I DNA topoisomerase** rectifies the situation behind by making a transient break in one strand of DNA. **Type II DNA topoisomerase** relaxes negative supercoiling during transcription by introducing a transient double stranded break in DNA (Wand, 1985).

Transcription in Eukaryotes

Transcription apparatus of eukaryotic cells is more complex than that of prokaryotes. In prokaryotes, there is only one species of RNA polymerase. There are 3 species of RNA polymerase in eukaryotes: **RNA polymerase I** - found in nucleolus and is responsible for transcription of genes for rRNA. **RNA polymerase II** - responsible for synthesis of heterogeneous nuclear (hn) RNA, the precursor of mRNA. **RNA polymerase III** - responsible for transcription of small nuclear RNAs and tRNAs.

Transcription and translation processes are integrated processes in prokaryotes. Translation on mRNA begins before transcription is complete. After translation, mRNA is released and degraded. In eukaryotes, there are certain post-transcriptional processes that occur before translation begins. These steps are known as processing of RNA.

RNAs, Their Processing and Functions

Messenger RNA

Thus RNA serves as a template on which a polypeptide is constructed contains an initiation codon (AUG or GUG), at least one of the termination codons (UAA/UGA/UAG) and base sequence in form of triplet codons that dictate the order of amino acid in a polypeptide chain. mRNA also includes certain trailer and leader sequences that are not translated.

Relationship between template DNA and its mRNA transcript is shown in **Fig. 16.6**. From a given piece of double helical DNA, mRNA that will be synthesized will have a predicted sequence (**Fig. 16.7**). RNA polymerase binds at the **promoter**. mRNA is synthesized using antisense strand ($3' \rightarrow 5'$) of DNA as a template. **Sense strand** in $5' \rightarrow 3'$ is not used as a template but this strand in $3' \leftarrow 5'$ direction may be a sense strand with respect to some other gene. Terms sense and antisense DNA strand are with reference to a particular gene. Adleman *et al.* (1987) have reported that two mammalian genes are transcribed from opposite strands of the same DNA locus.

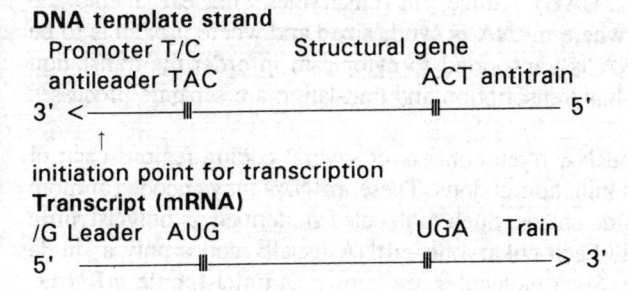

DNA template strand
```
  Promoter T/C        Structural gene
  antileader TAC         ACT antitrain
3' <————║——————————║————————— 5'
        ↑
initiation point for transcription
```
Transcript (mRNA)
```
 /G Leader   AUG            UGA  Train
5' ——————————║———————————————║————> 3'
```

Fig. 16.6. Relationship between DNA template strand and its transcript

276   Basic Genetics

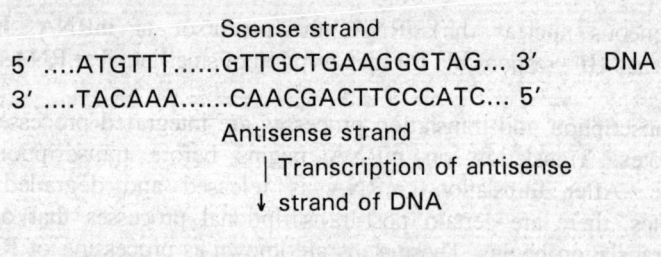

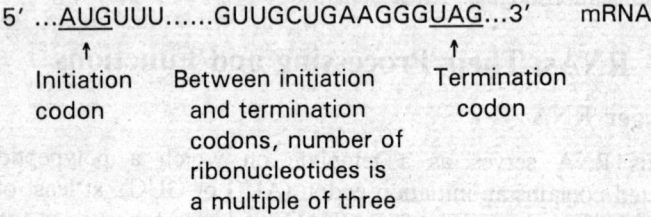

```
                 Ssense strand
        5' ....ATGTTT......GTTGCTGAAGGGTAG... 3'      DNA
        3' ....TACAAA......CAACGACTTCCCATC... 5'
                 Antisense strand
                      | Transcription of antisense
                      ↓ strand of DNA

        5' ...AUGUUU......GUUGCUGAAGGGUAG...3'     mRNA
             ↑                                ↑
          Initiation   Between initiation   Termination
          codon        and termination      codon
                       codons, number of
                       ribonucleotides is
                       a multiple of three
```

Fig. 16.7. Sequence of a transcript from a given double helical DNA molecule

Pre-mRNA Processing. RNA transcribed from genes of eukaryotic genes is called heterogeneous RNA (hn RNA) one of them is known as **pre-mRNA** which is processed into mRNA. Various steps are: removal of parts of leader and train, addition of "Cap" (5-m^7 Gppp) at 5' end, addition of poly(A) tail (200 nucleotides of adenine) at 3' end, splicing of noncoding sequences, called introns and retaining only some sequences, known as exons and methylation of 1 out of 400 adenines present. About 90 per cent of length of pre-mRNA is removed to produce mRNA. Exons are those sequence in DNA that are complementary to those present in finished mRNA. Exons may be translatable (e. g. sequence AUG......UAG). Since, in eukaryotes, nuclear membrane separates the site where mRNA is synthesized and where protein is to be synthesized, mRNA is transported to cytoplasm in order the translation process to start. Thus transcription and translation are separate processes in eukaryotes.

A bacterial mRNA often consists of several coding regions each of which has its own initiation codons. These mRNAs may encode for more than one polypeptide chains. Such molecules are termed as **polycistronic messenger RNAs**. Each eukaryotic mRNA usually codes only a single polypeptide chain. Such molecules are known as **unicistronic mRNAs**.

Processing of Leaders. It involves addition of 5' caps. Addition of cap (m_7 GpppN1) takes place at early stages of transcription in eukaryotes by polymerase II. Cap is suggested to be derived by endonucleolytic cleavage and subsequent addition.

Processing of Trains. **Polyadenylation** signal in the train is AAUAAA in pre-mRNA. Polyadenylation involves two steps: cleavage and addition of poly (A) tail. Multiple copies of **polyadenylation signal** is of common occurrence. Tropomyosin of *Drosophila* has a cluster of 5 poly (A) signals followed by pair of about 250 residues downstream. First of these was a dual structure AAUAAUAAA. Different signals are involved in addition of poly (A) tail in mRNA processing in different tissues.

Some times polyadenylation plays a role in generation of the translational **stop codons**. Other functions of polyadenylation are providing for **stability of mRNAs**, assisting in transport during processing and translation, and direct involvement in protein synthesis. Last activity is suggested by the observation that when mRNAs being translated, originally short poly (A) tail became longer and, on the other hand, some other mRNAs underwent depolyadenylation when they were no longer being translated.

One ribonucleoprotein (RNP) particle that contains a small RNA, U_7, is necessary in processing the nonpolyadenylated histone H_3 mRNA. Still another RNP that contains small RNA, U_1, is involved in the formation of the poly (A) tail.

Removal of Introns. Introns are removed by a two step reaction. (i) The first covalent modification of the substrate RNA is cleaved at a precise 5' site followed by covalent bond formation between 5' end of the intervening sequence and adenosine residue near the 3' splice site to produce a **lariat molecule** containing intervening sequence and 3' exon. (ii) The **bipartite intermediate** is subsequently resolved by cleavage at 3' splice site followed by ligation of 5' and 3' exons via 3', 5'-**phosphodiester bond**. The phosphates at the 5' splice site becomes incorporated into 2'-5' phosphodiester bond in the branch structure and phosphate at 3' splice site is used to form 3', 5' phosphodiester bond linking the two exons in the produced RNA. The introns are deleted very precisely. The process of intron removal is shown diagrammetically (**Fig 16. 8**). During introns removal, a **spliceosome** is formed. Involvement of spliceosome in **intron removal** is shown in **Fig. 16.9**. Partial sequence, complementarily and function of five small RNAs is given in **Table 16.1**.

278 *Basic Genetics*

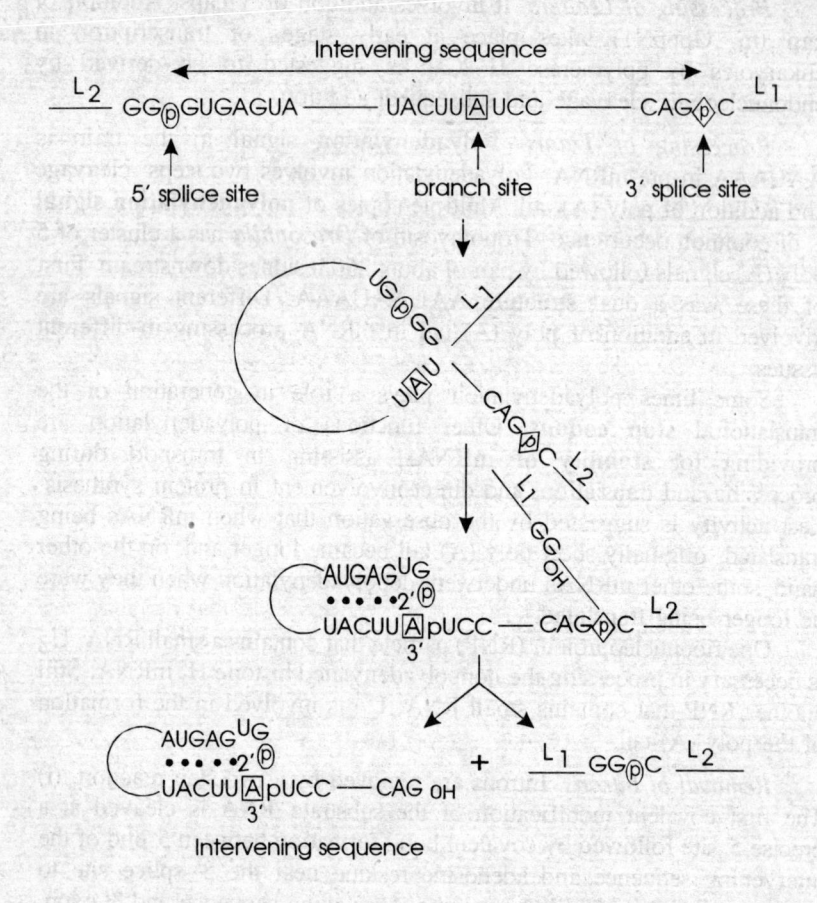

Fig. 16.8. Process of intron removal from pre-mRNA

RNA Editing. RNA editing is a process that results in changes in the nucleotide sequence of mitochondrial transcripts such that the RNA sequence differs from the DNA template from which it is transcribed. In this process, uridine (U) residues are added or deleted at multiple precise sites. RNA editing also involves C to U change, multiple C to U or U to C changes, addition of G or C residues, A to G modification, and conversion of U to A, U to G and A to G. RNA editing takes place with the help of guide RNAs (gRNAs). mRNAs are edited most frequently, whereas edits of tRNAs and rRNAs are extremely rare.

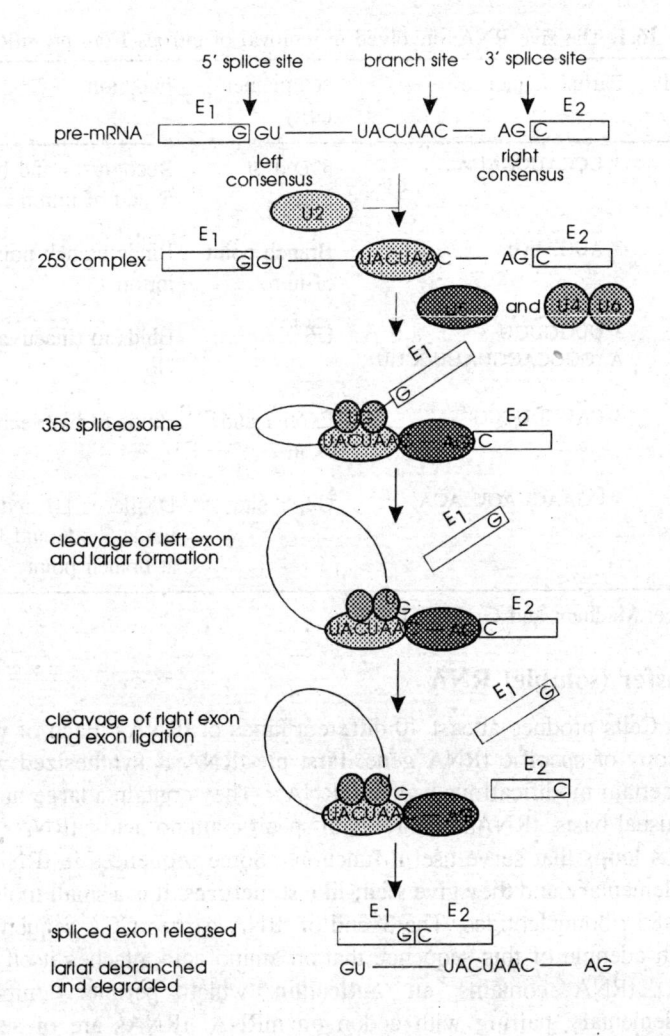

Fig. 16.9. Spliceosome formation involves the interaction of small U RNAs with consensus sequences

Guide RNAs. These are short RNA molecules that can form perfect hybrids with mRNA to be edited and they posses nucleotide seqquences at their 5' ends that are complementary to the sequences of the mRNAs immediately downstream of the **pre-edited regions (PER)**. gRNAs are of about 60 nucleotide length (Church *et al.*, 1979; Blum *et al.*, 1991).

Table 16.1. The five RNAs involved in removal of introns from pre-mRNA

Small RNA	Partial sequence	complementarity	Function
U1	3'-UCCAUUCAUA	5' end of intron	Recognizes and binds 5' end of intron
U2	3'-AUGAUGU	Branch point of intron	Binds branch point of intron
U4	3'-UUGGUGU... AAGGGCACGUAUUCCUU	U6	Binds to (inactivates) U6
U5	3'-CAUUUUCCG	Exon 1 and exon 2	Binds to both exons
U6	3'-CGAACUAGU..ACA	U2, 5' site	Displaces U1 and binds 5' site and U2 at branch point

(Source: Madhani and Guthrie, 1994)

Transfer (soluble) RNA

Cells produce atleast 40 different kinds of tRNAs. Each of which is a copy of specific tRNA gene. First pre-tRNA is synthesized which after certain modifications become tRNA's. They contain a large number of unusual basis. tRNAs interact with specific amino acids. tRNAs have various **loops** that serve useful functions. Some sequences in tRNA are complementary and they give **stem-like structures**. It is a small molecule of 75-85 ribonucleotides. The 3'-end of tRNA carries CCA sequence. It is with adenine of this sequence that an amino acid attaches itself with tRNA. tRNA contains an **anticodon** which permits temporary complementary pairing with codon on mRNA. tRNAs are of several different kinds and each type attaches with, it a specific amino acid and brings it to be ribosome. Partial pairing of some complementary sequences gives tRNA a clover-leaf structure. tRNA for alanine has been studied in detail and its structure is given in **Fig. 16.10**.

Pre-tRNA Processing-Removal of Introns. Intervening sequences (IVS) are located in most of the tRNA precursors of eukaryotes and in

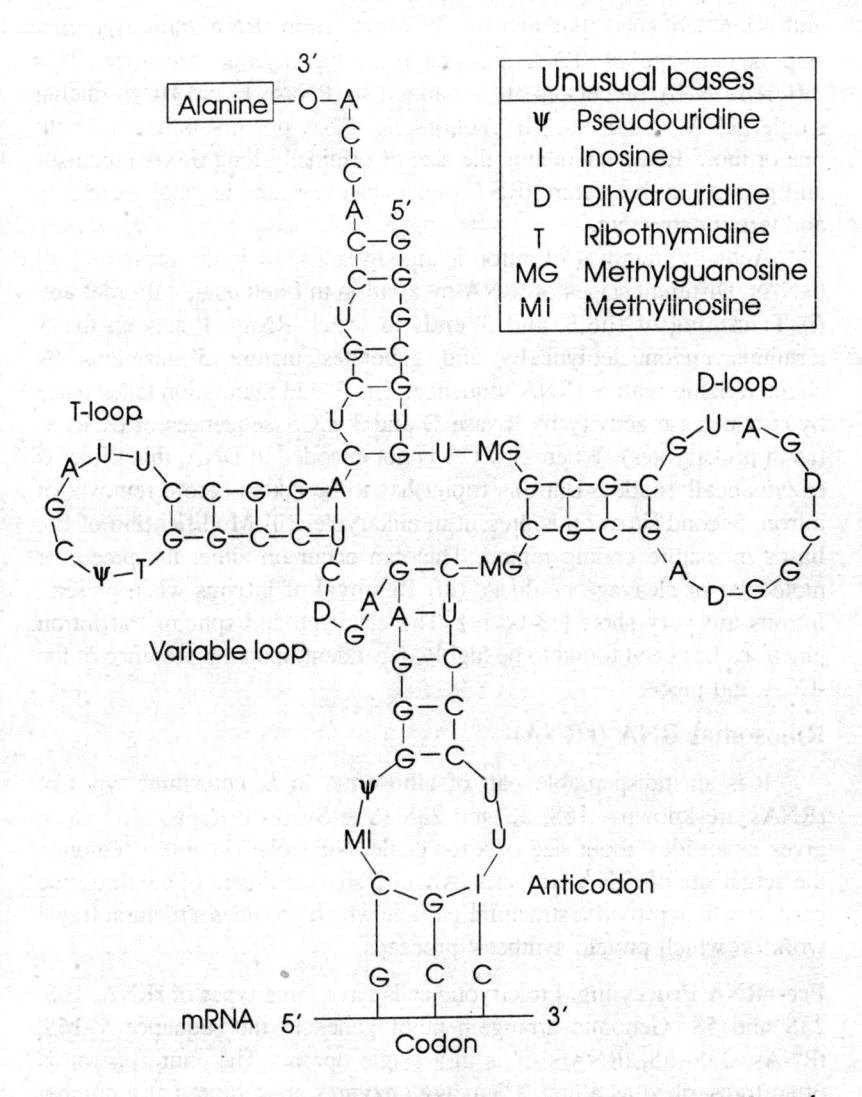

Fig. 16.10. Structure and sequence of alanine transfer RNA in yeast. Unusual bases are also defined. Anticodon (on tRNA and codon (on mRNA) relationship is shown

advanced genera of prokaryotes at a point one site removed downstream from anticodon. However in gene for tRNA leu the intron splits the anticodon sequence. For removal of intron from tRNA transcript, first step is cleavage of tRNA sections from the inserted sequence. This different tRNA precursors are separated by RNase E and III producing single tRNA precursors. The monomeric tRNA precursors interact with one or more RNases reducing the size of originally long tRNA precursor and permitting the mature tRNA precursor to assume its final secondary and tertiary structure.

Actually, removal of intron is among last steps in the processing of tRNAs. Different stages of tRNA maturation in fixed temporal order are: **(i) Trimming of the 5' and 3' ends** in which RNase P acts on the 5' terminus endonucleolytically and generates mature 5' terminus by characterizing mature tRNA sequence. The 3' end maturation takes place by exonuclease activity by RNase D and 3'-CCA sequences of tRNA C (as in prokaryotes). When -3'-CCA is not encoded in DNA, this triplet is enzymatically added. But this triplet has to be added before removal of intron. Second situation is present in eukaryotes. **(ii) Modification of the bases** in mature coding region. This can occur on either the precursor molecular or cleavage products. **(iii) Removal of introns** when present. Introns are very short (13-60 nt). The cleavage and splicing at intron junctions has been found to be highly dependent upon the presence of the -CCA and piece.

Ribosomal RNA (rRNA)

It is an indispensable part of **ribosome**. In *E. coli*, three types of rRNAs are known - 16S, 5S and 23S (S = Svedberg units). "S" value gives us an idea about size of a molecule. Nucleolar organizer region is the actual site of rRNA synthesis. An intrinsic constituent of the ribosome particle which provides structural particle which provides structural frame work on which protein synthesis proceeds.

Pre-rRNA Processing. Prokaryotic cells have three types of rRNA: 16S, 23S and 5S. Genomic arrangement of genes in the sequence 5'-16S, tRNAs, 23S, 5S, tRNA(s)-3' is in a single operon. The entire operon is often transcribed as a unit. **Cleavage enzymes** are required at a number of points even to separate several classes of pre-rRNAs. A typical **rRNA operon** is shown in **Fig. 16.11**.

Early Steps in Processing 5S rRNA. After transcription, the primary transcript is cleaved into two parts, one bears the genes for the minor rRNA and one or two tRNAs present on the intergenic spacer and other

```
P₁ P₂           16S      tRNA      23S                        5S
└┴┴─────────────┬───────++─++──────┬──────────────┬───────────++──────
              2,000    3,000    4,000          5,000        6,000
```

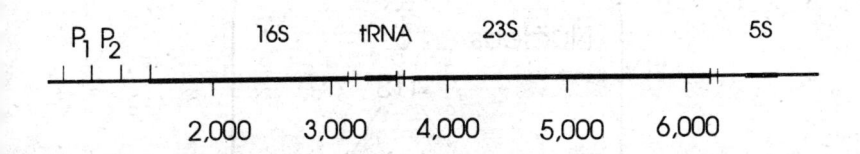

Fig. 16.11. A typical RNA operon. Genomic arrangement is in the sequence 5'-16S, tRNA(s), 23S, tRNA(s)-3'. The entire operon is transcribed as a unit

contains the major and the supplemental rRNAs, depending upon whether or not a tRNA occurs on the 3' end. Neither **RNase III** nor **E** is involved in initial enzymatic reaction.

Later Steps in Processing 5S rRNA. The 10S RNA contains 6 **stem-and-loop structures** generated by the action of RNase III. Then by action of **RNase P**, fragment 7S RNA is generated which contains **5S rRNA** and a termination stem. Removal of this stem is accomplished by **RNase E**.

Processing of other rRNAs. The primary transcript of the entire gene set is cleaved with RNase III or its equivalent, which breaks the precursor between the **minor and major coding sectors.**

Processing the Major Species of rRNA. The primary transcript of shorter operons, 30S pre-rRNA is in the form of two loops containing the minor and major species, respectively each closed at one side by a strongly base-paired stem. It is these stems that are cleaved by RNase III, which specifically acts on double stranded segments. The final trimming of **30S pre-rRNA** occurs at the ribosome to produce **23S RNA.**

Processing of the 5.8S rRNA. Processing of the supplemental species of rRNA is confined to the nucleolus and involved **small RNA U₃** and another type of **RNA, 7-1**. During much of its maturation, it is hydrogen bonded to the other pre-rRNAs. First product of this species is 12S RNA particle, followed by one sedimenting at 8S. Finally **5.8S rRNA** is produced. Maturation of rRNA in eukaryotic cells is described in **Fig. 16.12**.

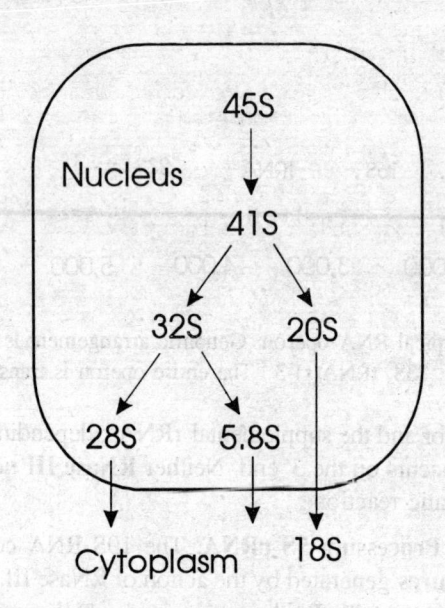

Fig. 16.12. Maturation of rRNA in eukaryotic cells

Relationship among three types of RNAs - mRNA, tRNA and rRNA - during protein synthesis is shown in Fig. 16.13. All three of them are found together at the ribosome during protein synthesis.

Genetic Code

The genetic code is concerned with the processes involved in translating or decoding the information contained in the primary structure of DNA genetic information flows from gene (DNA) to RNA to protein. RNA thus serves as an intermediate between DNA and proteins. The processes of genetic code are the basis of life or it may be considered as a fundamental secret of nature. The basic question that we try to answer is: Which DNA code words specify which specific amino acids? Answer to this comes from genetic code dictionary (**Table 16.2**)

There are only kinds 4 kinds of nucleotides in DNA/RNA and there are 20 different kinds of amino acids in proteins. So one nucleotide cannot code for one amino acid because only four amino acids can be coded in this way. Assuming that two nucleotides code for one amino acid, there will be 4^2 = 16 different doublets coding for only 16 amino

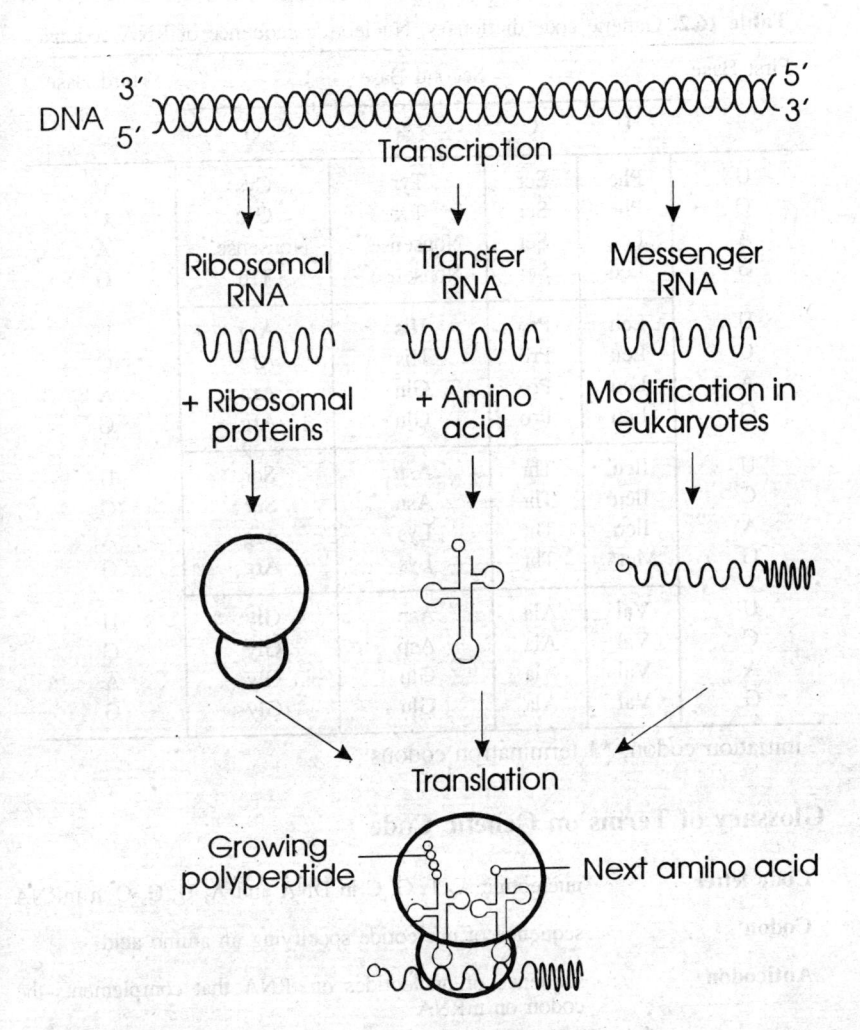

Fig.16.13. Relationship among tRNA, rRNA and mRNA during protein synthesis

acids. Four amino acids will still be left uncoded. A codon size of three nucleotides for one amino acids seems more likely, since it produces 4^3 = 64 possible triplets. Since only 20 amino acids need to be coded, 44 codons seem to be superfluous. If the coding ratio were 4:1 or 5:1, there will be excess of unnecessary codons. This aspect about length and nature of genetic code became somewhat clearer in 1961.

Table 16.2. Genetic code dictionary. Nucleotide sequence of RNA codons

First Base ↓	← Second Base →				Third Base ↓
	U	C	A	G	
U	Phe	Ser	Tyr	Cys	U
	Phe	Ser	Tyr	Cys	C
	Leu	Ser	Nonsense**	Nonsense**	A
	Leu	Ser	Nonsense**	Trp	G
C	Leu	Pro	His	Arg	U
	Leu	Pro	His	Arg	C
	Leu	Pro	Gln	Arg	A
	Leu	Pro	Gln	Arg	G
A	Ileu	Thr	Asn	Ser	U
	Ileu	Thr	Asn	Ser	C
	Ileu	Thr	Lys	Arg	A
	Met*	Thr	Lys	Arg	G
G	Val	Ala	Asp	Gly	U
	Val	Ala	Asp	Gly	C
	Val	Ala	Glu	Gly	A
	Val	Ala	Glu	Gly	G

* initiation codon; ** termination codons

Glossary of Terms on Genetic Code

Code letter   nucleotide A, T, G, C in DNA and A, U, G, C in mRNA

Codon   sequence of nucleotide specifying an amino acid

Anticodon   sequence of nucleotides on tRNA that complements the codon on mRNA

Coding dictionary   A table of all code words that specify amino acids (See Table 16.2)

Codon length   When only as many amino acids are coded as there are code words in end-to-end sequence. (e.g., UUUCCC has 2 code words; they code for only two amino acids)

Overlapping code	when more amino acids are coded for than there are code words represented in end-to-end sequence. (UUU CCC has two code words but in case codons may be UUU, UUC, UCC, CCC). But this actually is not found to be the case.
Nondegenerate code	When there is only one codon for each amino acid
Degenerate code	when there is more than one codon for a particular amino acid
Synonymous codons	different codons that specify the same amino acid in a degenerate code
Ambiguous code	when one codon can code for more than one amino acid
Commaless code	when there are no intermediary nucleotides (spacers) between code words
Reading frame	the particular nucleotide sequence that starts at a specific translation initiation point and is than partitioned into codons until the final word (in fact a termination codon) of that sequence is reached
Sense word	a codon that specifies an amino acid normally present in that position in a protein
Nonsense codon	a codon that does not produce an amino acid; also called a "termination codon"
Universality	utilization of the same genetic code in all organisms.
initiation codon	First codon in translation AUG is the initiation codon

Properties of Genetic Code

Genetic code is **triplet**, i.e., three code letters form one code word which code for one amino acid. Codon is **Commaless**, i.e., UUUCCCGGGAAA has four code words and upon translation we will have a tetrapeptide chain of phe-pro-gly-lys. So all the letters are used to code for one or other amino acid. Code is **degenerate**, there is more than one codon for a particular amino acid, e.g., there are 4 codons (GCU, GCC, GCA and GCG) for alanine. This is explained like this: the first two positions of the triplet codon on mRNA pair precisely with the first two nucleotides of anticodon of tRNA but pairing at third position may not be precise. Code is largely **unambiguous**, i.e. one code word codes for only one amino acid. In only one case the code seems ambiguous in the sense that codon AUG specifies two kinds of tRNA (tRNA met at initiation site and tRNA met in the reading frame).

Nonsense codons exist. e.g., UAA, UAG, and UGA do not code for any amino acid. They serve as **termination codons**. A synthetic ribonucleotide chains direct incorporation of the same amino acids into the polypeptides in cell free extracts derived from almost any organisms. This shows that code is **universal**. Even nonsense codons appear to be universal in prokaryote and eukaryotes. Code is **nonoverlapping**. There will be only two codons from six code letters if code is non-overlapping. **Initiation codon** is AUG or GUG.

Genetic Code at Work

If there is change in the nucleotide sequence of DNA, it would lead to a corresponding change in the mRNA. The change in nucleotide sequence of DNA may be brought about by a point mutation, insertion of nucleotide(s), deletion of nucleotide(s), etc. (**Fig. 16.14**). Thus a change in nucleotide sequence of DNA is reflected in protein via a change in the nucleotide sequence of mRNA.

Wobble Rules

There are thirty-two tRNAs (including one for initiation) in all. They can complement for all sixty-one sense codons. The genetic code is degenerate in that a given amino acid may have more than one codon. As can be seen from Table 16.2 eight of the sixteen boxes contain just one amino acid per box. Therefore, for these eight amino acid the codon need only be red in the first two positions because the same amino acid will be represented regardless of the third base of the codon.

These eight groups of codons are termed **unmixed families of codons**. An unmixed family is the four codons beginning with the same two bases that specify a single amino acid. For example, the codon family GUX codes for valine. **Mixed families of codons** code for two amino acids for stop signals and one are two amino acids. Six of the mixed families are split in half so that the codons are differentiated by the presence of a purine or a prymidine in the third base. For example, CAU and CAC both code for histidine; in both, codons the third base U or C is a prymidine. Only two of the families of the codons are split differently.

The lesser importance of the third position in the genetic code ties in with two facts about tRNAs. First, although there would seem to be a

```
     1    5    10   15
3' TACUUUCCCAAAGGG 5'              - DNA

5' AUGAAAGGGUUUCCC 3'              - mRNA

₂HN-met-lys-gly-phe-pro-COOH       - protein
```

Assume nucleotide A at position 10 changes to C.
New sequence will be:

```
     1    5    10   15
3' TACUUUCCCCAAGGG 5'              - DNA

5' AUGAAAGGGGUUCCC 3'              - mRNA

₂HN-met-lys-gly-phe-pro-COOH       - protein
```

Fig. 16.14. Genetic code at work. Effect of a mutation on amino acid sequence.

need for sixty-two tRNAs, there are actually only about thiry-two different tRNAs in *E. coli* cell. Second, a rare base such as inosine can appear in the anticodon, usually in the position that is complementary to the third position of the codon. These two facts lead to the concept that some kind of conservation of tRNAs is occurring and that rare bases may be involved.

Since the first position of the anticodon (5') is not as constrained as the other two positions, a given base at that position may be able to pair with any of the several bases in the third position of the anticodon. Crick characterized this ability as wobble. **Table 16.3** shows the possible pairings that would produce a tRNA system compatible with the known code. For example, if an isoleucine tRNA has the anticodon 3'-UAI-5', it is compatible with three codons for the amino acid: 5'-AUU-3', 5'-AUC-3' and 5'-AUA-3', i.e., inosine is the first (5') position of the anticodon can recognize U, C or A in the third (3') position of the anticodon, and thus one tRNA complements all three codons for isoleucine.

Hormones and Gene Expression

Study of sex differentiation and role of hormones shows that realization of genotypic capability depends upon and can be modified by

Table 16.3. Pairing combinations the third codon position

Number-one base in tRNA (5' end)	Number-three base in mRNA (3' end)
G	U or C
C	G
A	U
U	A or G
I	A, U, or A

hormones, while the production of hormones is itself guided by the genetic constitution of the individual. A further proof that hormones may influence gene expression is obtained by influence of sex hormones on (a) sex-influenced traits and (b) sex-limited traits.

Puffing in salivary chromosomes of dipteran flies are regarded as sites of gene activity. In Chironomus, to puffs A (on chromosome 1) and B (on chromosome 4) appear during molting of third and fourth instars. Ecdysone injected into larvae that are between molts produces puffs similar to normal molting. These puffs are not restricted to salivary chromosomes but are also formed in cells of other tissues. Temperature, oxygen, pressure, nutrition, etc. have acted as inducers of gene activity.

Genetic homeostasis pleiotropism, penetrance and expressivity are the genetic phenomenon that deal with effect of environment on gene expression.

Penetrance

Penetrance represents the proportion of genotypes that shows an expected phenotype. For example, a dominant mutation such as *Lobe* eye, in *Drosophila melanogaster* may, when heterozygous, show itself in only 75% of certain groups of flies although all individuals are carrying the *Lobe* gene. Under these conditions *Lobe* has 75% penetrance.

Expressivity

Expressivity deals with the degree to which a particular effect is expressed by individuals. The expression of the Lobe gene among various individuals that show its effect may vary considerably from complete absence of eyes to almost normal size (**Fig. 16. 15**). Although this gene is stable, its phenotypic consequence is variable.

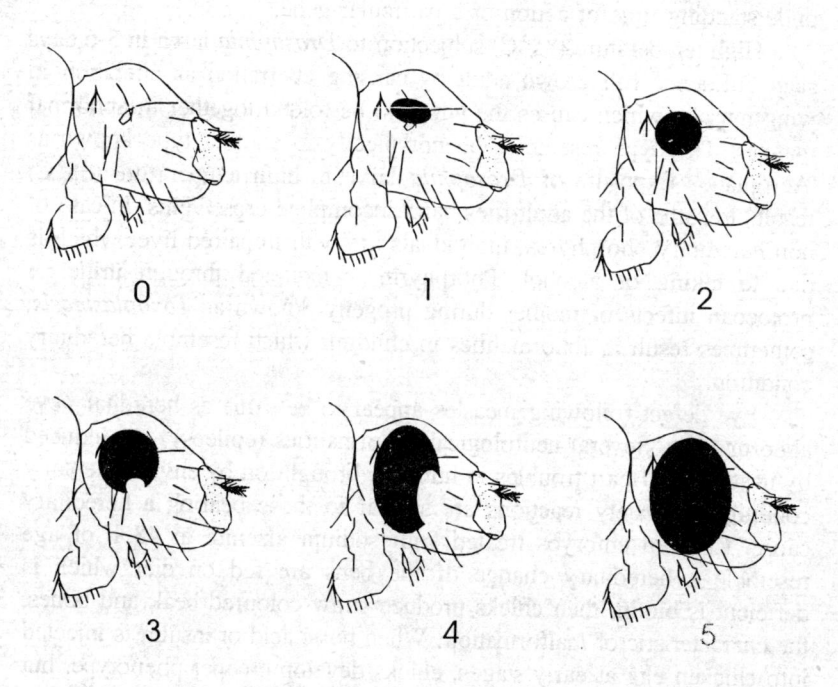

Fig. 16.15. Expression of *Lobe* gene in different individuals in *Drosophila melanogaster*. Eye sizes range from grade 0 with practically no eye facets to wild type phenotype, grade 5, with about 700 to 800 facets

Phenocopies

Phenocopies are non-hereditary phenotypic changes that are indistinguishable from or closely resemble known mutations. Phenocopies are of two types: The first type of phenotocpies are known as **phenocopying abnormal**. In this case phenocopy resembles a known mutation. For example, when *Drosophila* larvae are fed on silver nitrate mixed food, the adults emerged have straw body colour which resemble straw mutant flies. In 1960-61, it was noted that pregnant women who were kept on drug thalidomide produced children which had short limbs. This condition is known as **phecomelia**. The second of phenocopies are

phenocopying normal. In this case phenocopy resembles in normal individual whereas non-phenocopy individual is mutant. For example, the individuals suffering from diabetes when treated with insulin behave like normal individuals. Studies involving induction of phenocopies help in understanding time of action of a particular gene.

High temperature (35°C) subjection to *Drosophila* larva in 5-6 days stage influence full grown adult by causing aberration as alteration in wing muscle, which causes the wings to be folded together in abnormal manner. This type resembles phenotypically to another type known as 'Aeroplane'. Exposure of *Drosophila* larve to high temperature (35°C) results in some of the adult flies hairy incomplete cross veins. In case of non-hereditary phorphyrea, individuals are with impaired liver which is due to taking of alcohol. Phorphyrin is excreted through urine. A protocoan infects of mother during progeny, known as *Toxoplasmosis*, sometimes result is abnormalities in children which resemble hereditary condition.

Eye defect following measles appear to be same as hereditary eye abnormalities. Several neurological abnormalities (epilepsy) are induced by irradiation. Heart troubles in humans, brought on by environmentally conditioned anxiety reactions are similar to those bearing a hereditary cause. Chicken embryos treated with sodium arsenite at 48 h of age resemble a hereditary change. If the hens are fed on diet which is deficient is biotin, then chicks produce straw coloured beak and bones, the characteristic of malformation. When boric acid or insulin is injected into chicken egg at early stages, chicks develop creeper phenotype, but this get reverted when boric acid is not supplied. Patients having phenylketonuria can survive mental retardation when they are given did supplemented with carbohydrates and less of protein.

Some Useful Reading Hints

Brown (1981) gave a general view on gene expression in eukaryotes. Young (1991) gave current status of RNA polymerase II. Horikoshi *et al.* (1992) found that transcription factor TFIID induces DNA bending upon binding to the TATA element. Xu *et al.* (1992) reported that upstream box/TATA box order is the major determinant of the direction of transcription. Dreyfuss *et al.* (1993) described role of hnRNP proteins in biogenesis of mRNA. Zawel and Reinberg (1995) explained the common themes in assembly and function of eukaryotic transcription complexes. Adelman *et al.* (1987) observed that two mammalian genes transcribed from opposite strands of the same DNA locus.

Perry (1976) and Robertson (1988) dealt with processing of RNA. Sharp (1987) dealt with the process of splicing of messenger RNA precursors. Adhya and Gottesman

(1978) explained the mechanism for control of transcription termination. Nevins (1983) explained the pathway of eukaryotic mRNA formation. Buc (1986) described mechanism of activation of transcription by the complex formed between cyclic AMP and its receptor in *E. coli*. Brennan *et al.* (1987) reported that transcription termination factor rho is an RNA-DNA helicase. Das (1993) described control of transcription termination by RNA binding proteins.

Hoagland *et al.* (1958) described a soluble ribonucleic acid intermediate in protein synthesis. Hoagland (1959) reviewed the then status of the adaptor hypothesis. Brenner *et al.* (1961) reported an unstable intermediate carrying information from genes to ribosomes for protein synthesis. Chapeville *et al.* (1962) described the role of soluble ribonucleic acid in coding for amino acids. Altman (1978) dealt with structure and function of transfer RNA.

Crick (1961) dealt with general nature of the genetic code for proteins. Nirenberg and Matthaei (1961) reported the dependence of cell-free protein synthesis in *E. coli* upon naturally occurring or synthetic polyribonucleotides. Ochoa (1963) used synthetic polynucleotides to understand the genetic code. Nirenberg (1963) and Nirenberg and Leder (1964) observed the effect of trinucleotides upon the binding of sRNA to ribosome. Brenner *et al.* (1965) dealt with the nonsense triplets required for chain termination and their suppression. Sonneborn (1965) dealt with the extent, nature and genetic implications of degeneracy of the genetic code. Woese (1965) and Woese (1967a) discussed evolution of the genetic code. Woese (1967b) discusseed the genetic code as the molecular basis for gene expression. Crick (1966) and Crick (1968) dealt with the structure and origin of the genetic code. Crick (1970) gave the concept of central dogma of molecular biology. Fox (1987) dealt with natural variation in the genetic code. Ohno (1988) expresseed that codon preferene is but an illusion created by the construction principle of coding sequences. Gall (1991) dealt with spliceosomes and snurposomes, the structures that play important role in RNA processing. Gesterland *et al.* (1992) stated that recoding is a reprogrammed genetic decoding. Noller (1984) explained the structure of ribosomal RNA.

Temin and Mizutani (1970) explaind action of RNA-dependent DNA polymerase in virions of Rous sarcoma virus. Temin (1972) gave a general account of RNA-directed DNA synthesis. Landauer (1948) worked on hereditary abnormalities and their chemically-induced phenocopies to understand gene action.

Questions

1. What do you understand by central dogma of molecular biology and its reversal?
2. Define the term gene expression? Does this term have different meanings for different types of genes?
3. Define transcription. What are significant features of transcription in prokaryotes? What are minimum requirements for transcription? Describe structure of prokaryotic RNA polymerase. Differentiate between holo- and core-RNA polymerase.
4. Represent diagrammatically direction of transcription in relation to template DNA. Describe relationship between template DNA and its primary transcript. Describe the following events of transcription in prokaryotes: recognition of start signals and termination of transcription.

5. What are DNA topoisomerases? What is the role of these enzymes in transcription?
6. What are significant features of transcription in eukaryotes? How many different species of RNA polymerase are found in eukaryotes? Describe function of every one of them.
7. Describe the structure and functions of the three types of RNAs that make the protein synthesis machinery. Also describe relationship among the three types of RNAs.
8. Describe with appropriate diagrams production of messenger RNA, transfer (soluble) RNA, and ribosomal RNA from the processing of their respective precursor molecules.
9. Define the following terms along with appropriate examples:
genetic code, code letter, codon, coding dictionary, codon length, overlapping code, nondegenerate code, degenerate code, synonymous codon, ambiguous code, commaless code, reading frame, sense word, nonsense codon, termination codon, universality of code, initiation codon.
12. Describe various properties of genetic code giving appropriate examples.
13. Show through an example genetic code at work.
14. What are unmixed and mixed families of codons?
15. What is wobble rule? How does this rule explain economy in the use of tRNAs?
16. Describe role of hormones in gene expression with special reference to puffing in salivary chromosomes of dipteran flies.
17. What do you understand by genotype X environment interaction in relation to gene expression. What is genetic homeostasis?
18. Define the following concepts: penetrance; expressivity. What have these phenomenon contributed in understanding of gene expression?
19. What are DNA topoisomerases? What is the role of these enzymes in transcription?
20. What are phenocopies? What is usefulness of phenocopies in genetic studies?

17

TRANSLATION

Translation in Prokaryotes

Components of Translation

About 80% of cellular RNA is ribosomal RNA (rRNA), 15% is transfer RNA (tRNA) and 5% is messenger (mRNA). Out of all different types of RNAs, only mRNA is destined to serve as a template for production of protein which in turn performs structural or functional role. It is the ribosome on which the mRNA and tRNA in conjunction with amino acids aggregate during protein synthesis and function as the sites of **polypeptide synthesis**. The twenty amino acids found in proteins are shown in **Fig. 17.1**. These three types of RNAs are important components of the protein synthesizing machinery. Each species of tRNA has a specific anticodon. A specific tRNA attaches to a specific amino acid and then unites that amino acid with a specific codon on the mRNA. tRNA has an anticodon which is a sequence of three bases complementary to an mRNA codon. There is atleast one tRNA for each of the 20 amino acids. For binding of each amino acid and a specific tRNA, one specific enzyme called amino acyl synthetase is required. tRNA is an adaptor molecule with duel functions of being used to recognize both the codon and the amino acid.

Charging of tRNAs

Each tRNA can be charged only with the amino acid for which its anticodon is appropriate. The amino acid is linked by an ester bond involving its carboxyl group to one of the last base of the tRNA (which is always adenine). The meaning of tRNA is determined by its anticodon and not by its amino acid. **Activation** and **transfer** reactions are crucial for functioning of the tRNAs. The first important step in the initiation of

(a) Amino acids containing one amino and one carboxyl group:

L-Glycine (Gly,G)   L-alanine (Ala,A)   L-Valine (Val,V)

L-Isoleucine (Ile,I)   L-Leucine (Leu,L)   L-Serine (Ser,S)   L-Threonine (Thr,T)

(b) Amino acids containing one amino and two carboxyls: (c) Amides of dicarboxyl amino acids:

L-Aspartic acid (Asp,D)   L-Glutamic acid (Glu,E)   L-Asparagine (Asn,N)   L-Glutamine (Gln,Q)

(d) Basic amino acids (additional amino group):   (e) Imino or cyclic amino acid:

L-Lysine (Lys,K)   L-Arginine (Arg,R)   L-Histidine (His,H)   L-Proline (Pro,P)

(f) Aromatic amino acids (containing O group):

L-Tryptophan (Trp,W)   L-Phenylalanine (Phe,F)   L-Tyrosine (Tyr,Y)

(g) Sulphur-containing amino acids:

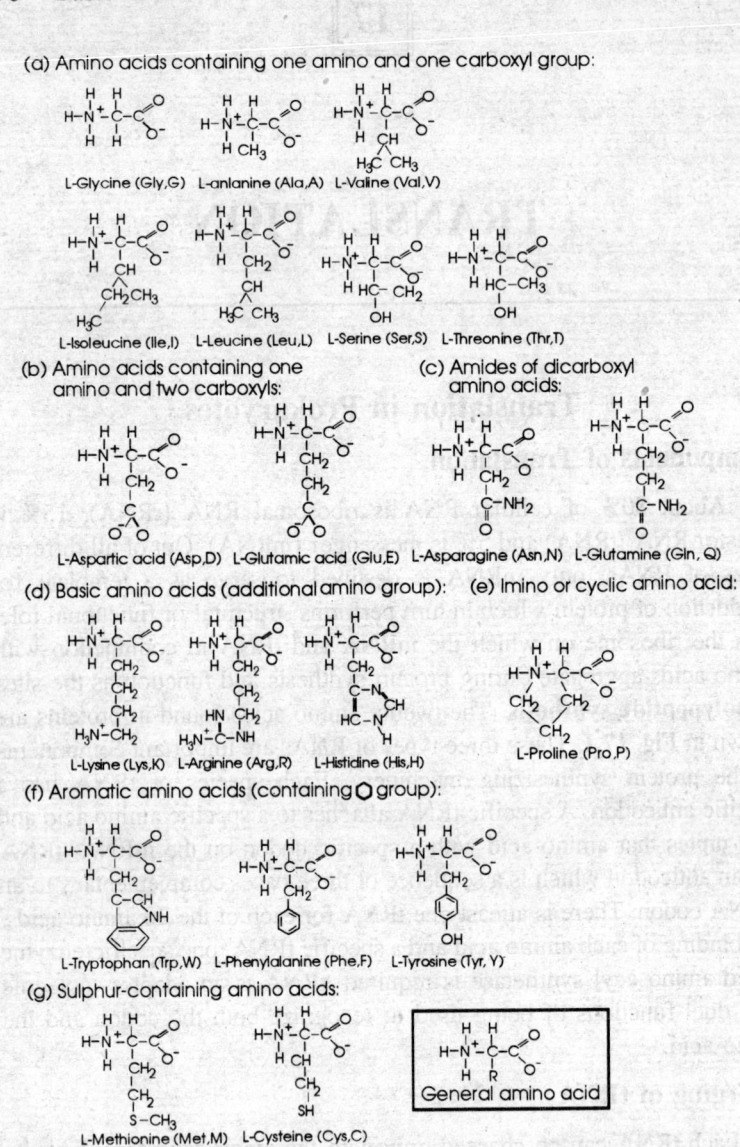

L-Methionine (Met,M)   L-Cysteine (Cys,C)   General amino acid

Fig. 17.1. The twenty essential amino acids found in proteins. At physiological pH, the amino acids usually exist as ions. Three- and one-letter abbreviations are also given. Asp and Glu are acidic, Lys, Arg and His are basic, Ala, Val, Pro, Leu, Phe, Trp, Met, LLe, Cys and Gly are nonpolar, and Ser, hr, Asn, Gln and Tyr are polar (uncharged) amino acids

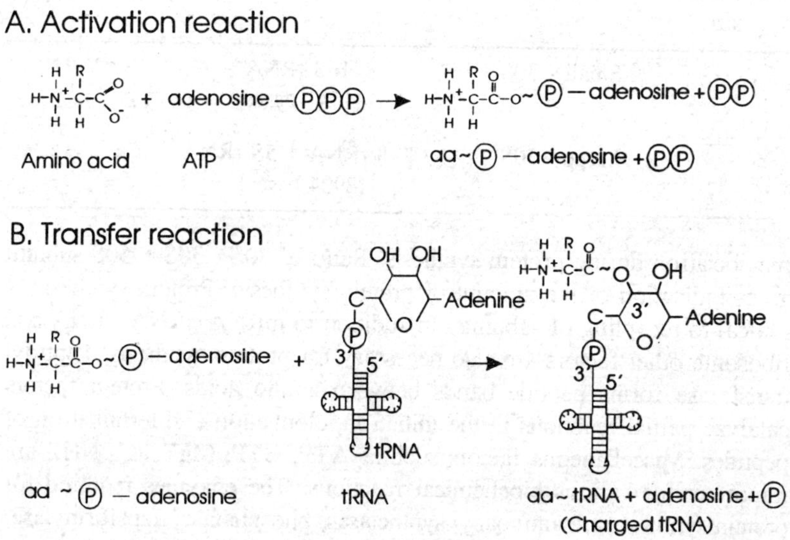

A. Activation reaction

Amino acid + ATP → aa~P—adenosine + PP

B. Transfer reaction

aa~P—adenosine + tRNA → aa~tRNA + adenosine + P (Charged tRNA)

Fig. 17.2. Charging of tRNA with an amino acid is a two step process: (A) Activation reaction in which an amino acid is attached to AMP with a high energy bond (~) and (B) Transfer reaction in which the activated amino acid is transferred to the RNA. Amino acyl-tRNA synthetase is used in these reactoins

translation is charging of tRNA$_f$ with methionine The reactions involved in this step are shown in **Fig. 17.2**.

mRNA is the blue print of genetic information encoded in the gene. There is an initiation codon AUG that signals the start of a polypeptide chain. After the initiation triplet, the bases are read in group of three in sequence, until a chain terminating codon is reached (UAA, UGA, UAG) at which time a polypeptide is released from the mRNA and ribosome.

Ribosomes of *E. coli*

Prokaryotic ribosomes are made up of rRNA and various proteins. Size of bacterial ribosome is 70S. It is made up of two subunits 30S and 50S. The composition of three subunits is given in **Table 17.1**. Ribosomes are the sites where protein synthesis takes place. rRNA contributes to the binding of tRNA and mRNA to ribosome during protein synthesis. Ribosomes undergo a cycle of dissociation and

Basic Genetics

Table 17.1. Subunit composition of bacterial ribosome

Ribosome size	Subunit size	rRNAs	Proteins
70S	Small - 30S	16S rRNA (1541 bases)	21
	Large - 50S	23S rRNA + 5S rRNA (2904 bases)	31

reassociation during protein synthesis. Ratio of 70S : 30S + 50S subunit gives indication of the amount of protein synthesis. Protein synthesis is critical to recycling of subunits. In addition to mRNA, tRNA, rRNA and ribosome other factors are also necessary for protein synthesis. Peptidyl transferase forms peptide bands between amino acids. Protein factors catalyze partial reactions in the initiation, elongation and termination of peptides. Miscellaneous factors such as ATP, GTP, Mg^{++}, K^+, NH_4^+ are important for various biochemical reactions. The enzymes required for protein synthesis are aminoacyl synthetases, phosphatase, transformylase, deformylase, methionine-specific peptidase.

Protein Synthesis

For sake of discussion the process of protein synthesis has been arbitrarily partitioned into three areas, initiation, elongation and termination, but they infact form a continuum. The simplified scheme gives below is concerned with *E. coli*.

Polypeptide Chain Initiation. Initiation involves reactions prior to forming the peptide bond between the first two amino acids of the protein which is a relatively slow step in protein synthesis. The **30S initiation complex** is first formed. It requires mRNA, 30S subunit of ribosome, a special initiating species of amino-acyl-tRNA (that apparently starts all polypeptide chain, and three initiating protein factors - IF_1, IF_2, and IF_3. The first amino acid incorporated is formylated at N-methionine (**Fig. 17.3**). At this stage, formation of 30S initiation complex is complete. Then **70S initiation complex (Fig. 17.4)** is formed. Two sites exist on the 50S subunit-one for the attachment of aminoacyl (amino acid tRNA) and the other for the growing peptide chain attached to tRNA, the former is known as acyl "A" site and later as peptidyl "P" site. F-met-tRNA$_{fmet}$ sits in the "P" site.

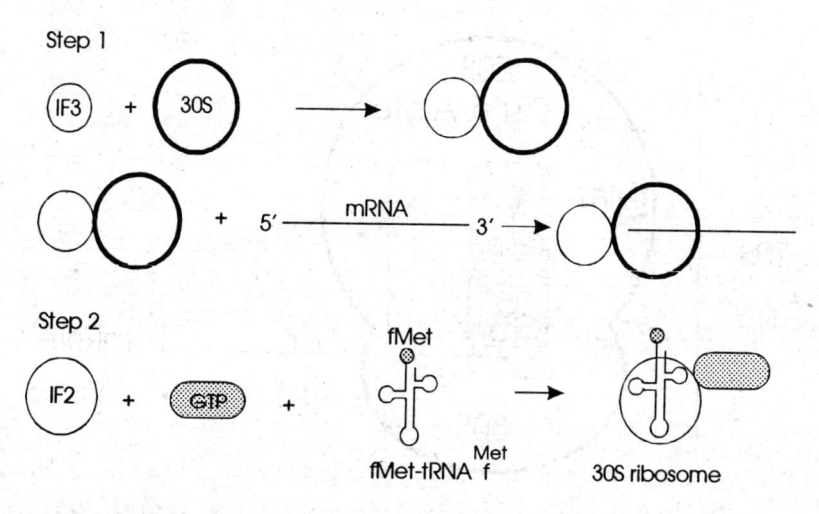

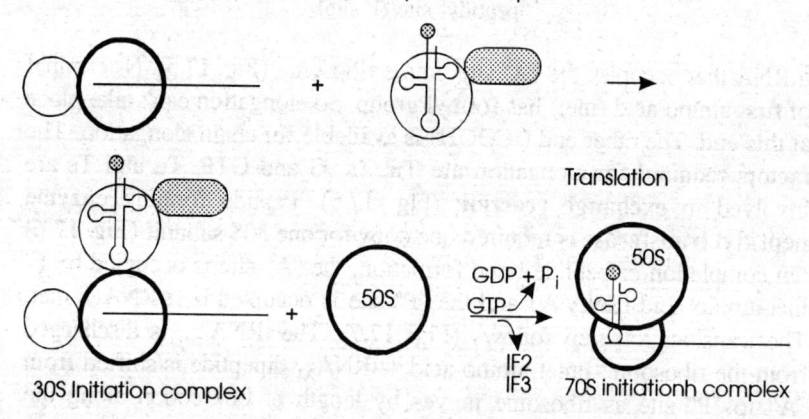

Fig. 17.3. Formation of prokaryotic 70S initiation complex is a three step process. In step 1, 30S ribosome and mRNA are combined. In step 3, initiator tRNA combines with IF2. In step 3, products of steps 1 and 2 are combines to form 30S initiation complex, followed by the formation of 70S complex

Polypeptide Elongation. This process starts when "A" site is occupied by an incoming amino-acid tRNA and peptide bond is formed between F-met and the incoming amino acid. The amino acid-tRNA that binds to the "A" site is specifically determined by the triplet codon of the

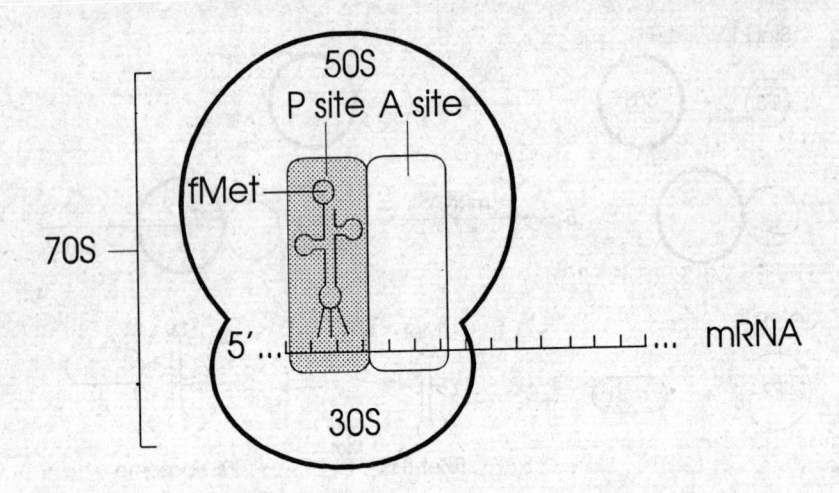

Fig. 17.4. The 70S ribosome contains an aminoacyl site (A site) and peptidyl site (P site)

mRNA that occupies the "A" site on the ribosome (**Fig. 17.5**). N-terminal of first amino acid (met) has formyl group. So elongation can't take place at this end. The other end (-COOH) is available for chain elongation. The factors required for elongation are Tu, Ts, G and GTP. Tu and Ts are involved in **exchange reaction** (Fig. 17.5). Peptide forming enzyme **peptidyl transferase** is required one copy for one 50S subunit (**Fig. 17.6**) On completion of peptide bond formation, the "A" site is occupied by F-met-amino acid-tRNA A1 and the "P" site is occupied by t-RNA F met. The translocation step follows (**Fig. 17.7**). The tRNA$_{fmet}$ is discharged from the ribosome. Fmet-amino acid 1-tRNA$_{A1}$ dipeptide is shifted from "A" to "P" site as ribosome moves by length of one codon along the mRNA in a 5′ → 3′ thus bringing a new codon to "A" site. Thus polypeptide assembly begins at the amino acid and ends at carboxyl end. mRNA is read in a 5′ → 3′ direction.

Polypeptide Termination. As polypeptide chain grows, it remains linked covalently to tRNA and bound to the ribosome. When complete, polypeptide is released from both the components. When **termination signal** (codon UAA, UGA or UAG) is reached at A site of the 50S subunit, chain terminates (**Fig. 17.8**). Release factors - RF1, RF2, RF3, GTP and nonsense codon(s) are required at this step. RF1 is required for

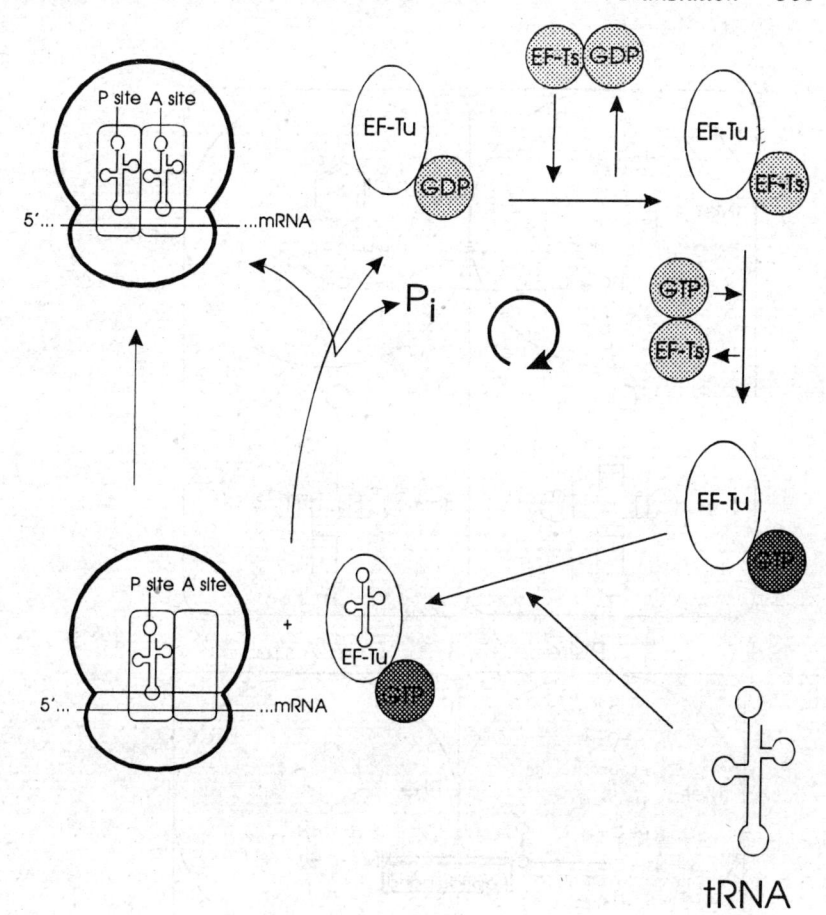

Fig. 17.5. Elongation of polypeptide chain. The step used EF-Tu and EF-Ts GTP provides energy for this step

UAG and RF2 for UGA. UAA can accept either RF1 or RF2. RF3 activates RF1 and RF2.

Polyribosomes

If only one ribosome is attached to a single mRNA, protein synthesis will not be efficient. Instead, a single mRNA usually has several ribosomes attached to it, as many as 1 for every 90 nucleotides. This mRNA with its attached ribosomes is called a **polyribosome** or **polysome** (**Fig. 17.9**). Ribosome move in a 5′ to 3′ direction, synthesizing

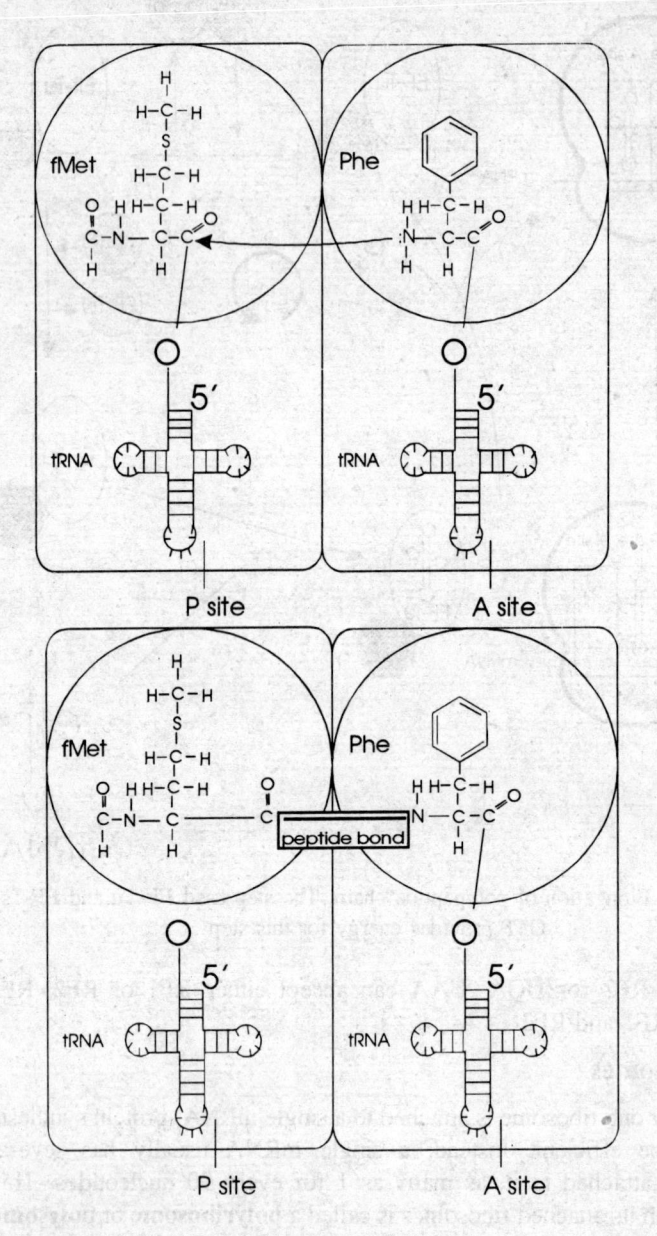

Fig. 17.6. Peptide bond formation during protein synthesis requires peptidyl transferase

Translation 303

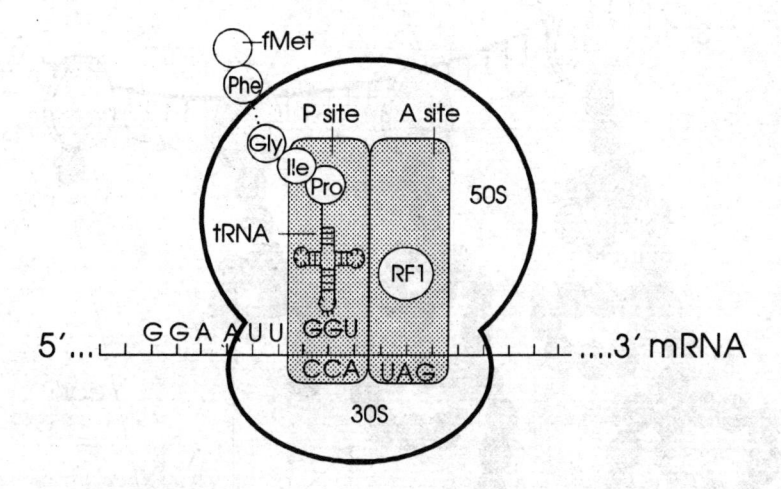

Fig. 17.7. Cycle of peptide bond formation and translocation on the ribosome

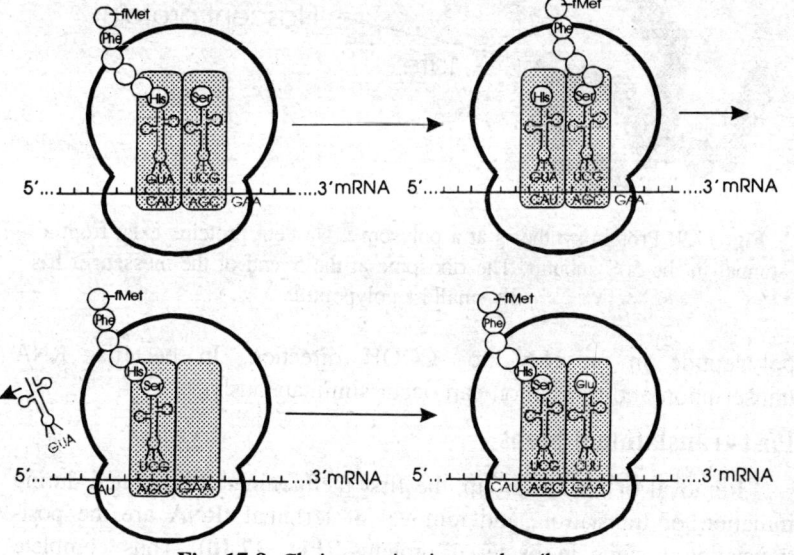

Fig. 17.8. Chain termination at the ribosome

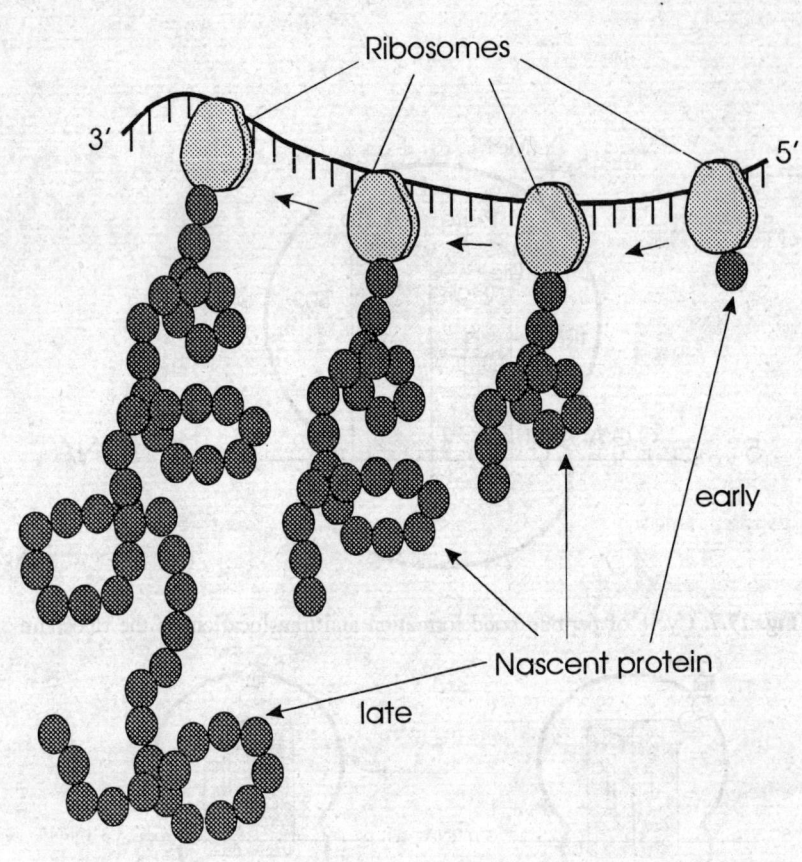

Fig. 17.9. Protein synthesis at a polysome. Nascent proteins exist from a tunnel in the 50S subunit. The ribosome at the 5' end of the messenger has the smallest polypeptide

polypeptide in an H$_2$N- to -COOH direction. In bacteria, RNA transcription and translation can occur simultaneously.

Post-translational Steps

Removal of formyl group, the first methionine incorporated during initiation of translation, and removal of terminal tRNA are the post-translational steps in bacterial proteins (**Fig. 17.10**). Thus complete polypeptide chain is obtained.

Deformylation of F-met
N-formyl-met- AA_1 - AA_2........AA_{n-1}-AA_n-tRNA$_{AAn}$
　　　　　　　↓　　　Deformylase
　　met - AA_1 - AA_2.........AA_{n-1}-AA_n-tRNA$_{AAn}$

Removal of first Amino acid (methionine)
　　methionine　　　　methionine-specific
　　removed　　　↓　　amino-peptidase
　　　　AA_1 - AA_2.........AA_{n-1}-AA_n-tRNA$_{AAn}$

Removal of terminal tRNA
　　　　　　　　↓　　terminal tRNA removed
　　　　AA_1 - AA_2.........AA_{n-1}-AA_n
　　complete protein becomes available

Fig. 17.10. Post-translation steps in bacterial protein biosynthesis

Translation in Eukaryotes

Ribosomes in eukaryotes are different from those of prokaryotes (**Table 17.2**). In eukaryotes, only AUG is recognized as initiation codon. Also methionine is not formylated in eukaryotes tRNA$_m$ is present in eukaryotes compared to tRNA$_f$ in prokaryotes. Newly synthesized bacterial proteins start with formyl-methionine and in eukaryotes, proteins start with methionine.

In eukaryotes, there are at least 8 **initiation factors** (eIF1, eIF2, eIF3, eIF4A, eIF4B, eIF4C, eIF4D, eIF5). Their roles are not yet clear. In eukaryotic system, only one release factor, eRF, has been found. GTP seems to be necessary for activity of eRF. The released peptidyl-tRNA comes in the cytoplasm. Some of the differences in translation machinery of prokaryotes and eukaryotes are listed in **Table 17.3**.

Table 17.2. Subunit composition of eukaryotic ribosome

Ribosome	Subunit	rRNAs	Proteins
80S	Small - 40S	18S rRNA (1900 bases)	33
	Large - 60S	28S rRNA + 5S rRNA (4700 bases+120 bases)	45

Table 17.3. Some differences between prokaryotic and eukaryotic translation

	Prokaryotes	Eukaryotes
Initiation Codon	AUG, occasionally GUG, UUG	AUG, occasionally GUG, CUG
Initiation Amino Acid	N-formyl methionine	Methionine
Initiation tRNA	$tRNA_f^{Met}$	$tRNA_i^{Met}$
Interior Methionine tRNA	$tRNA_m^{Met}$	$tRNA_m^{Met}$
Initiation Factors	IF1, IF2, IF3	Nine eIF fctors + CBP1, CBP2
Elongation Factor	EF-Tu	$eEF1\alpha$
Elongation Factor	EF-Ts	$eEf1\beta\gamma$
Translocation Factor	EF-G	eEF2
Release Factors	RF1, RF2	eRF

Fate of Completed Proteins

In both prokaryotes and eukaryotes, mRNAs that code for particular proteins may be translated at special locations and perhaps by special ribosomes. In bacterium *Bacillus*, synthesis of neutral protease (which is secreted into the medium) seems to occur on the ribosomes located at the cell periphery. In eukaryotes, ribosomes may be free or may be attached to the **endoplasmic reticulum** (ER). Membrane-bound ribosomes synthesize protein that are exported from the cell into the surrounding serum. 60S subunit of ribosome is attached to ER. The ER membrane is used to form a **vesicle** containing the protein. This vesicle them becomes secretary message whose membrane fuses with the cell membrane, emptying its contents outside the cell. On the other hand, free ribosomes synthesize specific nonserum liver proteins. It is possible that leader sequences on mRNA's determine which ribosomes shall translate them. Some short proteins are not synthesized on ribosomes. Such proteins are **antibiotic polypeptides**, such as gramicidin and tyrocidine.

Protein Secretion

Proteins that are secreted from the cell must pass through the plasma membrane to the exterior. These proteins start their synthesis in the same way as **membrane-associated proteins** but pass entirely through the system instead of halting at some particular point within it. Interaction between the membrane systems of the cell and parts of the protein structure (sequences) or groups added covalently to it give signals. By means of such signals a protein is recognized by receptors located within the membrane. A travelling protein may then be secreted through the membrane or may be passed on to another membrane. Free polysomes synthesize proteins that are released directly into the cytosol. Some proteins have signals for targeting to organelle, such as the nucleus or mitochondria. Proteins synthesized on membrane bound ribosomes pass into the endoplasmic reticulum, along to the Golgi, and them through the plasma membrane, unless they possess signals that cause retention at one of the steps on the pathway. **Co-translational transfer** is the process through which membrane-bound polysomes synthesized proteins enter the endoplasmic reticulum when they are being synthesized. For entry of these proteins into the ER, leader sequence is required at the N-terminal end of the protein. A protein that resides within a membrane is called an **integral membrane protein**. Secretion of proteins is explained by signal hypothesis.

Signal Hypothesis

Ribosomes are either free in the cytoplasm or associated with membranes. The choice is determined by the type of protein being synthesized. Membrane-bound ribosomes, polyosomes, synthesize proteins that enter membrane or in eukaryotes are passed into **membrane-bound organelles** (e.g., the Golgi apparatus, mitochondria, chloroplasts, vacuoles) or transported outside the cell membrane. The mechanism for membrane attachment is explained by the signal hypothesis, developed by G. Blobel and C. Milstein and their colleagues. The mechanism is applicable for both prokaryotes and eukaryotes. Here we describe it in mammals.

The signal for membrane insertion is coded into the first one to three dozen amino acids of membrane-bound proteins. This signal peptide takes part in a chain of events leading to membrane attachment by the ribosome and membrane insertion of the protein. The first step occurs when the signal peptide becomes accessible outside of the ribosome. It is recognized by a ribonucleoprotein particle called the **signal recognition particle** (SRP), which consists of six different proteins and a 7S RNA, which is about three hundred nucleotides long. The complex of signal recognition particle, ribosome, and signal peptide then passes, or diffuses, to a membrane in which the SRP binds to a receptor on the membrane, called a **docking protein** (DP) or **signal recognition particle receptor** (Fig. 17.11). During this time, protein synthesis is halted. The ribosome is brought in direct contact with the membrane. Other proteins of the membrane help anchor the ribosome. Protein synthesis then resumes, with the nascent protein passing directly into the membrane or passing into the membrane shortly after being synthesized. Once through the membrane, the **signal peptide** is cleaved from the protein by an enzyme called **signal peptidase**. A striking verification of this hypothesis came about through recombinant DNA techniques in which a signal sequence was placed in front of the α-**globin gene**, whose protein product is normally not transported through a membrane. Translation of this gene resulted in the ribosome becoming membrane bound and the protein passing through the membrane.

Since different proteins enter different membrane-bound compartments (e.g., Golgi apparatus), some mechanism must exist that directs a nascent protein to its proper membrane. This specificity seems to depend on the exact **signal-sequence receptors**. Apparently, after the ribosome binds to the docking protein, the signal peptide interacts with a signal-sequence receptor, which presumably determines whether that protein is specific for that membrane. If it is, the remaining processes continue. If not, the ribosome may be released from the membrane.

The signal peptide does not seem to have a consensus sequence like the transcription or translation recognition boxes. Rather, similarities (at least for the endoplasmic reticulum and bacterial membrane-bound proteins) include a positively charged (basic) amino acid (commonly lysine or arginine) near the beginning (N-terminal end) followed by about a dozen hydrophobic (nonpolar) amino acids, commonly alanine, isoleucine, leucine, phenylalanine, and valine. The signal peptide of the bovine prolactin protein reported by Sasavage *et al.* (1982) is:
NH2 - Met Asp Ser Lys Gly Ser Ser **Gln Lys Ser Arg Leu Leu Leu Leu Leu Val Val Ser** Asn Leu Leu Leu Cys Gln Gly Val Val Ser Thr Pro Val...Asn Asn Cys-COOH.

The aminco acids indicated in bold separates the signal peptide from the rest of the protein which consists of 199 amino acid residues.

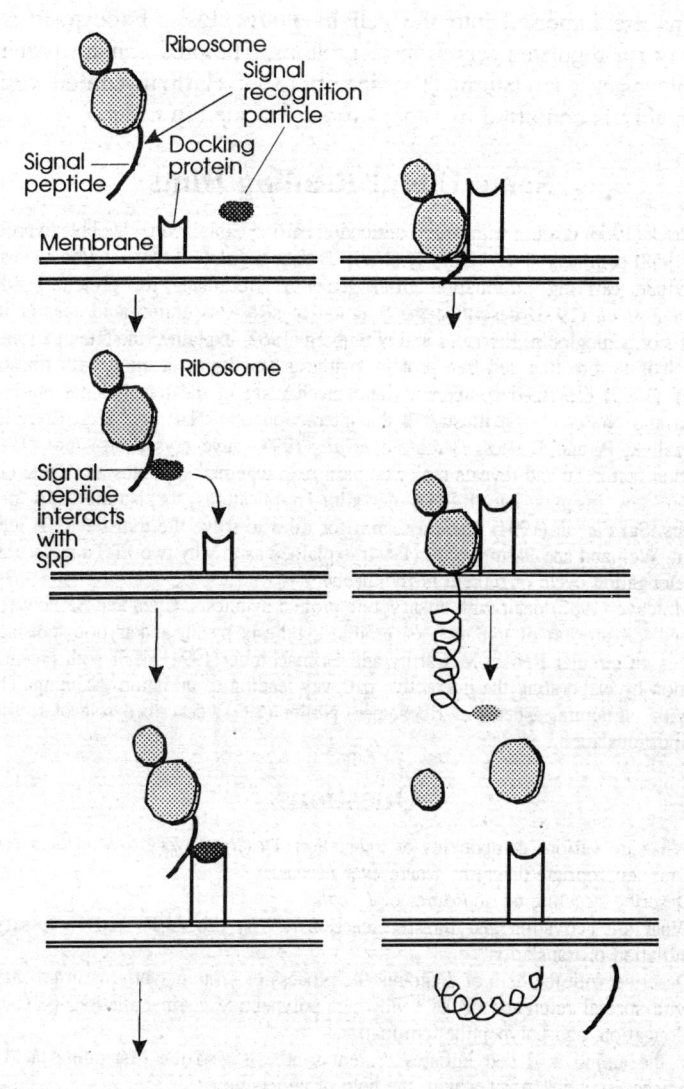

Fig. 17.11. The signal hypothesis

Secretory proteins are transported in coated vesicles. A vesicle is generated by pinching off from a membrane, it fulfills its function by fusing with another membrane to release its contents. Some proteins have **constitutive secretion** while others are secreted through **exocytosis**.

Proteins are imported into the cell by **endocytosis**. Exocytosis is the pathway for regulated secretion of proteins. The predominant protein in coat of vesicles is clathrin. The specificity of **clathrin-coated** vesicles presumably is conferred by other proteins located in coats.

Some Useful Reading Hints

Crick (1966) dealing with codon-anticodon pairing explained the wobble hypothesis. Crick (1958) commented on protein synthesis. Brenner *et al.* (1961) dealt with an unstable intermediate carrying information from genes to ribosomes for protein synthesis. Ehrenstein *et al.* (1963) described the function of sRNA as amino acid adaptor in the synthesis of hemoglobin. Barondes and Nirenberg (1962) explained the fate of a synthetic polynucleotide directing cell-free protein synthesis and its association with ribosomes. Hershey (1980) described components and mechanism of the translational machinery. Moazed and Noller (1989) illustrated the interaction of tRNA with 23S rRNA in the ribosomal A, P, and E sites. Berchtold *et al.* (1993) gave crystal structure of active elongation factor Tu and reveals major domain rearrangements. Santos and Tuite (1993) provided new insights into mRNA decoding implications for heterologous protein synthesis. Saks *et al.* (1994) made a search for rules to solve the transfer RNA identity problem. Weijland and Parmeggiani (1994) explained as to why two Ef-Tu molecules act in the elongation cycle of protein biosynthesis.

Moldave (1985) dealt with eukaryoteic protein synthesis. Chen and Sarnow (1995) explained the process of initiation of protein synthesis by the eukaryotic translational apparatus on circular RNAs. McCarthy and Brimacombe (1994) dealt with prokaryotic translation by elaborating the interactive pathway leading to initiation. Nanninga (1973) dealt with structural aspects of ribosomes. Noller (1991) described role of ribosomal RNA and translation.

Questions

1. What are various components of translation. Describe every one of them briefly. Draw appropriate diagrams where ever necessary.
2. Describe structure of ribosomes of *E. coli*.
3. What are activation and transfer reactions? Why are these steps necessary for initiation of translation?
4. Describe with the help of diagrams the process of protein synthesis in prokaryotes with special reference to the following: polypeptide chain initiation, polypeptide elongation, and polypeptide termination.
5. Is the amino acid that initiates protein synthesis also the first amino acid in a finished protein? Answer with the help of an example.
6. What are polyribosomes? What has nature evolved such a system as a part of protein synthesis machinery.
7. What are various post-translational steps in prokaryotes?
8. How does translation in eukaryotes differ from that in prokaryotes?
9. What is the fate of completed proteins?
10. What is cotranslation transfer?
11. Describe the the process of protein secretion on the basis of signal hypothesis.

18

GENE REGULATION

The Meaning of Gene Regulation

In *E. coli*, estimated number of genes is 3,000 to 4,000 and the number of proteins found at a given time is from 600 to 700. This means that all the genes are not always active. Only those genes are active at a particular time whose products are required by a cell. Further, in a bacterial cell there are about 10,00,000 protein molecules and number of copies of various protein molecules varies from 10 to 5,00,000. This suggests that all the active genes do not produce their products at the same rate. The genes synthesize their products in the quantities they are needed and at a time when they are needed by the cell/organism. This shows that activity of genes is regulated by some mechanisms. This is true not only for *E. coli* but for all the organisms - prokaryotes as well as eukaryotes. Why is regulation of gene expression necessary? It would be most efficient for the cell and in the interest of the organism to synthesize only those proteins it needs and in the needed proportions at a time and place it is needed. Some regulatory mechanisms have evolved to achieve above objective.

Terminology of Gene Regulation

Constitutive Versus Regulated Genes

All the gene are not subject to regulation. Depending upon whether a gene is regulated or not, it is classified as **constitutive gene** - a gene whose expression is not regulated or **regulated gene** - a gene whose expression is regulated. Product of a regulatory gene is synthesized regularly and is invariably present in the cell at roughly the same concentration regardless of whether substrate to that enzyme is ever presented to the cell or not. Such genes thus have **invariant control**. Constitutive genes encode for constitutive enzymes, like enzymes of glucose metabolism and those of monophosphate pathway. Expression of regulatory gene is subject to modification by chemical stimuli that appear in or disappear from cell's environment. Regulated genes produce regulated enzymes. These enzymes are not needed by the cell at all the times, like enzymes of metabolism of less common sugars such as lactose, arabinose and galactose, and the enzymes of biosynthetic pathways of amino acids, purines and pyrimidines.

Inducible Versus Repressible Gene Regulation

Substrates and end products play role in regulation of gene action. Accordingly, two types of controls are recognized. In **inducible control**, substrate acts to induce the production of the enzymes. Repression occurs in the absence of substrate. In **repressible control**, the end product, also known as **co-repressor**, acts to repress production of the enzymes needed for catabolism. Transcription in initiated in the absence of the end product.

Negative Versus Positive Control

Product of a regulatory gene also plays important role in regulation of gene expression. Accordingly, two types of controls are known to exist - **negative control**, in which case product of a regulatory gene acts to repress transcription and **positive control**, where product of a regulatory gene acts to stimulate transcription.

Based on whether control is inducible or repressible, or whether the control is positive or negative, four models of gene regulation can be conceived (**Fig. 18.1**). Examples of each **model of gene regulation** are also mentioned. Comparisons between negative and positive control systems have been drawn in **Table 18.1**.

Components of Gene Regulation in Prokaryotes

To understand working of the system, one is required to know the following terminology for the controlling elements. **Promotor** represents a sequence of bases in DNA that is recognized by DNA-dependent RNA polymerase. **Operator** is a sequence of bases in DNA which interacts with a regulator molecule to control transcription of a group of genes that are co-transcribed. **Operon** consists of a group of genes that are co-

Gene Regulation

	NEGATIVE CONTROL	POSITIVE CONTROL
INDUCTION	P inactive ↕ M P active genes ON ⇌ genes OFF M turns on transcription P is a repressor Examples: *lac, gal*	P inactive ↕ M P active genes ON ⇌ genes OFF M turns on transcription P is an activator Examples: *ara, mal*
REPRESSION	P inactive ↕ M P active genes ON ⇌ genes OFF M turns off transcription P is a repressor Examples: *trp, his*	P inactive ↕ M P active genes ON ⇌ genes OFF M turns off transcription P is an activator No example

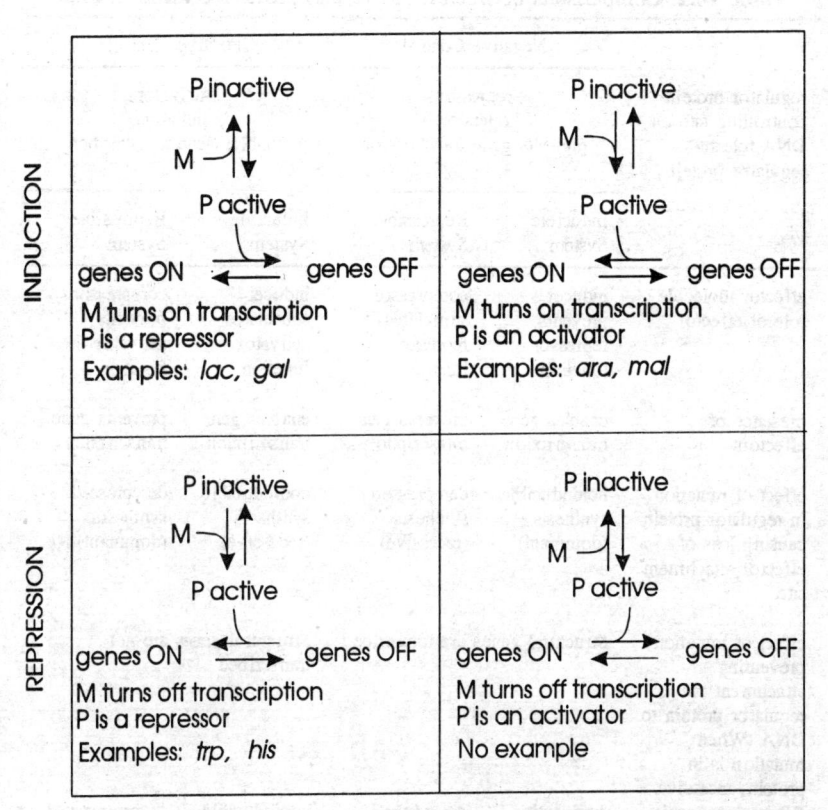

Fig. 18.1. Models of gene regulation based on whether control is inducible or repressible, or whether the control is positive or negative. P=repressor, M=metabolite, *lac*=lactose operon in *E. coli*, *gal*=galactose operon in yeast, *ara*=arabinose operon in *E. coli*, *his*=histidine operon in *Salmonella*

transcribed, alongwith a promotor and operator. **Effector molecule** is that molecule (a sugar, an amino acid or a nucleotide) that can bind to a regulator protein and thereby change the ability of the regulator molecule to interact with the operator. Effectors in inducible operons are called **inducers**.

Effectors of repressible operons are known as **co-regressors**. **Regulator molecule**, known as repressor, are the proteins which themselves are specified by regulatory genes. Regulatory genes have their own operator.

Table 18.1. Comparisons between negative and positive control systems

	Negative Control		Positive Control	
regulator protein controlling site on DNA role of regulator protein	repressor operator prevents gene transcription		Activator initiator enables gene transcription	
	Inducible System	Repressible System	Inducible System	Repressible System
effector molecule role of effector	inducer prevents repressor function	corepressor stimulates repressor function	inducer stimulates activator function	corepressor prevents activator function
presence of effectors	enables gene transcription	prevents gene transcription	enables gene transcription	prevents gene transcription
effect of mutation in regulator protein causing loss of effector attachment site	noninducible synthesis (dominant)*	derepressed synthesis (recessive)	noninducible synthesis (recessive)	derepressed synthesis (dominant)
effect of mutation preventing attachment of regulator protein to DNA (When mutation is in protein: recessive. When mutation is in DNA attachment site: dominant in *cis* position, recessive in *trans* position.)	Structural genes are transcribed constitutive synthesis	derepressed synthesis	Structural genes are not transcribed noninducible	superrepressed
effect of mutation that prevents regulator protein from leaving DNA (dominant)	Structural genes are not transcribed (repression continues) noninducible	superrepressed	Structural genes are transcribed (activation continues) constitutive synthesis	derepressed synthesis

Lactose Operon in *E. coli*

Structure of *Lac* Operon. Jacob and Monod (1961) studied regulation of lactose *(lac) operon* in *E. coli*. The *lac* operon in *E. coli* consists of promoter (*p*), operator (*o*) and three structural genes z, y and *a*. Gene z codes for enzyme β-**galactosidase**, y codes for enzyme β-**galactoside permease** and *a* codes for enzyme **thiogalactoside transacetylase**. These three cistrons (genes) are closely linked, co-transcribed and regulated coordinately as shown in **Fig. 18.2**.

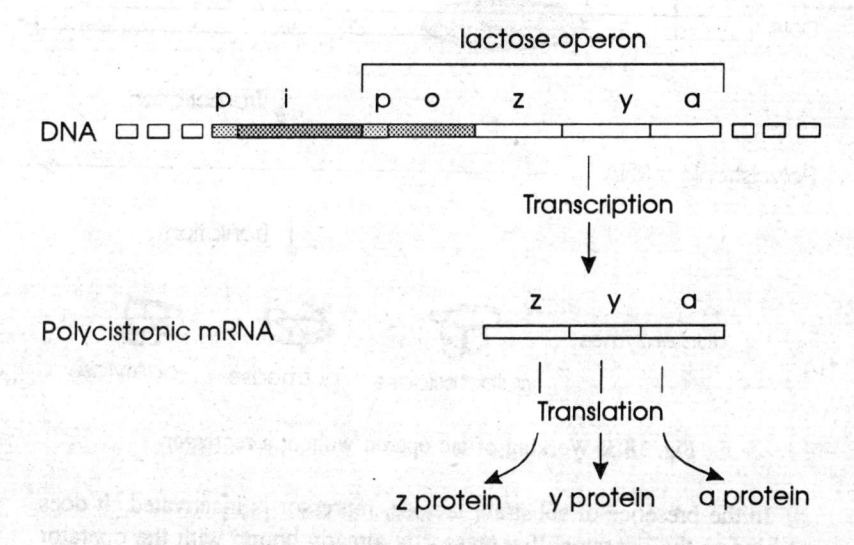

Fig. 18.2. The structure of lactose operon in *E. coli*. The operon is transcribed as a polycistronic mRNA. The operon and its regulator gene (i) have their own promoters (p). The operon has its operator (o), z, y and a are structural genes of the operon

Working of Lac Operon. The controlling region of lac operon is composed of three sites each of which serves to bind a specific protein. In absence of repressor protein, RNA polymerase begins transcription at the operator which is also the the site at which the repressor attaches (**Fig. 18.3**). When repressor, product of the regulator gene, binds to operator, RNA polymerase is physically blocked. Transcription is not allowed to take place. This happens when substrate lactose is absent (**Fig. 18.4**). *Lac* operon is thus an inducible operon. Since product of regulatory gene acts to repress transcription, it is undergoing a **negative control**. *Lac* operon, therefore, inducible negative control.

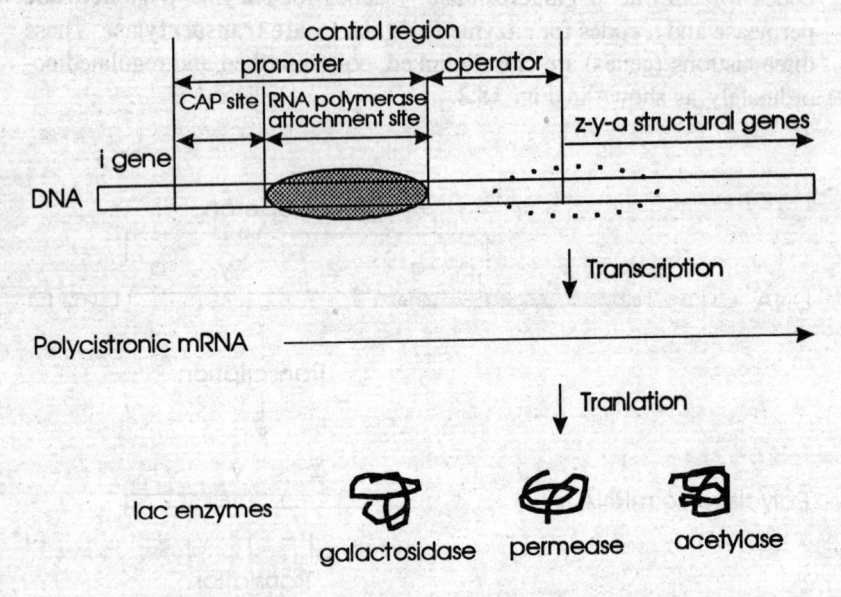

Fig. 18.3. Working of lac operon without a repressor

In the presence of substrate lactose, repressor is inactivated. It does not bind to the operator. If repressor is already bound with the operator it gets released. Now there is no physical block to stop activity of RNA polymerase (**Fig. 18.5**). All the three genes z, y, a are co-transcribed mRNA contains message from all the 3 genes. That is, it is a **polycistronic mRNA** which is then used as a template to produce rapidly and simultaneously all the three enzymes, β-galactosidase, galactoside permease and thiogalactose transacetylase which are used in lactose metabolism. β-galactosidase performs two functions - first, breakdown of disaccharide lactose into two monosaccharides, glucose and galactose, and second, formation of inducer by conversion of lactose into allolactose. Galactose permease facilitates entry of lactose into the cell. Thiogalactoside acetylase is involved at some point in lactose metabolism.

Understanding Gene Regulation Through Mutations. Understanding about regulation of lac operon came from the study of mutants that affected the production of these enzymes.

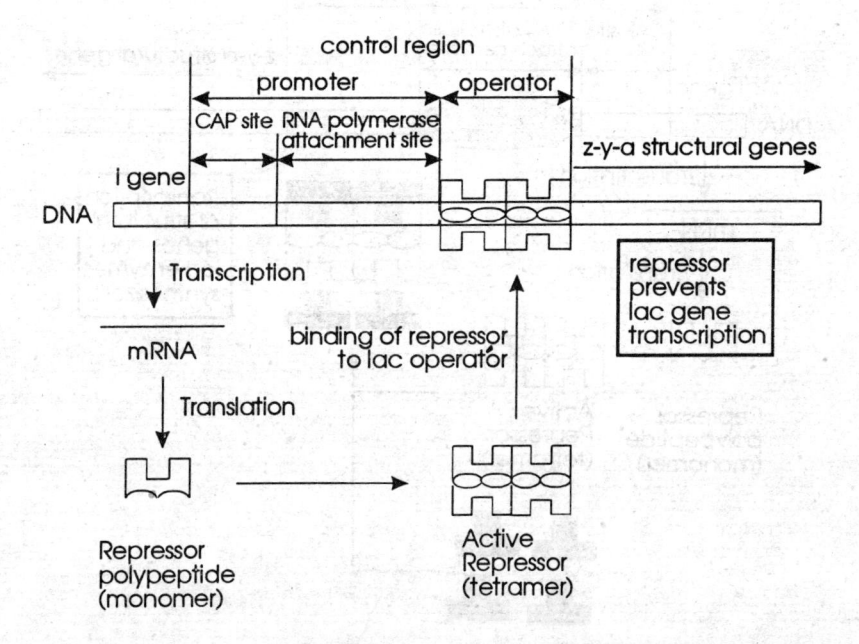

Fig. 18.4. Working of lac operon in presence of wild type repressor and the absnece of inducer

Mutations in Structural Genes. Mutation in any of the three structural genes - z, y, and a - led to amino acid changes in the protein. Of interest to Jacob and Monod were those mutations that determined how much of the three structural gene proteins were made, but which did not affect the structure of the three proteins themselves. These mutations were called regulatory mutations and were found to occur in the **regulatory elements**.

Mutations in Operator. The operator class of regulatory mutations controlled the three structural genes at the same time and in the same way. These mutants were termed as **operator-constitution mutants (o^c)**. These mutants synthesized all three proteins constitutively regardless of the presence or absence of lactose. The genotype was o^{c}z$^+$y$^+$a$^+$ and affected the activity of z-y-a genes in *cis*-arrangement and not in *trans*-arrangement. o^c allele is *cis*-dominant. This was shown by studies with **partial diploids (merodiploid** or **merozygotes)** constructed with the help

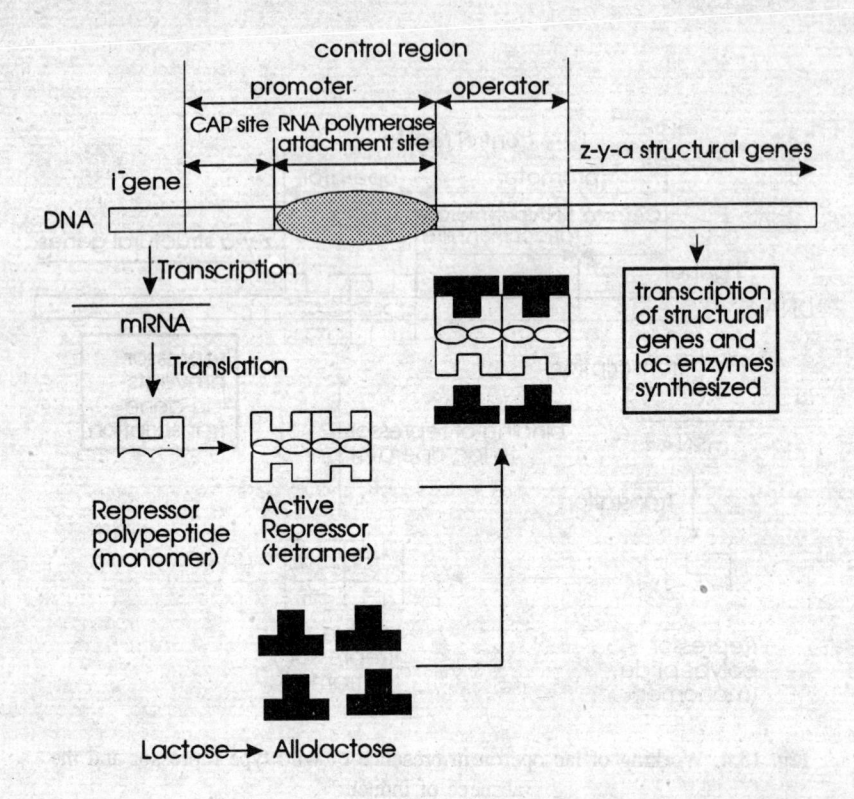

Fig. 18.5. Working of lac operon in presence or wild type repressor and inducer

of F' factor. This fact is clearly shown by data given in **Table 18.2**. The operator gene was mapped very close to z gene.

Mutations in Regulatory Gene. Second class of regulatory mutations were found to occur in the regulatory gene, *i*. These mutations mapped very close to the *o* gene of the *lac* operon. Mutations of *i* locus also resulted in **constitutive synthesis** of lac enzymes. Mutants of repressor gene were operationally different from operator mutants since they acted on z, y and a genes in the same way in both *cis*- and *trans*- arrangements in merodiploids. Two types of *i* mutants have been isolated. The *i*ˢ **mutants** produced **superrepressors** that prevented induction of the z, y, a proteins when lactose was present and the *i*⁻ **mutants** caused constitutive synthesis of the *i* **gene mutants** in haploids and merodiploids for the *lac* region in *E. coli* in shown in **Table 18.3**. The dominance-

Table 18.2. Genotype and phenotype of *E. coli* strains partially diploid for genes of the *lac* region and heterozygous for the *cis*-domonant o^c allele of the operator site. The second copy of the *lac* region is brought into cells by F'

Merodiploid genotypes	Phenotypes*	
	Constitutively synthesized proteins	Inducibly synthesized proteins
$\dfrac{o^+z^-y^-a^-}{F'\ o^c z^+y^+a^+}$	galactosidase permease acetylase	None
$\dfrac{o^+z^-y^-a^-}{F'\ o^c z^+y^+a^+}$	None	galactosidase permease acetylase
$\dfrac{o^+z^-y^-a^-}{F'\ o^c z^+y^+a^+}$	galactosidase permease	acetylase

*Allele o^c is *cis*-dominant to o^+, since o^c causes constitutive production only of proteins coded by structural genes on the same chromosome as itself; o^+ governs inducible enzyme synthesis only by structural genes on the same chromosome as itself.

recessiveness relationship for *i* gene mutants observed in merodiploids is: $i^s > i^+$; $i^+ > i^-$. Since *i* mutants acted in both *cis*- and *trans*-arrangement, this implied that *i* gene specified a product which could diffuse from site of synthesis to an altogether different region of the genome and there influence function of wild type operator. The action of i^- and i^s mutants is shown in **Fig. 18.6** and **Fig. 18.7**, respectively.

Interpretations that can be drawn from the studies on *i* and *o* mutants are: repressor and operator genes interact in the control of structural gene activity. Since operator controlled *z-y-a* genes in *cis*- arrangement, Jacob and Monod proposed that operator gene controlled **co-ordinated transcription** of *z-y-a* genes in the same strand. Operator determined whether or not the three structural genes would be transcribed. Repressor physically interacted with the operator site. When inducer was present, it combined with the repressor and the altered repressor could no longer bind with the operator site. Thus operator was available for RNA polymerase to start the transcription of polycistronic operon. In the

Table 18.3. Genotypes and phenotypes of haploids and merodiploids for the lac region in *E. coli* strains

Genotype	Phenotypes*		Interpretted *i* allele behaviour
	Constitutively synthesized proteins	Inducibly synthesized proteins	
Haploid			
$i^+o^+z^+y^+a^+$	None	All three	Wild type
$i^so^+z^+y^+a^+$	None	None	Superrepressor mutant
$i^-o^+z^+y^+a^+$	All three	None	Constitutive mutant
Merodiploid			
$\dfrac{i^so^+z^+y^+a^+}{i^+o^+z^-y^-a^-}$	None	None	i^s dominant over i^+
$\dfrac{i^so^+z^-y^-a^-}{i^+o^+z^+y^+a^+}$	None	None	i^s dominant over i^+
$\dfrac{i^-o^+z^+y^+a^+}{i^+o^+z^-y^-a^-}$	None	All three	i^+ dominant over i^-
$\dfrac{i^-o^+z^-y^-a^-}{i^+o^+z^+y^+a^+}$	None	All three	i^+ dominant over i^-

absence of inducer repressor could thus prevent transcription. Thus, in absence of inducer the three enzymes are not synthesized.

Confirmation of Jacob-Monod Model. Gilbert isolated in 1960's purified *i* gene product, repressor. Repressor was present in i^+ but not in i^- cells. Repressor protein binds to o^+ but not o^c DNA. This confirmed the model of Jacob and Monod.

Mutations in Promotor. The third kind of regulatory mutations in lac region of *E. coli* was found in another site, called promotor. This site was found to be present to the left of the operator. Promotor mutants coordinately reduced proteins of z-y-a cluster. There mutants responded to *i* gene (repressor) control. That is, the three proteins are made in low

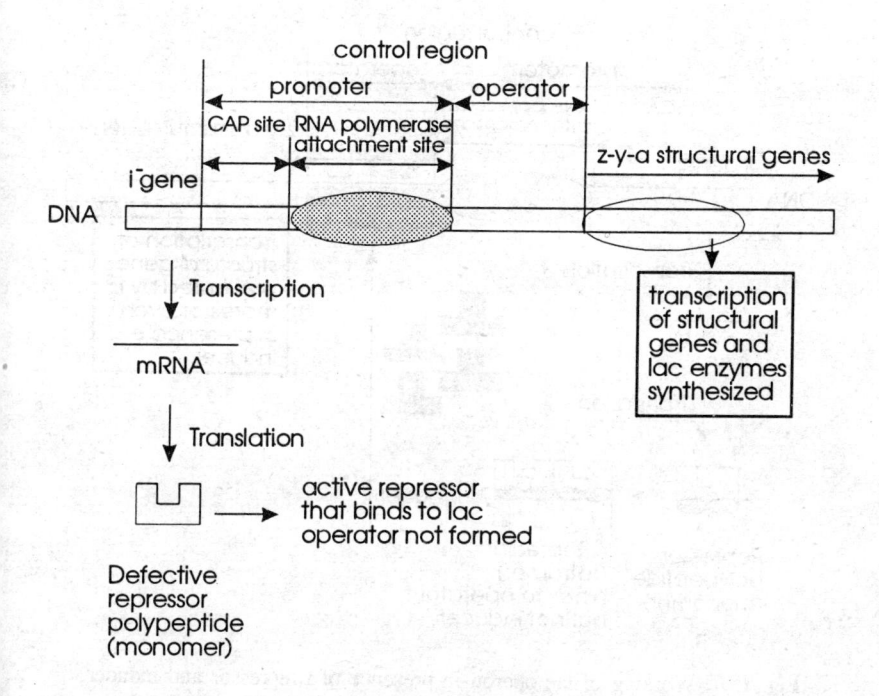

Fig. 18.6. The *i* gene mutation leads to constitutive synthesis of the three enzymes

amounts in a promotor mutant cell when lactose is present. This means repressor control has not changed. Promotor is the specific DNA site to which RNA polymerase binds. Polymerase can move along DNA if it is not blocked by a repressor bound to the operator between promotor site and *z-y-a* cluster. In mutant promotor cell, RNA polymerase does not bind to promoter tightly.

Polar Mutations. Direction of transcription was determined by studies on polar mutants. A polar mutation reduces all wild type activity distal to it, so that a polar mutation in *z* gene influences *z*, *y* and *a*; a polar mutation in *y* influences *y* and *a* genes only; and a polar mutation in gene *a* influences gene *a* only. These studies showed that direction of transcription is distally, from promotor to the operator and on to *z, y, a* in that order.

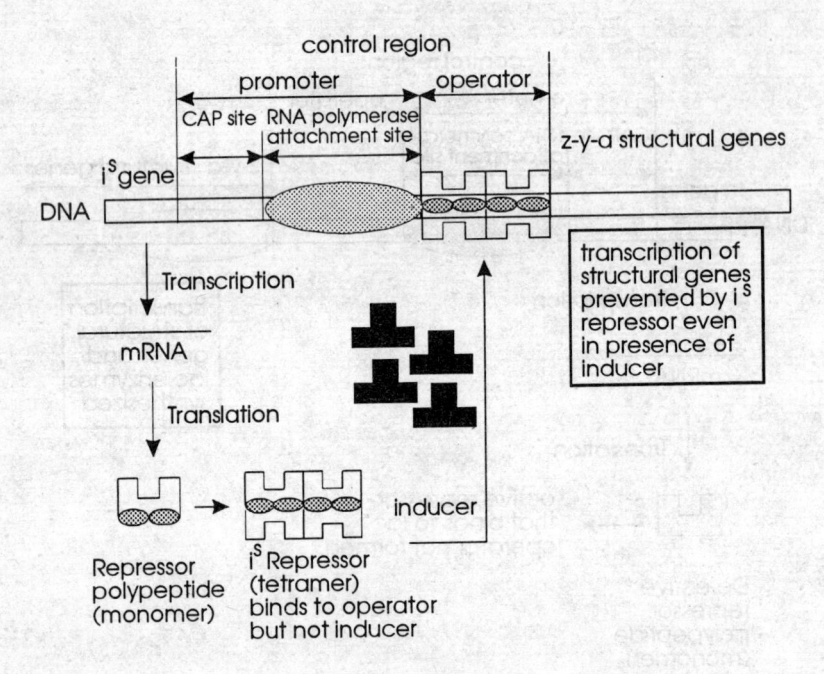

Fig. 18.7. Working of lac operon in presence of i repressor and inducer. Transcription of lac genes is prevented even in presence of inducer

Controlling Region of Lac Operon. Detailed base sequence of controlling region of *lac* operon of *E. coli* is given in **Fig. 18.8**. Sites of various components of the controlling region have been marked in this figure.

Simultaneous Control of Different Operons

Sometimes the products of regulator genes are affected by inducers in such a way they cross-regulate working of two operons simultaneously. For example, products of pegulator gene $_{II}$ are affected by inducer i_{II} but control **operon I**. Products of regulator i_I are affected by inducer i_I but control the **operon II** (Fig. 18.9).

Positive Control of *Lac* Operon. There also exists a positive control of gene action in lac operon. Positive control in *lac* operon of *E. coli* was studied by **glucose effect**. Cyclic AMP (cAMP) is the key metabolite

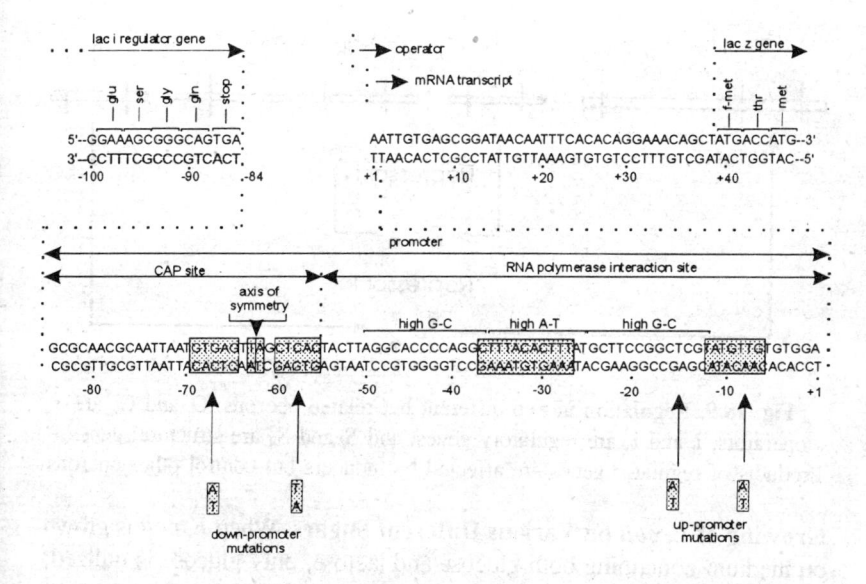

```
            lac i regulator gene                    ·  ▶ operator                          lac z gene
                                                    ·  ▶ mRNA transcript                 ▶
        glu  ser  gly  gln  stop                    ·                             f-met  thr  met
         |    |    |    |    |                      ·                               |    |    |
   5'--GGAAAGCGGGCAGTGA·                        AATTGTGAGCGGATAACAATTTCACACAGGAAACAGCTATGACCATG--3'
   3'--CCTTTCGCCCGTCACT·                        TTAACACTCGCCTATTGTTAAAGTGTGTCCTTTGTCGATACTGGTAC--5'
       -100      -90   ·-84                     +1·     +10      +20      +30      +40
```

                                                 promoter
       ◀─────── CAP site ────────▶◀──────── RNA polymerase interaction site ──────────▶
                       axis of
                      symmetry
                                        high G-C   high A-T   high G-C
   · GCGCAACGCAATTAAT GTGAGTTAGCTCAC TACTTAGGCACCCCAGG CTTTACACTTT ATGCTTCCGGCTCG TATGTTGT GTGGA ·
   · CGCGTTGCGTTAATTA CACTCAATCGAGTG AGTAATCCGTGGGGTCC GAAATGTGAAA TACGAAGGCCGAGC ATACAACA CACCT ·
       -80   -70   -60   -50   -40   -30   -20   -10   +1
                          ▲   ▲   ▲   ▲
                         A   T   A   A
                         T   A   T   T
                         down-promoter   up-promoter
                           mutations   mutations

Fig. 18.8. Control region of lac operon of *E. coli*

which influences glucose effect. cAMP is effective only when bound to catabolite activator protein (CAP). When cAMP-CAP complex binds to a specific site on operon, it increases the rate of transcription of *z-y-a* cluster. This system enables the cells to utilize a large number of different sugars as a source of carbon. CAP-deficient strains and adenylate cyclase deficient strains do not influence glucose effect. Adenylate cyclase-deficient strains make low amounts of inducible enzymes. CAP-deficient strains make low amounts of inducible enzymes. The positive and negative controls act together, but each is a separate regulation system for operon transcription.

cAMP-CAP Complex. Binding site of cAMP-CAP complex is located proximally to the *lac* promotor. When cAMP-CAP complex is bound at the promotor, binding of RNA polymerase to the operon is greatly enhanced. Local destabilization of double helical DNA is induced at RNA polymerase interaction site of the promotor by cAMP-CAP binding. RNA polymerase binds to the destabilized area. RNA polymerase then moves to the operator site and begins transcription. cAMP is synthesized from adenosine triphosphate (ATP) by adenylate cyclase (**Fig. 18.10**).

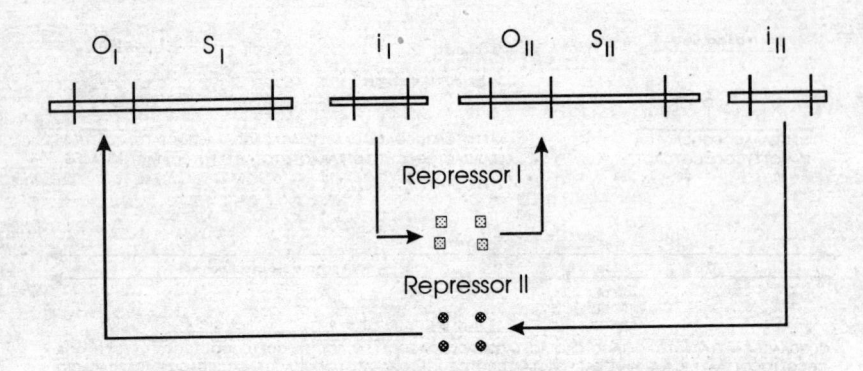

Fig. 18.9. Regulation in two different but related operons. O_I and O_{II} are operators, i_I and i_{II} are regulatory genes, and S_I and S_{II} are structural genes. Products of regulator genes are affected by inducers but control other operons

Growing of *E. coli* on Various Different Sugars. When *E. coli* is grown on medium containing both glucose and lactose, only glucose is utilized; the lac operon in inactive. The same situation holds good in case of glucose and galactose. The galactose operon remains inactive. The glucose blocks the activity of other operons through cAMP. When glucose is available as energy source, cAMP level decreases, which prevents the functioning of other operons involved in the metabolism of other sugars, like lactose, arabinose, galactose, maltose. When glucose levels decrease, cAMP levels increases, and other source of sugar can be utilized.

Control at Translational Level. For *lac* operon of *E. coli*, there is 10:5:2 ratio of *z, y,* and *a* gene products. The relative amounts of polycistronic mRNA depend upon the primary structure of mRNA. At the end of each cistron on mRNA there is atleast one chain termination codon followed by an intercistron segment prior to next translation initiation codon. The **intercistron nucleotide sequences** govern whether or not a ribosome "out of gear" will fall off the message before initiating the next polypeptide chain. Presumably, in *lac* operon, half (50%) of the ribosomes fall off between *z* and *y* and another more than half (60%) fall off between genes *y* and *a*.

Genes in prokaryotes are apparently continuously "turned on" and must be repressed if activity is to be controlled. In general, genes of

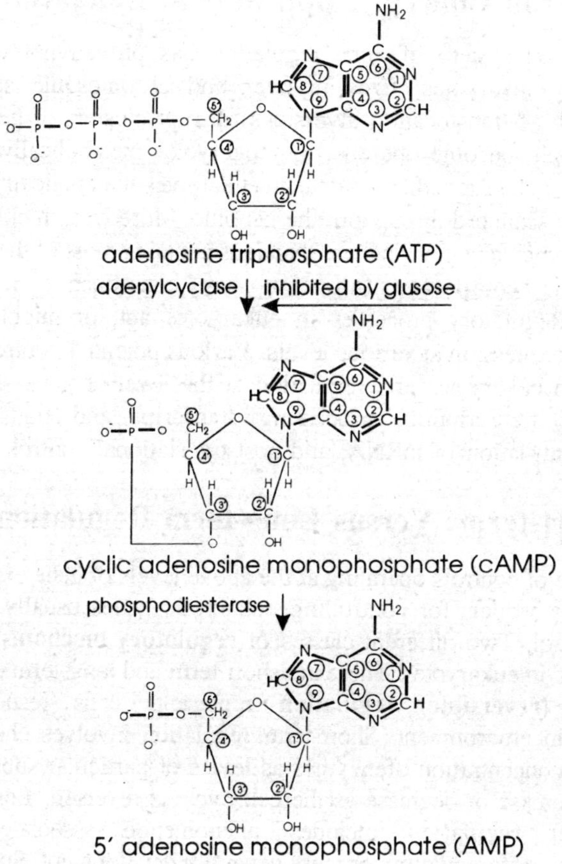

adenosine triphosphate (ATP)

adenylcyclase ↓ ← inhibited by glusose

cyclic adenosine monophosphate (cAMP)

phosphodiesterase ↓

5′ adenosine monophosphate (AMP)

Fig. 18.10. Conversion of ATP into cAMP and to AMP

eukaryotes are apparently "turned off" and must be activated. Differentiation and growth/development takes place only in eukaryotes. No cell division occurs in prokaryotes. At any one time, in a eukaryote cell, only 2-15 per cent of total genome is expressed. It means some sort of regulation is occurring. In chromatin, two types of proteins present are histones and non-histones. Histones are basic in nature and are involved in a structural role of chromosomes. Recently histones have been found to play regulatory role also. Non-histones are acidic in nature and are thought to act as controlling elements for gene transcription.

Is Operon Concept Applicable to Eukaryotes?

The operon system of gene regulation has proved not to be in applicable in eukaryotes. Crossing-over and chromosome structural changes (such as translocation inversions) disrupt clusters of genes that otherwise might become operons. In prokaryotes metabolically related genes are lying close together whereas in eukaryotes metabolically related genes may be scattered throughout the genome. Moreover, in eukaryotes cells, compartmentalization is there. Nucleus contains most of the genetic material. The **compartmentalization** has consequence for gene expression. Regulatory processes in eukaryotes act, or might act, to control gene expression at various levels. Various potential control points recognized in eukaryotes are: regulation at the level of gene structure, initiation of transcription, processing transcript and transport to cytoplasm, translation of mRNA, and post-translational control.

Short-terms Versus Long-term Regulation

Evidence of controls operating at the above levels of gene expression are available. Models for controlling eukaryotic genes usually assume positive control. Two different classes of **regulatory mechanisms** may be recognized in eukaryotes. These are short-term and long-term controls. **Short-term (reversible) regulation** represents cells response to fluctuations in environment. Short-term regulation involves changes in activities or concentration ofenzymes as levles of particular substrate or hormones increase or decrease as the cell cycle is reversed. **Long-term (irreversible) regulation** includes phenomenon associated with determination, differentiation, or more generally development. Short-term regulation is a feature of both developing and fully differentiated eukaryotic cells, i.e., even when a cell is undergoing differentiation, it responds to responding to its environment. During differentiation certain features of a short-term regulatory process may change dramatically. For example, DNA synthesis in early embryos occurs rapidly and cell cycle exhibits very short, if any, G_1 and G_2 periods. As differentiation proceeds, these periods lengthen and ultimately assume cell-specific proportions.

Mechanisms of Gene Regulation in Eukaryotes

A model designed to portray gene regulation in eukaryotes must account for the following observations: A simple external signal mediates a change in activity of a cell. Integrated activity of a large number RNA of noncontiguous genes is required for a particular state of differentiation. heterogeneous nuclear (hnRNA) is produced only in nuclei of higher organisms. .Genomes of higher organisms are very large as compared to bacterial genomes. A high proportion of DNA in higher organisms does not code for mRNA. Some DNA may be having regulatory and some protective functions. Repetitive DNA sequences are found in higher (but not in bacterial) cells. The repetitive DNA sequences are transcribed according to cell-specific patterns in differentiated cells.

Britten and Davidson Model

All these conditions require great flexibility in gene regulation. It is accepted fact that portions of DNA involved in regulatory functions would prevail over those carrying structural information. R.J. Britten and E.H. Davidson developed such a theoretical model in 1969 to explain enormous genome and possible modes of regulation in higher forms. This model was later elaborated by them in 1973. The model assumes the presence of the following four types of sequences:

Producer Gene. A producer gene is comparable to a structural gene in prokaryotes. It produces pre-mRNA which after processing becomes mRNA. Expression of a producer gene may be under the control of many receptor sites.

Receptor Site. Receptor site is comparable to operator in bacterial operon. One such receptor site is assumed to be present adjacent to each producer gene. A specific receptor site is activated when a specific **activator RNA** or **activator protein**, a product of integrator gene, complexes with it.

Integrator Gene. It is comparable to regulator gene and is responsible for synthesis of an activator RNA molecule that may or may not give rise to proteins before it activates the receptor site.

Sensor Site. A sensor site regulates activity of an integrator gene which can be transcribed only when the sensor site is activated. The sensor sites are also regulatory sequences that are recognized by **external stimuli (agents)**, like temperature or light, or internal **developmental signals** like hormones or chemical inducers after binding with some regulatory proteins.

328 Basic Genetics

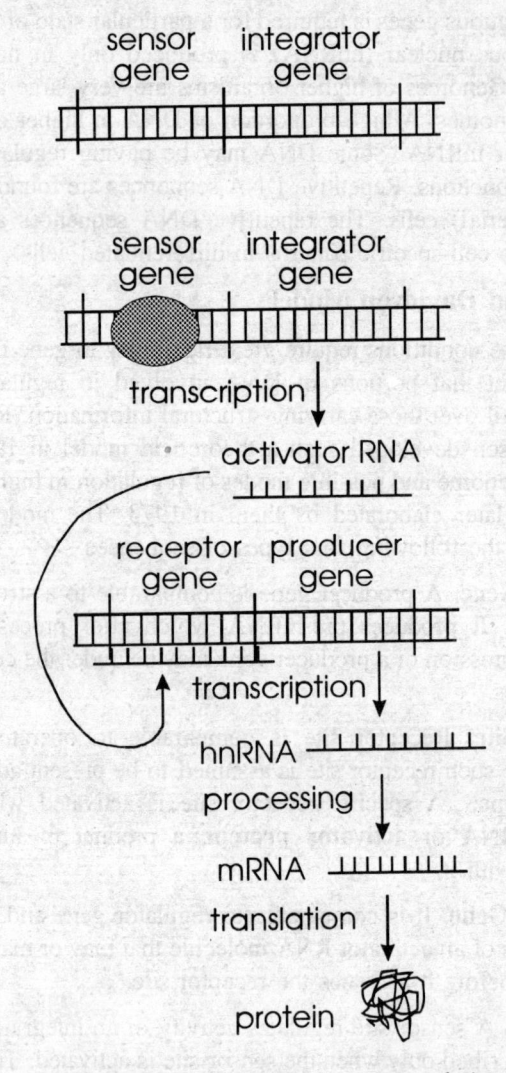

Fig. 18.11. Britten-Davidson model of eukaryotic gene regulation. Interactions among different components of the machinery are shown

The inter-relationship of four classes of molecules are illustrated in **Fig. 18.11**. Transcription of a producer gene could occur only if at least one of its receptor sites was activated by forming a sequence-specific complex with activator RNA (or activator protein). This RNA (or protein) would be synthesized by integrator gene in response to the signals by the sensor site that are sensitive to external or internal development signals.

In Britten-Davidson model, it is also proposed that receptor sites and integrator genes may be repeated a number of times so as to control the activity of a large number of genes in the same cell. Repetition of receptor ensures that same activator recognizes all of them and in this way several enzymes of one metabolic pathway are simultaneously synthesized. Transcription of the same gene may be needed in different developmental stages. This is achieved by multiplicity of receptor sites and integrator genes. Each producer gene may have several receptor sites, each responding to one activator (**Fig. 18.12**). Thus, a single activator though can recognize several genes but different activators may activate the same gene at different times. An integrator gene may also fall in cluster with the same sensor site (**Fig. 18.13**).

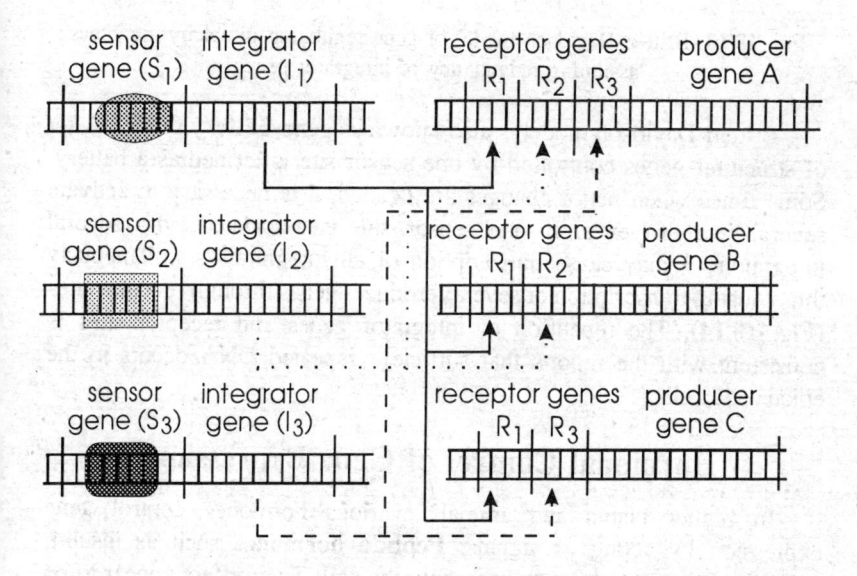

Fig. 18.12. Britten-Davidson model of gene regulation in eukaryotic genes assuming redundancy of receptor genes

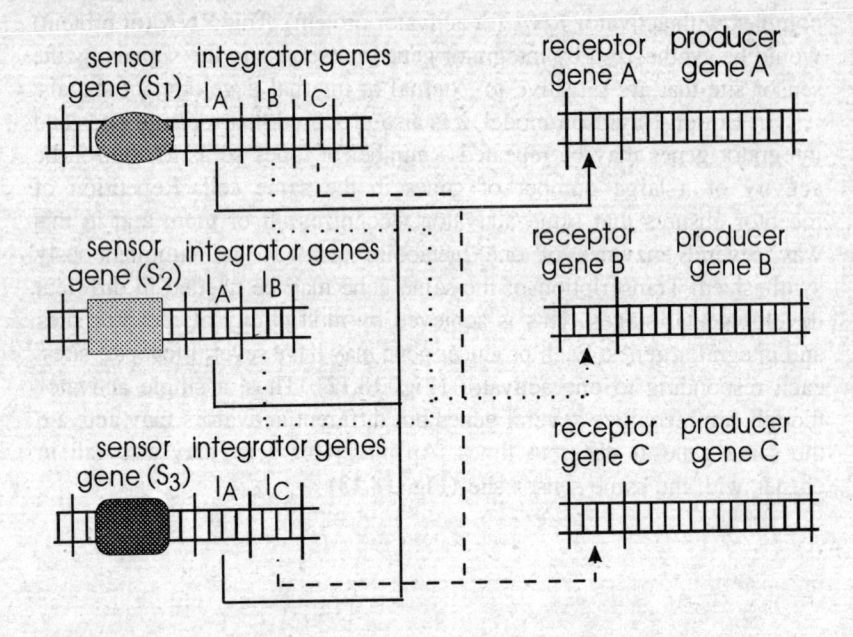

Fig. 18.13. Britten-Davidson model of gene regulation in eukaryotic genes assuming redundancy of integrator genes

Britten-Davidson model is also known as **gene-battery model**. A set of structural genes controlled by one sensor site is termed as a battery. Some times when major changes are needed, it is necessary to activate several sets of genes. If one sensor site is associated with several integrators, it may cause transcription of all integrators simultaneously thus causing transcription of several producer genes through receptor sites (**Fig. 18.14**). The repetition of integrator genes and receptor sites is consistent with the reports that sufficient repeated DNA occurs in the eukaryotic cells.

Hormonal Control of Gene Expression

In higher plants and animals, various hormones control gene expression by acting as signals. **Peptide hormones** such as insulin, epinephrine, are too large to enter into the cell. Their effect appear to be mediated by receptor proteins located in target cell membranes and by the

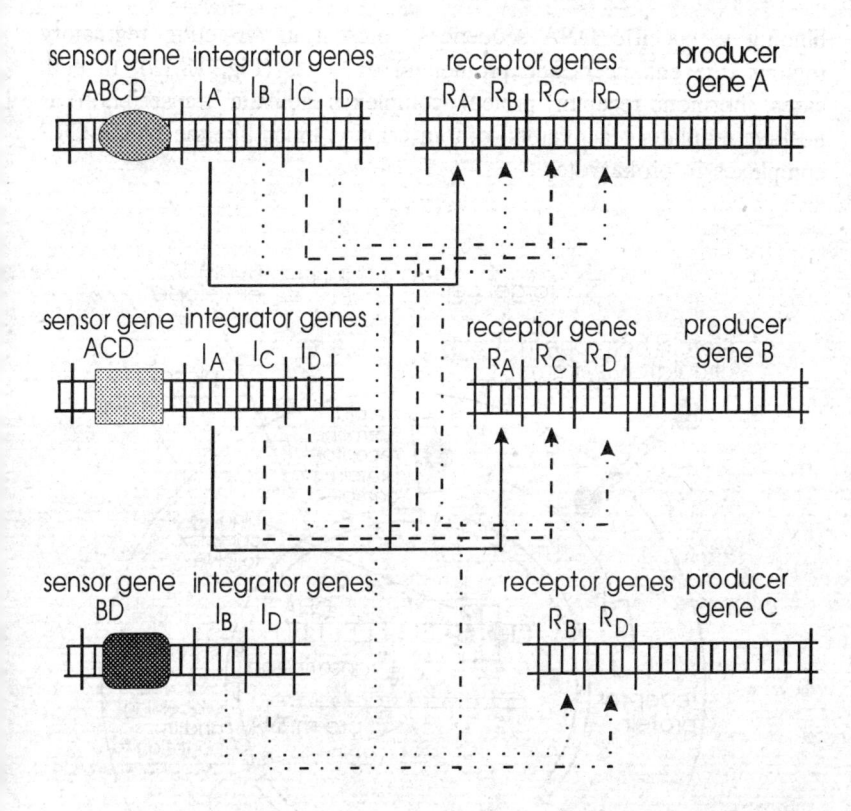

Fig. 18.14. Britten-Davidson model of gene regulation in eukaryotic genes assuming redundancy of receptor as well as integrator genes

intracellular levels of cyclic AMP (called **secondary messenger**). The cyclic AMP activates a **protein kinase** which phosphorylates and thus activates many specific enzymes. **Steroid hormones** such as estrogen, progesterone, testosterone are small molecules which readily enter through the plasma membrane. Once inside the appropriate target cells, the steroid hormones get tightly bound to specific receptor proteins which are present only in the cytoplasm of the target cells. The **hormone-receptor protein complexes** activate the transcription of specific genes by two possible ways: First, the hormone-receptor protein complexes interact with specific nonhistone chromosomal protein and this interaction stimulates the transcription of correct genes. Second, the hormone-receptor protein complexes activate transcription of target genes by binding to specific DNA sequences present in *cis*-acting regulatory regions (the enhancers and promoters) of genes (**Fig. 18.15**). In both cases, hormone-receptor protein complexes activate transcription as positive regulators (activators) of transcription much like the cAMP-CRP complexes in prokaryotes.

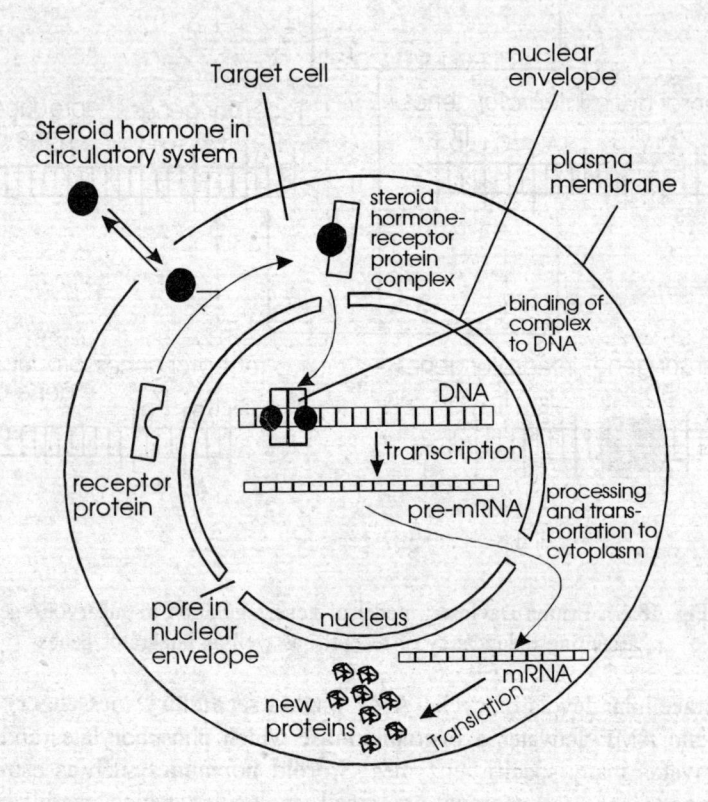

Fig. 18.15. Model action of steriod hormones control of gene expression

During development of dipteran flies (e.g. *Drosophila melanogaster*, *Chironomus tentans*) the steroid hormone **ecdysone** is released and triggers moulting. If larvae of these insects are treated with ecdysone at stages of development prior to or between moultings, patterns of chromosome puff occur that are identical to those occurring during natural moultings. Ecdysone tends to affect gene expression at the level of transcription.

Feedback Inhibition

This type of inhibition is also also known as **end-product inhibition**. Studies on isoleucine synthesis in *E. coli* have demonstrated that addition of isoleucine (the end-product of a five-step conversion of threonine) into a culture of bacterium resulted in immediate blocking of the isoleucine ← threonine pathway (**Fig. 18.16**). Addition of isoleucine inhibits the action of enzyme responsible for deamination of threonine to α-ketobutyrate. The end product binds to the enzyme so that it completes with substrate for a site on the enzyme molecule. The competition is not for the same site. The enzyme apparently has two specific recognition sites, one for its substrate and other for the inhibition end product. Attachment at inhibitor site affects the substrate site. It alters the ineraction between the substrate site and substrate. The term **allosteric interaction** is applied to such changes in enzyme activity due to a change in the shape of the enzyme molecule produced by binding of a second substance at a different and nonoverlapping site.

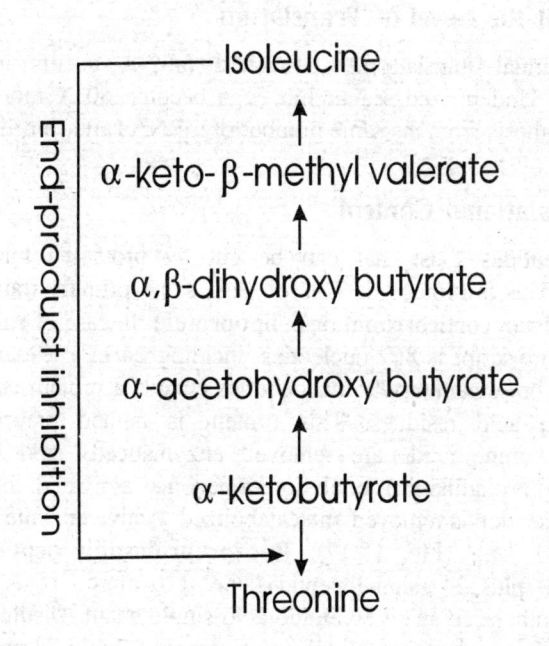

```
                Isoleucine
                    ↑
End-product     α-keto-β-methyl valerate
inhibition          ↑
                α,β-dihydroxy butyrate
                    ↑
                α-acetohydroxy butyrate
                    ↑
                α-ketobutyrate
                    ↓
                Threonine
```

Fig. 18.16. Post-transcriptional control by end-product (feed-back) inhibition

An other example is the coat protein of RNA phage R17. It binds to a hairpin that encompasses the ribosome binding site in the phage mRNA. Phage T_4 Reg A protein binds to the consensns sequence that includes the AUG initiation codon in several T_4 early mRNAs. T_4 DNA polymerase binds to a sequence in its over mRNA that includes the Shine-Dlgarno elements needed for ribosome binding. **Autogenous regulation** is formed among proteins that are incorporated into macromolecular assemblies. The assembled particle may be unstable as a regulator, because it is large, too numerous, or too restricted in its locations. If assembly pathway is blocked for some resaon, free subunits accumulate and shut off unnecessary synthesis of further components. Eukaryotic cells have this type of regulation, e.g., tubulin synthesis. Free tubulins bind to nascent polypeptide or mRNA. mRNA is degraded. The T7 early region is transcribed into a single RNA that is cleaved into five individual mRNAs by RNase III. Hairpins in precursor RNA are recognized by the enzyme RNase III. When cleavage is prevented certain coding regions are not translated. At least 12 ribonucleases exist in *E. coli*. Some are concerned with processing of tRNAs, some of rRNAs and some mRNAs.

Control at the Level of Translation

Differential translation of finished mRNA occurs in the cell cytoplasm. Undertilized sea urchin eggs become 50 X more active in protein synthesis from the same number of mRNAs after fertilization than before.

Post-translational Control

Polypeptides exist that can be cut by proteases into different activities. This is proteolytic control. e.g., same primary transcript may produce **adrenocorticotropin** or **β-lipoprotein**. In case of **rat killikrein gene**, the transcript is 867 nucleoties, inclding 24 in a leader and 48 in the train. The remaining 795 nucoeotides encode a protein consisting of 265 amino acid residues. This protein is named preprokallikrein. Seventeen amino acids are removed enzymatically and becomes a proenzyme (prokallikrein), but before it becomes activated, the 11-amino acid "pro" section is removed and catabolized. Active enzyme is only 237 amino acid long (**Fig. 18.17**). **Rat preproinsulin peptide** has 86 (proinsulin) plus 24 (signal peptide), i.e. 110 amino acids. Insulin is initially synthesized as an 86 amino acid single chain. Disulphide bonds are formed, one between cysteines at positions 7 and 72 and the other

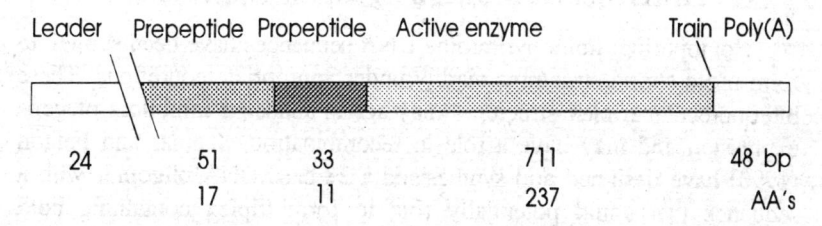

Leader	Prepeptide	Propeptide	Active enzyme	Train	Poly(A)
24	51	33	711	48	bp
	17	11	237		AA's

Fig. 18.17. Rat kallokrein gene serves as an example of post-tranlational gene regulation

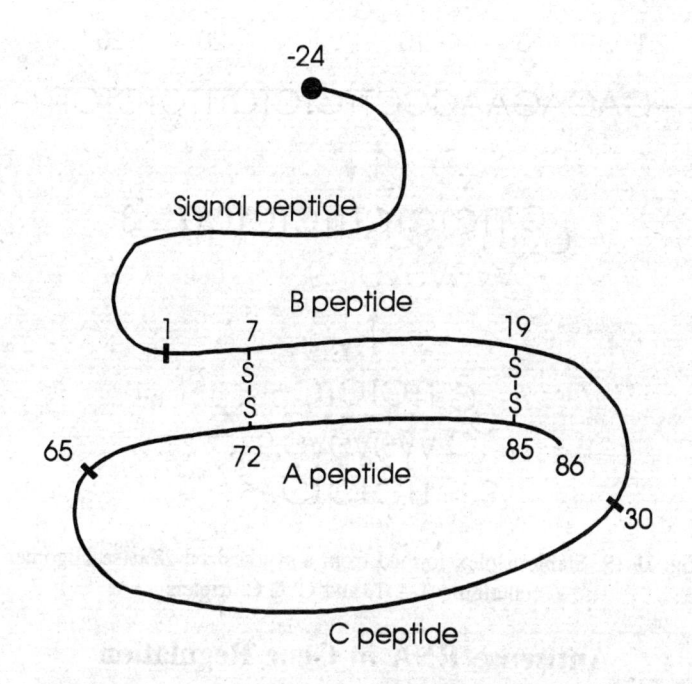

Fig. 18.18. Rat preproinsulin peptide undergoes post-translational modifications before yielding functional insulin

between cysteines at positions 19 and 85 (**Fig. 18.18**). The internal cuts excise 36 amino acids. Mature insulin hormone carries the A chain (30 amino acids) and B chain (21 amino acids).

Antisense Inhibition of Gene Expression

Homopurine-homopyrimidine DNA sequences have been shown to form triple strand structures readily under appropriate conditions. These intermolecular triplex structures may act as antisense inhibitors of gene expression and may have a role in recombination. Skelnar and Feigon (1990) have designed and synthesised a 28-base DNA oligomer with a sequence that could potentially fold to form triplex containing both T.A.T. and C+.G.C. triplets (Fig. 18.19). In case of formation of such a triple helix structure DNA looping is seen.

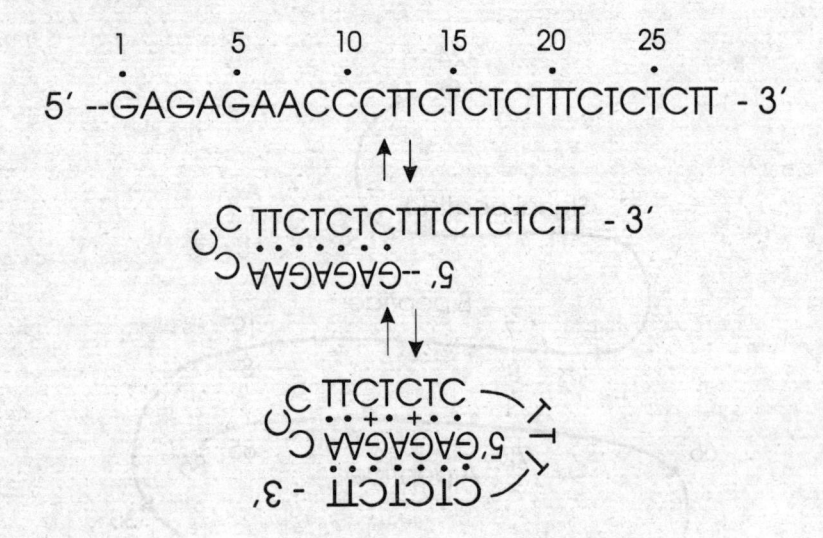

Fig. 18.19. Stable triplex formed from a synthesized 28-base oligomer containing T.A.T. and C⁺.G.C. triplets

Antisense RNA in Gene Regulation

Translation can be regulated by **RNA-RNA hybridization**. RNA complementary to the 5' end of a mRNA can prevent the tranlation of that messanger RNA (Itoh and Tomizawa, 1980). Several examples of this type of regulation are known. The regulating RNA is called **antisense RNA**. The messenger RNA from the *ompF* gene in *E. coli* is prevened from being translated by complementary base pair binding with an antisense RNA, called *micF* RNA (*mic* stands for **mRNA-interfering complementary RNA**). The *ompF* gene codes for a memebrane component called a **porin**, which, as the name suggests, provides pores in the cell mmebrane for transport of materials. Surprisingly, a second porin gene, *ompC*, seems to be the source of the *micF* RNA. Transcription of the opposite DNA strand (the one not normally transcribed), near the promoter of the *ompC* gene yield the antisense RNA. One porin gene thus seems to be regulating the expression of another porin gene, for reasons that are not completely understood. **Antisense RNA** has also been implicated in such phenomena as the control of plasmid number and the conrol of transposon TN10 transposition. Control by antisense RNA is a fertile field for gene therapy because antisense oligonucleotides can be artifically synthesized and then injected ito eukaryotic cells (Cazenave and Helene, 1991).

Molecular Zippers in Gene Regulation

Many gene regulating proteins "zip" together into pairs. This linkage is critical to their ability to bind to DNA. The "teeth" that join the molecules almost always consist of amino acid leucine. Landschlutz *et al.* (1988) have called these toothed regions the **leucine zippers**. Detailed account of the action of these leucine zippers has been given by McKnight (1991).

These workers concentrated on a protein that recognized a motif called CAT (short for CCAAT) found in many promoters of viral and mammalian genes. This protein had ability not only of binding to CAT motif in the promoters but also had affinity for the core homology common to many enhancers. This substance was thus named C/EBP, for CAT/Enhancer-Binding Protein. This protein was not made in all the tissues. it was abundant in the liver, lungs, small intestine and placental tissues as well as in fat. C/EBP was made by hepatocytes but not by cells that form secretory duct work of liver. This observation suggested that this substance may participate in **selective gene expression**. This protein was found to be 359 amino acid long. Sixty-amino acid segment of C/EBP was quite similar to segments of two other proteins: products of *myc* and *fos* proto-oncogenes which are also regulators. A 35-amino acid segment revealed amino acid leucine which is extremely hydrophobic and protrudes relatively far from the backbone of the helix, occupied every seventh position thus forming a continuous ridge. The regulatory protein molecules join at their zipper regions, forming a dimer. Each dimer subunit interacts with one of the dyad halves of the motif. This was also observed to be the case for regulatory proteins that are products of genes *GCN4* and *jun*. Leucine zippers create dimers needed for DNA binding, the zipper region cannot itself bind DNA.

A region rich in positively charged amino acids - specially arginine and lysine reside close to the zipper. This lysine/arginine rich region is the DNA-binding region. Studies on eleven gene regulatory proteins from plants, fungi and mammals that all regulatory proteins were designed so (**Fig 18.20**). The comparison of these protein showed that certain amino in the protrusion acids were conserved suggesting that they aid binding and touch the DNA. Asparagine, a potential helix breaker was found at a fixed position in every DNA-binding region of every protein. This amino acid may allow the protruding section to bend and thus contact the DNA continuously. When monomers were assembles into dimers, the resulting unit would adopt the shape of a "Y". The **zipper region** would form a stem and arginine/lysine regions would form the DNA-binding arms.

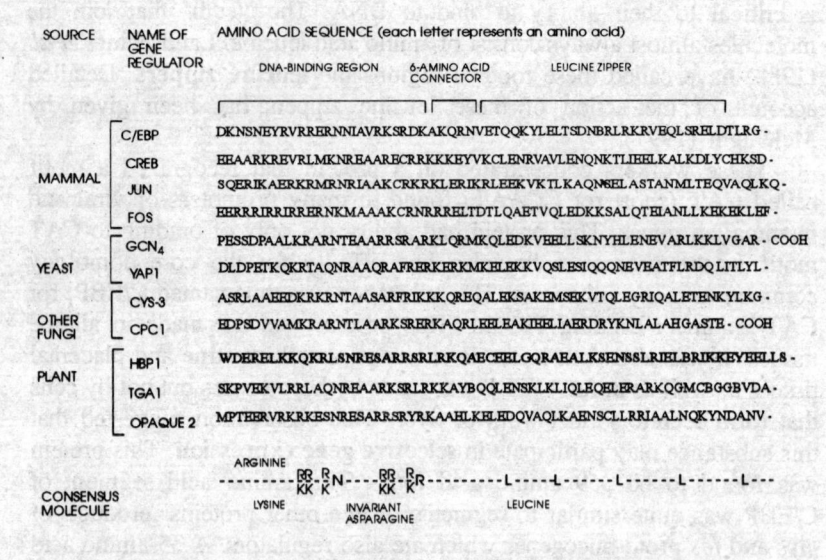

SOURCE	NAME OF GENE REGULATOR	AMINO ACID SEQUENCE (each letter represents an amino acid)
		DNA-BINDING REGION / 6-AMINO ACID CONNECTOR / LEUCINE ZIPPER
MAMMAL	C/EBP	DKNSNEYRVRRERNNIAVRKSRDKAKQRNVETQQKYLELTSDNBRLRKRVEQLSRELDTLRG -
	CREB	EEAARKREVRLMKNREAARECRRKKKEYVKCLENRVAVLENQNKTLIEELKALKDLYCHKSD -
	JUN	SQERIKAERKRMRNRIAAKCRKRKLERIKRLEEKVKTLKAQNSELASTANMLTEQVAQLKQ -
	FOS	EERRRIRRIRRERNKMAAAKCRNRRRELTDTLQAETVQLEDKKSALQTEIANLLKEKEKLEF -
YEAST	GCN$_4$	PESSDPAALKRARNTEAARRSRARKLQRMKQLEDKVEELLSKNYHLENEVARLKKLVGAR - COOH
	YAP1	DLDPETKQKRTAQNRAAQRAFRERKERKMKELEKKVQSLESIQQQNEVEATFLRDQLITLYL -
OTHER FUNGI	CYS-3	ASRLAAEEDKRKRNTAASARFRIKKKQREQALEKSAKEMSEKVTQLEGRIQALETENKYLKG -
	CPC1	EDPSDVVAMKRARNTLAARKSRERKAQRLEELEAKIEELIAERDRYKNLALAHGASTE - COOH
PLANT	HBP1	WDERELKKQKRLSNRESARRSRLRKQAECEELGQRAEALKSENSSLRIELBRIKKEYEELLS -
	TGA1	SKPVEKVLRRLAQNREAARKSRLRKKAYBQQLENSKLKLIQLEQELERARKQGMCBGGBVDA -
	OPAQUE 2	MPTEERVRKRKESNRESARRSRYRKAAHLKELEDQVAQLKAENSCLLRRIAALNQKYNDANV -
CONSENSUS MOLECULE		ARGININE / --------RR-R N---- RR-R R---------L------L-------L------L------L--- / KK K / KK K / LYSINE / INVARIANT / LEUCINE / ASPARAGINE

Fig. 18.20. Amino acid sequence of gene regulators of 11 protein

Dimeric structure of regulatory proteins suggests that dimers can be heterodimers or homodimers. Heterodimers increase number of usable motifs and also increase variety of protein combinations.

Some Useful Reading Hints

Jacob and Monod (1961) explored genetic regulatory mechanisms in the synthesis of proteins. Gilbert and Müller-Hill (1966) Gilbert and Müller-Hill (1967) isolated lac repressor and found that *lac* operator is DNA. Beckwith and Zipser (1970) dealt with the *lactose* operon. Englesberg and Wilcox (1974) described regulation by positive control. Steitz et al. (1974) gave the details of molecular shape, subunit structure, of *lac* represor protein and proposed model for operator interaction based on structural studies of microcrystals. Goldberger (1974) described autogenous regulation of gene expression. Dickson et al. (1975) explained genetic regulation in the *lac* control. Beyreuther (1978) gave chemical structure and functional organization of *lac* represor from *Escherichia coli*. Barkley and Bourgeois (1978) described the mechanism of tepressor recognition of operators and effectors. Savageau (1979) detailed out a general theory and experimental evidence of autogenous and classical regulation of gene expression. Slilaty and Little (1987) demonstrated that Lysine-156 and Serine-119 are required for LexA repressor clevage. Gold (1988) dealt with posttranscriptional regulatory mechanisms in *Escherichia coli*. Lilley (1991) discussed the events that follow when the CAP fits bent DNA. Schultz et al. (1991) gave crystal structure of a CAP-DNA complex and reveal that the DNA is bent by 90°. Sharp (1992) considered that TATA-binding protein was a classless factor. Kim (1993) gave crystal structure of a yeast TBP/TATA-box complex. Busby and Ebright (1994) dealt in detailed with promoter structure, promoter recognition, and transcription activation in prokaryotes.

Simons and Kleckner (1988) explained biological regulation by antisense RNA in prokaryotes. Weintraub (1990) discussesed scope of antisense RNA and DNA. Cazenave and Hélène (1991) discussed nature, structure and applications of antisense oligonucleotides. Eguchi et al. (1991) gave details of antisense RNA technology. Aggarwal (1992) discussed antisense oligonucleotides as antiviral agents. Richardson (1997) discussed the blocking of adenosine with antisense RNA.

Darnell et al. (1973) dealt with biogenesis of mRNA and genetic regulation in mammalian cells. Darnell (1982) discussed with examples variety in the level of gene control in eukaryotic cells. Britten and Kohne (1968) dealt with repeated sequences in DNA. Britten and Davidson (1969) gave a theory of gene regulation for higher cells. Stein et al. (1975) gave the nature of chromosomal proteins and their role in gene regulation. Klukas (1976) described role of non-histone proteins in transcription. O'Malley et al. (1977) gave general theory of regulation of gene expression in eukaryotes. Robertson (1979) gave the regulatory sequences involved in the promotion and termination of transcription. Davidson and Britten (1979) described regulation of gene expression and explain possbile role of repetitive sequences in this process. Long and Dawid (1980) described repeated genes in eukaryotes.

Questions

1. What is the meaning of gene regulation. What is the need of gene regulation for a cell?
2. Differentiate between the following pairs of terms: constitutive versus regulated genes, inducible versus repressible gene regulation and negative versus positive control

340   Basic Genetics

3. Based on whether control is inducible or repressible, or whether the control is positive or negative, what models of gene regulation can be conceived? Also state examples of each model of gene regulation.
4. Draw comparisons between negative and positive control mechanisms of gene regulation.
5. Define the following terms with reference to gene regulation in prokaryotes: Promotor, Operator, Operon, Effector, Inducer, Co-repressors, Regulator molecule, Repressor and Regulatory gene.
6. Describe structure and working of *lac* operon.
7. Mutations have been very helpful in understanding gene regulation. In what way the following types of mutations been useful? mutations in structural genes, mutations in operator, mutations in regulatory gene, and mutations in promotor.
8. How was the model of Jacob-Monod confirmed?
9. How did the polar mutations reveal the direction of transcription in *lac* operon in E. coli?
10. Draw a diagram showing the controlling region of lac operon of *E. coli*.
11. Comment on the statement:: "*lac* operon of *E. coli* has inducible negative control".
12. How is simultaneous control of different operons achieved?
13. Describe positive control of *lac* operon. What is role of cAMP-CAP complex in this positive control?
14. Suppose growing *E. coli* are grown on various different sugars, including glucose, which sugar will it prefer and why?
15. Are the the products of z, y and a structural genes of *lac* operon of *E. coli* found in the cell, in presence of lactose, in the same quantity. If not, why and at what level such a control operates?
16. A mutational change can lead to the transformation of an inducer into a corepressor. On the other hand, the conversion of a corepressor can also occur into an inducer. What do these observations show?
17. What will the effect of a chromosomal aberration when it puts structural genes of an operon next to another operator affected by a different repressor?
18. What is antisense RNA? How is antisense RNA able to regulate gene expression at translation level. What is the practical use of antisense RNA technology?
19. Is operon concept applicabile to eukaryotes?
20. What are the basic requirements of a mechanism for explaining gene regulation in eukaryotes?
21. Describe various components of Britten and Davidson model of gene regulation in eukaryotes. How do these components interact?
22. How is flexibility and simultanous regulation of metabolically related genes provided in the Britten-Davidson model?
23. What is the mechanism of action of steroid hormones in controlling gene expression?
24. What do you understand by antisenese inhbition of gene expression? How is it achieved through triple helix DNA structure?
25. What role do the leucine zippers and arginine/lysine regions of regulatory proteins play in their to DNA? Are different regulatory proteins related?

19

MUTATIONS

Definition

Mutation is defined as a change in the DNA at a particular locus in an organism. Mutation is also defined as the process that produces a gene or a chromosome set differing from the wildtype. The gene or chromosome set that results from such a process is also known as mutation. Mutation is thus a process as well as an effect. Mutation is thus a change in the genetic material of an organism that is detectable and inherited by offspring and is not caused by recombination or segregation. The term includes nucleotide substitutions, insertions, ommissions, duplications, translocations within a gene, and large chromosodmal changes also.

Mutagenesis is the use of physical or chemical agents to increase the rate of mutations. In nature, different alleles of a gene, arise through spontaneous mutations.

Characteristics of Mutations

Mutations are more or less permanent changes in kind, number or sequence of nucleotides in genetic material. The making or breaking of hydrogen bonds in duplex DNA is a temporary, hence non-mutational change. Mutations are uncoded or unprogrammed changes. Methylation of cytosines or adenines is not a mutation because it is a programmed change as it occurs due to instructions coded by DNA. The mutations are relatively rare and unusual changes. Genetic material tends to be protected against occurrence of uncoded and unusual changes by double helical structure and its bonds to proteins. Mutations are regarded as accidental, unintended, undirected, random events as we have no way of knowing when a mutation will occur, which individual, cell, cell organelle, chromosome or gene will mutate. Mutations are un-oriented with respect to purpose. Mutation is a result of deviation from a natural process.

Classification of Mutations

Size of Mutations

Microlesions. These mutations are also known as **point mutations** or **gene mutations**. These are those mutations in which only one nucleotide pair is replaced. These changes being of very small size are not observable even under a microscope. Such mutations can be detected by comparing nucleotide sequence of wild type and mutant DNA/RNA. Due to such a substitution, the number of nucleotide pairs does not change in a gene. Such mutations are of two typpes. **Transitions** are the change from purine to purine (A → G or G → A) or pyrimidine to pyrimidine (T ⇌ C or C → T) whereas **transversions** are the change from a purine to a pyrimidine (A → T, A → C, G → T, G → C or or vice-versa (T → A, C → A, T → G, C → G).

Intermediate Lesions. In this case either there is addition (insertion) or deletion (removal) of one or few nucletotide pairs from a gene. This leads to a change in number of nucleotide pairs in a gene. Thus changes may cause a shift in the translational reading frame and are called as **frame shift mutations**. Write the mRNA sequence AUGCCCAAAGGGUUU .... and study the effects on translational reading frame when (a) one base is deleted, (b) two bases are deleted, (c) three bases are deleted, (d) one base is added, (e) two bases are added, and (f) three bases are added. Take the help of genetic code dictionary for this exercise. Unless a frameshift mutation cccurs close to the C-terminus, they knock down function of the gene.

Macrolesions. Macrolesions are also known as **Chromosome mutations** which include changes in chromosome structure and changes in chromosome number. These changes at high resolution can be observed under a light microscope.

Changes in Chromosome Structure

These mutations include duplications, deletions, inversions and translocations. A **deletion** or addition may include of a part of a gene, whole gene, a group of genes, complete chromosome, more than one

A. Deletion

a b c d e f g h i j
─────────────
a b c d e f g h i j

A pair of homologous chromosomes

a b c d e f g h i j
─────────────
a b e f g h i j

Deletion heterozygote

a b ⌒c⌒d⌒ e f g h i j
─────────────
a b e f g h i j

Deletion loop at pachytene

B. Duplication

a b c d e f g h i j
─────────────
a b c d e f g h i j

A pair of homologous chromosomes

a b c d e f g h i j
─────────────
a b c d c d e f g h i j

Deletion heterozygote

a b c d e f g h i j
─────────────
a b ( ) c d e f g h i j
   c d

Duplication loop at pachytene

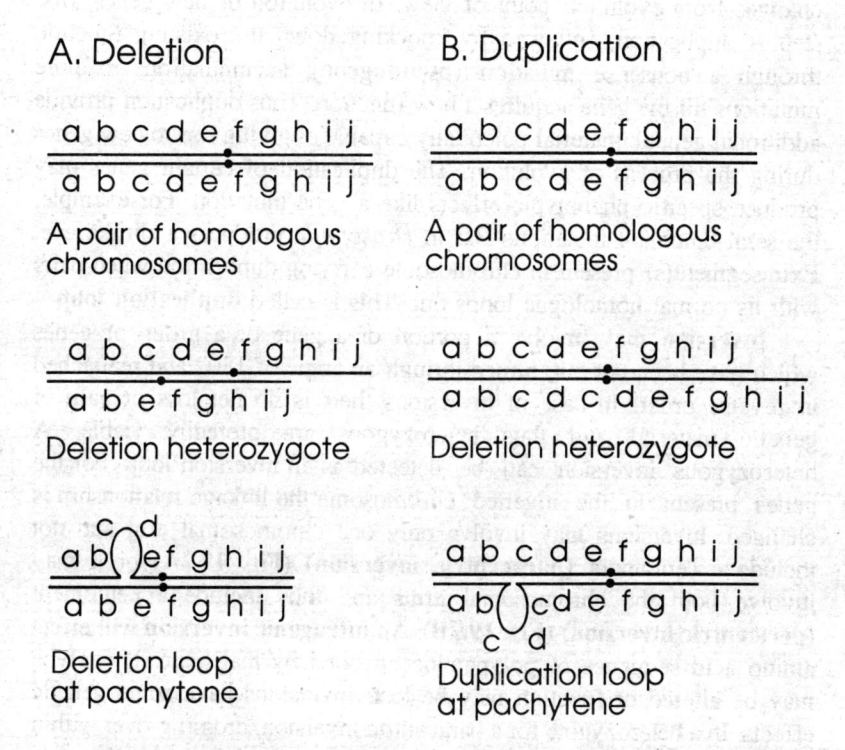

Fig. 19.1. Deletion and duplication loop. Each line represents a chromosome

chromosome or even complete set of chromosomes. Deletions are usually lethal in homozygous state. Cytologically, deletions can be detected by failure of a segment of a chromosome to pair properly (**Fig. 19.1A**).

The segment present in normal chromosome but absent in mutant homologue loops out during pachytene synapsis. This is called **deletion loop**. Chromosome with a deletion can nevert rever to a wildtype condition. Deletions may show **pseudodominance** because in this case a recessive gene present in a single dose expresses itself. Some deletions may have phenotypic effects. For example, deletion of a part of short arm of human chromosome 5 leads to Cri-du-chat syndrome in which case the individual survives only upto age of 30.

The reciprocal of a deletion is called **duplication** which at pachytene gives a loop-like structure (**Fig. 19.1B**). There may be a **tandem duplication** of a gene or whole of gene. Dulpications are very important changes from evolution point of view. In evolution of new genes, first step is duplication, followed by knocking down the existing function through a nonsense mutation (**pseudogene**), accumulation of more mutations till the gene acquires a new function. Thus duplication provide additional genetic material potentially capable of giving rise to new genes during the process of evolution. The duplication of certain genes may produce specific phenotypic effects like a gene mutation. For example, the semi-dominant mutations Bar in *Drosophila* produces a slitlike eye. Extrasegment(s) present in chromosome carrying duplication when pairs with its normal homologue loops out. This is called **duplication loop**.

Inversion may involve a portion of a gene or a group of genes which have been cut out, turned thriugh an angle of 180° and reattached in reverse order. In case of inversions their is no net loss or gain of genetic material and thus heterozygous are prefectly viable. A heterozygous inversion can be detected as an inversion loop. For the genes present in the inverted chromosome the linkage relationship is changed. Inversions may involve only one chromosomal arm and not include a centomere (**paracentric inversion**) (**Fig. 19.2A**) or it may involve both the chromosomal arms and thus include a centomere (**pericentric inversion**) (**Fig. 19.2B**). An **intragenic inversion** will effect amino acid sequence of polypeptide encoded by that gene. Phenotype may be altered or function may be lost. Inversions have some genetic effects. In a heterozygote for a paracentric inversion, crossing-over within the inversion, crossing-over within the inversion loop produces **dicentric** and **acentric chromosomes**. Individuals producing crossover chromatids do not survive and thus do not produce recombinant gametes. Thus crossovers are not detected. In a heterozygote for a pericentric inversion also, crossover products are not recovered.

Translocations involve transfer of a part of chromosome to the same chromosome (**intercalary translocation**) (**Fig. 19.3A**), or to a nonhomologous chromosome (**simple transloction**) (**Fig. 19.3B**), or two nonhomologous chromosome may mutually exchange their parts and in this case in reality two simple translocations are simultaneously achieved (**reciprocal translocation**) (**Fig. 19.3C**). If break is within a gene, the

A. Paracentric inversion B. Pericentric Inversion

```
a b c d e f g h i j        a b c d e f g h i j
─────────────────          ─────────────────
a b c d e f g h i j        a b c d e f g h i j
```

A pair of homologous chromosomes

A pair of homologous chromosomes

```
a b c d e f g h i j        a b c d e f g h i j
─────────────────          ─────────────────
a d c b e f g h i j        a b c f e d g h i j
```

Inversiontion heterozygote

Inversion heterozygote

```
   b c d
  ╱     ╲                      d e f
a (b d) e f g h i j        a b c (d e f) g h i j
─────────────────          ─────────────────────
a       e f g h i j        a b c         g h i j
```

Inversion loop at pachytene. Crossing-over within inverted loop

Inversion loop at pachytene. Crossing-over within inverted loop

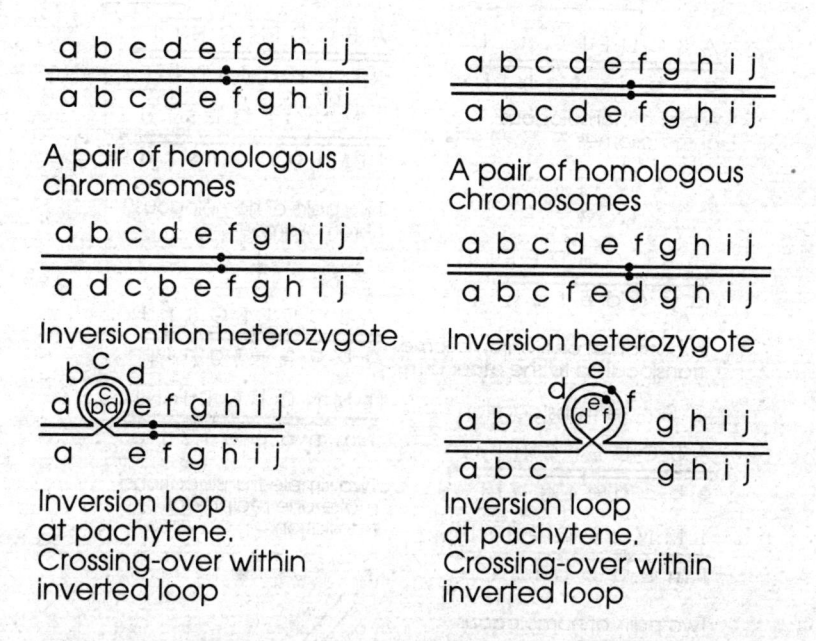

Fig. 19.2. Paracentric and pericentric inversion loops. Pairing between only one normal and one inverted chromosome is shown

base sequence the gene will be changed and its function may be lost. Translocations, when in homozygous from change linkage relationsips. This is the genetic effect. They may drastically alter size of a chromosome as well as the position of its centromere. Translocation is used as important tools in gene transfers and production of duplications and deficiencies. Philadelphia chromosome responsible for chronic myeloid leukemia in man is due to translocation of a part of long arm of chromosome 22 to chromosome 9.

Translocations, inversions and deletions produce semisterility by generating unbalanced meiotic produts that may themselves be lethal or that may result in lethal zygotes. Various genetic diseases in humans have been identified to be due to chromosomal rearrangements.

On the Basis Changes in Chromosome Number.

Monoploid is a cell that contains one chromosome set in a kind of

A. Intercalary translocation

```
A B C D E F G H I J
─────────────────────
a b c d e f g h i j
```
A pair of homologous chromosomes

↓

```
A B C G H D E F I J
─────────────────────
a b c d e f g h i j
```
A portion of one chromosome translocated to the other arm

B. Simple translocation

```
A B C D E F G H I J
─────────────────────
a b c d e f g h i j

L M N O P Q R S T U
─────────────────────
l m n o p q r s t u
```
Two pairs of homologous chromosomes

↓

```
A B C D E F G H I J S T U
─────────────────────────
a b c d e f g h i j

L M N O P Q R
─────────────────────
l m n o p q r s t u
```
A portion of one chromosome is translocated to a non-homologous chromosome

C. Reciprocal translocation

```
A B C D E F G H I J
─────────────────────
a b c d e f g h i j

L M N O P Q R S T U
─────────────────────
l m n o p q r s t u
```
Two pairs of homologous chromosomes

↓

```
A B C D E F G S T U
─────────────────────
a b c d e f g h i j

L M N O P Q R H I J
─────────────────────
l m n o p q r s t u
```
Two simple translocations make one reciprocal translation

↓

Pachytene configuration of reciprocal translocation heterozygote

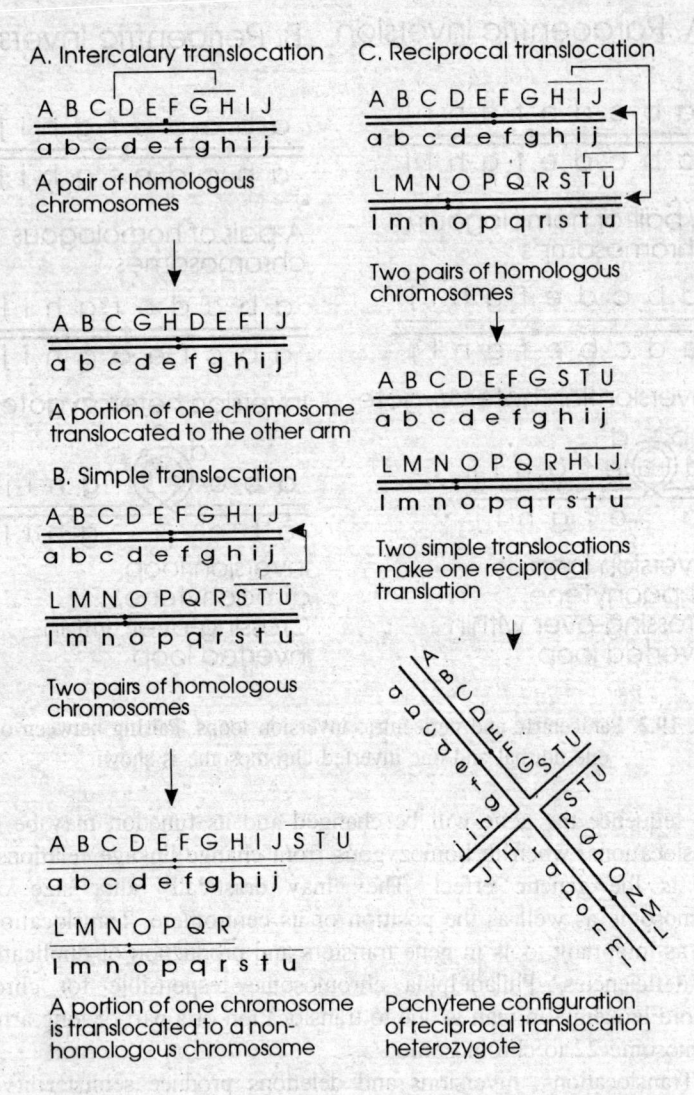

Fig. 19.3. Different types of translocations

organism usually associated with diploid type of life cycle. Monoploids usually arsie due to mutations. In wasps, bees and ants, males are monoplods. **Haploids** exist as a single set of chromosomes during almost entire part of their life and thus monoploids are distinguishable from true haploids. Monoploids may be derived from the products of meiosis. Monoploids have a major role to play in modern approaches to animal and plant breeding.

Change in Chromosome Number

Change in chromosome number may involve incomplete set of chromosomes (**aneuploids**) or may involve a complete set of chromosomes (**euploids**).

Aneuploids. Anuploids may involve, in a disomic (2n), deletion of one chromosome (2n-1, **monosomic**), deletion of two nonhomologous chromosomes (2n-1-1, **double monosomic**), or deletion of a pair of homologous chromosomes (2n-2, **nullisomic**). Aneuploids may also arise due to the addition of one chromosome (2n+1, **trisomic**), addition of two nonhomologous chromosomes (2n+1+1, **double trisomic**), or addition of a pair of homologous chromosomes (2n+2, **tetrasomic**). **Non-disjunction** (failure of homologous chromosome to separate during meiosis) is the cause of aneuploids. Trisomics play important role in gene location programmes and nullisomics and monosomics are useful in locating newly found genes on specific chromosomes in plants. Trisomy is the cause of various genetic disorders in humans. (**Table 19.1**).

Table 19.1. Human disorders due to aneuploidy

Name of the syndrome	Chromosome(s) present or absent
Down's syndrome	21 extra
Klinefelter's syndrome	X extra
Klinefelter's syndrome	XY extra
Turner's syndrome	Only one X present
Patau's syndrome	13 extra
Edward's syndrome	18 extra

Euploids. Euploids may contain more than two complete sets of chromosomes. Individuals having three, four, five, six... sets of chromosomes are called **triploids, tetraploids, pentaploids, hexaploids**... and so on. In animals, polyploids are restricted to groups that reproduce asexually or are **hermaphroditic** (an individual with both male and female reproductive organs) and **parthenogenetic** (individual developed from an unfertilized egg). Examples are earthworms, some types of shrimps, and parthenogenetic species of insects and lizards.

Polyploids which are formed from sets of chromosomes from one single species are known as **autopolyploids** and those that are formed from sets of chromosomes of different species are called **allopolyploids** or **amphidiploids**. Alfalfa, coffee, potato and peanut are autotetraploid plants. Example of allopolyploids are cabbage and wheat (*Triticum aestivum*). Triploids are characteristically sterile. A practical application of sterility associated with triploidy is the production of seedless varieties of watermelons, bananas, etc. These varieties are more palatable to consumers. **Allopolyploidy** seems to have played as an important force in speciation of plants. An alkaloid, **colchicine**, is used to produce polyploids as it acts by dissolving the spindle apparatus thereby doubling the number of chromosomes in the cell.

Forward And Reverse Mutations

Forward mutation. Any mutation away from the standard or wild type.

$$y^+ \rightarrow y \text{ or } R \rightarrow r$$

Reverse or back mutation. Any mutation towards the standard or wild type.

$$y \rightarrow y^+ \text{ or } r \rightarrow R$$

Reverse or back mutations are classified further as follows:

(a) **Exact reversion** In this case same codon is reconstituted.

```
DNA   AAA                    GAA                    AAA
                 forward            reverse
RNA   UUU (Phe)  ──────> CUU (Leu) ──────> UUU (Phe)
Wild type                   mutant                  Wild type


DNA   TTT                    CTT                    TTT
                 forward            reverse
RNA   AAA (Lys)  ──────> GAA (Glu) ──────> AAA (Lys)
Wild type                   mutant                  Wild type
```

Equivalent reversion In one type of equivalent reversion a different codon coding for same amino acid is reconsitututed.

```
DNA  AGG              ACG                    TCG
              forward       reverse
RNA  UCC (Ser) ————> UGC (Cys) ————> AGC (Ser)
     Wild type        mutant              Wild type
```

In second type of equivalent reversion a different codon coding for similar amino acid is reconstituted.

```
DNA  GCG              GGG                    GTG
RNA  CGC (Arg) ————> CCC (Pro)  ————> CAC (His)
     (Basic)         (Non-polar)              (Basic)
     Wild-type        Mutant            Pseudo-wildtype
```

Same Sense, Missense, and Non-sense Mutations

Same Sense or Neutral Mutations. A muation which the changed codon codes for the same amino acid as in wild type. For example,

```
DNA         AGA           AGG  AGC or AGT
RNA         UCU    ————>  UCC, UCA or UCG
Amino acid  (Ser)              (Ser)
```

Mis-sense Mutation. A point mutation that changes meaning of a codon such that new codon codes for a different amino acid. For example,

```
DNA         TCA           TCT  TCC
RNA         AGU    ————>  AGA or AGG
Amino acid  (Ser)              (Arg)
```

Non-sense Mutation. When a point mutation creates one of the three-termination codons, it is called non-sense mutation. It causes premature termination of protein synthesis.

```
DNA         ATG           ATT  ATC
RNA         UAC    ————>  UAA or UAG
Amino acid  (Tyr)         ochre   Amber
                          (Terminator)
```

Spontaneous Mutation

The only operational definition of "spontaneous" mutations is that these mutations are produced in natural or laboratory populations under normal growth conditions. The frequency may, therefore, depend on the growth medium, temperature, pH, etc., which are normally employed. These are those mutations whose cause is not known. These may orginate due to errors during DNA replication, due to environmental mutagens, tautomerism or transposable genetic elements.

Some well known examples of spontaneous mutations are: (a) A male lamb with short, lowed legs was found in the stock of seth Wright of New England, (b) hornless cattle, (c) pacing horses, (d) double-toed cats, (e) mutants in *Oenothera lamarkiana* (evening primrose) reported by de Vries, and (f) *white*-eyed *Drosophila melanogaster* reported by Morgan.

Changes in Electronic Structure of Bases

Some spontaneous mutations must come about by mistakes in the normal duplication of DNA or RNA. Due to tautomeric shift (or ionization) rare base pairs occur spontaneously. Transitions may arise due to **proton migration** within bases (**Fig. 19.4**) and they show unusual pairing (**Fig. 19.5A**). **Tautomers** are rare and short lived. Tautomeric adenine (A^*) pairs with cytosine and this mispairing leads to A.T $\rightarrow$ G.C transition. Tautomeric guanine (G^*) may pair with thymine. This mispair leads to G.C $\rightarrow$ A.T transition. Similarly, $T^* \equiv G$ forbidden base pairing leads to T.A $\rightarrow$ C.G transition and $C^*=A$ forbidden base pairing leads to C.G $\rightarrow$ T.A transition (**Fig. 19.5B**). Proton migration within base pairs can set off a chain of modification that end in DNA mutation. Transversuins may arise by mistake pairing between syn adenine produced by proton migration and rotation and common adenine, and between neoguanine produced by shifting if glycosydic bond from normal from normal to extracyclic nitrogen and common guanine Methylated guanine undergoes tautomeric shifts and pairs with thymine (**Fig. 19.6**) and leads to G.C $\rightarrow$ T.A transversions. Transversions occur by apposition of two errors - proton migration in one and $180°$ rotation around glycosydic bond (syn) in the other (**Fig. 19.7**). Tautomeric syn-adenine (A^*) may pair with syn A which leads to A.T $\rightarrow$ T.A transversion. Similarly, tautomeric neo-guanine (G^*) may also pair with syn G which leads to G.C $\rightarrow$ C.G transversion.

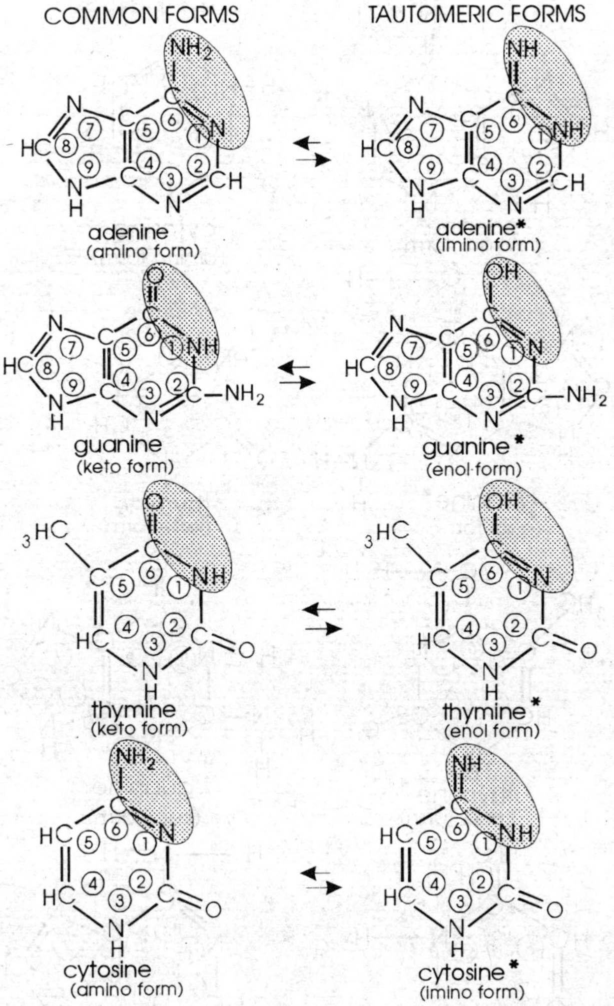

Fig. 19.4. Tautomerism in bases of DNA. The tautomeric forms differ from commom forms by rearrangement of protons and electrons

Rare Base Pairing Between Two Purines or Two Pyrimidines

Rare base pairing between two purines or two pyrimidines can cause spontaneous production of transversion (**Fig. 19.8**). These A.G and T.C

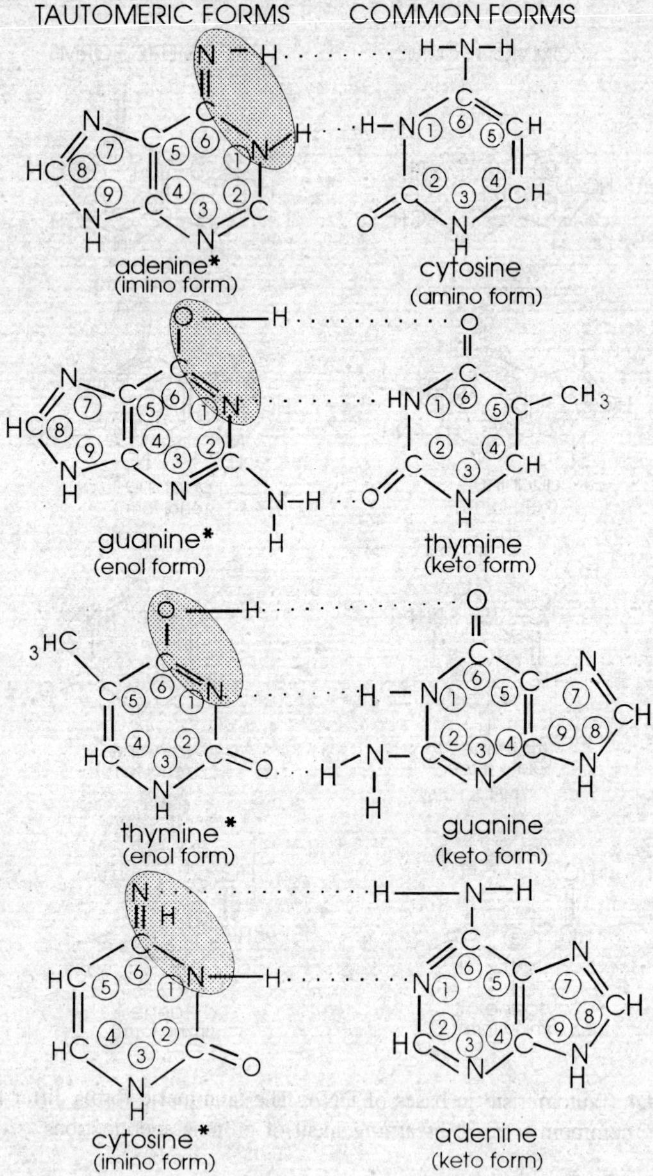

Fig. 19.5. Forbidden base pairs between tautomeric and common bases of DNA. A*=C, G*=C, T*=G, C*=A.

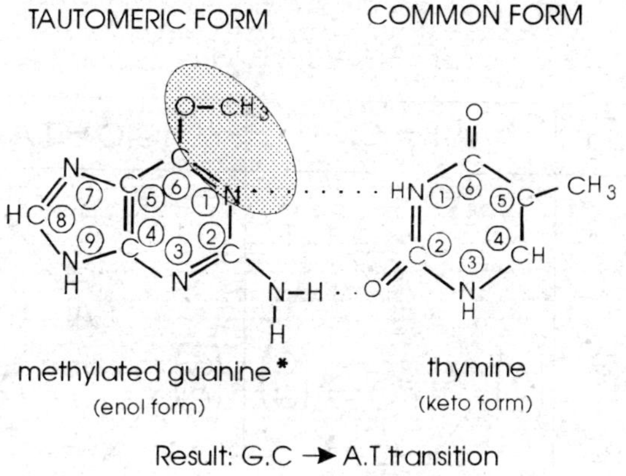

TAUTOMERIC FORM   COMMON FORM

methylated guanine*   thymine
(enol form)   (keto form)

Result: G.C → A.T transition

Fig. 19.6. Forbidden base pairs between tautomeric methylated guanine and common thymine

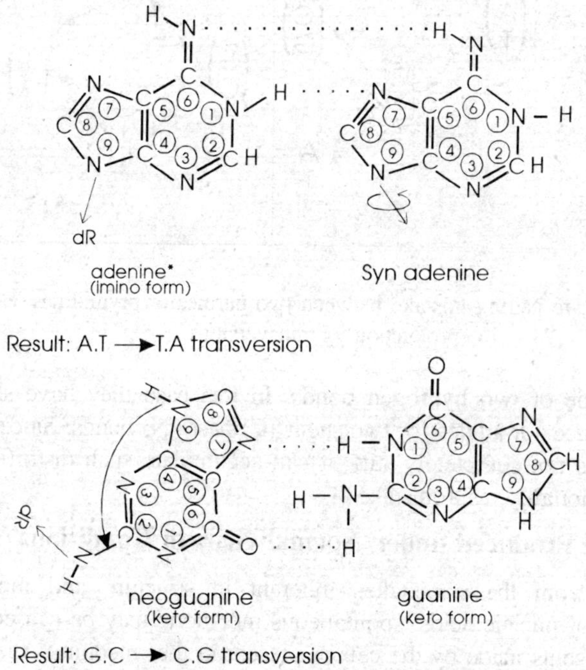

adenine*   Syn adenine
(imino form)

Result: A.T → T.A transversion

neoguanine   guanine
(keto form)   (keto form)

Result: G.C → C.G transversion

Fig. 19.7. Transversions may arise by mistake pairing

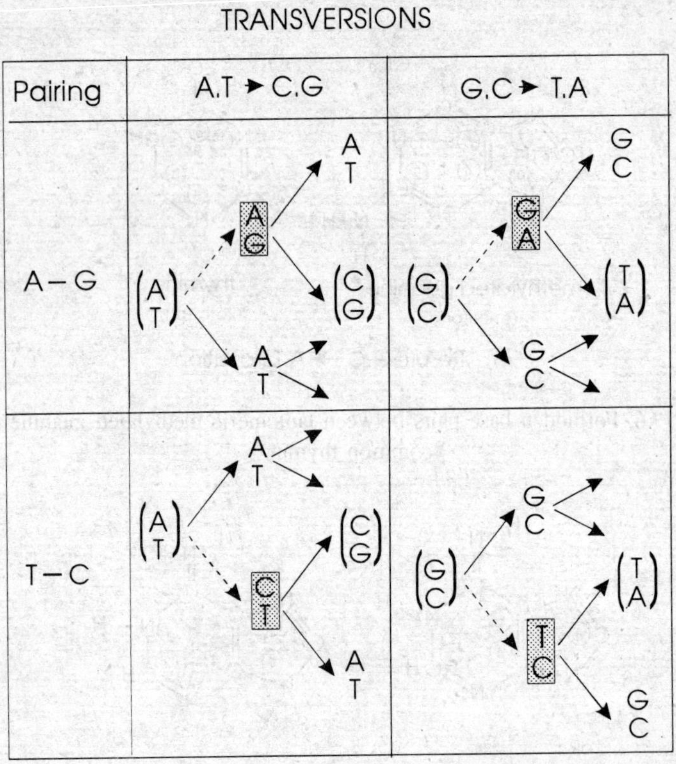

354 *Basic Genetics*

TRANSVERSIONS

Pairing	A.T → C.G	G.C → T.A
A–G	(figure)	(figure)
T–C	(figure)	(figure)

Fig. 19.8. Rare pairing mistake between two purines or prymidines leads to production of transversions

pairs use one or two hydrogen bonds. In this case they have slightly wrong distances or angles for two sugar (C)-base (N) bonds. Since DNA back-bone is not completely stiff, it can accumulate such distortions at least occassionallly, as a mistake.

Chemicals Produced under Normal Growth Conditions

Apart from these mistakes inherent in structure and mode of replication of nucleic acids, spontaneous mutations may be caused by a variety of agents made by the cell or present in the medium. Agents of this kind are base analogues, peroxides, nitrous acid, etc., which are produced under some growth conditions. These agents are often inactivated by special enzymes like catalase, peroxidase, etc. Some of these agents inside the cell act on duplicating DNA others on resting DNA. Mode of action of such agents will be discussed later in this chapter under chemical mutagenesis. Processes like methylation, depurination, gross base distortion and formation of bulky adducts are quite common in celular metabolism.

Induced Mutations

These are the mutations that are produced under changed growth conditions, i.e., as a result of exposure of an organism to some agent that does not form normal growth condition for the organism. An agent that has the ability to produce mutations is called **mutagen**. A mutagen can be a physical agent (UV light, X-rays, γ-rays, α-particles, β-particles, etc.) or a chemical agent (nitrogen mustard, hydrazine, hydrogen peroxide, nitrogus acid, base analogues, alkylating agents, dyes, etc.). An individual possessing a mutation is termed as **mutant**. These are those whose cause is known.

History of Induced Mutations

Morgan (1914) reported unsuccessful attempts to induce mutations in *Drosophila* with alcohol or ether. Mann (1923) failed to induce mutations in *Drosophila* with morphine, quinine and a number of metal salts. Muller (1927) was the first scientist to successfully induce mutations in *Drosophila* with X-rays. he devised a method to detect sex-linked recessive lethal mutations in *Drosophila*. His two methods, ClB and Muller-5, are even in use today to test the ability of an agent to produce mutations in *Drosophila*. He was awarded a Nobel prize for his work in the year 1949. In 1928, Stadler showed that γ-rays were able to induce mutations in barley. The mutagenic effects of ultravoilet light (UV) was first discovered by Altenberg (1934) in polar cap cells of *Drosophila*. Thom and Steinberg (1934) found during World War II nitrogen acid was effective in causing mutation in *Aspergillus*. Auerbach and Robinson (1946) found that mutations could be induced in *Drosophila* by nitrogen mustard and sulpher mustard. Since then a large number of physical and chemical agents have been found to induce mutations in diverse organisms.

Physical agents responsible for induced mutations include UV light, X-rays, γ-rays, α- and β-particles. The mechanism of induction of mutations by these agents is decribed below.

Physical Mutagens

Radiations are difined as electromagnetic energy having a specific wavelength. There are different kinds of radiations like X-rays, γ-rays, μv, infrared etc. X-Ray and γ-rays are called ionizing radiations, as they cause ionisation in matter from which they pass. Ionisation results in the production of free radiclas and cause mutations. Wavelengths of some of the radiations are:

Radiation		Wavelength
Radiowaves	=	$10^{-4} - 5 \times 10^{4}$ m
Infrared	=	$10^{-6} - 5 \times 10^{-3}$ m
visible light	=	$7 \times 10^{-7} - 4 \times 10^{-7}$ m
UV light	=	$5 \times 10^{-7} - 5 \times 10^{-9}$ m
X-rays	=	$5 \times 10^{-8} - 5 \times 10^{-15}$ m
γ-rays	=	$5 \times 10^{-11} - 4 \times 10^{-15}$ m

UV light, X-rays and γ-rays have very short wavelength and penetrate deep and hence high ionising capacity is there. Radiations can change the genetic material. Radiations are used for three purposes: visibility, carrying radio and T.V. signals, and creating new variability.

Radiations are of two types - non-particulate and particulate. **Non-particulate radiations** are electromagentic in nature, i.e., they are generated by using electricity, e.g., X-rays and γ-rays. X-rays of shorter λ are called **hard X-rays** and others are called **soft X-rays**. Soft X-rays are less penetrating. Wavelength is .001 - 10 nm and it will depend upon potential voltage. γ rays are produced by using high voltage, that is not possible to handle it. So radioisotoes are used e.g. Co^{60}, Cs^{137}. Radioisotoes are not stable, they keep on decaying. For example, Co^{60} —> Ni^{60} + γ-rays. Wave length of γ-rays is shorter than X-rays.

Particulate radiations are of four types: (i) β **particles** - these penetrate only upto mm. Source is radioactive isotopes (S^{35}, N^{15}, P^{32}) when radioisotopes decay, they loose electrons and ultimately become stable. (ii) **Neutrons** - Fusion and fission cycle generate neutrons. These are of two types: (a) **Fast neutrons**, whose energy is 0.5 -2.0 Mev and (b) Thermal or **slow neutrons** whose energy is 0.025 ev neutrons. (iii) α-**particles** - Source is radioisotopes, These are Helium nuclei. Their energy is 2-9 Mev. Penetration of α-particles is small. (iv) **Protons** - Energy of protons is higher than α-particles source is acclrators and nuclear reactors.

Radiations with which we are concerned are the α, β, γ-radiations of radioactive substances, X-rays, protons, neutrons, etc. These may be grouped together as ionizing radiations.
Ultraviolet (UV) light is a non-ionizing radiation. It is an electromagnetic radiation. UV light for biological experiments is obtained from a mercury lamp. Effects of UV light are several. It kills bacteria and it also induces mutations among the survivors. The first photochemical lesion found in the DNA of irradiated bacteria was thymine dimer. UV light produces thymine dimers mainly by linking adjacent thymine bases in the same srand of DNA via C-C bonds. Dimers containing thymine block DNA replication *in vitro* and *in vivo* and are responsible for an unimportant fraction of lethal effects of low doeses of UV light in some strains of bacteria. It also causes hydration, i.e., it acts on 'C' of DNA in water solution, adding a water molecule across a double bond.

UV light can temporarily delay cell divisions. It also produces chromatid type of aberration. It also disrupts the H-bonding between strands of natural DNA upon exposure of viral suspensions to UV light. This results in inactivation of virus. In yeast reversion to prototrophy from auxotrophy. The deleterious effect of UV light of sunlight on skin is present in certain individuals with a rare inherited disorder called

Xeroderma pegmentosum. These people are homozygous for a recessive mutation. These rays injure the surface cells of cornea making the latter opaque. This causes blindness for a few days untill the injured cells are replaced by new cells.

Ultraviolet Light. .Mutations are produced if the UV light-indued lesions are not repaired by reapir mechanism of the cell. Unrepaired pyrimidine dimers (**Fig. 19.9**) cause mutations by inducing the formation of daughter strand gaps. One gap is formed opposite each lesion. These gaps which are lethal, if not repaired cannot be mended by repair replication, because each is located opposite a non-coidng lesion. These are repairable only by a mechainsm which probably involves some form of genetic recombination between complementary daughter strands. Most UV light-induced mutations are errors introduced into the DNA in the courses of inaccurate recombination repair of single strand gaps.

γ-irradiation. The γ-rays used in biological work are those of radium and its produts or sometimes radon and its products, sealed in a cloned container with sufficient thickness of wall to absorb α and β rays. These γ-rays are measured in terms of roentgens. When growing plants are

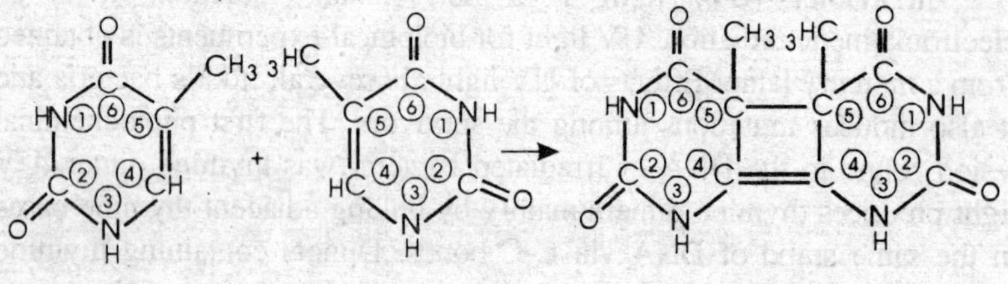

Fig. 19.9. Formation of thymine dimers from the adjacent molecules in DNA due to ultraviolet light

subjected to chronic and acute γ-irradiation, several morphoological changes are induced such as reduction in growth, thickness of plant parts induction of somatic mutations and tumours. One of the striking effects observed in chronic γ-irradiated plants was the development of epidermal hairs on leaves and other parts following treatment. In many systems, exposure to γ-irradiations gives rise to, among other effects, a temporary disappearance of mitotic figures from the cell-population. This phenomenon is called **mitotic inhibition**. The cause of inhibition is a block in the cell-division cycle immediately prior to prophase I; neither the site of the radiation lesion nor its mechanism of repair is known.

X-rays. It is an electromagnetic readiation emetted in quanta. 21's wavelength is (0.05 - 10Å). Two rare chromosomal aberrations induced by X-ray irradiation were recorded on modifying factors in induced mutagenesis. First is the occurance of ring like metaphse chromosome at mitosis in the root-tips from the dessicated seeds of *Oryza sativa*. This condition airses due to the failure of separation of the ends. The phenomenon has been explained on the basis of the alteration of the cycle of separation of the four regions with in a chromosome. There is occurrence of an interlocked chromosomes or bridge at mitosis in the root tips. This has been explained on the basis of translocation following breakage of two chromosomes, leading to the formation of a dicentric chromosome. The oridge arises due to a full turn of the chromatids over one another at Anaphase.

The biological effects of X-rays, γ-rays, α-particles, protons, high energy neutrons, and electrons have been studied extensively. The radiation can have a direct effect on the chromosome e.g. directly break it or alter one of the DNA bases and an indirect effect, since along the track of each particle a large number of ions or radicals remain which can initiate a chain of chemical reactions. Chromosomes are extensively sensitive to breakage in meiotic prophase.

The inactivation of organisms follows a first-order reaction equation: $dN/dt = -\beta N$ or $N = N_o e^{-\beta t} = N_o e^{-KD}$. β is proportional to the intensity of the radiation. D is the dose, i.e., the energy absorbed/gram of material. K is a dose-independent constant describing the sensitivity of the organism. The frequency of mutants per viable organism often increases with the dose. $M = \alpha D$. Timofeeff-Ressovsky et al. (1935) interpretted these equations in terms of a 'target theory' which states that a single 'hit' of the particle on the 'target', i.e., the genetic material inactivates or mutants it if the inactivation follows a first order equation. One calls $\beta t = KD = n$, the number of lethal hits because then the survival is given by the zero order Poisson term $N/N_o = e^{-n}$ meaning of the word 'target' depends on the chemical and metabolic life time of the radiation produced, i.e., whether it can get to any place or is united to the ionizing surrounding of the track of the inoizing particles. If the life time is long, the radiation produces mutagenic chemicals whose conc extration extration depends on the number of ionisation and the amount of the absorbed. Energy is the proper measure for the 'dose'. The word **single 'hit'** can mean a **single chemical event** or a **single ionisation** but it can also mean a **single cut** by a densely ionizing particle. Deletions arise by single hits and ring dicentrics and exchange aberration occurs by two hits.

Reed and Thoday found that low oxygen concentration reduced the frequency of chromosome breaks induced by X-radiation. This O_2 efect has been extensively studied and it is probable that radiation in the presence of oxygen causes the formation of some peroxide radical which in turn affects the frequency of chromosome breaks, and may be other mutations also. The formed peroxides may cuase an increase in the frequency of chromosome breaks either because they induce breaks themselves or because they prevent the rejoining of chromosome cut by the radiation.

Radiation Units

Roentgen (R). It is unit of exposure of X-γ-rays based on ionisation, that these radiation produce in air: 1R = 1 esu/cc of air.

Radiation Absorbed Dose (RAD). It is a unit of absorbed dose and expresses the energy absorbed/gram of a substrate from any radiation.

Rem. It is a unit of dose equivalent. By this we can compare different types of radiations. Rem is equal to dose in rads multiplied by appropriate modifying factor. For example, for X-rays and γ-rays QF = 1.0, for Fast neutrons and protons, QF = 10.0, and for α-particles QF = 10.0, i.e., if we are using α-particles, 10 times more dose is given than X-rays to observe the same biological efffect.

Curve. Unit of radioactivity and is defined as the rate at which the atoms of a radioactive substance decays.

Relative Biological Efficiency (RBE). It compare different type of radioactive substance with respect to their biological activity.

Chemical Mutagens

A variety of chemicals identified to be capable of inducing mutations are the inhibitors of nucleic acid precursors, base analogues, dyes, nitrous acid, hydroxylamine, hydrazine, low pH, heat, hydrogen peroxide, maleic hydrazide, antibiotics, azides, alkylating agents, chemicals used in industries, pesticides, cosmetics, food preservatives, etc.

The chemical agents that induce mutations are classified into two types depending upon the physiological state of DNA affected by these agents. In first group fall those chemicals which **affect replicating nucleic acids**. Examples of such chemicals are **inhibitors of nucleic acid precursors** (e.g., 5-aminouracil, azaserine, benzimidazole, caffeine, 8-ethoxycaffeine, 8-mercaptopurine, 5-nitroquinoxaline, paraxanthine, tetramethyl uric acid, theobromine), **base analogues** (e.g. 5-bromouracil, 5-chlorouracil, 5-iodouracil, 2-aminopurine, 2, 6-diaminopurine), and **dyes** (e.g., proflavin, acridine orange, 2-aminoacridine, methylene blue, toluidine blue ethidium bromide, 10-methyl acridinium phenanthridine, propidium diiodide). 5-aminouracil and 8-ethoxycaffeine inhibit thymine while azaserine, benzimidazole, caffeine, 8-mercaptopurine, 5-nitroquinoxaline, paraxanthine, tetramethyl uric acid and theobromine inhibit purines. The frequency of mutations increases because the lack of one base either cause a chromosomal break and thus a large alteration or it increases frequency of pairing mistakes. It is possible that an inhibtor causes the increased formation of another base analogue which is actually mutagenic since it is incorporated in place of a normal base. The substances that are incorporated into DNA without hindering its replication losely related to adenine, guanine, cytosine or thymine.

Base Analogues and Dyes. Since analogue differs from normal base in certain substituents, its electronic structure is modified and occassional errors may occur in specificity of hydrogen bonds during incorporation into DNA. 5-bromouracil, 5-chlorouracil and 5-iodouracil replace thymine in DNA and they thus pair with adenine. 5-bromouracil (rare enol state) replaces cytosine. It pairs with guanine and thus causes G-C to A-T transitions and vice-versa (**Fig. 19.10**). 2-aminopurine can pair in its normal tautomeric form with two bases, with thymine by two hydrogen bonds and with cytosine by one hydrogen bond (**Fig. 19.11**). More frequently, 2-aminopurine pairs with thymine by two hydrogen bonds. Thus, 2-aminopurine is incorporated most frequently in place of adenine which itself would not be mutagenic. When 2-aminopurine pairs with cytosine it is highly mutagenic. It is expected to induce base pair transitions from A-T to G-C and vice-versa. Base analogues, 5-bromouracil and 2-aminopurine do not cause transitions. Since both these base analogues cause transitions in both the directions A-T to G-C in both the directions the mutatins induced with them can be reversed by use of these substances. The the normal base that pairs with base analogues, 2-aminopurine and 2, 6-diaminopurine, is not known. Caffeine owes its

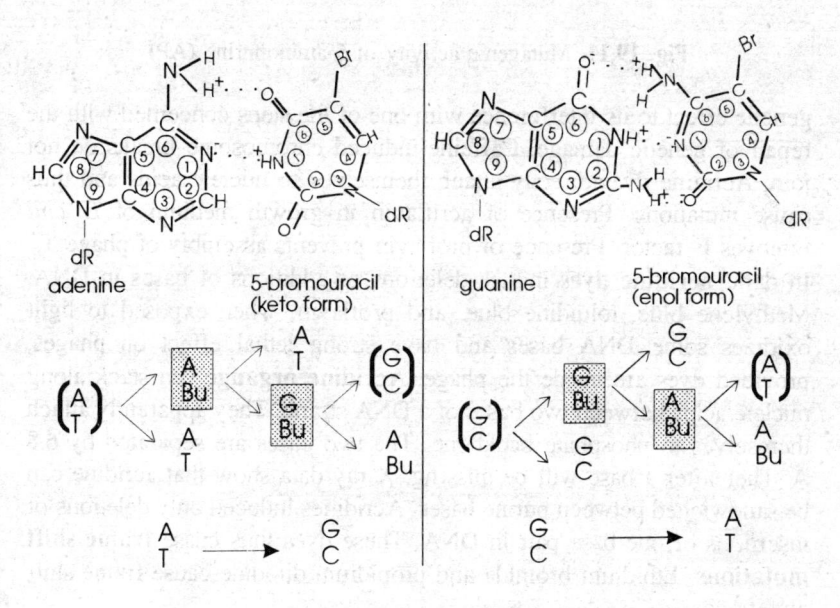

Fig. 19.10. Mutagenic activity of 5-bromouracil (5-Bu)

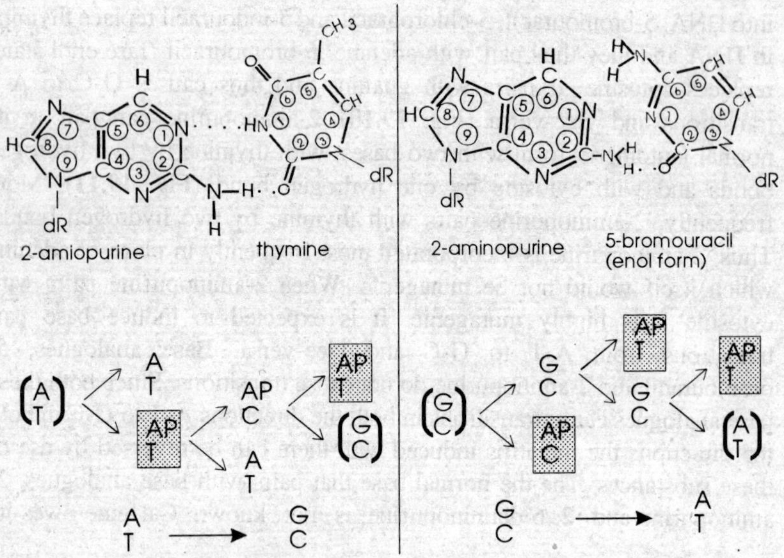

362 Basic Genetics

Fig. 19.11. Mutagenic activity of 5-aminopurine (AP)

genetic effect to its interference with one of the steps concerned with the repair of genetic damage. Caffeine-induced chromosome breaks do not join. Acridine dyes directly attach themselves to nucleic acids and thus cause mutations. Presence of acriflavin in growth medium of *E. coli* removes F' factor. Presence of proflavin prevents assembly of phage T_4. In dark, **acridine dyes** induce deletions or additions of bases in DNA. Methylene blue, toluidine blue, and proflavin, when exposed to light oxidizes some DNA bases and have strong lethal effect on phages, provided dyes are inside the phage. **Acridine organge** can stack along nucleic acids between two bases of a DNA strand. They apparently attach themselves to phosphate backbone. The two bases are separated by 6.8 Å. Thereafter a base will be missing. X-ray data show that acridine can be sandwiched between purine bases. Acridines induced only deletions or insertions of one base pair in DNA. These dyes thus cuase **frame shift mutations**. Ethidium bromide and propidium diiodide cause frame shift mutations.

In second group fall those chemicals that **affect resting nucleic acids.** Examples of such chemicals are nitrous acid, hydroxylamine, hydrazine, low pH, heat, hydrogen peroxide, maleic hydrazide, antibiotics, azide and alkylating agents.

Nitrous Acid. Nitrous acid (HNO$_2$) oxidatively deaminates with decreased frequency the bases guanine, cytosine and adenine in DNA and RNA. In this reaction, amino group in these bases is replaced by oxygen. HNO$_2$ induces point mutations in T$_4$ phage and yeast. Its mutagenic effect is not reported for higher plants. Adenine is deaminated to hypoxanthine which then pairs preferentially with cytosine in place of thymine. The result is transition of A-T to G-C (**Fig. 19.12**). Cytosine is deaminated to uracil which now pairs with adenine in place of guanine. The result is transition of G-C to A-T. Guanine is deaminated to xanthine which still pairs with cytosine but by only two hydrogen bonds. Thus deamination of guanine is not mutagenic. Nitrous acid thus causes transitions A-T to G-C in both the directions. Thymine which does not contain any amino

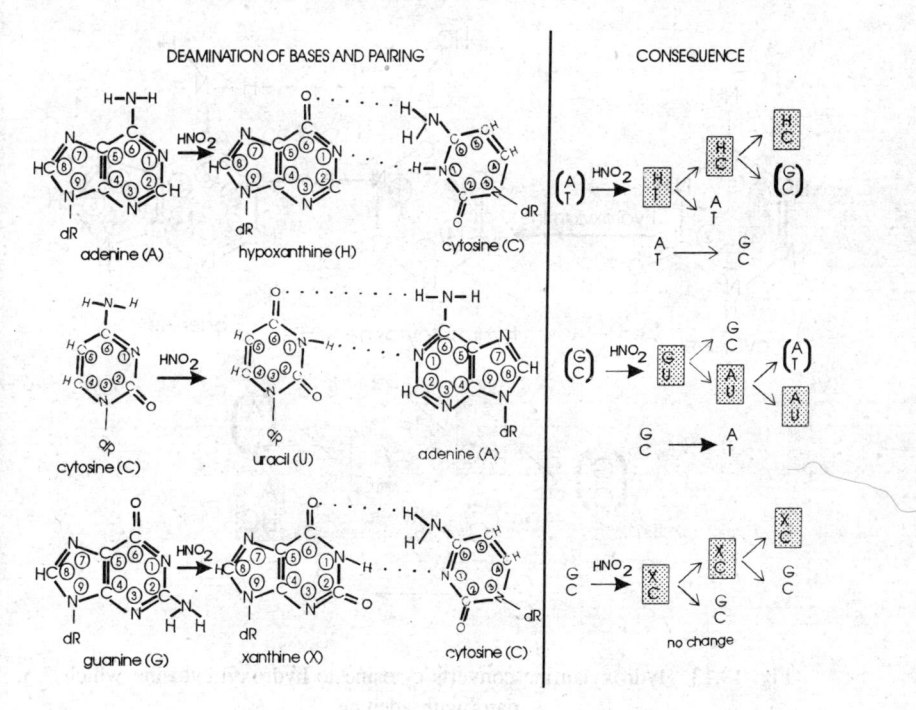

Fig. 19.12. Mutagenic activity of nitrous acid (HNO$_2$)

364 Basic Genetics

group is not altered by nitrous acid. Nitrous acid also causes **interstrand cross-linking** of DNA. The DNA strands fill to separate and there is no DNA duplication which is lethal for phages. Blockage of replication may result into deletions. One cross link is formed for four deaminations.

Hydroxylamine. Hydroxylamine (NH_2OH) reacts with pyrimidine bases. Its effect is strong on cytosine while on uracil the effect is less. It acts at pH 6.0 and at concentrations higher than 0.1M. The toxicity of this chemical is high but it is a weak mutagen. Hydroxylamine breaks and removes pyrimidine ring of uracil thus producing phosphoribosyl urea and 5 isoxasolone (**Fig. 19.13**). With cytosine, hydroxylamine finally produces hydroxyl amino (-NOH) derivative which might be responsible for base pair change. The new form of cytosine cannot pair with guanine but can form one or two hydrogen bonds with adenine. Thus, hydroxylamine causes base pair transitions from G-C to A-T only. Mutations induced with hydroxylamine cannot be reversed with hydroxylamine.

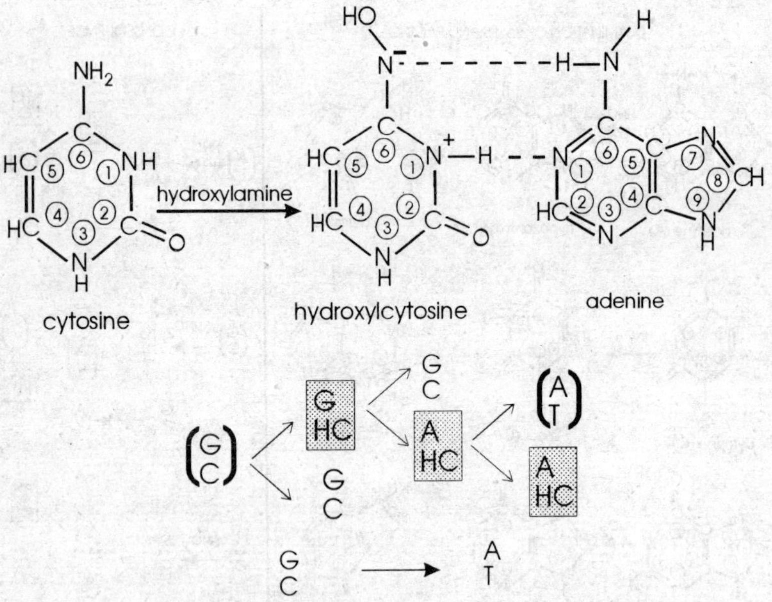

Fig. 19.13. Hydroxylamine converts cytosine to hydroxyl cytosine which pairs with adenine

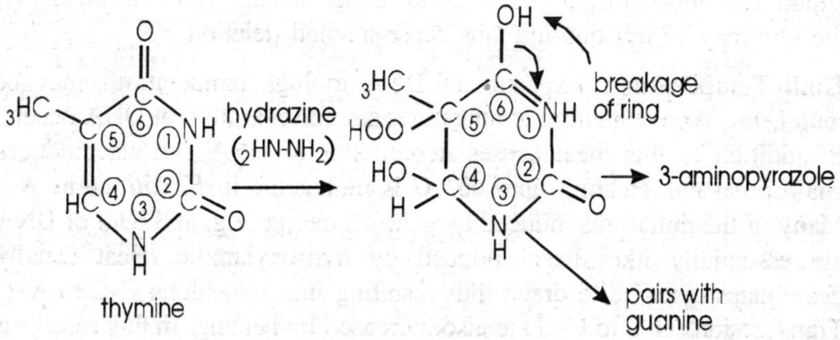

Fig 19.14. Hydrazine breaks rings of cytosine giving rise to 3-aminopyrazole.

Hydrazine. Hydrazine ($_2$HN-NH$_2$) breaks rings of uracil, cytosine and thymine (**Fig. 19.14**). Hydrazine breaks rings of cytosine giving rise to 3-aminopyrazole. Similarly uracil is converted into pyrazolone. Remaining urea remains attached to the sugar and in the presence of water is subsequently hydolyzed off Hydrazine acts at pH 8.5 with uracil producing pyrazolone and urea. With cytosine, hydrzine produces 3-aminopyrazole and urea. The urea thus produced remains attached to sugar-phosphate backbone of DNA and subsequently hydrolyzes it off. The new β-glycosidic bond becomes unstable and sensitive to breakage. Breakage of prymidine is much more frequently lethal than mutagenic. Mutagenic effect of hydrazine is A-T to G-C.

pH. When DNA solution is exposed to pH below 4.0, some bases become positively charged and **strand spearation** occurs which may be lethal. At pH 4.0 and above this effect is negligible. At pH 4.0, two additional reactions occur which should be active at pH above 4.0 also. Adenine and guanine bases are removed; this effect is known as **depurination.** At pH 1.6, **apurinic acid** is formed from purines. When it is exposed to solutions at high pH, all sugar phosphate bonds which have no base attached to sugar, apparently break. Even at neutral pH such breaks occur, although very slowly. In T4 phage, at pH 4.2, point mutation are observed. It seems these are result of depuination not leading to backbone breakage. A gap left by purine removal might be filled again before DNA replicates; no such mechanism exists. Alternately, the gap could remain empty untill the DNA replicates. If DNA replication stops at this site, it would be merely lethal. If any one of the four DNA bases could get incorporated, into new complementary strand, transition or transversion or no change would occur. Alternatively, the gap may be left out and thus cause a small deletion.

High Temperature. Exposure of DNA to high temperatures may be mutagenic. Above melting temperature, strand separation of DNA occurs. In addition to this, heat causes depurinatin of DNA and also induces chain breakage. Heating upto 60° C is mutagenic in *E. coli* strain W6. Many of the mutations induced by simple heating (e.g. at 37°C) of DNA are essentially like those induced by hydroxylamine. Heat usually deaminates cytosine to uracil thus resulting into transitions G-C to A-T. Transversions G-C to C-G are also increased by heating. In this case heat converts guanosine to neoguanosine which leads to guanine-guanosine base pairing. Heat can thus cause both transitions (G-C to A-T) and tranversions (G-C to C-G).

Hydrogen Peroxide. Hydrogen peroxide and organic peroxides are decomposed into OH radicals which may react with pyrimidines at 5-6 double bonds thus attaching -OH group at position 6, initially (**Fig. 19.15 A and 19.15 B**). In presence of O_2; O_2 attaches itself at position 5. In presence of a reducing agent (RH) -OOH is at position 5. Hydroperoxide is thus produced which leads to transitions G-C to A-T. An other OH may attach itself at position 5. The -NH2 group at position 4 may be replaced by OH ; or OH at position 6 may be removed. The compound formed is 5-hydroxycytosine. 5-hydroxycytosine is an analogue of thymine. Results in G-C to A-T transition. Base-sugar (β-glycosidic) bond is sensitive to attack of OH radicals. This bond may get broken and sugar and base may get separated.

Maleic Hydrazide. Maleic hydrazide (MH) is a herbicide, fungicide, growth inhibitor and growth regulator. It is structural isomer of uracil (**Fig. 19.16**). It induces chromosomal breaks frequently in heterochromatic segment near the centromere of M chromosome of *Vicia faba*. Mode of action of maleic hydrazide is not known.

Antibiotics. Antibiotics, like actinomycin D, Mitomycin, alfatoxin B and streptomycin, have chromosome breaking properties. They affect the biosynthesis of DNA, RNA or proteins. Antibiotics can also enhance or reduce mutagenic effects of other mutagens depending upon the physiological state of the cell.

Azide. Azide is a potent mutagen which induces gene mutations in high frequency and chromosome aberrations in low frequency.

Fig. 19.15(A). Hydrogen peroxide is converted into hydroxyl radicals which react with cytosine at 5-6 double bond to produce 5-hydroxycytosine. β-glycisydic bond is also susceptible to the attack of hydroxyl radicals

Fig. 19.15(B). Hydrogen peroxide is converted into hydroxyl radicals which react with thymine in presence of oxygen and RH to produce hydroperoxide thymine that pairs with guanine

Alkylating Agents. Aklylating agents carry one, two or more alkyl groups in reactive form which are capable of being transferred to other molecules where electron density is high. Accordingly, such aklylating are called mono-, bi- or polyfunctional aklylating agents (**Fig. 19.17**). For example in sulphur mustard, one or both alkyl groups may be halogenated. In case of nitrogen mustard, one, two or three alkyl groups

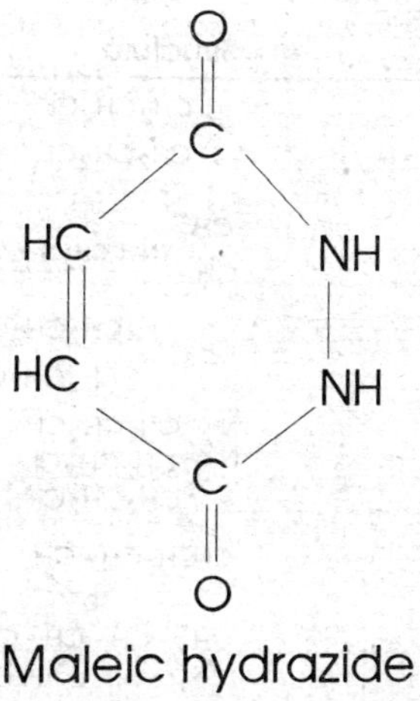

Maleic hydrazide

Fig. 19.16. Structure of maleic hydrazide

may be halogenated. Although dialkyl sulphates, may look bifunctional but it can not aklylate more than once so they are monofunctional aklylating agents. Similarily, aklylalkane sulphonates are **monofunctional** aklylating agents. Epoxides and ethylene amines are monofunctional aklylating agents. Some of the sulphates and sulphonates are monofunctional (dimethyl sulphate, diethyl sulphate, methyl methanesulphonate, ethyl methanesulphonate) while others (e.g. myleran) are **bifunctional**. Diazomethane, an diazoalkane, is a monofunctional agent. Most of the nitroso compounds, namely N-methyl-N-nitrosourethane, N-ethyl-N-nitrosourea, 1-methyl-3-nitro-1 nitroso guanidine, are monofunctional aklylating agents.

Alkylating agents may cause **cross-linking** of DNA and thus cause lethal effect by inhibiting DNA duplication because due to cross-linking DNA strand do no separate. Alkylating agents may readily react with phosphate groups and may thus break sugar phosphate backbone of DNA. This may induce larger alterations.

Most alkylating agents have same mutagenic effect. Some of the base of DNA is alkylated. For example, dimethylsulphate upon reaction with DNA may most frequently produce 7-methyl guanine which may pair with thymine. The result is transition G-C to A-T. Other products of this reaction may be 1-methyl adenine, 3-methyl adenine, 1,3-dimethyladenine or 7-methyladenine. The relative reactivities of various bases for

Group/Name	Structure	Nature
Sulphur mustards	$S\begin{smallmatrix}\diagup CH_2CH_2Cl\\ \diagdown CH_2CH_2Cl\end{smallmatrix}$	bifunctional
Nitrogen Mustards	$\begin{smallmatrix}CH_3\\ CH_3\end{smallmatrix}\!\!>\!N\text{-}H_2CH_2Cl$	monofunctional
	$CH_3N\begin{smallmatrix}\diagup CH_2CH_2Cl\\ \diagdown CH_2CH_2Cl\end{smallmatrix}$	bifunctional
	$N\!\begin{smallmatrix}\diagup CH_2CH_2Cl\\ \leftarrow CH_2CH_2Cl\\ \diagdown CH_2CH_2Cl\end{smallmatrix}$	trifunctional
Epoxides	$Cl\,CH_2CH\text{-}CH \atop \diagdown O \diagup$	monofunctional
	$CH_2\text{-}CH\text{-}CH\text{-}CH_2 \atop \diagdown O \diagup \quad \diagdown O \diagup$	bifunctional
Ethylene-imines	$HN\!\begin{smallmatrix}\diagup CH_2\\ \;\;\;\;\;\;|\\ \diagdown CH_2\end{smallmatrix}$	monofunctional
Sulphates and sulphonates - SO OR$_2$		
Dimethyl sulphate	$SO_2(OCH_3)_2$	monofunctional
Methyl methanesulphonate	$CH_3SO_2OCH_3$	monofunctional
Ethyl methanesulphonate	$CH_3SO_4O_2CH_5$	monofunctional
Myleran	$(CH_3)_2\!\!\begin{smallmatrix}\diagup CH_2OSO_2CH\\ \diagdown CH_2OSO_2CH_3\end{smallmatrix}$	bifunctional
Diazoalkanes		
Diazomethane	$N\!\equiv\!N^+\text{-}CH_2$	monofunctional
Nitroso compounds		
N-methyl-N-nitrosourethane (MNU)	$\begin{smallmatrix}CH_3\\ ON\end{smallmatrix}\!\!>\!N\text{-}COOC_2H_5$	monofunctional
N-ethyl-N-nitrosourea (ENH)	$\begin{smallmatrix}CH_2H_5\\ ON\end{smallmatrix}\!\!>\!N\text{-}CONH_2$	monofunctional
1-methyl-3-nitro-1 nitroso guanidine (MNNG)	$\begin{smallmatrix}CH_3\\ ON\end{smallmatrix}\!\!>\!N\text{-}C\text{-}NHNO_2$	monofunctional

Fig. 19.17. Structure and nature of various alkylating agents

methylation are 100 (7-methyl guanine), 50 (1-methyl adenine), 20 (1,3-dimethyl-adenine) and 2 (7-methyladenine). Diethyl sulphate with DNA produces 7-ethylguanine. N_1 position of cytosine is also akylated but to a very less degree. This reaction produces 1-methylcytosine which can pair with adenine. The result is transition G-C to A-T. The alkylation of bases in DNA does not seen to be direct but by **transalkylation** of alkyl group changing from phosphate to base. The alkylated base might inhibit DNA duplication thus causig lethal effect or casue base pairing mistakes during DNA duplication thus causing mutagenic effect. Alkylation of thymine may produce 6-O, ethyl thymine which pairs with guanine thus resulting into transition T-A to C-G. Alkylation of guanine produces 6-O, ethyl guanine less frequently and 7-alkyl guanine more frequently. Alkylation of purines in ^{7}N-position gives rise to quaternary nitrogens (N^+) that are unstable. Either alkyl group itself hydrolyzes away from the purine or the alkylated purine separates from deoxyribose leaving it "depurinated". The gap thus left may interefere with DNA duplication. At high pH depurinated DNA may occassionally break. This may induce larger alterations or be lethal but probably does not induce point mutations. From point of view of inducing mutations, methylating and ethylating agents have been most extensively used. Alkylated guanine in DNA is liberated slowly because of slow hydrolysis at glycosidic linkage. It may lead to loss of a base pair (deletion), or replacement of missing base by any available purine or pyrimidine base thus causing a transition, transversion or a return to the original state. Bifunctional agents have toxicity greater than monofunctional agents.

Detection of Mutations

Some of the important methods used in detection of mutations are described below:

Replica-plating Technique

Lederberg and Lederberg (1952) introduced replica plating technique to indirectly select bacterial mutants (**Fig. 19.18**). Master plate showing diffused bacterial growth of phage-sensitive *E. coli* on a nonphage medium. Replicas are made by pressing a velvet-covered wooden block against the master plate, and then pressing this, in the same oriented direction, to the surface of petri dishes containing culture medium mixed with phage T_1. One master plate possesses sufficent bacteria to start colonies on a number of replica plates. Replica plates show occurrence of resistant colonies in identical locations, indicating that resistnace to phage T_1 must have been present at each of these positions in the master plate.

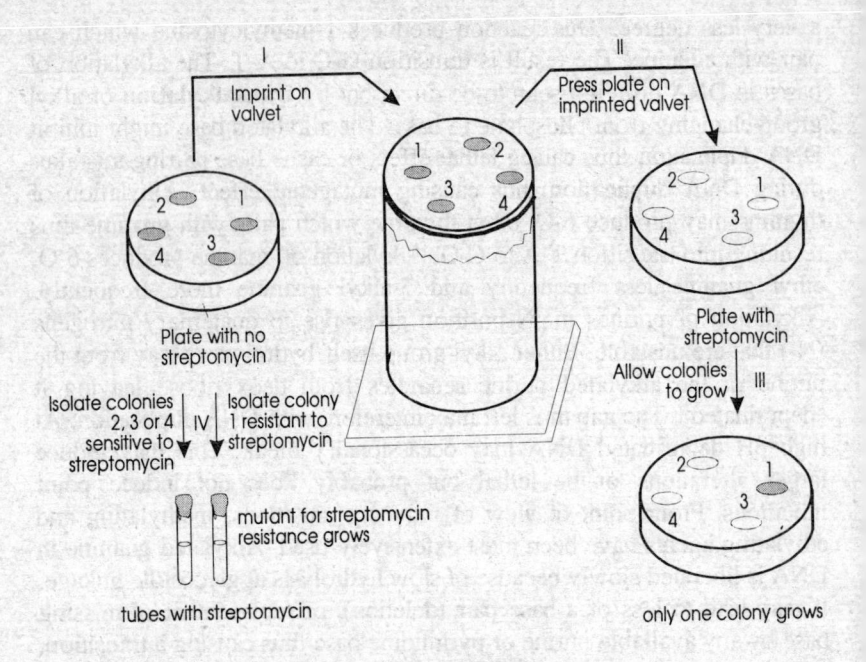

Fig. 19.18. Replica plating technique

Ames Test

To detect the mutagenic effect of various commercial and pharmecutical products, Ames devised a simple inexpensive test that relies on the reverse mutation of histidine-requiring (his^-) **auxotorphs** in *Salmonella typhimurium* to wild-type (his^+) **prototrophs** (Fig. 19.19). Ordinarily, such **reversions** are quite rare, no more that about once in 100 million (10^8) cells; meaning that very few, if any, prototrophic colonies will be found when 10^8 his^- cells are plated on minimal agar. However, should a tested substance be mutagenic, then increased densities of his^+ revertants will occur in those areas of his^- plates to which the substance is applied. To improve the efficiency of the test, strains of bacteria are used that are defective in DNA repair (improving the ability to detect mutagenicity) and that also have increased permeability (allowing tested substances to enter the cell more easily). In addition, tested substances are preincubated with liver extracts in order to simulate mammalian metabolic activity, and thus to discover whether such substances, as metabolized in the body, produce mutagenic acitvity. The Ames test and its various

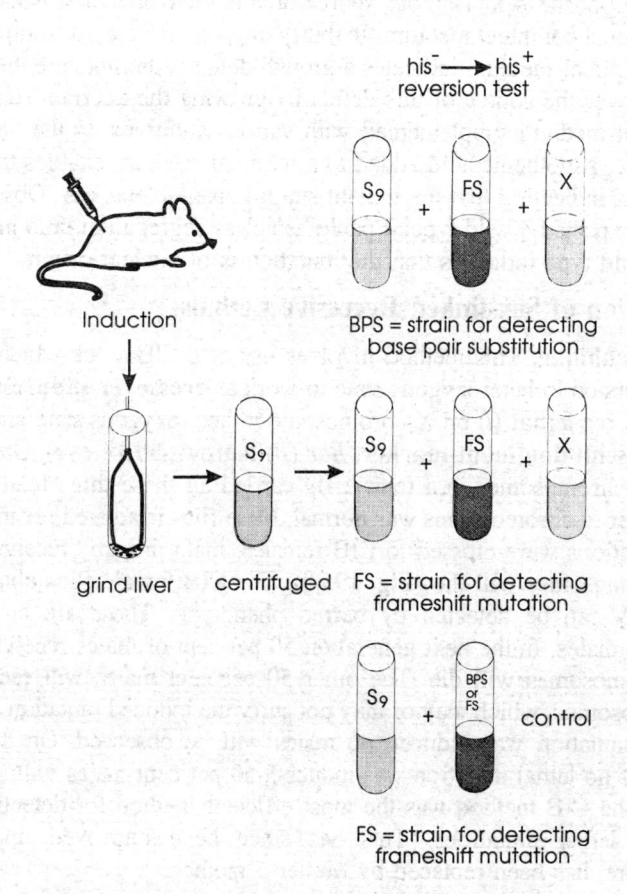

Fig. 19.19. A summary of procedure for Ames test

derivatives are therfore being extensively used to obtain information on both mutagenicity and carcinogenicity of a large number of chemicals. Two types of *his⁻* auxotrophs strain are used: Strain TA100 is highly sensitive to reversions by base-pair substitutions whereas strains TA1535 and TA1538 are sensitive to reversions by frameshift mutations.

Nutritional Mutants

Conidia of a particular strain of *Neurospora crassa* causing mutation (X-rays or ultraviolet, see Chapter 16 for details) and crossed to wild type of the opposite sex. Haploid spores of this cross are then isolated and grown on a complete medium. Inability of such an isolated strain to grow on a minimal medium indicates a growth defect. Attempts are then made to discover the source of this defect by growing the aberrant strain on a minimal medium supplemented with various additives. In the illustrated example, panothenic acid added to a minimal medium enables the strain to grow, indicating that the mutant strain is pantohtenicless. Observation of the expected 4 wild type: 4 pantothenicless segregation ratio in a cross with wild type indicates that the mutation is of nuclear origin.

Detection of Sex-linked Recessive Lethals.

ClB technique. This method involves use of a ClB stock which carries an inversion in heterozygous state to work as **crossover suppressor** (*C*), a **recessive lethal** (*l*) on X-chromosome in heterozygous state and a sex-linked **semi-dominant marker,** *Bar* (*B*)(narrow slitlike eye). One of the two X-chromosomes in a female fly carried all these three features and the other X-chromosomes was normal. Male flies irradiated for induction of mutations were crossed to ClB females. male progeny receiving ClB X-chromosomes will die (**Fig. 19.20**). The ClB female flies obtained in progeny can be detected by barred phenotype. These are crossed to normal males. In the next generation 50 per cent of males receiving ClB X-chromosomes will die. The other 50 per cent males will receive X-chromosomes, which may or may not carry the induced mutation. In caqe lethal mutation was induced no males will be observed. On the other hand, if no lethal mutation was induced, 50 per cent males will surivive. Thus, the ClB method was the most efficient method for detecting sex-linked lethal mutations. This has since been improved upon and, therefore, has been replaced by Muller-5 method.

Muller-5 technique. Muller (1927) evolved a method by which certain *Drosophila* mutation could be detected. One of such techniques is Muller-5 technique which is utilized to detect sex-linked recessive lethals induced in males of *D. melenogaster*. Muller-5 stock carries an X chromosome that contains *Bar* (*B*) and *white-apricot* (w^a) and two *scute* inversions. The precise genetic transcription of the stock is $In(1)sc^{S1L} sc^{8R} + S, sc^{S1} sc^8 w^a B$. If a lethal is induced in the wild type male being tested, F_1 hetrozygous female will carry the lethal on the X chromosome derived from the male parent. If this is so, wild type males will not appear in F_2 (**Fig. 19.21**). Frequency of lethal mutations can be accurately scorded in large samples.

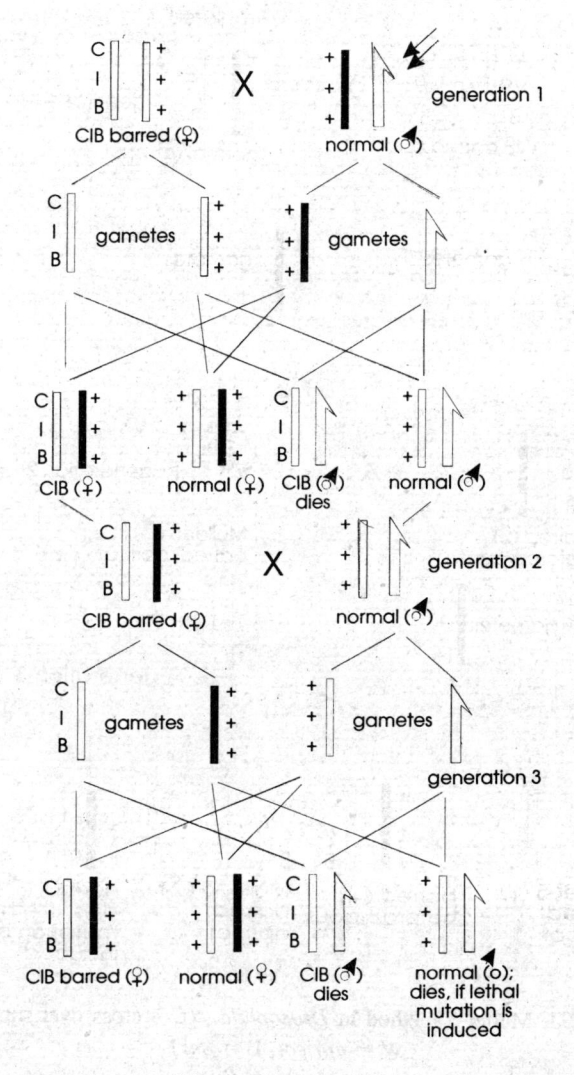

Fig. 19.20. ClB method in *Drosophila*. (*C* = cross over suppressor, *l* = recessive lethal, *B* = Bar)

Detection of Autosomal Recessive Lethals

Detection of mutations on autosomes (chromosomes other than sex chromosomes) makes use of a balanced lethal stock. Balanced lethal stock

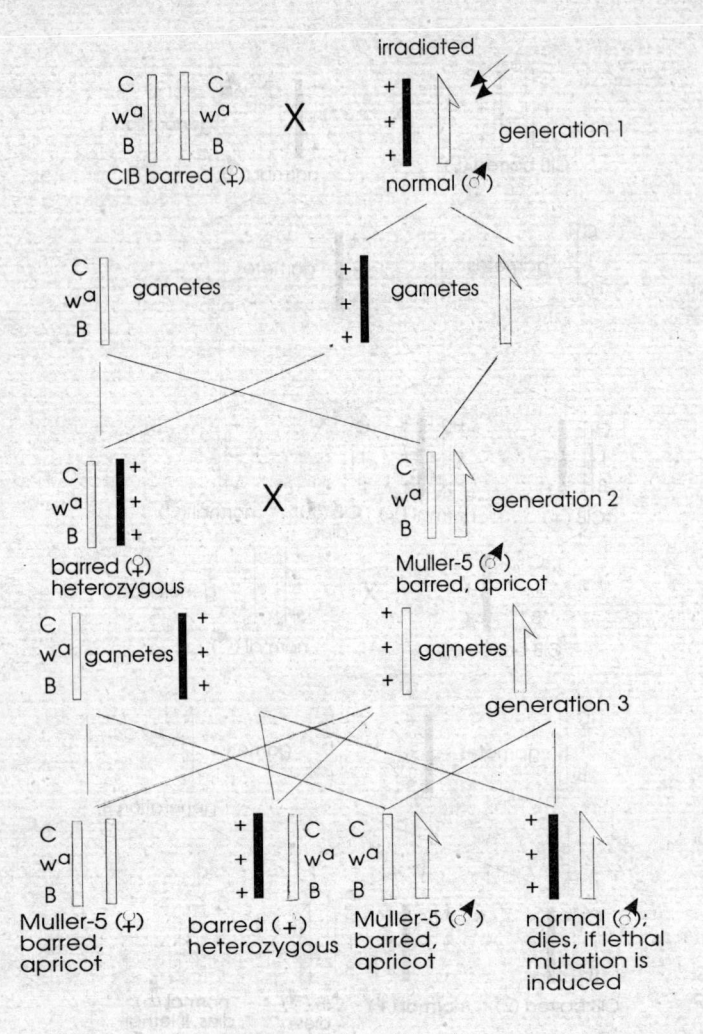

Fig. 19.21. Muller-5 method in *Drosophila*. (C = cross over suppressor, w^a = *apricot*, B = *Bar*)

is one which carries recessive lethal alleles on two specific homologous chromosomes, their normal allele being present on its homologue; consequently, homozygous individuals will not survive. The concept of balanced lethal would be better understood using a specific example. For detection of visible mutations in *Drosophila*, a stock carried the dominant genes *Cy* (*curly* wing) and *L* (*lobed* eye) on one chromosome and *Pm* (*plum*, i.e., brownish eye) on the other homologue so that the organism can be disgnated as *Cy L/Pm*. Phenotype of the stock which was heterozygous for *Cy*, *L* and *Pm*, was *Curly*, *Lobe* and *Plum*. If such individuals are crossed among themselves, progeny is always heterozygous, because each of the two homologous chromosomes carried recessive lethal genes (**Fig. 19.22**). Through the presence of inversion, crossing over was also suppressed, so that *Cy* and/or *L* may not be transferred to the chromosome carrying *Pm* and vice-versa.

For detection of an autosomal mutation *Cy L/Pm* stock is crossed to irradiated fly. From F_1 generation *Cy L* flies were again backcrossed to *Cy L/Pm* in order to obtain *Curly, Lobe* (heterozygotes) male and female flies carrying the same autosome to be tested. Such *Cy L* heterozygotes when crossed among themselves would give 1 *Cy L/Cy L* : 2 *CyL/++* : 1 *++/++*. Since *Cy L* homozygotes are lethal, phenotypic ratio will be 2 curly : 1 wildtype. However, if lethal mutation was induced, only curly will appear. If a recessive **visible mutation** was induced this will appear in homozygous state in one-third of the progeny (Fig. 19.22). Thus balanced lethal system also detects visible mutations.

Host-mediated Assay

In this test, microorganisms (bacteria or yeast cells) are introduced into a host organism such as a rat (or a mouse). The host is then treated with the substance being tested. After a period of time, the microorganisms are recovered, checked for mutations and compared to the control. It is presumed that experimental micro-organism comes in contect with the substance after it has been metabolized by the host. The host-mediated assay was originally developed by Gabridge and Legator (1969) and was later improved by Fahrig (1974, 1977).

A recent method for detection of mutation in man and other organisms (plants and animals) is to screen various enzymes (Glucose-6-phosphate dehydrogenase or proteins (hemoglobin) for slight variation. The technique used is known as **electrophoresis** which separates DNA and protein molecules in an electric field on the basis of mass/charge ratio. Relative positions occupied by different electrophoretic products (bands) of an individual under test are compared with a wild-type (standard) individual. The molecules showing variation can be shown to be heritable, and their transmission can be traced between various pedigree generations. Not all variant forms of a protein can be detected by this method.

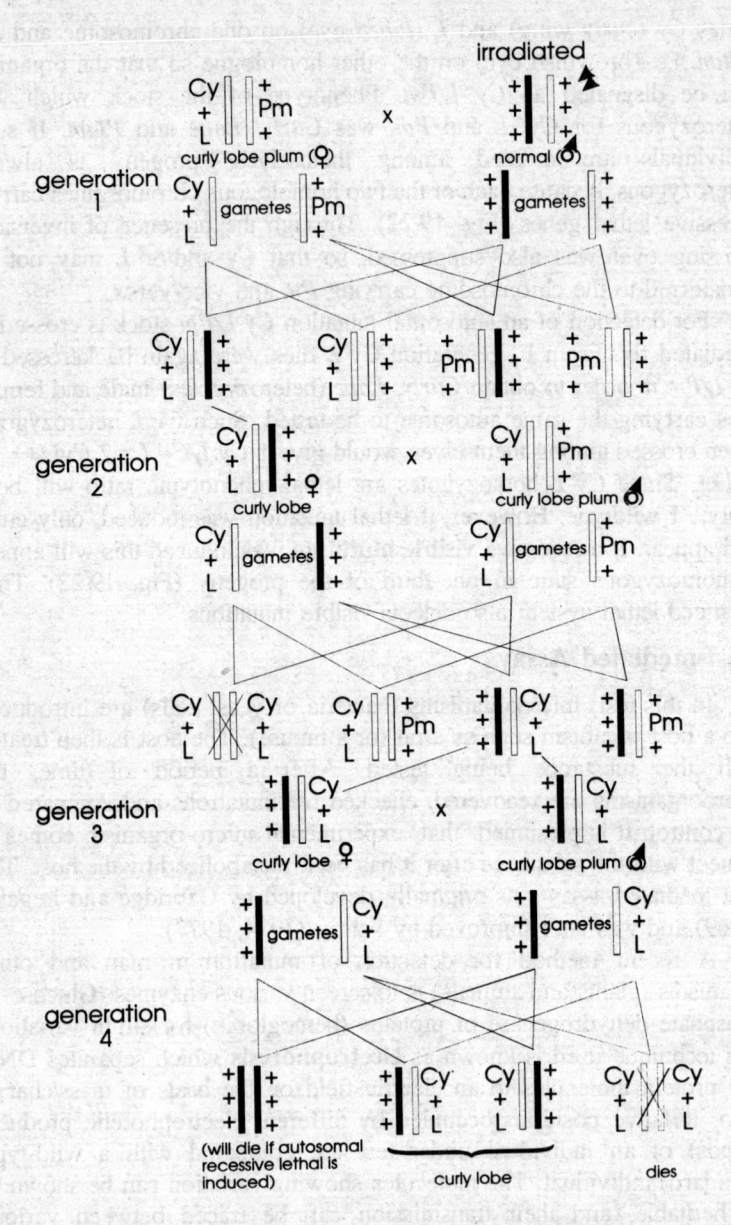

Fig. 19.22. Balanced lethal system in *Drosophila*. (*Cy* = *Curly*, L = *Lobe*, Pm = *Plum*)

Applications of Induced Mutations

Mutations are mostly deleterious and recessive and therefore majority of them have no practical value. However, a small frequency (one out of 1,000) of induced mutations have been successfully used for production of improved varieties.

Nilsson-Ehle and Gustafson (1947) induced mutations in barley by use of X-rays. They were able to produce varieties with stiffer straw, a more dense head and higher yield than parents. Mexican wheat varieties though high yielding were red grained. Treatment with γ-rays converted them to amber colour. For example, Sharbati Sanora was evolved from Sonora 64 and Pusa Lerma was evolved from Lerma Rojo 64A. In wheat, several mutants like lodging resistant, high protein and lysine contant have been obtained and utilized in plant breeding programmes.

Semi-dwarf variety of rice was evolved from tall rice variety with the help of X-rays. Mutants were also obtained for increased protein and lysine content. In some mutants isolated, duration of crop was reduced from 120 to 95 days. Reduction in duration of crop plants has been important contribution of mutations. For example, in sugar cane, growing period was reduced from 18 months to less than 10 months and in castor, duration has been reduced from 9 months to 4.5 months. In California, peanuts with thick shells were produced through induced mutations. Several muuants of *Penicillium* have helped in increasing the yield of pencillin.

Environmental Mutagenic Hazards

We are using a variety of chemicals in our industry and to these chemicals million of workers are being exposed regularly. Even our agriculture is being industrialized and here also we are using large number of chemicals in form of fertilizers, insecticides, fungicides, bactericides, etc. The persons not working in industry and agricultural expose themselves to chemicals in other ways, e.g., medicines, food dyes, food preservatives, dyes. It will not be an exaggeration to state that chemicals have become an important part of our life.

A dangerous consequence of modern industrialized society is the effects its products can have on the genetic material. Our technologically oriented society has greatly increased the amounts of various gases, viz., CO, CO_2, SO_2, NO, NO_2, H_2S that can be quite hazardous. For example, sulphurous gases enter the atmosphere through industrial and automobile combustion.

$$SO_2 + H_2O \longrightarrow H_2SO_3 \longrightarrow H^+ + HSO_3$$

Both SO_2 and H_2SO_3 have been shown to be mutagenic in a number of organisms. (2) NO and NO_2 are produced during combustion of fossil fuels.

$$NO_2 + H_2O \longrightarrow HNO_2$$

HNO_2 is a potent mutagen as such; also

$$NO_2 \xrightarrow{light} NO + O.$$

All these compounds are mutagenic. Combustion of fossil fuels in industry and automobiles produces another class of hazardous chemcials known as hydrocarbons. Some of the hydrocarbons are directly mutagenic, while others are metabolically converted into mutagens. Ozone is produced around electric utility plants where sparking is common, and in polluted areas exposed to intense sunshine. It enters into free-radical reaction that mimics the effects of ionizing radiations on the genetic material.

Now extensive lists of pesticides, drugs, industrial chemicals, food additives are available that have been tested in one or more systems and they have been shown to be mutagenic in one or the other test system, while information on many is lacking. It has been emphasized that all of our environmental chemicals should be thoroughly tested. It has been recommended that the "high risk" persons should be offered routine check up for any cytogenetic abnormality.

Even after observing that a particular chemical is mutagenic in some test system, it does not mean that it will be mutagenic in all the system. For example, a chemcial found mutagenic in Ames test many not be mutagenic in *Drosophila* or man. Such observations lead to controversy over the safe use of the particular chemicals. e.g., controversy prevailed for a long time on the safe use of DDVP, sachharrine, etc. Before a chemical is declared to be genetically hazardous, testing of the chemicals must be throughly done in various test systems. Basically the genetic material is same in bacteria, *Drosophila*, wheat, and man. A chemical reacting with DNA is bacteria is also expected to do so in the other organism, provided it penetrates into the cells. The negative results in one test system do not rule out genetic activity in other organisms.

Some Useful Reading Hints

DeVries (1901) gave theory of mutation. Sturtevant (1921) reported presence of inversions by studying rearrangement of genes in *Drosophila*. Sturtevant (1926) and Sturtevant and Beadle (1936) reported a crossover reducer in *Drosophila melanogaster* due to inversion of a section of the third and X chromosome, respectively. Altenburg (1934) reported artificial production of mutations by ultraviolet light. Bridges (1936) reported that the *Bar* "gene" in *Drosophila* was due to a duplication. Demerec (1937) estimated frequency of sontaneous mutation in certain stocks of *Drosophila melanogaster*. Auerbach and Robson (1946) and Auerbach (1949) worked on chemical induction of mutations. Lederberg and Lederberg (1952) and Lederberg (1989) used replica-plating for indirect selection of bacterial mutants. Hollaender (1954) gave a complete account of radiation biology. Lea (1955) described actions of radiations on living cells. Brenner *et al.* (1961) gave the theory of mutagenesis. Singleton (1962) gave an excellent account of various methods of detection of mutations in different test organisms. Muller and Oster (1963) introduced some mutational techniques in *Drosophila*. Hollaender (1971) and Hollaender and DeSerres (1982) dealt with principles and methods for detection of chemical mutagens. Auerbach (1976) dealt with the problems, results and perspectives of mutation research. Ehling (1991) reviewed principles and methods of genetic risk assessment. Foster and Cairns (1992) reviewed the machanisms of directed mutations. Kibley (1977) published a handbook of mutagenicity test procedures. Miglani (1986) reviewed methodologies for genetic risk evaluation of environmental chemicals. Nagao *et al.* (1979)

Ames *et al.* (1975) gave new methods for detecting carcinogens and mutagens with *Salmonella*/mammalian - microsome mutagenicity test. Ames (1979) used these methods for identifying environmental chemicals causing mutations and cancer. Brusick (1987) gave complete account of principles of genetic toxicity. Heslot (1965) gave a comprehensive review of the nature of mutations. Drake (1969) reported comparative rates of spontaneous mutation. Freese (1963) and Freese (1971) gave an excellent review on the molecular mechanism of mutation. Drake (1970) reviewed the molecular basis of mutation.

Questions

1. What do you understand by the terms mutation, mutagen and mutant?
2. Describe various characteristics of mutations.
3. Classify mutations on the basis of the following crireria: size of mutations, changes in chromosome structure, changes in chromosome number, direction of mutations, and effects on the sense of genetic code.
4. What are spontaneous mutations? How are the following processes responsible for their origin: spontaneous changes in electronic structure of bases, rare base pairing between two purines or two pyrimidines, chemicals produced under normal growth conditions.
5. What are induced mutations? Give brief history of induced mutations.
6. What are different types of physical agents that are capable of inducing mutations?
7. What are particulate and non-particulate radiations? How do they differ in their biological effects?

8. Describe mechanism of action of the following radiations: Ultraviolet light, X-rays and γ-rays
9. What is target theory of mutagenic action of X-rays?
10. Define the following terms related with radiation genetics: Roentgen (R), Radiation absorbed dose (RAD), Relative biological efficiency (RBE), Rem, and Curve.
11. What are different classes of chemical agents that are known to induce mutations?
12. Describe the mechanism of action of the following chemical mutagens: inhibitors of nucleic acid precursors, base analogues, dyes, nitrous acid, hydroxylamine, hydrazine, low pH, heat, hydrogen peroxide, maleic hydrazide, antibiotics, azides, and alkylating agents.
13. What is the usefulness of various organisms in detection of mutations?
14. Describe the following methods of detection of mutations in bacteria: Replica-plating technique and Ames test.
15. Describe the method of detection of nutritional biochemical mutations in *Neurospora crassa*.
16. Describe the following methods of detection of mutations using *Drosophila* as a test organism: sex-linked recessive lethals and autosomal recessive lethals.
17. What is host-mediated assay? What is the advantage of this method over others?
18. How is electrophoresis useful in detection of mutations?. What type of mutations can be detected through this technique?
22. Describe various applications of induced mutations.
23. What are environmental mutagenic hazards? What classes of agents pose threat to genetic health of man. Why a battery of test system is recommended for testing mutagenicity of an environmental chemical?

20

GENES IN POPULATIONS

Laboratory and Natural Populations

Population genetics is a study of the frequencies of genes in a population. Upto now we have been dealing with laboratory populations. From genetics point of view laboratory experiments that natural populations differ from each other. The relative frequency of alleles at a locus in a laboratory population are usually fixed at some convenient ratio by the design of the experiment, e.g., when two homozygous parents are crossed (*TT* X *tt*) to begin an experiment, the alleles are therefore introduced in equal frequency. In natural populations, on the other hand, the relative frequency of alleles may vary greatly. In laboratory populations, system of mating is usually sharply defined, e.g., crossing F_1 X F_1 or F_1 to one of the parents. Thus, breeder entertains a wide variety of crosses. In natural populations, much more is generally left to chance. Any genotype may mate with any genotype.

Binomial Distribution of Genotype Frequencies

Gene frequencies can be computed in decimals and can be treated and manipulated as probabilities. One chromosome carries one allele and its homologous chromosome carries the other allele. The probabilities of different genotypes in the population can be calculated by combining the chromosomes by chance. In natural populations, the relative frequencies of alleles may vary greatly.

Hardy-Weinberg Law

G.H. Hardy (from England) and W. Weinberg (from Germany) independently reported in 1908 basic principles of population genetics.

Their discovery is known in form of Hardy-Weinberg Law which states that in a large population mating at random, after one generation of such mating the frequencies of alleles and of genotypes remain constant generation after generation provided the following conditions are met: The population must be large enough so that sampling errors can be disregarded. This implies that there should be no **random genetic drift**. There can be no mutations, either $A \to a$ or $a \to A$. If there are, $A \to a$ mutation its frequency must be the same as the $a \to A$ frequency. If this is not the case the A and a gene frequencies will change over time. Mating must be random; that is all mating combinations must be equally probable. For examples, let us consider three genotypes - AA, Aa, aa. Various mating combinations $AA \times AA$; $AA \times Aa$; $AA \times aa$; $Aa \times Aa$; $Aa \times aa$; and $aa \times aa$ should be in same proportion. All the genotypes should be equally viable and equally fertile. No selection should be operative. All gametes must have equal probabilities of forming zygotes. The population must be isolated to avoid individuals, and hence genes, from migrating into and out of the population.

If these condition are met with, **genetic equilibrium** will be established; if not, gene frequencies will change. The changing of gene frequencies over time is the entire basis of **evolution**.

Principle of Genetic Equilibrium

The **principle of genetic equilibrium** states that if the a fore mentioned conditions are met, gene frequencies and phenotypic ratio in a population will remain stable. Example: A gene has two alleles, A and a, only in a population so that total number are $A + a$. Proportion of allele

$$A = \frac{A}{A + a} = p \text{ and that of}$$

$$a = \frac{a}{A + a} = q$$

$$p + q = \frac{A}{A + a} + \frac{a}{A + a} = \frac{A + a}{A + a} = 1$$

Note that $p + q = 1$.

Let us suppose a cross F_1 (Aa) X F_1 (Aa). Both the parents are assumed to form the two types of games in same proportion (A) and (a).

♀\♂	(A)	(a)
(A)	AA p x q = p^2	Aa p x q = pq
(a)	Aa p x q = pq	aa q x q = q^2

Note that $p^2 + 2pq + q^2 = 1$

Let us consider a hypothetical population composed of 490 individuals of *AA*, 420 *Aa* and 90 *aa* genotype. Total number of individuals in the population is 1000. Frequency of different genotypes is: $AA = p2 = 490/1000 = 0.49$, $Aa = 2pq = 420/1000 = 0.42$, and $aa= q^2 = 90/1000 = 0.90$.

Frequency of allele $A = p = 0.49 + 0.21 = 0.7$

Frequency of allele $a = q = 0.09 + 0.21 = 0.3$.

Let us now consider whether these frequencies change from one generation to next if all the conditions of Hardy-Weinberg Law are met with.

Genotypic Frequencies (♀)	Genotypic frequencies (♂)		
	AA (0.49)	Aa (0.42)	aa (0.09)
AA (0.49)	AA x AA (0.240)	AA x Aa (0.206)	AA x aa (0.044)
Aa (0.42)	Aa x Aa (0.206)	Aa x Aa (0.176)	Aa x aa (0.038)
aa (0.09)	aa x AA (0.044)	aa x Aa (0.038)	aa x aa (0.008)

Let us now calculate frequencies of *A* and *a* alleles in the progress of these matings.

Mating	Mating probability	Offspring ratio	Allele ratio in offspring A	a
AA x aa	0.240	1AA	0.240	0.000
AA x Aa	0.206	1AA:1 Aa	0.155	0.051
AA x aa	0.044	1Aa	0.022	0.022
Aa x AA	0.206	1AA:1Aa	0.155	0.051
Aa x Aa	0.176	1Aa:2Aa:1aa	0.088	0.088
Aa x aa	0.038	1Aa:1aa	0.009	0.029
aa x AA	0.044	1Aa	0.022	0.022
aa x Aa	0.038	1Aa:1aa	0.009	0.029
aa x aa	0.008	1aa	0.000	0.008
Total	1.000		0.700	0.300

Note that gene frequencies among the first generation are identical those among the parents. This is a characteristic of an equilibrium population and would hold all the time, under defined set of conditions, no matter how many generation were studied.

Equilibrium in Mixed Populations

Let us consider two populations A and B which differ in genotypic constitution. Mix populations A and B to get new Population C. The genotypic make up of this population will be as under.

```
Population A       Population B       Population C
AA  Aa  aa    +   AA  Aa  aa    =   AA   Aa   aa
640 320 40    +   250 500 250   =   890  820  290
```
Frequency of genotypes = AA 0.445, Aa 0.410, aa 0.145
Frequency of allele A = p = 0.445 + 1/2(0.410) = 0.65
Frequency of allele a = q = 0.145 + 1/2(0.410) = 0.35

Expected genotypic frequencies	Expected individuals
$AA = p^2 = (0.65)2 = 0.42$	840
$Aa = 2pq = 2(0.65)(0.35) = 0.46$	920
$aa = q^2 = (0.35)^2 = 0.12$	240

Genotype	Observed	Expected
AA	890	840
Aa	820	920
aa	290	240

The calculated chi-square value (24.26) is greater than theoretical value (3.841) at one degree of freedom and 0.05 probability. Observed and expected thus do not agree because population C is not a product of random mating. Genetic equilibrium will, however, be reached after one generation of random mating.

Determination of Gene Frequency

Dominant Inheritance

In cases in which one allele is completely dominant over the other, two of the genotypic classes are phenotypically identical. In these cases we can use the square root of the homozygous recessive class, if we assume Hardy-Weinberg equilibrium. For example, **phenylthiocarbamide (PTC) tasters** have a dominant gene T. Thus tasters are of TT or Tt genotypes. The nontasters are tt. In a population of 320 individuals, 218 tasters and 102 nontasters were found. Frequency of tasters, $tt = q^2 = 102/320 = 0.32$. Therefore, frequency of recessive allele, $q = \sqrt{0.32} = 0.56$. Frequency of dominant allele, $p = 1 - q = 1 - 0.56 = 0.44$ (since $p + q = 1$).

From frequencies of alleles T and t, genotypic frequencies can be determined. By multiplying each frequency with total number of individuals in the population, expected number of individuals belonging a particular genotype can be computed.

Genotype	Frequency	Expected numbers
TT	$p^2 = (0.44)^2 = 0.19$	0.19 X 218 = 41
Tt	$2pq = 2 \times 0.44 \times 0.56 = 0.49$	0.49 X 218 = 107
tt	$q^2 = (0.56)^2 = 0.32$	0.32 X 218 = 70

In order to test the population for genetic equilibrium, chi-square analysis is employed.

Phenotype	Observed	Expected
Tasters	116	148
Nontasters	102	70

The calculated chi-square value (21.546) is greater than theoretical value (3.841) at one degree of freedom and 0.05 probability. Therefore, population is not in genetic equilibrium.

Co-dominant and Incompletely Dominant Inheritance

In MN blood group system in man, M and N blood antigens are inherited in Mendelian fashion, and are co-dominant, so that genotype *MM* results in M blood type, *MN* results in the MN blood type, and *NN* results in the N blood type. In case of incomplete dominance, like co-dominance, homozygous dominants, heterozygotes and homozygous recessives give different phenotypes. So the procedure used for calculation of gene frequency in both these cases is the same. Half the frequency of heterozygotes is added to frequency of homozygotes for the allele for which the frequency is to be determined. To illustrate the method let us consider the following example. In a given population, frequency of *MM* individuals = 0.512, that of *MN* individuals = 0.400, and that of *NN* individuals = 0.088. Frequency of allele $M = p = MM + 1/2\ MN = 0.512 + 1/2\ (0.400) = 0.712$ and frequency of allele $N = q = NN + 1/2\ MN = 0.088 + 1/2(0.400) = 0.288$.

Expected genotypic frequency	Expected number
$MM = p^2 = (0.71)^2 = 0.505$	505
$MN = 2pq = 2(0.71)\ (0.29) = 0.412$	412
$NN = q^2 = (0.29)2 = 0.083$	83

Phenotype	Observed	Expected
M	512	505
MN	400	412
N	88	83

The calculated chi-square value (0.748) is less than theoretical value (3.841) at one degree of freedom and 0.05 probability. Expected class size agree so well with the observed class size that we can conclude that population is in Hardy-Weinberg equilibrium.

Multiple Allelic Traits

The ABO blood group series in humans is determined by three alleles I^A, I^B, i^o. The i^o allele is recessive to I^A and I^B. The I^A and I^B alleles are co-dominant (Chapter 7).

Genotypes	Blood groups	Expected genotypic frequency
$I^A I^A$, $I^A i^o$	A	$p^2 + 2pr$
$I^B I^B$, $I^B i^o$	B	$q^2 + 2qr$
$I^A I^B$	AB	$2pq$
$i^o i^o$	O	r^2

If we assume that A', B', AB' and O' represent phenotypic frequencies of A, B, AB and O blood types, the following formulae are used to calculate frequency of the three alleles.

Frequency of allele i^o = r = $\sqrt{O'}$

Frequency of allele I^A = p = 1 - $\sqrt{(B' + O')}$

Frequency of allele I^B = q = 1 - $\sqrt{(A' + O')}$

This procedure can be better understood with the help of the following example. A sample of 500 people had the following distribution of blood types: A 195, B 70, AB 25, and O = 210. Assume p = frequency of allele

I^A, q = frequency of allele I^B, and r = frequency of allele i^o. Genotype frequencies of four blood types are:

A	O	AB	B
$p^2 + 2pr$	r^2	2pq	$q^2 + 2qr$
195/500	210/500	25/500	70/500
0.39	0.42	0.05	0.14

To calculate gene frequencies:
$r^2 = 0.42$, $r = \sqrt{0.42} = 0.65$
Frequencies of A and O phenotypes =
$p^2 + 2pr + r^2 = (p+r)^2 = 0.39 + 0.42 = 0.81$
$p + r = \sqrt{0.81} = 0.9$
$p = 0.9 - r = 0.9 - 0.65 = 0.25$
Frequency of the three alleles, $p + q + r = 1$
$0.25 + q + 0.65 = 1$
$q = 1 - 0.65 - 0.25 = 0.10$

In order to test the population for genetic equilibrium, expected frequencies and expected numbers for different genotypes are calculated.

Genotypes	Expected frequency	Expected numbers
$I^A I^A$	$p^2 = (0.25)^2 = 0.06$	30
$I^A i^o$	$2pr = (2)(0.25)(0.65) = 0.33$	165
$I^A I^B$	$2pq = (2)(0.25)(0.10) = 0.05$	25
$i^o i^o$	$r^2 = (0.65)2 = 0.42$	210
$I^B i^o$	$2qr = 2(0.10)(0.65) = 0.13$	65
$I^B I^B$	$q^2 = (0.10)^2 = 0.01$	5

Blood type	Observed number	Expected number
A	195	195
B	70	70
AB	25	25
O	210	210

The calculated chi-square value (0.0) is less than theoretical value (3.841) at one degree of freedom and 0.05 probability. In fact there is no deviation. The population is in Hardy-Weinberg equilibrium.

Sex-linked Inheritance

In organisms such as humans and *Drosophila*, where males are heterogametic (XY) and females are homogametic (XX) sex, procedure for determination of gene frequency must take into account the fact that females carry twice as many X-linked genes as do the males. There are only two genotypic classes, A and a, for males; their total genotypic frequency being $p + q = 1$. However, there are three genotypic classes, AA, Aa and AA, for females; their total genotypic frequency being $p^2 + 2pq + q^2 = 1$. First of all frequencies of the two alleles p and q are separately calculated in males and females. Then keeping in mind the fact that two-thirds of a sex-linked allele is found in females and one-third in males, frequency of p and q alleles in the population in calculated.

Colourblindness trait in humans and white eye colour in *Drosophila* are sex-linked characters. Let us consider an example of colourblindness in humans. In a sample of 2,000 individuals (1,000 men and 1,000 women), 90 males are colourblind. In this example, we are interested to calculate only the frequency of carriers. If p be the frequency of normal allele R^g and q be the frequency of colourblind allele r^g.

Sex-linked genes being in **hemizygous** state, phenotypic frequencies and genotypic frequencies in heterogametic males are same. Frequency of allele for colourblindness, $q_m = 90/1000 = 0.09$, and frequency of wild type allele, $p_m = 1 - 0.09 = 0.91$.

Assuming a population with random mating proportions, the expected numbers of females expressing the trait and carrying it

392 Basic Genetics

heterozygous fashion can be estimated as follows considering the frequencies of p and q same in both the sexes.

Genotypes	R^gR^g	R^gr^g	r^gr^g
Genotypic frequency	p^2	$2pq$	q^2
	$(0.91)^2$	$2(0.91)(0.09)$	$(0.09)^2$
	0.83	0.16	0.01

Frequency of females carriers of colourblindness = $2pq$ = 0.16. Number of carrier females = 0.16 X 1000 = 160.

Let us now consider a slightly different example where number of mutants for the sex-linked trait in males as well as females is known. In a sample, of 2,000 (1,000 females and 1,000 males), 17 females and 40 males were hemophilic. In this example, we will calculate frequency of (H) normal and hemophilic (h) alleles and find out whether or not population is in genetic equilibrium.

In males, $q_m = 40/1000 = 0.04$

$p_m + 40/1000 = 1$

$p_m = 1 - 0.04 = 0.96$

In females, $q^2_f = 17/1000 = 0.017$

$q_f = \sqrt{0.017} = 0.13$

$p_f = 1 - 0.13 = 0.87$

In order to calculate frequency of p and q in the population one must remember that females carry twice as many X-linked genes as do the males.

$p = 2/3\ p_f + 1/3\ p_m\ = 2/3\ (0.87) + 1/3\ (0.96) = 0.90$

$q = 2/3\ q_f + 1/3\ q_m\ = 2/3\ (0.13) + 1/3\ (0.04) = 0.10$

With this frequency of p and q, expected frequencies of different genotypes and phenotypes in males and females can be calculated.

	Normal		Carrier females	Hemophilic	
	Males	Females		Males	Females
Frequency	0.9	0.81	0.18	0.1	0.1
Expected Number	900	810	180	100	10

Phenotype	Observed	Expected
Normal	1943	1890
Hemophilic	57	110

The calculated chi-square value (31.022) is greater than theoretical value (3.841) at one degree of freedom and 0.05 probability. We find that the given population is not in Hardy-Weinberg equilibrium.

Some Useful Reading Hints

Hardy (1908) and Weingerg (1908) did poineering work on Mendelian proportions in a mixed population. Wright (1948) discussed the roles of directed and random changes in gene frequency in the genetics of populations. Lush (1951) reviewed concepts of genetics in relation to animal breeding. Li (1955) and Lewontin (1985) presented concepts and principles of population genetics. Mather (1973) described genetical structure of populations. Crow (1963) did lot of theoretical work to understand behavior of genes in populations. Crow (1986) gave deataied of basic concepts in population, quantitative, and evolutionary genetics.

Questions

1. What are salient differences between laboratory and natural populations?
2. How and why is binomial distribution used in determination of genotype frequencies?
3. Define as completely as possible Hardy-Weinberg law describing all the assumptions considered there in.
4. Explain the principle of genetic equilibrium?
5. Do we observe equilibrium in mixed populations in the initial generation? Give reason for your answer.
6. Describe the basis and method for determination of gene frequency in the traits showing: (a) dominant inheritance, (b) co-dominant inheritance, (c) incompletely dominant inheritance, (d) multiple allelic traits, and (e) sex-linked inheritance.

Numericals

1. In a population of 200 individuals, if 100 individuals have M-blood group, 80 have MN and 20 have N blood group, find out whether the population is in equilibrium or not?
2. In a population of 320 individuals, 215 are tasters and 105 are non tasters. What are the frequencies of alleles T and t and different genotypes?
3. What is the frequency of heterozygote tasters in a random mating population, if the frequency of non-tasters is 0.09?
4. What is the frequency of heterozygotes Tt in a random mating population in which the frequency of all the tasters is 0.19?
5. A sample of 500 people had the following distribution of blood types: A-195, B-70, AB-25, 0-120. Find out whether the population is in equilibrium or not.
6. The following are the observed ABO blood group phenotypes among a group of 600 Canadian Indians: 0-288, A-28, B-19, AB-13. (a) Calculate the estimates of A, B and O gene frequencies (b) How does the observed distribution of phenotypes agree with those expected on the basis of calculated gene frequencies?
7. In testing the blood types of 2047 Guernsey cattle, Stormont found genotypes for the co-dominant alleles, Z and z, in the following frequencies: 542 ZZ, 1043 Zz, and 462 zz, Using the Chi-square method, calculate how far the observed frequencies depart from the expected equilibrium frequencies.
8. Among 569 Egyptians tested for the ability to taste phenylthiocarbamide (PTC), 442 tasters were found. If the allele for tasting has a dominant effect over the nontasting allele: (a) What proportion of mating will be of the type taster x taster, assuming that mating is purely random? (b) What is the ratio of taster to nontaster children expected from these taster x taster matings? (c) What is the proportion of taster to nontaster children expected from taster x nontaster matings in this population?
9. The following are the observed ABO blood group phenotypes among a group of 600 Canadian Indians: O 288, A 280, B 19, AB 13. How well does the observed distribution of phenotypes agree with those expected on the basis of the calculated gene frequencies.
10. Waaler tested 9049 school boys in Oslo for red-green colourblindness and found 8324 of them to be normal and 725 colorblind. He also tested 9072 school girls and found normal and 40 colorblind. Assuming that all forms of red-green colourblindness are caused by effects of the same sex-linked recessive allele, c, estimate whether this sample demonstrates equilibrium for this locus.

QUANTITATIVE INHERITANCE

Quantitative Traits

Quantitative inheritance in study of inheritance of those trait that exist in a range of phenotype for a specific character. Such characters differ by degree rather than by distinct qualitative differences. **Phenotypic range** of quantitative trait appears to be continuous without distinct steps that could be accounted for by distinct genes. Most quantitative traits tend to follow a **normal distribution** curve.

Galton's Interest in Quantitative Traits

Galton, a cousin of Charles Darwin, noted that certain physical and mental characteristics which showed continuous variation appeared to be inherited, so that taller individuals, for example, produced taller children, on an average. For such traits, although segregation and assortment of individual hereditary factors could not be determined, the biometricians were able to demonstrate statistically liklinesses between relatives with respect to numerous continuously distributed traits.

Most of the actual variations among organisms in natural population are quantitative. Some examples of quantitative traits are stature in man, weight in man, egg-laying in chickens, yield in crop plants, milk production, intelligence quotient (I.Q.), metabolic activity, reproductive rate, behaviour. All these characters are continuous over a range. The continuity of phenotypes in quantitative traits is result of two phenomenon: Each genotype does not have a single phenotype but a wide phenotype range. **Phenotypic differences** between genotypic classes become blurred and it is not possible to assign a particular phenotype to a particular genotype and (b) Expression of quantitatrve traits depends on action and interaction of several genes each having a small additive effect on the phenotype. Such characters are subject to considerable phenotypic modifications by environment.

Measurement of Quantitative Traits

Quantitative characters cannot be sharply classified. Quantitative traits are measured in terms of various **statistical parameters** like (a) arithmetic mean, (b) mode, (c) median, (d) variance, (e) standard deviation, (f) coefficient of variation. Data of a quantitative trait fit a **normal distribution**.

Multiple Factor Hypothesis

East and Nilsson-Ehle (1910) put forward 'multiple factor hypothesis'. They explained quantitative inheritance on the basis of action and segregation of a number of allelic pairs having duplicate and accumulative effects without complete dominance. Studies on ear length in corn, seed, weight in beans, and grain colour in wheat supported multiple factor hypothesis.

Ear Length in Corn

R.A. Emerson and E.M. East (1913) worked on ear length in corn. They crossed two varieties - one was Thom Thumb Popcorn (Parent 60) and other was Black Mexican sweet corn (Parent 54). Details are given in **Fig. 21.1**.

	Parent 60		Parent 54
Range	: 5-8 cm	X	13-21 cm
Mean	: 6.63 cm		16.80 cm
Variance	: 0.816		1.887

↓
F_1

Range	:	9-15 cm
Mean	:	12.116 cm
Variance	:	1.519

F_1 X F_1
↓
F_2

Range	:	7-19 cm
Mean	:	12.888 cm
Variance	:	2.252

Fig. 21.1. Results of the crosses conducted by Emerson and East (1913) for ear length in corn

The following conclusions can be drawn from the above data: (1) mean of F_1 is approximately intermediate between the means of the two parental varieties, (2) mean of F_2 is equal to the mean of F_1, (3) variance in F_2 is greater than that in F_1, and (4) extreme measurements in the F_2 overlap well into the distribution of parental values. The experiment of Emerson and East (1913) thus provided evidence to multiple factor hypothesis.

A hypothetical model to explain multiple factor hypothesis has been given in **Fig. 21.2.**

Mendelian Genes and Quantitative Inheritance. After rediscovery of Mendelism, the view emerged that genes controlled inheritance of discontinuous variation. It was found difficult to explain inheritance of quantitative traits on the basis of particulate genes. Were the genes controlling quantitative inheritance different from those controlling qualitative inheritance? Answer to this question came from detailed analysis of quantitative characters in beans (*Phaseolus vulgaris*).

Seed Weight in Beans

Johannsen's technique, based on inbreeding, was to separate the beans into distinctive lines on the basis of weight of seeds. Since beans are highly self-pollinated, it was easy to establish inbred stocks derived initially from single seeds. He established 19 **pure lines**. Each pure line produced average seed weights that were individually distinct (**Table 21.1**). Line 1 was the heaviest (64.2 cg) and Line 19 was the lighest (35.1 cg). Genetic differences between lines could be seen when comparing the offspring of different pure lines with same parental seed weight. A study of Table 21.1 shows that different parental sizes within a particular line produced offspring with uniform mean values. Some degree of variability existed within each pure line that was apparently caused by environment, as can be seen in a variety of seed weights produced by a single sample line, that of number 13 in **Table 21.2**, for example.

On the basis of Johannsen's experiment, a population of individuals that varied in respect to a quantitative characteristic was conceived to consist of a number of genetically different groups. Within each group, the quantitative characteristic would have a range of measurements. Some degree of overlapping would occur between the ranges of different groups, so that the separation between them was not distinct. Johannsen's experiment explained that (a) continuous variation noted for quantitative

398 Basic Genetics

True-breeding Tall is 78"　　X　　True breeding short is 60"
　　XXYYZZ　　　　　　↓　　　　　　xxyyzz
F$_1$:　　　　　　　　　XxYyZz
　　　　　　　　　　　　　69"

F$_2$:

Frequency	Genotype	Height (inches)
1	XXYYZZ	78
2	XXYYZz	75
2	XXYyZZ	75
2	XxYYZZ	75
4	XXYyZz	72
4	XxYYZz	72
4	XxYyZZ	72
8	XxYyZz	69
1	XXYYzz	72
2	XXYyzz	69
2	XxYYzz	69
4	XxYYZz	66
1	XXyyZZ	72
2	XXyyZz	69
2	XxyyZZ	69
4	XxyyZz	66
1	xxYYZZ	72
2	xxYyZz	69
2	xxYyZz	69
4	XXYyZz	66
1	XXyyzz	66
2	Xxyyzz	63
2	xxyyZz	63
1	xxYYzz	66
2	xxYyzz	63
1	xxyyzz	60

No. of active alleles	0	1	2	3	4	5	6
Height (inches)	60	63	66	69	72	75	78
Frequency	1	6	15	20	15	6	1

Fig. 21.2. A model explaining multiple factor hypothesis

Table 21.1. Johannsen's experiments on 19 pure lines. Each pure line produced average seed weights that were individually distinct

Pure line	Yield of the parents						Average weigt of offspring
	20x20	30x30	40x40	50x50	60x60	70x70	
1					63.1	64.9	64.2
2			57.2	54.9	56.5	55.5	55.8
3				56.4	56.6	54.4	55.4
4				54.2	53.6	56.6	54.8
5			52.8	49.2		50.2	51.2
6		53.5	50.8		42.5		50.6
7	45.9		49.5		48.2		49.2
8		49.0	49.1	47.5			48.9
9		48.5		47.9			48.2
10		42.1	47.7	46.9			46.5
11		45.2	45.4	46.2			45.4
12	49.2			45.1	44.0		45.5
13		47.5	45.0	45.1	45.8		45.4
14		45.4	46.9		42.9		45.3
15	46.9			44.6	45.0		45.0
16		45.9	44.1	41.0			44.6
17	44.0		42.4				42.8
18	41.0	40.7	40.8				40.8
19		35.8	34.8				35.1
Mean of all lines =							47.9

traits was result of both genotype and environment, (b) although basically genetic, the distinction between Johannsen's pure lines could not be ascribed to particular genes since none were individually identified. On the basis of above experiments it was concluded that quantitative variation is also governed by Mendelian genes.

Table 21.2. Number of offspring of different seed weights produced in Johannsen's pure line, number 13

Weight of parents	Weight classes of offspring centigrams									Total
	17.5	22.5	27.5	32.5	37.5	42.5	47.5	52.5	57.5	62.5	
27.5			1	5	6	11	4	8	5		40
32.5				1	3	7	16	13	12	1	53
37.5		1	2	6	27	43	45	27	11	2	164
42.5	1		1	7	25	45	46	22	8		155
47.5				5	9	18	28	19	21	3	103
52.5		1	4	3	8	22	23	32	6	3	102
57.5			1	7	17	16	26	17	8	3	95
	1	2	14	38	104	172	179	140	53	9	712

Yule (1906) suggested that continuous quantitative variation might be produced by a multitude of individual genes, each with a small effect on the measured character.

Grain Colour in Wheat

Nilsson-Ehle (1909) demonstrated an actual segregation and assortment of genes with quantitative effect. According to Nilsson-Ehle, there were three individual gene pairs involved in determination of grain colour in wheat i.e., A,a; B,b; and C,c with genes for red (A, B, C) dominant over genes for white (a, b, c). Each of these three gene pairs segregated in predictable Mendelian fashion, so that the products of heterozygotes for any one pair produced offspring in 3:1 when two gene pair differences were segregating at the same time, the results also followed Mendelian principles, producing 15 red : 1 white. Similarly, a cross between heterozygotes for 3 gene pairs produced a close fit to 63 red : 1 white. Not all red phenotypes showed identical shading, suggesting that some quantitative effect was involved in determining the degree of redness. A variety of red genotypes occur containing different numbers of the three red genes, A, B, and C. If we imagine that, instead of dominance, the action of each red gene is to add a small equivalent degree of redness to the plant, then the range of red phenotypes corresponds to the range of red genotypes. The extreme phenotypes

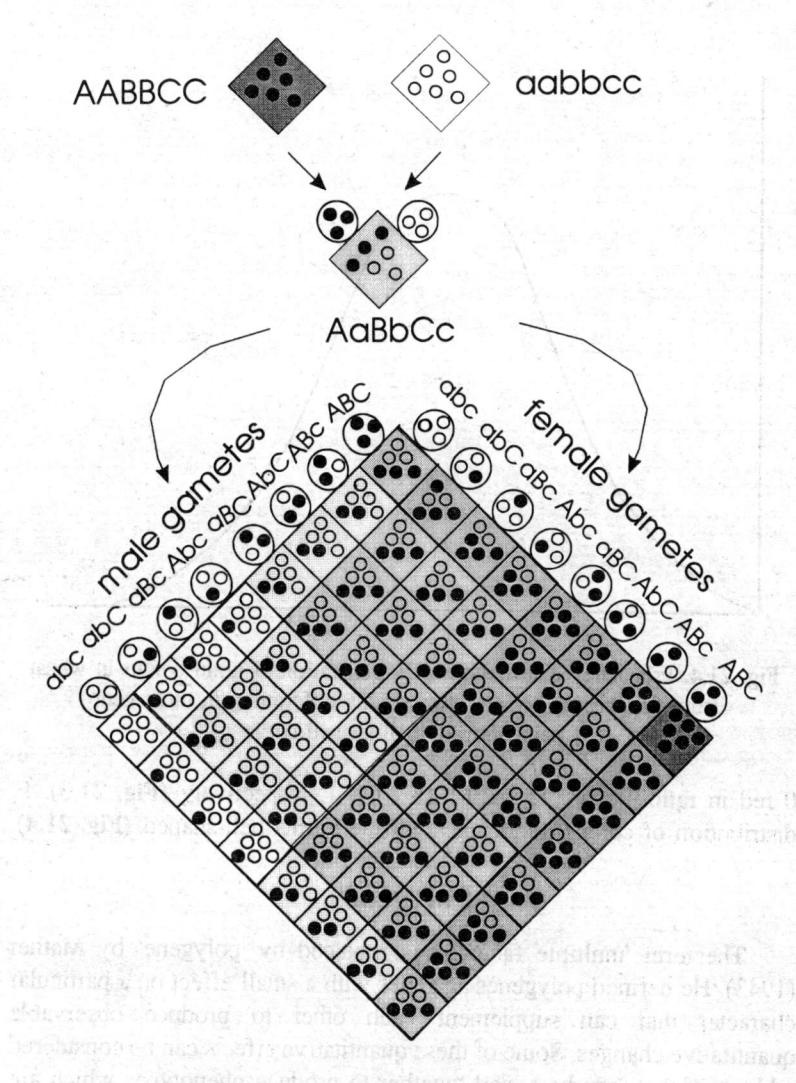

Fig. 21.3. Inheritance of a quantitative trait

among the F_2 are expected to be quite rare, while the intermediate types should be more frequent. In wheat, three genes determining grain colour segregate independently. Consequently, there will be seven classes with respect to number of red genes - 6 red, 5 red, 4 red, 3 red, 2 red, 1 red,

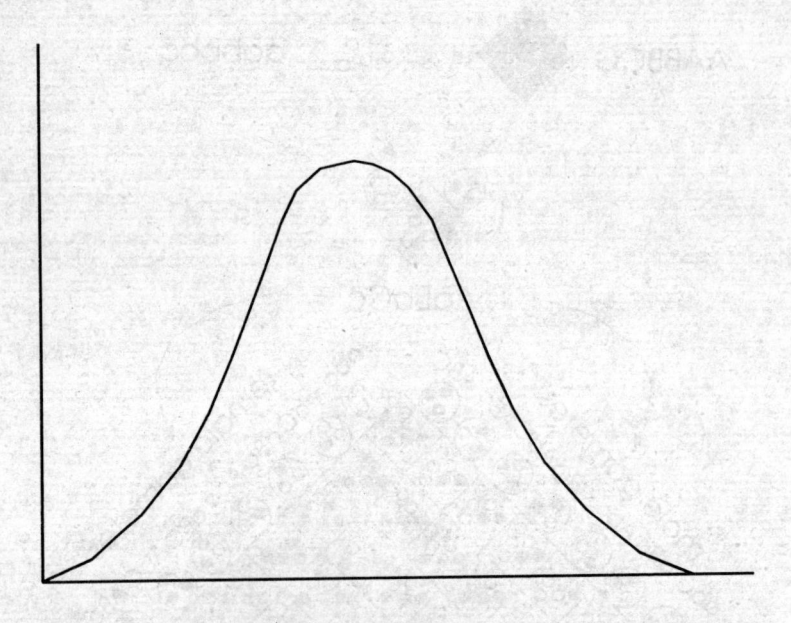

Fig. 21.4. Frequency distribution in F_2 generation of grain colour in wheat from the crosses between two strains differening in three gene pairs shows a bell-shaped curve

0 red in ratio 1 : 6 : 15 : 20 : 15 : 6 : 1, respectively (**Fig. 21.3**). F_2 distribution of colour frequencies will be norma bell-shaped (**Fig. 21.4**).

Polygenes

The term 'multiple factor' was replaced by 'polygene' by Mather (1943). He defined polygenes as genes with a small effect on a particular character that can supplement each other to produce observable quantitative changes. Some of these quantitative effects can be considered additive if they can be added together to produce phenotypes which are the sum total of the negative and positive effects of individual polygenes.

Environmental effects, although not inherited, may also produce modifications in the expected phenotypes of any generation. This factor is also operating to produce continuity of measurement for quantitative characters.

Genetic and Environment Components of Continuous Variation

The **phenotypic variations** observed between individuals in a population may be the result of genetic differences, environmental differences and/or interaction between the genotype and the environment. Mathematically, $V_P = V_G + V_E + V_{GE}$, where, V_P = total phenotypic variance observed, V_G = genetic variance, i.e., all genetically based differences, V_E = environmental variancem i.e., all non-genetically based differences, and V_{GE} = variance due to genetic and environmental interactions. Genetic expression varies as a function of the environment in which the genotype is placed. This component is difficult to analyze and quantify, so it is usually ignored.

Heritability

It is useful to know relative importance of heredity and environment in expression of a trait. When genotype completely determines the phenotype, heritability is 100 per cent. Heritability specifies the proportion of total variability that is due to genetic causes. Moderately low heritability or moderately large number of genes will result in more or less normal distribution.

Heritability can be expressed quantitatively by the ratio of genetic variance to the total variance. Mathematically,

$$H = \frac{\sigma G^2}{\sigma G^2 + \sigma E^2} \tag{1}$$

where, σG^2 = genetic variance and σE^2 = environment variance

The genotypic variance among families and various interaction components can be obtained from the mean-square expectations. Various components of variance are:

σa^2 = A genetic component arising from genetic differences among families.
σal^2 = A component arising from interaction of families and locations.
σay^2 = The family X year component of variance.
σaly^2 = The family X location X year component of variance.
σe^2 = The error variance.

The relationship between above components of variance is depicted as: $\sigma A^2 = \sigma a^2 + \sigma al^2/L + \sigma ay^2/Y + \sigma aly^2/LY + \sigma e^2/LYR$.

$$H = \frac{\sigma a^2}{\sigma A^2} \tag{2}$$

Additive and Dominance Components of Genetic Variance

Considering only the A_a locus, the values associated with the genotype are:
(i) AA = d,   (ii) Aa = h,   (iii) aa = -d.
Since all the values are measure from AA-aa mid plants (M) as the origin, the difference between the AA and aa homozygotes has the value of 2da.

```
      aa              M         Aa              AA
    ──┬───────────────┬──────────┬───────────────┬──
                         ← h →
    ──┬───────────+ d ─────┬───── - d ───────────┬──
```

The constitution of the ompulation as regards the locus Aa is expected to be:

Genotype	¼ AA	½ Aa	¼ aa
Phenotype	+d	h	-d

Letting $D = \sum(d)^2$, $H = \sum(h)^2$, and E = environmental variance, the variance of an F_2 segregating for many genes can be written as:

$$V_{F2} = 1/2 D + 1/4 H + E \tag{3}$$

By similar method, summed variance of the backcrosses to the parents is found to be:

$$V_{B1} + V_{B2} = 1/2 D + 1/2 H + 2E \tag{4}$$

Using the mean variance of non segregating generations,

$$E = \frac{V_{P1} + V_{P2} + V_{F1}}{3}$$

Heritability in broad sense (H_b)

If is defined as ratio between total genetic variance and total phenotypic variance. It is calculated by using the formula:

$$H_b = \frac{V_{F2} - E}{V_{F2}} = \frac{\sigma G^2}{\sigma A^2}$$

Heritability in narrow sense (H_n)

It is defined as ratio between additive genetic variance and total phenotypic variance by using the following formula:

$$H_n = \frac{2V_{F2} - (V_{B1} + V_{B2})}{V_{F2}} = \frac{\frac{1}{2}D}{V_{F2}}$$

Degree of Dominance

Degree of dominance is expressed as $\sqrt{(H/D)}$. In order to find out value of ½D, substract $(V_{B1} + V_{B2})$ (equation 4) from 2 X V_{F2} (equation 3). Substitute values of ½D and E in equation (3) or (4) and calculate H. If value or degree of dominance is = 1, it indicates complete dominance, if value is > 1, it shows over dominance, and if value is < 1, it is a case of incomplete dominance. Thus degree of dominance reflects nature of dominance.

The Meaning of Heritability

A trait with a heritability of one is not influenced by environment. For example, if you inherit the blood alleles I_A and I_B, your phenotype will be AB, irrespective of the environment. Some traits, such as bristle number in *Drosophila*, have a heritability of 0.5. A trait with heritability value of zero has no genetic basis.

Estimation of heritability gives a clear cut pointer towards contribution of different components towards the total phenotypic variance and is thus used as an index to the effect of selection before having actually performed it.

Some Useful Reading Hints

Yule (1906) worked on the theory of inheritance of quantitative characters on the basis of Mendel's laws. Nilsson-Ehle (1909) conducted basic studies to understand the action of genes responsible for continuous variation. Davenport (1913) worked out heredity of skin color in Negro-White crosses. East (1916) gave a Mendelian interpretation of variation that is apparently continuous. Charles and Smith (1939) distinguished between various types of gene action in quantitative inheritance. Mather (1943) worked on polygenic inheritance and natural selection. Burton (1951) studied quantitative inheritance in pearl millet. Carlson (1959) worked on comparative genetics of complex loci. Allard (1960) has dealt in depth the principles of quantitative inheritance. Falconer (1960) has illustrated various concepts and principles in quantitative genetics. Tanksley (1993) described a method for the mapping of polygenes. Cheverud and Routman (1993) worked out individual gene effects on quantitative characters. Cheverud and Routman (1995) discusses epistasis and its contribution to genetic variacne components.

Questions

1. Define properties of quantitative traits.
2. What are Galton's observations on inheritance of quantitative traits?
3. What did Yule's work suggest about the inheritance of quantitative traits?
4. In what ways is the measurement of quantitative traits done?
5. Describe multiple factor hypothesis.
6. What do you understand about the inheritance of a quantitative trait?
7. How have experiments of Emerson and East contributed towards understanding of the inheritance of quantitative traits?
8. Do the Mendelian genes explain inheritance of quantitative inheritance?
9. What did the experiments of Johannsen on seed weight in beans add to our knowledge of inheritance quantitative traits?
10. What is the contribution of Nilsson-Ehle's work on grain colour in wheat in understanding inheritance quantitative traits?
11. What are polygenes? How do polygenes differ from Mendelian genes?
12. What are various genetic and environment components of continuous variation?
13. Define: (a) heritability (H), (b) heritability in broad sense (H_b), and heritability in narrow sense (H_n). What do these terms signify?
14. Give the steps for calculating heritability in broad sense (H_b), and heritability in narrow sense (H_n).
15. What is the meaning of heritability in applied sense?
16. What does degree of dominance mean? How is this parameter calculated? Does success of selection depend on degree of dominance?

Numericals

1. Two full-grown plants of a particular species are given to you which have extreme phenotypes for a quantitative character such as height, e.g., 1 foot tall and 5 feet tall. (a) If you had only a single set of environment conditions in which to conduct your experiment (e.g., one uniformly lighted greenhouse), how would you determine whether plant height is environmentally or genetically caused? (b) If genetically caused, how would you attempt to determine the number of gene pairs that may be involved in this trait?

2. A cross between two inbred plants that had seeds weighing 20 and 40 centigrams, respectively, produced an F_1 with seeds that weighed uniformly 30 centigrams. An $F_1 \times F_1$ cross produced 1000 plants; four had seeds weighing 20 centigrams, four had seeds weighing 40 centigrams, and the other plants produced seeds with weights varying between these extremes. How many gene pairs would you say are involved in the determination of seed weight in these crosses.

3. If skin color is caused by additive genes, give your answers to the following: (a) Can matings between individuals with intermediate colored skins give birth to lighter-skinned off-spring? (b) Can such matings produce dark-skinned offspring? (c) Can matings between individuals with light skins produce dark-skinned offspring?

4. Assume that two pairs of genes with two alleles each, Aa and Bb, determine plant height additively in a population. The homozygote $AABB$ is 50 centimeters high, the homozygote $aabb$ is 30 centimeters high. (a) What will be the F_1 height in a cross between these two homozygous stocks? (b) After an $F_1 \times F_1$ cross, what genotype in the F_2 will show a height of 40 centimeters? (c) What will be the F_2 frequency of these 40-centimeter plants?

5. If three independently segregating genes, each with two alleles, determine height in a particular plant, e.g., Aa, Bb, Cc, so that the presence of each capital-letter allele adds 2 centimeters to a base height of 2 centimeters: (a) Give the heights expected in the F_1 progeny of a cross between homozygous stocks $AABBCC$ (14 centimeters) x $aabbcc$ (2 centimeters). (b) Give the distribution of heights (phenotypes and frequencies) expected in an $F_1 \times F_1$ cross. (c) What proportion of this F_2 progeny would have heights equal to the parental stocks? (d) What proportion of the F_2 would breed true for the height shown by the F_1?

6. The components of variance obtained from analysis of data for total yield and seed yield from 2 locations in 2 years and 3 replications are given below:

Character	σ_a^2	σ_{al}^2	σ_{ay}^2	σ_{aly}^2	σe^2
Total yield	2480	0	2057	3099	16578
Seed yield	317	27	101	2.9	1061

Calculate H eritability for both the traits.

408 Basic Genetics

7. The variance for two traits for various generations are given below:

Generation	Variance for trait 1	Variance for trait 2
P_1	11036	315
B_1	17352	495
F_1	5237	150
F_2	40350	1140
B_2	34288	995
P_2	10320	250

Calculate (i) heritability in narrow sense, (ii) heritability in broad sense, and (iii) degree of dominance.

22

EXTRANUCLEAR AND EXTRACHROMOSOMAL INHERITANCE

Extranuclear Inheritance

Certain characters are controlled by the genes present in the cell organelles such as mitochondria and chloroplasts. These characters do not reveal Mendelian pattern of inheritance. Extranuclear inheritance is the term used to describe inheritance of this type and is characterized by differences in reciprocal crosses (Fig. 22.1) and **maternal inheritance**, i.e., progeny shows the characters of the female parent. Cytoplasmically controlled characters show no linkage with characters controlled by nuclear genes. Segregation of cytoplasmic genes is of non-Mendelian type, i.e., segregation reveals no predictable ratio. Even if nucleus is substituted these cytoplasmic characters are not affected.

Organelle DNA

Cytoplasmic DNA is found in cell organelles, namely, mitochondria, chloroplasts, kinetoplasts and centriole.

Mitochondrial DNA. All living beings cells except bacteria, blue green algea and mature erythrocytes contain mitochondria. Mitochondria are centre of aerobic respiration. Mitochondrial DNA (mt DNA) replicates independent of nuclear DNA. In many multicellular animal cells, mt DNA is double stranded, has a circular structure (similar to bacterial DNA) with a molecular weight of $9 \times 10_6$ to 10×10^6 and length 5 µm; number of base pairs being 14,000.

Chloroplast DNA. Characteristics of chloroplast DNA (cp DNA) are: molecular weight $55 \times 10_6$ to 97×10^6; circular DNA; length 419 µm; number of base pairs 1.3×10^5 to 1.5×10^5. cp DNA differs from nuclear

Reciprocal Crosses

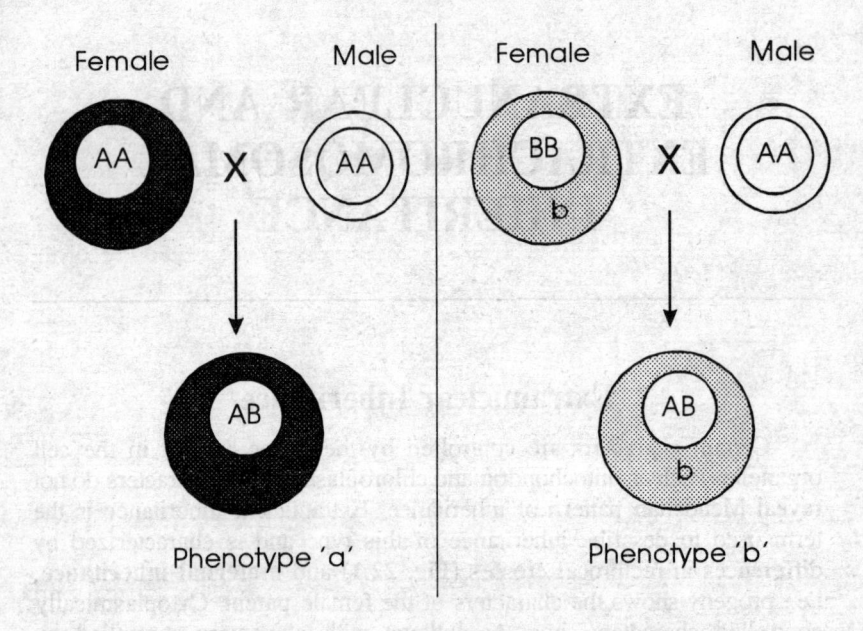

Fig. 22.1. Inheritance of a trait through a cytoplasmic gene

DNA of same species in nucleotide composition. Plastid DNA is shown to replicate through **semi-conservative mode** in *Chlamydomonas*. It higher plant species, cp DNA replicates by **rolling circle method**.

Kinetoplast DNA. Kinetoplasts are modified mitochondria located near to the base of flagella. Each protozoan contains a single kinetoplast which may be larger than the nucleus. The kinetoplast DNA consists of a large number of interlocked "minicircles". They are comparable with small bacterial plasmids.

Centriole DNA. Presence of DNA in centriole is still disputed with different experiments giving different results. They, however, seem to contain double stranded DNA.

Evolutionarily, both mitochondria and chloroplast DNA's seem related to bacterial DNA.

Cytoplasmic Inheritance

Cytoplasmic inheritance of those characters whose genes are located in the cytoplasmic cell organelles rather than in the nucleus.

Plastid inheritance in *Mirabilis jalapa*. Plastid inheritance in four o' clock plant *Mirabilis jalapa* provides an excellent example of cytoplasmic inheritance (Correns, 1909). Leaves are of three types with respect to chrloroplasts - pale, green and variegated. Controlled pollination of flowers borne on three kinds of branches gave the results shown in **Fig. 22.2**. These observations reveal that type of pollen used is not important. The factor determining phenotype. Factor determining phenotype of the progeny is female's contribution. The results can be interpreted on the basis of two different kinds of plastid transmitted only through the female with the demonstration that DNA occurs in plastids, it is established that plastids include material with known potential for replication and plastids store genetic information. In case of variegated phenotype, the egg type contains both green and white chloroplasts which produces a zygote that also contains both types of chloroplasts. In the subsequent mitotic divisions, some form of cytoplasmic segregation occurs that produces chloroplast types into pure cell lines thus producing a variegated phenotype. This process is known as **cytoplasmic segregation and recombination** (CSAR) (**Fig 22.3**). Thus, CSAR is the process in which

Pollen from branch of type	Pollinated flowers on branch of type	Progeny grown from seeds
Pale	Pale Green Variegated	Pale Green Pale, Green, Variegated
Green	Pale Green Variegated	Pale Green Pale, Green, Variegated
Variegated	Pale Green Variegated	Pale Green Pale, Green, Variegated

Fig. 22.2. Results of experiments conducted by Correns (1909) on *Mirabilis jalapa*

412 Basic Genetics

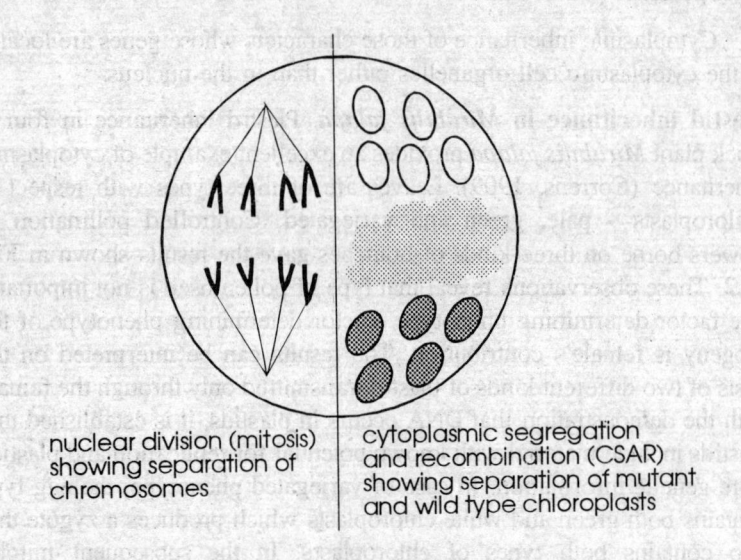

Fig. 22.3. Comparison between nuclear division, mitosis, and analogous hypothetical cytolpasmic segregation and recombination, CSAR

organelle-based genes assort and recombine. This is common behaviour of extranuclear genome. CSAR is, however, a hypothetical process.

Streptomycin-resistance in *Chlamydomonas*. Sager (1954) isolated a streptomycin sensitive mutant of *Chlamydomonas*. It has two mating types, mt$^+$ and mt$^-$. Streptomycin-sensitive mutants have a peculiar inheritance pattern as shown in the crosses depicted in **Fig. 22.4**. Hure all progeny cells show the streptomycin phenotype of the mt$^+$ parent. This is a kind of **uniparental inheritance**. Uniparental genome in this case is

Cross	Progeny
sm-r mt$^+$ X sm-s mt$^-$	sm-r
sm-s mt$^+$ X sm-r mt$^-$	sm-s

Fig. 22.4. Results of cross in streptomycin sensitive and resistant strains of *Chlamydomonas* differing in mating types (mt$^+$ and mt$^-$)

chloroplast (cp DNA). Buoyant densities of **cp DNA** in progeny from various *Chlamydomonas* crosses are shown in **Table 22.1**. The results show that cp DNA of the mt⁻ parent is lost, inactivated or destroyed.

Table 22.1. Buoyant densities of cp DNA in progeny from various *Chlamydomonas* crosses

Crosses	Buoyant density of zygote cpDNA
^{14}N mt⁺ X ^{14}N mt⁻	1.69
^{15}N mt⁺ X ^{15}N mt⁻	1.70
^{15}N mt⁺ X ^{14}N mt⁻	1.70
^{14}N mt⁺ X ^{15}N mt⁻	1.69

Mapping of Uniparently Inherited Genes. *Chlamydomonas* has a large number of uniparentally inherited genes. To map genes, we need to study recombination. For this, both types of parental cp DNAs need to be present. We know that mt⁻ parental cp DNA is eliminated. There is way out of this dilemma. In crosses of mt⁺ *sm-r* X mt⁻ *sm-s*, about 0.1 per cent the progeny zygotes were formed to contain both *sm-r* and *sm-s* alleles. This means segregation of two alleles presumably occurs from a CSAR process in the zygote at a later stage of division. Such zygotes are called **biparental zygotes** and their genetic condition is known as **cytohet**. Cytohet is cytoplasmically heterozygous conditions. This means the inactivation of mt⁻ cpDNA fails in certain cases. Mpas in Chlamydomonas have been constructed using such cytolets. Eventually circular map of *Chlamydomonas* cpDNA was derived from genetic analysis (**Fig. 22.5**). *Chlamydomonas* has only one chloroplast which contains 500-1500 cpDNA molecules, commonly observed to be packed in **nucleoids**. Each chlorplasts may have 4-18 nucleoids and each nucleoids may contain 4-8 cp DNA molecule.

Poky *Neurospora*. Mitchell (1952) isolated a mutant strain of *Neurospora* that she called *poky*. *Poky* is slow growing, it shows **maternal inheritance** and has abnormal amounts of cytochromes. Cytochromes are mitochondrial electron transport proteins necessary for the proper oxidation of foodstuffs to generate ATP energy. There are three types of cytochromes a, b and c. In poky, there is no cytochrome a or b but there is an excess of cytochrome c. *Neurospora* is a haploid organism, How

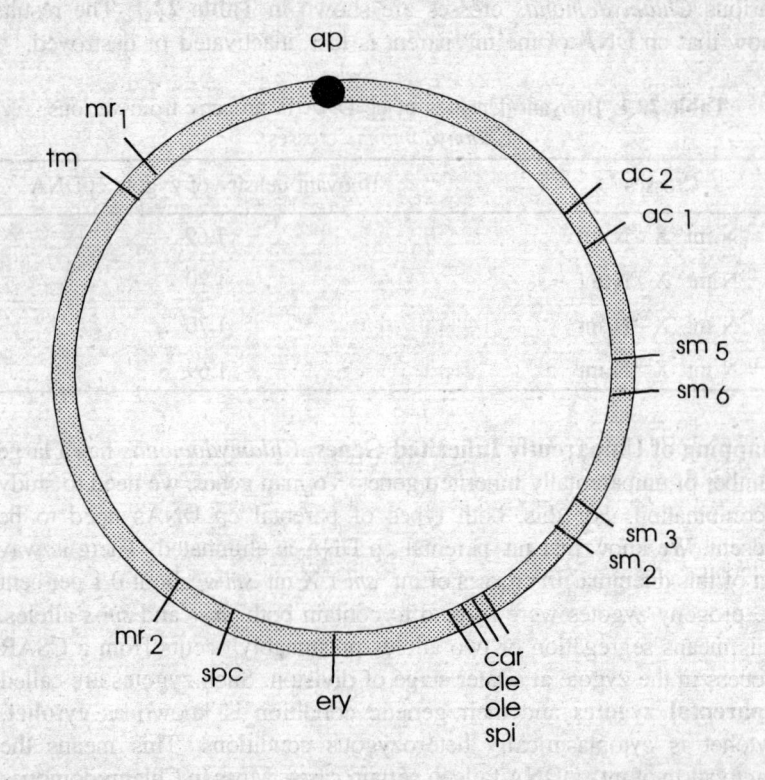

Fig. 22.5. Tentative linkage map for chloroplast genes of *Chlamydomonas reinhardtii*. Symbols: *ap* = attachment point, *ac* = acetate requirement, *sm* = streptomycin resistance, *cle* = cleosine resistance, *ole* = oleandomycin resistance, *spc* = strptomycin resistance, *mr* = multiple resistance, *tm* = temperature sensitivity (Singer *et al.*, 1976)

can maternal inheritance be expressed in a haploid organism is shown by the following crosses:

Cross	Progeny
poly ♀ X wild type ♂	All poky
wild type ♀ X poky ♂	All wild type

Thus poky does not behave like a nuclear mutation. It was termed as extranuclear mutation or a cytoplasmic mutation because it seems to be carried in the cytoplasm of the female parent. Poky mutants in *Neurospora crassa*

Drug Resistant Mutants. Drug resistant mutants in yeast, *Neurospora Paramecium*, mammalian cells were found to be due to mitochondrial genes.

Minute Mutants in *Chlamydomonas*. Minute mutants in *Chlamydomonas* can be induced with acriflavine or ethidium bromide. These mutants live only for 8-9 cell divisions in light after induction, and then die. If minutes are crossed with normal cells during interval between induction and death, the zygotes show 4:0 or 0:4, mutants normals. showing that minutes are not due to nuclear genes. Unlike *sm-2* mutants, minutes can be transmitted equally readily through mt^+ or mt^- gametes. Thus these mutants are extranuclear but have **biparental pattern of inheritance**.

CO_2 **sensitivity in *Drosophila melanogaster*.** Normally, *Drosophila* is immobilized rapidly by exposure to high concentration of CO_2 but recovers fully within a few minutes of its removal. L'Heritier and Tessier (1937) discovered a strain that was permanently paralyzed by CO_2. This was named as CO_2 **sensitive strain**. CO_2 sensitivity in *Drosophila melanogaster* is due to the presence of virus, called **sigma**, in the cytoplasm. This virus is transmitted in or association with, the germ cells of *Drosophila*. When a CO_2 sensitive female is crossed with a normal male (cross a), all the progeny is CO_2 sensitive. In the reciprocal cross (cross b), almost all the progeny is normal but a small proportion is CO_2 sensitive. This suggests extranuclear inheritance. Under an electron microscope, this virus was found attached to a cytoplasmic membrane. The trait is usually transmitted by the female because of the large volume of cytoplasm in the egg. This condition is also transmitted through male, although less frequently, because there is so little cytoplasm in sperm cells.

Variants are known for sigma. Recombination between different **sigma particles** has been demonstrated. Certain *Drosophila* nuclear genes inhibit or slow down sigma multiplication. A semi-dominant gene on chromosome 2 affects multiplication of sigma. The *ref-2/ref-2* homozygotes do not allow multiplication of virus although virus itself is not destroyed The *ref-2 gene* allele acts on the synthesis in viral RNA. Sigma mutant P' grows even in *ref-2/ref-2* homozygotes.

416 Basic Genetics

Sex-ratio Strains in *D. willistoni*. Sex-ratio (SR) strains produce progenies that are predominantly or exclusively females. This is due to mortality of males during embryonic development. Occasionally, a son of an SR female survives and reciprocal crosses can be made. These crosses show that SR condition is transmitted only through the female gamete.

Incompatibility in Mosquitoes. Various strains of mosquito *Culex pipiens* show various incompatibilities when hybridized. A cross between Hamburg (Ha) female and Oggelshausen (Og) male might produce viable progeny as about 87 per cent eggs develop into adults whereas its reciprocal cross might produce progeny that is sterile, viable, inviable or progeny having reduced proportion of viable individuals as 17 per cent of the eggs laid develop into adults which have mother characteristics. These adults are considered to be of parthenogenetic origin. This situation is due to a cytoplasmic factor which may be a **rickettsiae**-like microorganism. It has been shown that union of male and female nuclei does not take place during fertilization in certain crosses.

Males lacking rickettsiae were found to be compatible with all the females with or without rickettsiae while males containing rickettsiae were unable to produce viable offspring with females lacking rickettsiae. It seems that rickettsiae in males is responsible for incompatibility while inheritance of rickettsiae occurs through the females. Each strain of mosquito contains its own co-adapted strain of rickettsiae. When a male contains rickettsiae of a type different from those in the female, or when female has none, the action of the sperm is changed in such a way to prevent the formation and development of diploid eggs. Genes of any one strain are not required for maintenance of its own cytoplasmic rickettsiae.

Killer Strain of Yeast. A cytoplasmic factor which is double stranded RNA is involved in this trait. Yeast strains are of three types: **killer** - the yeast cells contain a cytoplasmic factor, **sensitive** - cytoplasmic factor is absent, and **neutral** - these cells are neither killed by the killer strain nor kill sensitive strains. Neutrals also contain a cytoplasmic factor which is shown by the following crosses:

Parents	Diploid hybrids	Tetrads
Killer X sensitive	Killer	4 Killer : 0 Sensitive
Neutral X sensitive	Neutral	4 Neutral : 0 Sensitive
Killer X Neutral	Killer	Various ratios from 4:0 to 0:4 killer: neutrals; No sensitives

Dominant gene *M* (Later designated as $+^{mak-1}$) is required for maintenance of both killer and cytoplasmic factors. Later on two nuclear genes, kex^1 and kex^2, were identified which convert killers into neutrals. Killer and neutral strains contain two (a larger and a smaller) species of double stranded RNA which was separately encapsulated into virus like particles. Smaller RNA containing particles are necessary for production of toxins. Function of smaller RNA-containing particles is dependent upon $+^{mak-1}$ allele (nuclear) and larger RNA-containing particles. The larger RNA-containing particles are inherited cytoplasmically and independently of host genotype. Smaller and larger RNA-containing viruses have been observed under the electron microscope.

Milk Factor in Mice. A low cancer incidence individual embryo if nursed by a female of high cancer incidence strain develops high rate of cancer. This is due to an infective agent that is transmitted through milk. It has characteristics of a virus.

Cytogenic Inheritance

Those biological properties that are controlled by cytoplasmic genes in conjunction with nuclear genes are referred to as cytogenic inheritance.

Male Sterility in Plants. Male sterility in plants (e.g., maize, sorghum, bajra) represents an example of cytogenic inheritance where inheritance of a character is controlled by the interaction of cytoplasmic and nuclear genes. This is because of the fact that male sterility can be of genetic and/or cytoplasmic origin. **Genetic male sterility** controlled by nuclear genes shows Mendelian inheritance (**Fig. 22.6A**). **Cytoplasmic male sterility** is controlled by mitochondrial genes and shows non-Mendelian inheritance (**Fig. 22.6B**). **Chimeric genes** formed in mitochondrial genome are also responsible for male sterility.

Male sterility is of great use in hybrid seed production since male flowers (anthers) need not be removed. It thus saves lot of time and money. For hybrid seed production maintainer line is required for maintaining the male sterile line.

Maintenance of Male Sterile Lines. Male sterile lines are pollinated by the pollen from the maintainer plants. Maintainer plants are those which when used as a pollen parent for pollinating cytoplasmic male sterile progeny (**Fig. 22.7**).

Hybrid Seed Production. Male sterile plants are used as female parent. Desirable pollen parent is used which restores the male fertility in the progeny. Such pollen parent is known as **restorer** (**Fig. 22.8**).

418 Basic Genetics

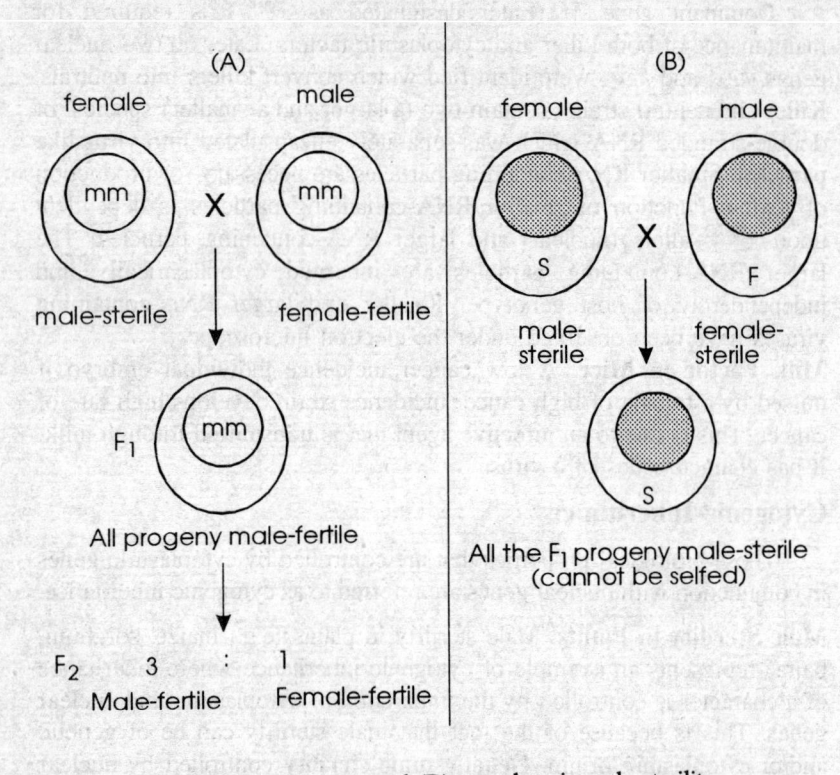

Fig. 22.6. (A) Genetic and (B) cytoplasmic male sterility

Cytoplasmic male sterility (cms) is clearly having cytogenic inheritance. More than 80 species of angiosperms have been reported to be producing strains with cytoplasmically inherited pollen sterility. Cytoplasmic male sterility is due to a cytoplasmic factor known as **cms factor**. Cytoplasmic male sterility could be transmitted in corn through cytoplasm for 20 generations continuously. During these generations, alongwith male sterile plants, some partially or completely male fertile plants were obtained. The latter type of plants were shown to contain certain **"restorer" genes** derived from fertile plants.

Restorer genes are dominant nuclear genes. These genes responsible for restoration of fertility in cms lines are designated as Rf_1, Rf_2, Rf_3, etc. These genes do not permanently modify or eliminate cms factors but merely suppress their expression. Different strains of maize may have different types of cms factors (**Table 22.2**). Different restorer genes have

Male sterile plant (female parent)   maintainer line (A-line) (male parent)

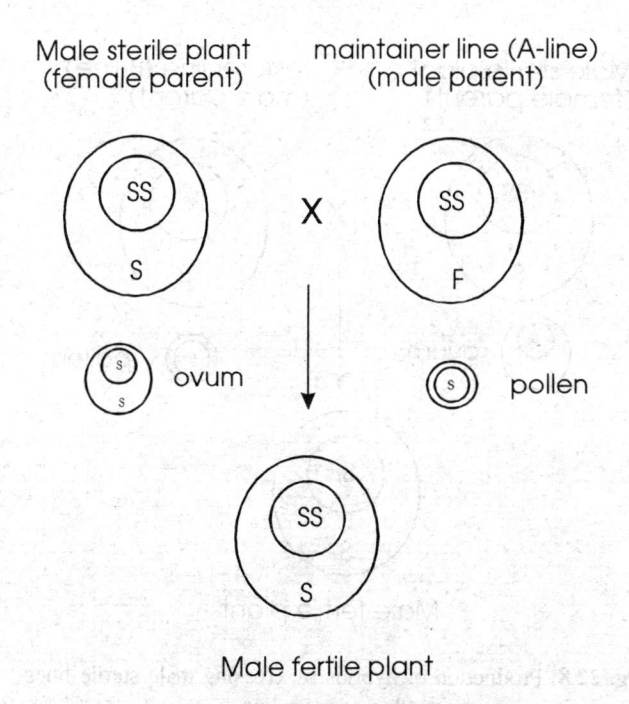

Fig. 22.7. Maintenance of male sterile line.

different time of action, e.g., restorer gene Rf_1 interacts with cms-S cytoplasm in diploid (sporophytic) generation whereas gene Rf_3 interacts with cms-T cytoplasm only in the haploid (gametophytic) generation. Environment factors play a modifying effect on the action of a restorer gene.

Cytoplasmic male sterility factors (cms factors) exert their effect by acting on mitochondria. For example, cms-T plants have decreased respiration, oxidative phosphorylation and electron transfer when treated with fungus pathotoxin, as compared to mitochondria for normal plants.

Nature of CMS Factors. Mitochondria seem to be the most suitable candidates for the position of bearers of the cms particles. This view is supported from the observations that a mt DNA of cms-T maize differs from the normal. mt DNA of cms-S maize contains two additional DNA fragments, possibly **episomes**. The episomes may confer a new property to the carrier organism. Even a plastid mutation is known to affect pollen

Male sterile plant (female parent) × **restorer line (B-line) (male parent)**

SS / S ovum (ss)   FF / F pollen (f)

↓

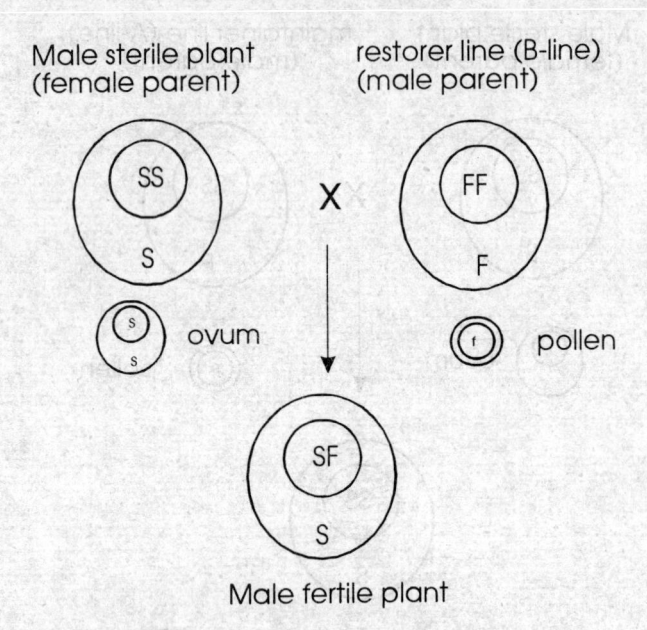

SF / S

Male fertile plant

Fig. 22.8. Production of hybrids by crossing male sterile lines with a restorer line

Table 22.2. Types of cms factors in different strains of maize

Inbred line	A158	Ky21	K55	CE1
Restorer gene	-	Rf_1+Rf_3	Rf_1	Rf_3
Cytoplasmic type				
USDA (cms-S)	Sterile	Fertile	Sterile	Fertile
Texas (cms -T)	Sterile	Fertile	Fertile	Sterile

fertility in *Epilobium hirsutum*. Some workers propose that cms is caused by the alternative steady states of a **regulator-operator system**. This proposal has little support.

Yeast Mitochondrial Genome

This map shows large mt rRNA genes small rRNA gene, ATPase gene, cytochrome oxidase - I, II and III genes, cytochrome b gene in

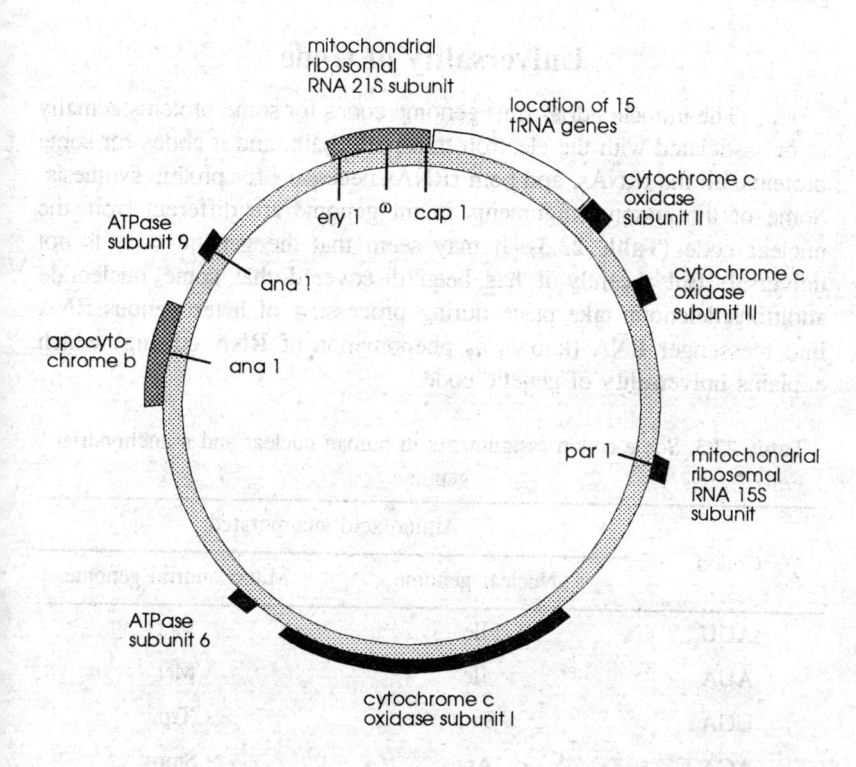

Extranuclear and Extrachromosomal Inheritance 421

Fig. 22.9. Map of yeast *Saccharomyces cerevisiae* including loci for some drug resistance genes. *ana* = antimycin, *cap* = chloramphenicol, *ery* = erythromycin, *oli* = oligomycin, *par* = paromycin, ω = gene affecting recombination between mitochondrial genes

yeast and man (Fig. 22.9). The map also contains some surprises too. Occurrence of introns in several mt genes, e.g., subunit I of cytochrome oxidase contains one intron. In yeast nuclear genes, introns are relatively rare. Occurrence of long **unassigned reading frames** (URFs) has been reported. URFs, are those sequences that have correct initiation codons and are uninterrupted by stop codons. Probably these genes are in search of a function. Some URFs occur within an **intron** may be involved in specifying proteins important in splicing out of introns. There seems to be much lesser **spacer DNA** between the mt genes and URFs. Inactive

pseudogenes are detectable in the nucleus, showing homology with mt genes.

Universality of Code

The mitochondrial (mt) genome codes for some proteins actually in or associated with the electron transport chain, and it codes for some proteins, all the tRNAs, and both rRNAs necessary for protein synthesis. Some of the codon assignments in mt genome are different from the nuclear code (**Table 22.3**). It may seem that the genetic code is not universal. But recently it has been discovered that some nucleotide additions/deletions take place during processing of heterogenous RNA into messenger RNA (known as phenomenon of **RNA editing**) which explains universality of genetic code.

Table 22.3. Some codon assignments in human nuclear and mitochondrial genome

Codon	Amino acid incorporated	
	Nuclear genome	Mitochondrial genome
AUU	Ile	Met
AUA	Ile	Met
UGA	Stop	Trp
AGA	Arg	Stop
AGG	Arg	Stop

Mitochondrial Genes and Diseases

In human beings there are approximately diseases caused by improper functioning of the mitochondria. One of the diseases Leber optic atrophy has been traced to mt DNA. It is maternally inherited. This disease causes blindness around the age of 20-24 years. A mutation in the mitochondrial gene encoding for NADH dehydrogenase subunit 4 affects the optic nerve. The disease also does some damage to the heart.

Maternal Effects

Direction of Coiling in Snails. In this case a nuclear gene product in the cytoplasm determines the phenotype of the organism. Direction of coiling in snail, *Limnea peregra* illustrates the influence of nuclear genes acting through effects produced in the cytoplasm. Coiling in snails is of two types: (i) dextral - coiling is to left and (ii) sinistral - coiling is to left. Direction of coiling is determined by pair of nuclear genes dextral (*D*) and sinistral (*d*). Gene for dextral, being dominant to that for sinistral.

Reciprocal crosses between the two types (dextral and sinistral) are shown in **Fig. 22.10**. The salient observations made from these reciprocal

Cross I

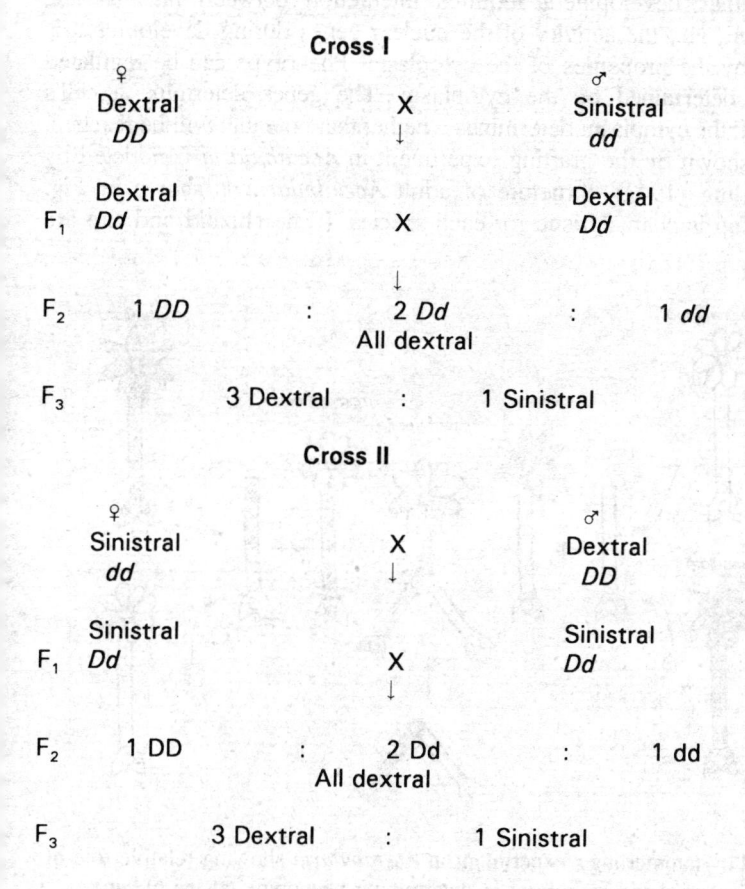

	♀ Dextral *DD*	× ↓	♂ Sinistral *dd*
F₁	Dextral *Dd*	× ↓	Dextral *Dd*
F₂	1 *DD* :	2 *Dd* All dextral	: 1 *dd*
F₃	3 Dextral	:	1 Sinistral

Cross II

	♀ Sinistral *dd*	× ↓	♂ Dextral *DD*
F₁	Sinistral *Dd*	× ↓	Sinistral *Dd*
F₂	1 DD :	2 Dd All dextral	: 1 dd
F₃	3 Dextral	:	1 Sinistral

Fig. 22.10. Reciprocal crosses between dextral and sinistral snails

crosses are: In F_1, both crosses have the Dd genotype, but reflect mother's genotype in respect to coiling. DD mothers produce dextrally coiled offspring whereas dd mothers produce sinistrally coiled offspring. The F_2 generation in both cases is identical because genotype of the F_1 mothers are identical. The F_3 generation in both cases is identical because genotypes of the F_2 mothers are identical. Thus, it is concluded that the coiling of a snail's shell is determined by its mother's genotype.

Nucleo-cytoplasmic Interactions

Grafting Experiment in *Acetabularia*

Cellular development requires interaction between nucleus and cytoplasm, i.e., the activity of the nuclear genes during development is limited by the properties of the cytoplasm. Phenotype can be regulated but not determined by the cytoplasm. The genes determine a cell's potential; the cytoplasm determines whether that potential will be reached. This is shown by the grafting experiment in *Acetabularia* conducted by Hammerling (1943). Structure of adult *Acetabularia* is shown in **Fig. 22.11**. Cap is characteristic for each species. If the **rhizoid** and **cap** are

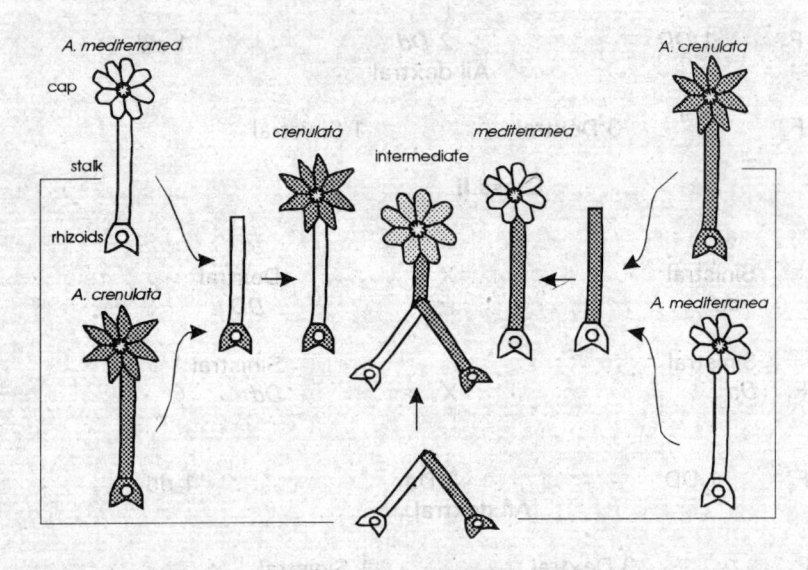

Fig. 22.11. Hammerling's experiment in *Acetabularia* showing relative role of nucleic acid and cytoplasm in determining phenotype (shape of cap)

removed leaving only the enucleated stalk, a new cap will be generated but this will not occur the second time if that new cap is removed. After removal of the nucleus, gene products that are contained in the stalk are sufficient to regenerate only one cap.

If a stalk of *A. crenulata* from which the cap, rhizoidal hold fast, and nucleus have been removed is grafted onto a rhizoid base and nucleus of *A. mediterranea*, the regenerated cap will be intermediate between *A. crenulata* and *A. mediterranea* (**Fig. 22.12**). The first cap thus shows interaction of the gene products of first nucleus (which are retained in the stalk) with those of the second nucleus (which was transplanted). If this intermediate cap is removed, a second cap is regenerated that is identical to *A. mediterranea*. Moreover, in the reciprocal experiment, in which an *A. mediterranea* stalk is grafted onto an *A. crenulata* rhizoid base and nucleus, the second regenerated cap is that of *A. crenulata*. The second cap thus reflects gene products of the transplanted nucleus only; the gene products of the first nucleus having been used up in regeneration of the first cap. Thus the nucleus determines the phenotype.

The important point that this experiment brings forth is that the nucleus controls the synthesis of products at points far removed from its own location. At molecular level these results are interpreted in terms of mRNA being transported from nucleus to cytoplasm.

Killer Trait in *Paramecium*

Killer trait in *Paramecium aurelia* depends on the presence of a cytoplasmic factor called **kappa**, 0.4 microns long, which has its own DNA and which in turn is depended on a dominant chromosomal gene *K*, *S1*, or *S2* and possibly other loci to keep kappa particles in cytoplasm. A killer strain, which possesses a kappa, destroys sensitive strains, which have no kappa particles. Killer strain liberates toxin **paramecin** which kills sensitive strains. The killer cells are immune to their own toxin in the process of conjugation.

The inheritance of the killer trait does not follow a Mendelian pattern. The heterozygous cells produced by brief conjugation between homozygous killer and sensitive strains are not identical phenotypically (**Fig. 22.13**). When a heterozygote undergoes autogamy, the heterozyotes segregate with some organisms *AA* and some *aa*. All sensitive heterozygotes produce only sensitive offspring and all killer heterozygotes produce only killer offspring. When conjugation between a killer and a sensitive strain was prolonged allowing cytoplasm to be exchanged, all

426　Basic Genetics

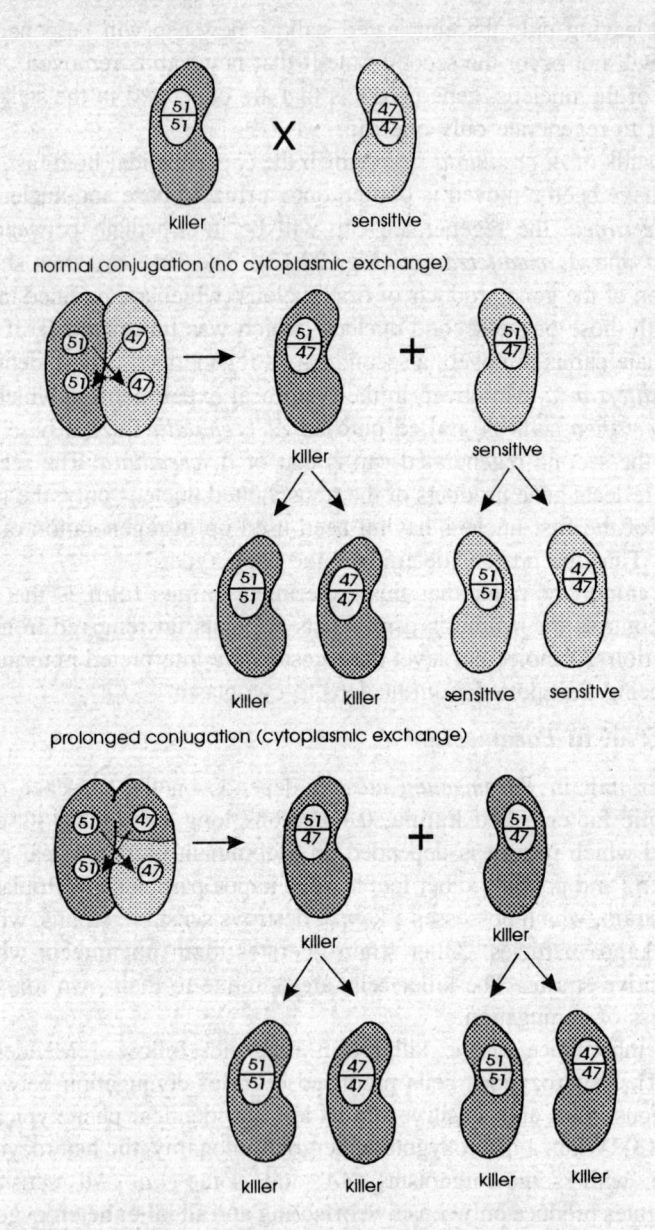

Fig. 22.12. Results of crosses between killer and sensitive strains of *Paramecium aurelia* with and without cytoplasmic exchange

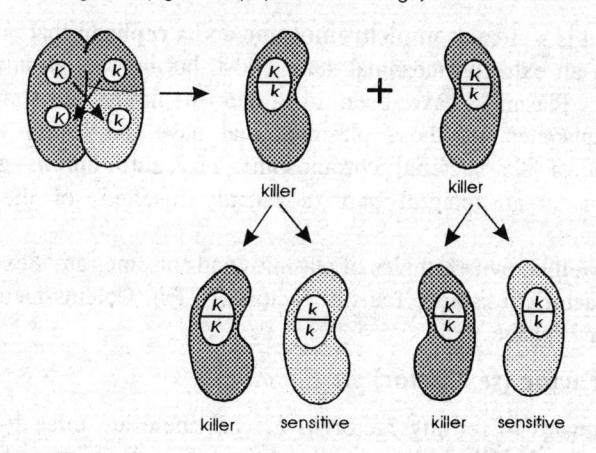

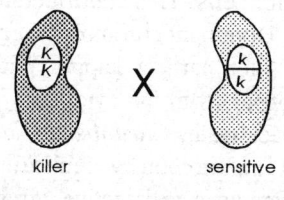

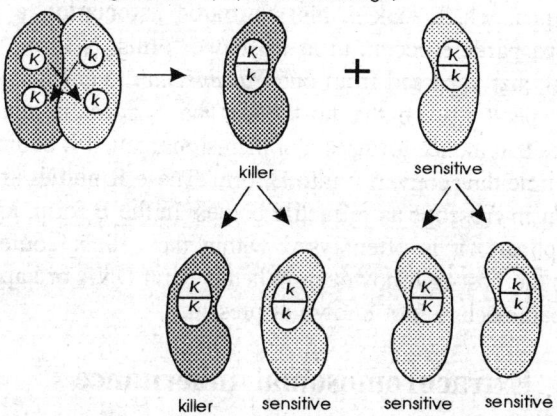

Fig. 22.13. Results of crosses between killer and sensitive strains of *Paramecium aurelia* with and without cytoplasmic exchange illustrating role of a dominant gene *K* in survival of kappa particles

heterozygotes were then killer (Fig. 22.13). Autogamy in these heterozygotes produced all killer cells. This confirmed that the killer trait is determined cytoplasmically. Dominant chromosomal genes are required to maintain the cytoplasmic killer particle kappa. Without a dominant gene this particle would disappear from the cytoplasm of the host.

Kappa appears to be a bacterium, *Caedobacter laeniospiralis*, that contains its own DNA and autonomously replicating. It has been suggested that the toxic substance is a defective virus particle housed inside the kappa, which makes this **symbiotic association** a complex triangle. Kappa particle occurs in at least two forms. N form, which is infective form, that is passed from one *Paramecium* to another, does not confer killer specificity on the host cell. The N form is attacked by bacteriophages that induce formation of inclusions, called R bodies, inside the kappa particle thus convert it into B form. These **R bodies** are visible under the light microscope as refractile bodies. In the B form, kappa can no longer replicate; it is often lysed within the cell. it confers killer specificity on the host cell, however. Whether viral DNA or kappa DNA codes toxin paramecin is not known at present.

Extrachromosomal Inheritance

Plasmids and Episomes

Plasmid is a circular **minichromosome** and a **replicon** that is stably inherited in an extrachromosomal state. Most, but not all, plasmids are dispensable. Plasmids have been identified in almost all strains of bacteria. Episomes are those plasmids that have the ability to live independent of the bacterial chromosome, i.e., **autonomous genetic element**, or as an integral part (covalently inserted) of the main chromosome.

Some well known examples of plasmids and episomes are: Sex-factor (F and F′ factor), resistance transfer factors (RTFs), Colcins factor (col factor), and λ phage.

Fertility Factor (sex-factor) in *E. coli*

With regard to fertility factor in *E. coli*, there are three types of bacterial cells. F cells lack a fertility factor, F⁺ cells possess fertility factor in **autonomous state**, and Hfr (high frequency recombination) cells possess fertility factor in integrated form in the bacterial chromosome. F factor is a circular piece of DNA that is self- reproducing. It is a kind of parasite. It can exist in autonomous or **integrated state**. About 1 in every 10,000 F^+ cell, undergo a significant change such that F factor becomes associated with the bacterial chromosome. **Integration** of F-factor into and its **excision** from a bacterial chromosome has been explained in Chapters 10 and 13.

Sex-duction

Under some conditions, the F factor has been shown to carry a variable number of *E. coli* genes as part of its own circular construction. Such fertility factors are designated as F'. When F'-factor is transferred to a recipient cell becomes partially diploid for some of the bacterial genes brought about by F'. The partially diploid individuals are called **merozygotes** or **merodiploids**. Recombinants may arise as a result of crossing-over between these pairs of homologous genetic regions. The process of F'-mediated creation of merodiploids is called **sex-duction, F-duction** or **F-mediated transduction**. The merodiploids have been instrumental in study of dominance-recessiveness and *cis-trans* effects of mutations.

λ Phage of *E. coli*

λ phage has **linear genome** which is composed of double stranded DNA except at the ends where 3' terminus of each strands extends beyond the 5' terminus of opposing strand by 12 nucleotides. The base sequence of these extended single-stranded regions are complementary to one another. Such ends are called **cohesive ends** or **sticky ends**. When viral DNA is injected into a sensitive host cell or upon infection the linear DNA assumes **circular form** (Fig. 22.14) Covalent closure of each strand in completed by DNA ligase. The viral DNA may enter a **lytic cycle** during which circular λ genome undergoes rapid multiplication, possibly involving a rolling circle mechanism of DNA replication. Nicks are enzymatically produced in daughter circular DNA of phage to generate **linear genophore**. Linear λ genophores are packed into newly synthesized head capsules subsequent lysis release 50-100 progeny phage particles. The phenomenon is called lysis.

Alternatively, **circular viral genome** may be covalently inserted into the genome of the host. This phenomenon is called **lysogeny**. Once integrated, viral DNA is called **prophage**. Prophage replicates with the host chromosome and is transmitted to daughter bacterial cells as a stable genetic entity. In the prophage state the **lytic genes** of virus are repressed.

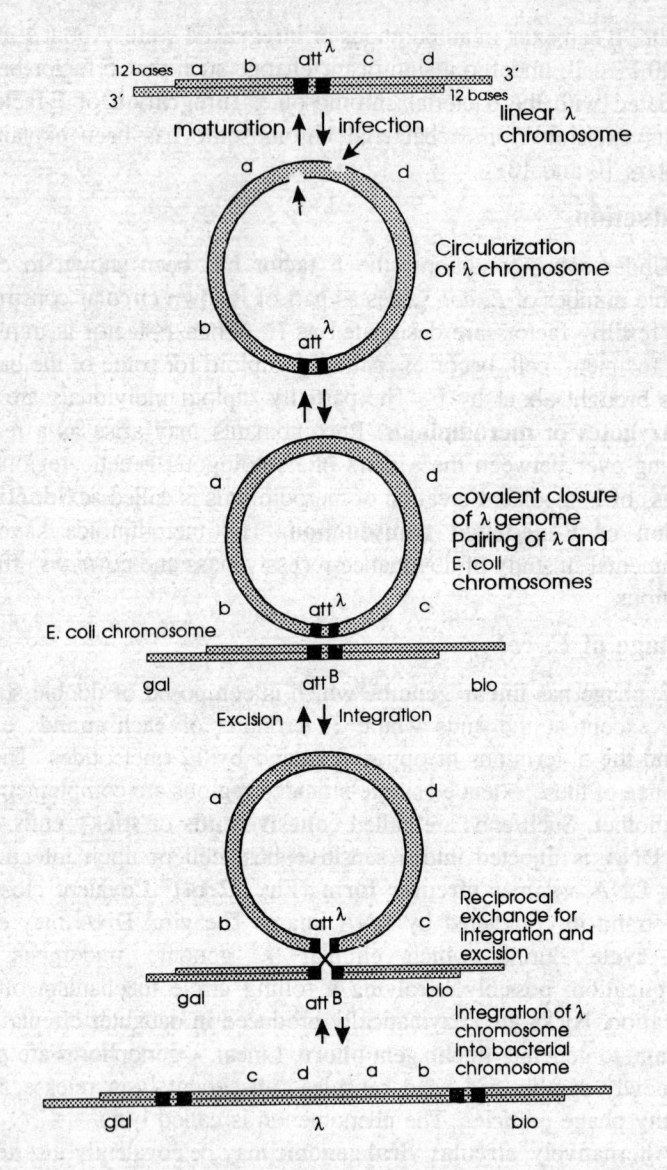

Fig. 22.14. Life cycle of bacteriophage λ. Integration leads to lysogeny and excision leads to lysis. Each of the attachment sites contain base pair sequence GCTTTTTTTATACTAAA/CGAAAAAATATGATT (*gal* = gene for gaplactose, *bio* = gene for biotin)

cl gene of λ phage synthesize repressor protein which when synthesized in sufficient quantity blocks the synthesis of those viral proteins that cause lysis. A bacteria harbouring a prophage is said to be **lysogenic**. A lysogenic bacterial cell is immune to secondary infections by the same virus.

Col Factors

Various types of colicin factors are known. Colicins are strain-specific. E_1 colicin (Col E1) appears to affect membrane bound oxidative phosphorylation and interfere with ATP production, E_2 colicin (Col E_2) degrades DNA or prevents DNA synthesis whereas E_3 colicin (Col E_3) causes ribonuclease activity leading to ribosomal alterations and inhibition of protein synthesis. **Colicinogenic cells** (Col$^+$) are cells carrying colicinogenic factor in inhibited or suppressed form and, as in lysogenic cells, are themselves immune to death by colicins of own type. Such colicinogenic cells may, however, transfer col factor to Col$^-$ bacterial recipients through conjugation or transduction, making them Col$^+$. Col factors may help to bring about bacterial chromosome transfer from their hosts to recipients. There is no evidence to indicate that col factors ever become stably integrated into bacterial chromosome.

Col factor may have originated in temperate bacteriophages that became defective. As evidence, some of the bacteriocidal proteins produced by col factors closely resembles the tail structures of bacteriophages. Chromosome transfer activity has been noted for a group of plasmids called resistance factors (R). Simultaneous resistance to four or more antibiotics (chloramphenicol, streptomycin, sulfonamide, tetracycline etc.) of *Shigella* was noted. Such resistance could be transferred extrachromosomally by cell to cell contact both within and between species such as *Shigella* and *E. coli*. Such transfer was independent of whether the cells were F^+ or F^-. This indicated that a plasmid other than F factor was involved. R plasmids are circular cytoplasmic **self-replicating DNA molecules** which specify sex pili and bear F-like nucleotide sequences for transfer genes. But R factor lacks in ability to integrate stably into bacterial chromosome.

Resistance Transfer Factors

Some of the **transposon** mechanisms exist that have enabled R plasmids to accumulate genes for antibiotic resistance. Plasmid vector **Th system** that allows the isolation of many replicate copies of a particular

DNA sequence is greatly dependent on the kind of vector used to carry the DNA insert into the cells where it can be replicated or "cloned". Plasmid vectors are sought to have small size to have more stability. Relaxed control on replication is an other requirement. Replication independent of host chromosome is a must to have a multiple copies of a plasmid within a cell. Non-conjugation transmission is a safety feature that makes them unable to transfer themselves from one cell to another but instead puts the cellular introduction of **plasmid DNA** ("transformation") under control of the experimenter.

The incorporation of DNA fragments into plasmid vectors not only allows foreign DNA to be replicated in cloned cells for later isolation and identification but can also be designed so that cells transcribe and translate this DNA into proteins. For expression, foreign genes must be inserted into plasmids at positions downstream to **transcriptional promoters**.

Plasmids and episomes are added to a genome through an external source. They do not arise by mutation or rearrangement of the existing genome. This introduction of external genetic material is infectious. Episomes are passed on not only among individuals of same species but also among individuals of different species. F-factor of *E. coli* can be transferred during conjugation to bacteria of other genera within the family Enterobacteriaceae. Plasmid from *Pseudomonas aeruginosa* (R 1822) has an extraordinary broad host range as it can be transferred to many different families of Gram negative bacteria. This makes the study of episomes vital to our programmes to control bacterial diseases. Plasmids are a very useful tool to transfer genetic material from one species to another. In genetic engineering programmes, plasmids are very essential. To suit one's needs plasmids have been even artificially synthesized.

Importance of Plasmids

Because of their properties, plasmids have become of central importance in problems of medicine. The infectious nature of these particles has been not merely from one genetic material not merely from one strain to another but even between widely separated taxonomic groups. F factor transferred from *E. coli* into the *Salmonella* cell, may now mobilize the *Salmonella* chromosome for the transfer. More than 1,000 different plasmids have been identified in bacteria. All their possess are colicinogenic factors that occasionally cause death of some cells with subsequent release of proteins (colicins) capable of killing susceptible srains.

Some Useful Reading Hints

Boycott and Diver (1923) worked out inheritance of sinistrality in *Limnaea peregra*. Sturtevant (1923) deatt with inheritance of direction of coiling in *Limnaea*. Sonneborn (1947) put together recent advances in the genetics of *Paramecium* and euplotes. Sonneborn (1950) discussed methods in the general biology and genetics of *Paramecium aurelia*. Sonneborn (1959) discussed kappa and related particles in *Paramecium*. Sonneborn (1963) discussed a question: Does preformed cell structure play an essential role in cell heredity? Srb (1963) described extrachromosomal factors in the genetic differentiation of *Neurospora*. L'Heritier (1958) reported a hereditary virus of *Drosophila*. Jinks (1963) worked on cytoplasmic inheritance in fungi. Jinks (1964) reviewed various instances of extrachromosomal inheritance. Beisson and Sonneborn (1965) studied cytoplasmic inheritance of the organization of cell cortex in *Paramecium aurelia*. Beale and Knowles (1978) Attardi et al. (1979) studied the organization of the gene in the human mitochondrial genome and their mode of identification. Landis (1987) discussed factors determining the frequency of the killer trait within populations of the *Paramecium aurelia* complex. Preer (1950) detected microscopically visible bodies in the cytoplasm of the "killer" strains of *Paramecium aurelia*.

Campbell (1962) and Campbell (1969) reviewed structure and usefulness of episomes. Hardy (1986) reviewed bacterial plasmids. Ephrussi (1953) studied nucleocytoplasmic relations in micro-organisms. Hammerling (1953) did poineering work on nucleo-cytoplasmic relationships in the development of *Acetabularia*. Wright and Lederberg (1957) reported work on extranuclear transmission in yeast heterokaryons. Wettstein (1961) described nuclear and cytoplasmic factors in development of chloroplast structure and function. Sanger and Ramanis (1963) worked out the particulate nature of nonchromosomal genes in *Chlamydomonas*. Wilkie (1964) described the role of cytoplasm in heredity. Gillham (1965) conducted experiments on linkage and recombination between non-chromosomal mutations in *Chlamydomonas reinhardi*. Levine and Goodenough (1970) described the genetics of photosynthesis and of the chloroplast in *Chlamydomonas reinhardi*. Wettstein et al. (1971) worked on the genetic control of chloroplast development in barley. Sager (1972) reviewed cytoplasmic genes and organelles. Singer et al. (1976) reviewed chloroplast genetics of *Chlamydomonas*. Preer (1974) wrote on Kappa and other endosymbionts in *Paramecium aurelia*. Grun (1976) dealt with cytoplasmic genetics and evolution. Preer et al. (1971) worked on isolation and composition of bacteriophage-like particles from kappa of killer Paramecia. Walbot and Coe (1979) concluded that a nuclear gene *iojap* conditions a programmed change to ribosome-less plastids in *Zea mays*. Clayton (1982) dealt with replication of animal mitochondrial DNA. Elisson (1978a) worked on non-Mendelian female sterility in *Drosophila melanogaster* and variation of chromosomal contamination when caused by chromosomes of various inducer efficiencies. Picard (1978b) did further work on non-Mendelian female sterility in *Drosophila melanogaster* and provided further data in chromosome contamination.

Rhoades (1946) worked on plastid mutations in plants. Rhoades (1955) studied interaction of genic and non-genic heredity units and the physiology of nongenic function. Gyllensten et al. (1991) reported paternal inheritance of mitochondrial DNA in mice.

Questions

1. What is difference between (a) extranuclear and extrachromosomal inheritance and (b) between cytoplasmic and nuclear inheritance?
2. Describe characteristics of mitochondrial DNA, chloroplast DNA, kinetoplast DNA and centriole DNA.
3. Explain cytoplasmic inheritance through plastid inheritance in *Mirabilis jalapa*. What do you understand by cytoplasmic segregation and recombination?
5. Explain inheritance of streptomycin-resistance in *Chlamydomonas*. What is difference between uniparental and biparental inheritance.
5. How are the uniparently inherited genes mapped?
6. Differentiate between genetic and cytoplasmic male sterility in plants
7. What do you understand by cytogenic inheritance?
8. What are maintainer lines? How are these lines used to maintain the male sterile lines? What are restorer lines? How are these lines used in hybrid seed production?
9. Describe the nature of cytoplasmic male sterility (cms) factors. In what respects do various cms factors differ?
10. What do you know about the inheritance of *poky* trait in *Neurospora*?
11. What are various surprises in yeast and human mitochondrial genome?
12. Some differences have been observed in nuclear and mitochondrial codons. What are your comments on the universality of genetic code
13. What is relationship between mitochondrial mutations and diseases?
14. What is genetic basis of drug resistant mutants in yeast, *Neurospora Paramecium* and mammalian cells?
15. Describe the inheritance of *minute* mutants in *Chlamydomonas*
16. What is the nature of the factor responsible for CO_2 sensitivity in *Drosophila melanogaster*?
17. How are the sex-ratio strains in *D. willistoni* recognized?
18. What is the genetic basis of incompatibility in mosquitoes?
19. How is the killer strain of yeast genetically controlled?
20. What is the significance of milk factor in mice in incidence of cancer? How is this factor genetically controlled?
21. What are maternal effects? How are these effects differentiated from cytoplasmic and nuclear inheritance? Illustrate maternal effects with the help of direction of coiling in snails
22. What are nucleo-cytoplasmic interactions? What is the importance of these interactions in genetic analysis? What phenomenon does grafting experiment in *Acetabularia* conducted by Hemmerling (1943) describe?
23. What type of inheritance is exhibited by killer trait in *Paramecium*? What are differences between killer and sensitive strains of *Paramecium*? Killer trait in *Paramecium* is said to demonstrate a complex triangle, comment.
24. What are differences between plasmids and episomes? List a few examples.
25. What is the role of fertility factor (sex-factor) in *E. coli* as an episome?
26. What do you mean by sexduction/F-duction/F-mediated transduction?
27. Describe lysogenic and lytic phases in the life of λ phage of *E. coli*.
28. What are *Col* factors? What is their significance?
29. What is the role of resistance transfer factors in bacteria in conferring resistance to various drugs?
30. What is practical importance of plasmids?

23

SEX DETERMINATION AND DOSAGE COMPENSATION

Sex-determination

Sexual reproduction is essential for the production of male and female gametes. It is useful in producing genetic variability and has played an important role in evolution. In higher animals, male and female individuals are separate (**dioecious**). Most of the plants are having bisexual flowers (**monoecious**). In dioecious forms, male and female individuals exist in equal numbers. So a very precise mechanism of sex-determination should be involved.

Chromosomal Sex-determination

In most of the animals and in some plants a pair of chromosomes is responsible for the determination of sex. This pair of chromosomes is called **sex chromosomes** or **allosomes**. The chromosomes other than the sex chromosomes are called as **autosomes**. In most diploid individuals there are 2n-2 autosomes and 2 sex chromosome while in some there are 2n-1 autosomes and 1 sex chromosome. In one of the sexes two chromosomes are identical or homomorphic (**homogametic sex**). In other sex these are different or heteromorphic (**heterogametic sex**). Different sex- determination systems, names of the organisms, genotypes of males and females, genotypes of sperms and eggs produced are summarized in Table 23.1.

Based on Chromosome Number (XO type). McClung (1901, 1902) observed that somatic cells in female grass hopper carry a different number of chromosomes than do the corresponding cells in the male. He called the extra chromosome in the female as X body.

Table 23.1. Various sex-determination systems

System	Organism	Males	Females	Sperms	Eggs
XY ♂	Drosophila Humans, Coccinia indica Melandrium album	XY+2A	XX+2A	X+A Y+A	X+A
ZW ♀	Birds, Poultry, some fishes, Butterflies, Fragaria sp.	WW+2A	ZW+2A	W+A	W+A Z+A
XO ♂	Pronetor (insect) Dioscorea sp.	X+2A 2N=13	XX+2A 2N=14	X+A A	X+A
WO ♀	Fumea (insect)	WW+2A	W+2A	W+A	W+A A

Wilson observed different number of chromosomes in diploid cells of male and female protenor; males carried 13 chromosomes and females carried 14 chromosomes in unreduced cells. Males formed two types of gametes (heterogametic sex) some gametes carry 6 chromosomes while others carry 7 chromosomes. All the eggs carried 7 chromosomes (homogametic sex). Eggs fertilized with sperm carrying 7 chromosomes produced females. It was thus hypothesized that female carried 12 autosomes and two X chromosomes whereas male carried 12 autosomes and one X chromosome. In domestic fowl, fumea (a moth) and butter flies female is heterogametic sex (XO) and male is homogametic sex (XX).

Based on Differences in Sex Chromosomes (XY type). XY system of sex-determination is found in most higher animals and some plants (*Malandrium album*). In humans, presence of Y chromosome determines maleness (**Table 23.2**). Number of chromosomes in somatic cells of male and female is same (2n = 46). Female carries 22 pairs of autosomes and two X chromosomes (homogametic sex) whereas male carries 22 pairs of autosomes and one X and one Y chromosome (heterogametic sex).

During development, all X chromosomes except one become inactive, highly coiled and heterochromatic. This heterochromatic body

Table 23.2. Y chromosome determines maleness in man

Sex chromosomes constitution	Sex
XX	Female
XY	Male
XXY	Male
XO	Female

found in the nuclei of normal females and absent in the nuclei of normal males is known as **Barr body** after the name of discoverer M. Barr. The Barr body is found attached with the nuclear membrane.

X and Y chromosomes have a short homologous segment which causes pairing. This region is called **pseudoautosomal region** and the genes located in this region show **pseudoautosomal inheritance**. Genes located in the differential region of X chromosome show **X-linked**, or also known as **sex-linked** inheritance while differential region of the Y chromosome shows **Y-linked**, or also known as **Holandric inheritance**. This aspect of inheritance has been dealt with Chapter 6. Y chromosome in man has two pseudoautosomal regions that allow synapsis between X and Y chromosomes. *ANT3* (ADP/ATP translocase gene), *CSF2RA* (*colony-stimulating factor* gene), *MIC2* (T cell adhesion gene) are located in pseudoautosomal region of the short arm of Y chromosome. *SRY*, *RSP4* (ribosomal protein gene) and *ZFY* are located in the differential region in the same arm.

Table 23.3. Y chromosome is not required for maleness in *Drosophila*

Sex chromosomes constitution	Sex
XX	Female
XY	Male
XXY	Female
XO	Male

In fruit fly *Drosophila* Y chromosome is not required for maleness (**Table 23.3**). In *Drosophila*, female determiners are located in the X chromosome and male determiners are located in the autosomes. Ratio of

X chromosomes to sets of autosomes (**X/A ratio**) determines sex in *Drosophila*. Change in this ratio leads to changed sex phenotype (**Fig 23.1**). The *Drosophila* Y chromosome is known to carry genes for at least six sterility factors, two in the short arm (*ks-1* and *ks-2*) and four in the long arm (*kl-1, kl-2, kl-3,* and *kl-4*). In addition, Y chromosome carries two other known genes, *bobbed*, which is a locus of ribisimal RNA genes (the nucleolar organizer) and *suppressor of Stellate* (*Su/Ste/*), a gene required for **RNA splicing**. The fertility factors code for proteins needed during spermatogenesis. For exampe, *kl-5* codes for the dynein motor needed for sperm flagellar movement.

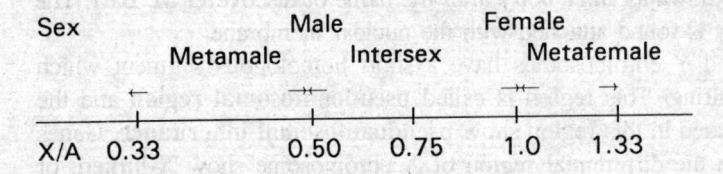

Sex	Metamale	Male Intersex	Female Metafemale
X/A	0.33	0.50 0.75	1.0 1.33

Fig. 23.1. Ratio of X chromosomes to sets of autosomes (X/A) determines sex in *Drosophila*

Balance theory of sex-determination put forward by Bridges (1925) explains sex-determination in *Drosophila*. The X chromosome is believed to carry **female determining genes** while autosomes carry the male determining genes. There has to be a balance between these sets of genes for an individual to become a male or a female. Y chromosome of *Drosophila* carries genes for female fertility. Chromosomal studies in gynadomorphs of *Drosophila* also suggest that Y chromosome does not play any role in the determination of sex in *Drosophila*. Gynadomorphs are the flies in which one half of the body shows male characteristics while the other half shows female characteristics. Such flies develop due to loss of an X chromosome from a cell at first mitosis of zygote. In birds, moths, reptiles and some plants, females are heterogametic sex (XY) and males are homogametic sex (XX).

Compound Sex chromosomes. In nematode, *Ascaris incurva* there is large difference in the number of chromsomes between males and females. The number of autosome pairs in both sexes is 13. Number of compound X chromosomes is 16 in female and 8 in male. In addition, male contains a Y chromosome.

Genic Control of Sex

One or two gene differences may control sex in some plant species. *Asparagus* is dioecious with separate male and female plants. Here the maleness is determined by a dominant gene (*P*) while the femaleness is attributed to recessive homozygous condition (*pp*). Maize (*Zea mays*) is monoecious with male inflorescence (tassel) and female inflorescence (silk) located on the same plant. Two gene pairs (*Ts, ts* and *Sk, sk*) have been identified to play role in sex-determination in maize as shown in Table 23.4. When a female *tsts sksk* (having seeds on tassels but silkless) is crossed with a male *Tsts sksk* (with normal tassels but silkless), it produces male and female plants in 1:1 ratio. This will be a stable dioecious condition.

Table 23.4. Genes controlling sex in *Zea mays*

Genotype	Sex
TsTs SkSk	Monoecious with male and female inflorescence
tsts SkSk	Tassel seed; tassel converted into seed bearing inflorescence
TsTs sksk	Silkless; effectively male plants; female flowers rudimentary
tsts sksk	Female plants; male flowers rudimentary

Sex-determination Through Egg Fertilization

In hymenoptera, sex appears to be determined by the number of sets of chromosomes a bee receives. Fertilized eggs produce diploid females while unfertilized eggs undergo an asexual process of embryonic development, called **parthenogenesis**, to produce haploid but fertile males. Male produce sperm by mitosis rather than meiosis, yielding one functional sperm per spermatocyte.

Effect of Environment on Sex-determination

Extracts made from the female proboscis influence young worms towards maleness. Any young worm released from a single isolated egg became a female. If newly hatched larval are released in water containing mature females, some of the larval get attached to the female proboscis. They were transformed into males.

Hormonal Control of Sex

In *Bonellia viridis*, a marine worm, all the larval are genetically and cytologically similar. If a particular larva settles near proboscis (suctorial mouth part) of an adult female, it becomes a male individual. On the other hand, if it has to develop free in water, it becomes a female. The proboscis secretes some substance suppressing femaleness.

Crew (1923) reported a case of complete **sex reversal** in fowl. A fertile hen (female) changed to a fertile cock (male). This resulted due to damaged ovary in female. The ovary in the female is secrets a male suppressing hormone, therefore in absence of ovary, the testis could develop.

In cattle, when twins are produced, one male and other female, the female is sterile and male is normal. These are known as **Free-Martin**, and would be produced only when there is a vascular connection between two embryos. The male hormone perhaps suppresses the development of the ovary in the female.

Human Sex Anomalies

Various **syndromes** are produced due to addition or loss of sex chromosomes but following are some important cases. (i) Genetically determined **sex reversal** in which case XY is reversed to female. (ii) 46 XY females have vestigial gonads, normal females tall, normal sex organs, normal uterus, normal fallopian tubes but did not conceive. (iii) **Gonadal dysgenesis** is due to an X-linked recessive allele which did not allow differentiation of gonads to testes in 46 XY embryos. It was presumed that Y chromosome product integrates into the X-linked product to produce testis differentiation. (iv) **Hermaphroditism** is a condition in which an individual has both male and female sexes. True hermaphroditism (TH) is rare. As many as 108 cases of TH were studied in USA. A 25 year old hermaphrodite reported pregnancy and delivered after 30 weeks a still born child. This individual spent early years in sexual activity as a male but preferred to live as a female partner. The genitalia of some individuals are masculinized but male and female sex characters vary. The 59 individuals were 46 + XX. They were normally females but expressed some traits of males. As many as 21 individuals which were also 46 + XY, expressed traits of both the sexes, 28 were mosaics.

Sex Correction

Normal individuals with chromosome constitution 46 + XX were produced from loss of Y from a 47, XXY. The genetic sex, somatic sex and socio-psychological sex of such individuals is different form normal XX females and normal XY males.

Dosage Compensation

Barr Bodies

M. Barr E. Bertram identified Barr body as X chromosome in 1949. They observed a darkly stained body in the nuclei of cat nerve cells which were not present in males. It was later called Barr body (**Fig. 23.2**). Human embryos could be differentiated at early stages of development. Number of Barr bodies observed in a cell was one less than the number of X chromosomes (**Table 23.5**).

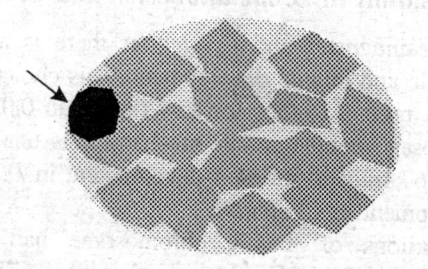

Fig. 23.2. Barr body (arrow) in the nucleus of a cheek mucosal cell of a normal cell

Table 23.5. Number of Barr bodies in individuals with different chromosome constitution

Chromosome constitution	No. of Barr bodies
XX	1
XY	0
XO	0 (Turner syndrome)
XXX	2

Lyon's Hypothesis

Mary Lyon suggested that Barr body represented an inactive X chromosome which in females becomes tightly coiled into heterochromatin, a condensed, and therefore visible, form of chromatin. One of the X chromosomes in homogametic sex becomes inactive to regulate the production of gene products controlled by X-linked genes. The activation is random and occurs through **heterochromatization** (permanent repression). In females one X-chromosome shows late replication and this X becomes inactive.

Dosage compensation is the process whereby total gene expression from a single X in one sex is made equal to total expression from the two X chromosomes in the other sex. This can be achieved either by **upgrading regulation** in the single X in one sex or down regulation of the two X chromosomes in the other, or by a combination of these mechanisms. Failure to establish correct dosage compensation is usually lethal in both *D. melanogaster* and *C. elegans*.

Dosage Compensation in *D. melanogaster* and *C. elegans*

In both *D. melanogaster* and *C. elegans* there is a dichotomous choice between male and female development. This choice is ultimately determined by X/A ratio. At ratios too low (less than 0.03) or too high (more than 1.5) dosage compensation mechanism is unable to prevent death. X/A ratio 0.67 leads to intersex development in *D. melanogaster* but to male development in *C. elegans*.

Partial duplications of the X chromosome had a significant feminizing effect in a triploid 2X : 3A animal (ratio 0.67). This implied the existence to multiple numerator elements which contribute additively to the X part of the X/A ratio. Two major numerator elements *sis-a* and *sis-b* (sisterless) have been defined so far. The two elements seem to be related but not identical. Both *D. melanogaster* and *C. elegans* appear to have more than one **numerator element** (but possibly not a large number). **Denominator elements** constituting the autosomal part of the X/A ratio have not yet been identified in flies or worms; such elements must exist as autosomal genes encoding the general machinery of the cell. A comparison of sex-determination and dosage compensation mechanisms in *D. melanogaster* and *C. elegans* is presented in **Table 23.6**.

Interfering with dosage compensation in *C. elegans* can affect the reading of the X/A ratio. Mutations of DCD genes, which lead to increased X transcript levels, transform 2X : 3A animals from male into

Table 23.6. Comparison of sex-determination and dosage compensation systems in *D. melanogaster* and *C. elegans*

Parameter	*D. melanogaster*	*C. elegans*
Similarities		
X/A ratio as basic signal	XX and X(Y) sexes	XX and XO sexes
Multiple numerator elements on X	*sis-a, -b...*	Octamers?
Master regulators of sex and compensation	*Sxl*	*xol, sdc*
Dosage compensation genes	*MSL* set	*DSD* set
Regulatory cascade for somatic cell	*tra, ix, dsc*	*her, fem, tra*
Germline specific controls	*ovo*?	*fog*?
Differences		
Sexes	Female, male	Hermaphrodite, male
Compensation mechanism	Positive	Negative
Soma and germline	Different genes	Same genes
Cascade interactions	Positive	Negative
Interaction mechanisms	Alternate splicing	Transcriptional and post-transcriptional

hermaphrodite. All the **master regulators** that respond most directly to the X/A ratio are themselves located on the X chromosome in both fruitfly and nematode. Most of the other regulatory genes are autosomal.

In *D. melanogaster,* dosage compensation acts by a two-fold increase of sex-linked gene transcription in male flies, relative to the basal level for an autosomal gene. This is achieved by *cis*-acting elements which act as two-fold enhancers.

In nematode, four autosomal genes affecting dosage compensation act as negative regulators for nematode genes for the *msl* type. In both nematode and fruitfly, somatic sexual phenotype is controled by a small number of autosomal genes, organized in an ordered cascade of interactions.

In *C. elegans* there are seven major autosomal regulatory genes: *her-1*, three *fem* genes and three *tra* genes. The activity of the last gene in the cascade, *tra-a*, determines somatic sexual phenotype. It is both necessary and sufficient to direct all aspects of female development. In fruitfly, somatic sex-determination genes have no effect on determination of germ-line, but in nematode the same set of genes functions in determining sexual phenotype in both soma and germ line.

A dominant **gain-of-function mutation** of *tra-1* leads to female development of an XO animal irrespective of the stage of nine upstream regulation genes. Absence or low activity of *tra-1* is both necessary and sufficient for male somatic development. The *fem-1* gene encodes a predicted protein of 67K (M 67,000); about one-third of the protein is composed of six repeats of sequence but its significance is presently unknown. Both genes *dsx* in *D. melanogaster* and *tra-1* in *C. elegans*, act as global regulators, affecting a large number of different developmental functions in different ways. These genes act in a cell autonomous manner, so they are likely to encode hormones.

Absence of Dosage Compensation

In sex-determination system where ZZ is male ZW is female, no dosage compensation is observed for Z-linked genes. Normally the genes whose products are required for early entry into oogenesis are compensated. In females giving heterogametic constitution, the eggs condition large amounts in excess of used for early embryogenesis. *Xenopus* larva 2n is female. The egg contains 10,000 X the RNA proteins and lipids than what is required for embryogenesis.

Some Useful Reading Hints

McClung (1902) reported the accessory chromosomes as sex determinants. Allen (1940) worked on the genotypic basis of sex-expression in angiosperms. Rick and Hanna (1943) discussed determination of sex in *Asparagus officinalis* L. Rothenbuhler *et al.* (1968) deal with genetics of sex detemination in bees. Byers *et al.* (1972) reported ethylene as a natural regulator of sex expression of *Cucumis melo*. Bull and Vogt (1979) dealt with temperature-dependent sex determination in turtles.

Lyon (1962) worked on sex chromatin and gene action in the mammalian X chromosome. Russell (1963) explained that mammalian X-chromosome inactivation is limited in spread and region of origin. Lyon (1992) discusses some milestones in the history of X-chromosomes inactivation. Baker *et al.* (1991) reviewed mechanism of dosage compensation in *Drosophila*. Gillis (1994) described the mechanism of turning off the X chromosome. Parkurst and Meenley (1994) dealt with sex determination and dosage compensation and draw out lessons form flies and worms. Gorman and Baker (1994) discussed the question of how flies make one equal two in dosage compensation in *Drosophila*. Williams (1995) gave a fresh look at dosage compensation and discuss how males and females achieve equality. Lyon (1996) reviewed current status of X-chromosome inactivation. Heard *et al.* (1997) discussed the process of X chromosome inactivation in mammals.

Ohno (1979) discusses major sex determining genes. Simpson (1982) discusses causes of abnormal sexual differentiation in humans. Palmer *et al.* (1990) conducts comparison of human ZFY and ZFX transcripts. O'Neill (1990) talks about the gene that makes a man of you. Hodgkin (1990) compares sex determination in *Drosophila* and *Coenorhabditis*. Gubbay *et al.* (1990) find a gene mapping to the sex determining region of the mouse Y chromosome is a member of a novl family of embryonically expressed genes. Sinclair *et al.* (1990) report that a gene from the human sex-determining region encodes a protein with homology to a conserved DNA binding motif. Villeneuve and Meyer (1990) discuss the role of *sdc-1* gene in the sex determination and dosage compensation decisions in *Caenorhabditis elegans*. Harley *et al.* (1992) report DNA binding activity of recombinant SRY from the normal males and XY females. Hodgkin (1992) discusses genetic sex determination mechanisms and thier evolution.

Questions

1. Enumerate and define various basis of sex determination.
2. What do you understand by sex-determination, dosage compensation and sex-differentiated. Are these process related?
3. What do you mean by chromosomal sex-determination mechanisms?
4. Describe sex determining mechanisms based on chromosome number (XO type).
5. Describe sex determining mechanisms based on differences in sex chromosomes.
6. Compare the structures of Y chromosome in man and *Drosophila melanogaster*.
7. With the help of an example explain sex determination through compound sex chromosomes.
8. Describe genic control as a mechanism of sex determination.
9. How is sex-determination controlled through fertilization.
10. Describe effect of environment on sex-determination.
11. Describe involvement of hormones in control of sex.
12. What do you understand by sex reversal and sex correction.
13. What are various events in sex determination?
14. Write a short note on human sex anomalies.
15. What is dosage compensation? What are Barr bodies? Where are these bodies located? How, when and why are they formed in mammalian somatic cells?
16 Describe Lyon's hypothesis of dosage compensation.
17. Describe various similarities and differences in dosage compensation mechanisms in *D. melanogaster* and *C. elegans* and nematode.
18. How does dosage compensation differ in mammals and *D. melanogaster*?

24

HUMAN GENETICS

Genetics and Man

Everything found out and understood about genetics has application to human welfare. Researches in area of genetics are conducted to understand various principles of genetics and to find out ways and means to use them for human welfare. Principles of genetics can be used for the improvement of human race. This branch of genetics is known as **eugenics**. Identical twins and multiple births provide the only human groups that approach a genotype standard. If identical twins are separated early in life and raised in different environments, it is possible to compare effects of different environments on similar genotypes. There are a large number of human disorders that are inherited in a simple Mendelian fashion. Thus genetics has contact with medicine. The branch of genetics that deals with study of causes, identification, cure/treatment of human disorders is known as **medical genetics**.

Principles of genetics can be applied in preventing, diagnosing, and treating illness. Maupertuis, an eighteenth century biologist, was one of the first to apply genetics to man. He collected pedigrees of families in which polydactyly and albinism occurred, analyzed them, and predicted for the future by applying theory of probability. Garrod (1908) described certain diseases, like alkaptonuria, phenylketonuria, as "inborn errors of metabolism". He proposed that several biochemical diseases were dependent on recessive genes. By 1910, it was possible to predict the probability of expression of some hereditary diseases. With the development of techniques for enzyme analysis it became possible to identify biochemical genetic disorders. Some 170 human disorders are now identified with a genetically determined enzyme deficiency.

Disease/disorder for the present discussion is any condition in which bodily or mental well-being is seriously impaired. One-quarter to one-half of all human diseases is of genetic origin. In the general population, about one in every ten human gametes carries a bit of misinformation introduced by a new mutation. Most of the cost of this "**mutational load**" is paid during the early stages of development. About 3 per cent of newborns suffer from a recognizable genetic defect.

Genetics has helped us in understanding role of genes in development of man from zygote to adult, maintenance and human health. Physicians should have enough knowledge of genetics. Degenerative diseases such as cancer, heart disease and arthritis pose difficult problems that require basic insights into heredity and environment. Importance of genetics in medical practice is going to be major because vaccines, antibiotics and sanitation have decreased the importance of infectious diseases and brought genetic diseases into prominence.

About 20 per cent of all congenital effects are predominantly inherited about 20 per cent are due to environmental factors and 60 per cent are due to combinations of inherited and environmental factors.

Genetic Basis of Disorders

Physicians must learn to deal with approximately 1,500 known genetic diseases. Controlled crosses cannot be made in humans. So we cannot deduce mode of inheritance of human traits from mathematical ratios in segregating progeny as demonstrated in peas and *Drosophila*. In case of humans, mode of inheritance can be traced, when sufficient family information is available, through the technique of **pedigrees analysis**.

Hereditary diseases can be due to defect in a single gene or in multiple genes. It is difficult to understand genetic basis of latter type. Single gene disorders are usually due to point mutations. Pedigree analysis is helpful in finding out genetic basis of such disorders. Alkaptonuria, phenylketonuria, tyrosinosis, tyrosinemia, haemophilia are human disorders due to a point mutation.

Pedigree is a "family tree", drawn with standard genetic symbols, showing inheritance pattern for specific phenotypic characters. See **Fig. 24.1** for symbols. **Propositus** or **proband** is the number a family who first comes to the attention of geneticist. Usually phenotype of the propositus is unusual or exceptional in some way. The investigators then traces the history of the character in the propositus back through the history of the family. A few examples given below explain the procedure

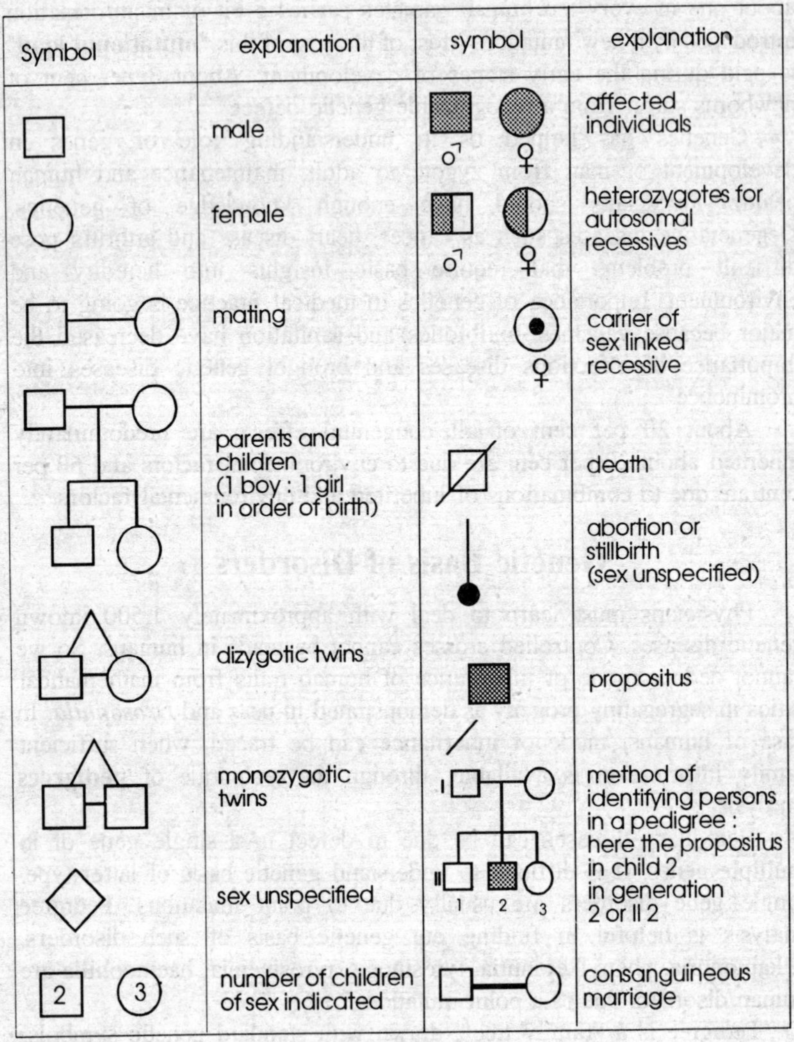

448 Basic Genetics

Symbol	explanation	symbol	explanation
□	male	▨ ⊙ ♂ ♀	affected individuals
○	female	▤ ◐ ♂ ♀	heterozygotes for autosomal recessives
□—○	mating	⊙ ♀	carrier of sex linked recessive
□—○ with children	parents and children (1 boy : 1 girl in order of birth)	⌀	death
△ ○ twins joined	dizygotic twins	• (stillbirth symbol)	abortion or stillbirth (sex unspecified)
△ □ joined	monozygotic twins	▨	propositus
◇	sex unspecified	/ □—○ pedigree with 1, 2, 3	method of identifying persons in a pedigree: here the propositus in child 2 in generation 2 or II 2
2 ③	number of children of sex indicated	□—○	consanguineous marriage

Fig. 24.1. Symbols used in human pedigree analysis

of determining whether the exceptional characters is governed by an autosomal or sex-linked, or a recessive or dominant gene. Other cause of human disorders is changes in chromosome number and structure. In pedigree analysis, genetic basis of a disorder can be worked out with certainty when **penetrance** is complete and **expressivity** is cent per cent. Pedigree analysis is possible for any simply inherited genetic trait.

Changes in Chromosome Number or Structure

Some human disorders caused due to aneuploidy are described in Table 24.1. Some patients with Down syndrome have a total of only 46

Table 24.1. Some human disorders caused due to aneuploidy

Chromosome Nomenclature	Chromosome Formula	Clinical Syndrome	Main Phenotypic Characteristics
2n + 21	2n + 1	Down	Short broad hands with simian-type palmar crease, short stature, hyperflexibility of joints, mental retardation, broad head with round face, open mouth with large tongue, epicanthal fold.
2n + 13	2n + 1	Edward's syndrome	Mental deficiency and deafness, minor many organs, low set, malformed ears: receding mandible, small mouth and nose with general elfin appearance; mental deficiency; horseshoe or double kidney: short sternum. 90% die in the first 6 months.
2n + 18	2n + 1	Trisomy-18	Female with retarded sexual development, usually sterile, short stature, webbing of skin in neck region, cardiovascular abnormalities, hearing impairment.
15,X	2n − 1	Turner	Male, subfertile with small testes, developed breasts, feminine pitched voice, long limbs, knock knees, rambling talkativeness.
17, XXY 18, XXXY 18, XXYY 19, XXXXY 20, XXXXXY	2n + 1 2n + 2 2n + 2 2n + 3 2n + 4	Klinefelter	Female usually normal genitalia and limited fertility. Slight mental retardation.
n, XXX	2n + 1	Triple X	

instead of 47 chromosomes but in such cases a translocation has joined the long arm of chromosome 21 with another chromosome in the same complement, mostly chromosome 14. With assurance that the translocated chromosome t(14q.21q) will remain intact and assuming that one parent has a normal chromosome complement, the theoretical risk is 1/3 that each additional child will have the Down syndrome (**Table 24.2**). Down syndrome can also be due to partial chromosome 21 deletion. Cri du chat syndrome is due to partial group B chromosome deletion. Philadelphia syndrome is due to deletion of long arm of chromosome 21. When a large part of long arm of chromosome 22 is translocated to chromosome 9, it leads to chronic myeloid leukemia.

Table 24.2. Calculating theoretical risk of Down syndrome when caused by mating between a chromosomally normal individual and a heterozygote for chromosomal translocation t(14q.21q)

Gamete of Second parent	Gametes of First Parent					
	14.21	14q21q	t(14q21q), +21	14	t(14q21q), +14	21
(14.21)	14.21 14.21	(14q21q) 14.21	t(14q21q), +21 14.21	14 14.21	t(14q21q), +14 14.21	21 14.21
Phenotypes of Zygotes	Normal	Normal translocation carrier	Trisomy-21 Down Syndrome	Monosomic-21 lethal	Trisomic-14 lethal	Monosomic-14 lethal

Point Mutations

Autosomal Recessive Mutations. The pedigree of a person showing a disorder due to autosomal recessive condition (**Fig. 24.2**) has the following salient features: Condition appears in the progeny of two unaffected (heterozygous; carrier) individuals. Abort one-fourth of the children are affected. Two affected individuals can not have an unaffected child. Condition occurs with equal frequency in males and females. Autosomal recessive disorders are frequently revealed in cousin marriages. For example, myoclonic epilepsy, Tay-Sachs disease, cystic fibrosis, phenylketonuria, alkaptonuria, tyrosinosis, albinism, limb girdle muscular dystrophy.

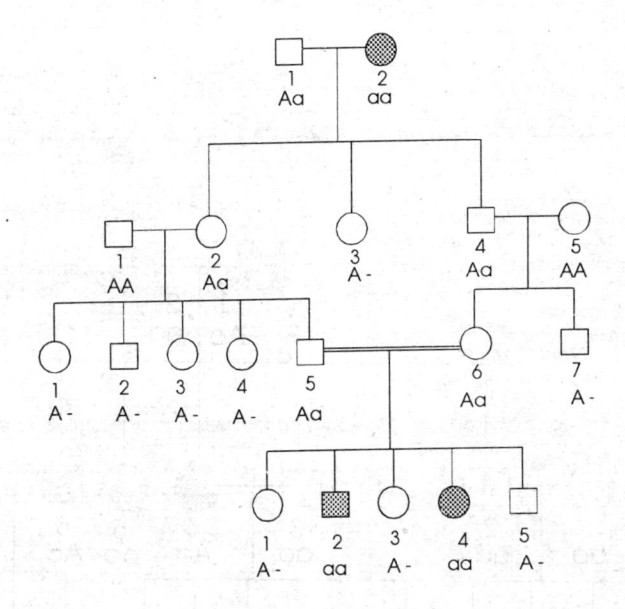

Fig. 24.2. A typical pedigree of a trait due to autosomal recessive gene

Autosomal Dominant Mutations. Following are salient features of a pedigree (**Fig. 24.3**) having a disorder due to autosomal dominant mutation. The condition typically occurs in every generation. Unaffected individuals never transmit the conditions to their offspring. Two affected parents may have unaffected children. One-fourth children of parents heterozygous for such a disorder are normal. When one of the two parents is affected, the condition may pass on to one-half of the children of affected individual. For examples, brachydactyly, achondroplasia (a kind of dwarfism), white forelock, Huntington's chorea, retinoblastoma.

Sex-linked Recessive Mutations. Pedigree pattern of a X-linked recessive disorder (**Fig. 24.4**) tends to be oblique because of transmission of such a condition from a male to his grandson is through his carrier daughter. In female, a sex-linked recessive condition expresser only in homozygous conditions for whereas in males it expresses in hemizygous from as Y chromosome lacks the genes present on the X chromosome. Father-to-son transmission of this condition is absent. In man, more than 90 pathological traits that are X-linked recessive are known. A few examples are: colourblindness, haemophilia, hyphosphatemia, glucose-6-

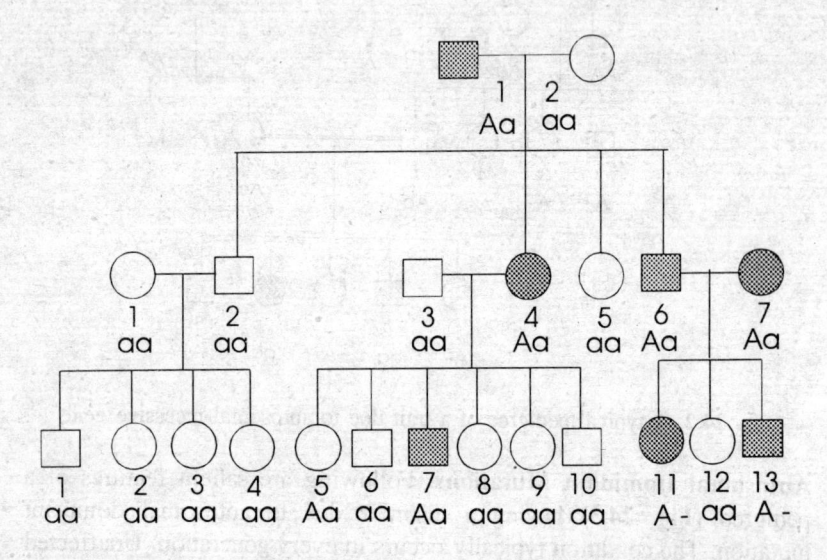

Fig. 24.3. A typical pedigree of a trait due to autosomal dominant gene

phosphate deficiency, ichthyosis (scaly skin), phosphoglycerate deficiency, Lesch-Nyhan syndrome, Hunter syndrome.

Sex-linked Dominant Mutations. In case of a condition due to sex-linked dominant gene, both males and females are affected and both males and females transmit the disorder to their offspring, just as in autosomal dominant condition (**Fig. 24.5**). Affected males transmit the condition to none of his sons but to all of the daughters. For examples, vitamin D-resistant rickets, low blood phosphate.

Y-linked disorders. Any trait controlled by differential region of Y-linked gene will show father to son inheritance whereas that controlled by homologous regions of X and Y chromosomes will pseudoautosomal inheritance.

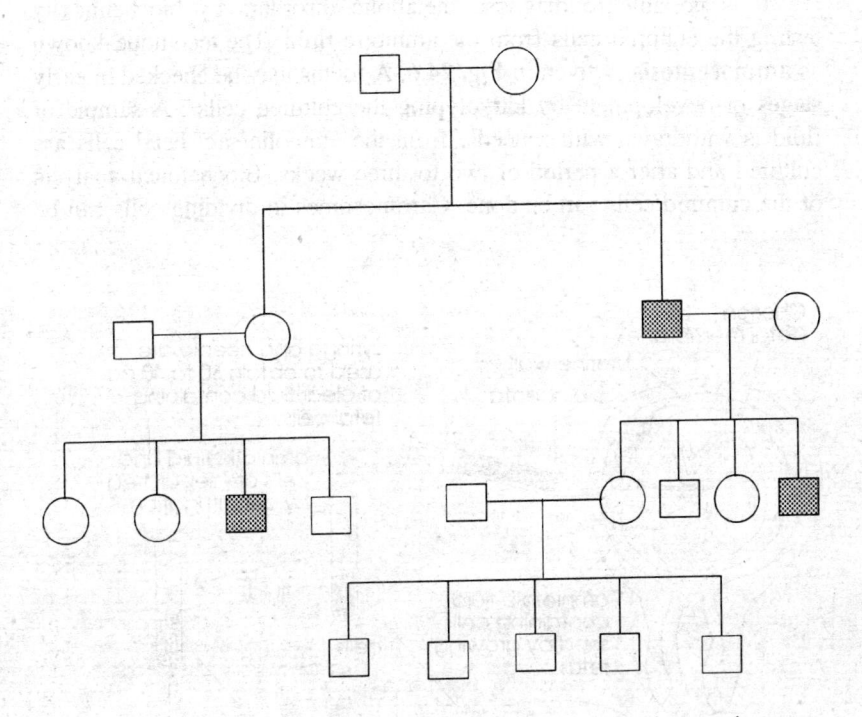

Fig. 24.4. A typical pedigree of a trait due to sex-linked recessive gene

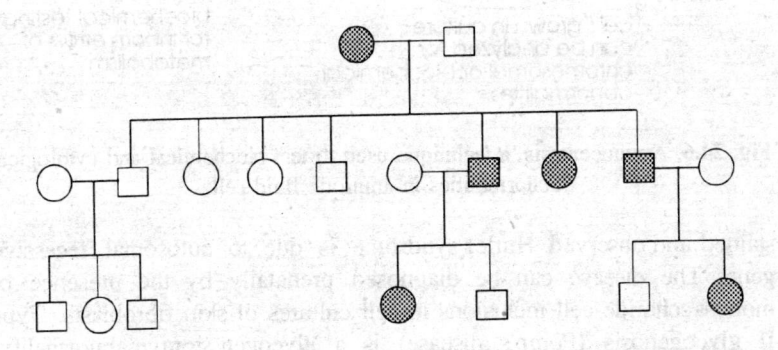

Fig. 24.5. A typical pedigree of a trait due to sex-linked dominant gene

Diagnosis and Carrier Detection
Prenatal Diagnosis

It is possible to diagnose metabolic disorders by biochemically testing the cultured cells from the amniotic fluid. The technique known as **amniocentesis** is given in **Fig. 24.6**. A foetus may be checked in early stages of development by karyotyping the cultured cells. A sample of fluid is withdrawn with a needle from the amniotic sac. Fetal cells are cultured and after a period of two to three weeks. Biochemical analysis of the cultured cells can be done. Chromosomes in dividing cells can be

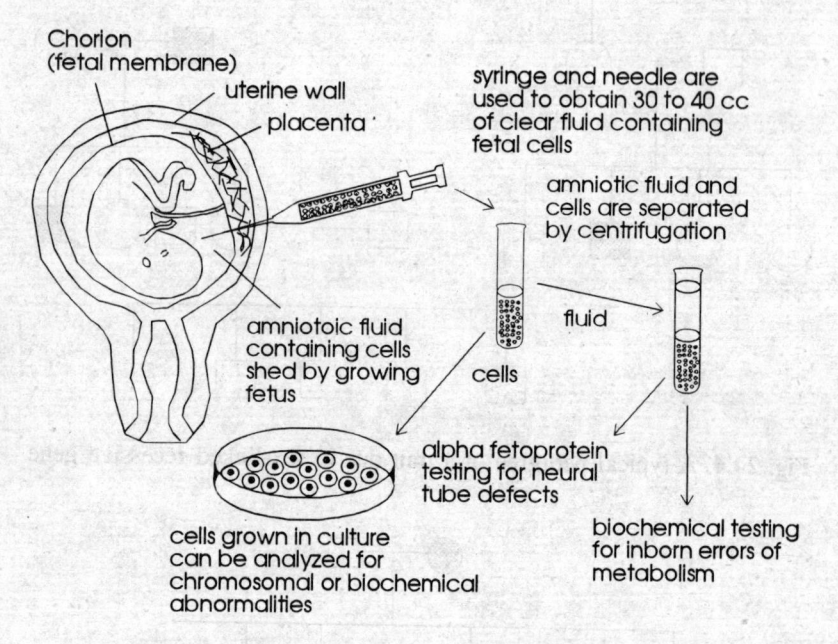

Fig. 24.6. Amniocentesis, a technique used to test biochemical and cytological abnormalities in amniotic fluid cells

stained and observed. Hurler syndrome is due to autosomal recessive gene. The disease can be diagnosed prenatally by the presence of monosaccharide cell inclusions in cell cultures of skin fibroblasts. Type II glycogenosis (Pomp's disease) is a glycogen storage abnormality.

Activity of enzyme α-1,4-glucosidase is deficient in amniotic fluid cells due to autosomal recessive gene. Amniocentesis is also a useful method for detecting aneuploidy which is responsible for almost all chromosomal diseases.

Postnatal Diagnosis

Cystic fibrosis, a condition due to autosomal recessive gene, is characterised by abnormal functioning of the exocrine glands. Staining cultured cells from cystic fibrosis patients has led to the identification of two or possibly three kinds of cystic fibrosis. Phenylketonuria (PKU), caused by autosomal recessive gene, can not be detected prenatally. In USA, routine screening at birth is done for PKU. A simple change in colour in the treated urine identifies PKU. Infants showing positive tests are placed on particular diet for first a few years of life. Thus patients are treated. PKU patients excrete excessive phenylpyruvic acid in the urine. Now we know that PKU positive test is obtained for many other metabolic disorders connected with defects in tyrosine metabolism.

Human genome mapping is progressing at a very fast rate. With completion of this project it would become much easier to get to a gene responsible for a disorder. Once gene-specific probes are available it become very easy to identify an abnormal gene through *in situ* hybridization.

Restriction Fragment Length Polymorphisms (RFLPs)

In recent years RFLPs have been used as molecular genetic markers in human beings. RFLPs are revealed by digesting genomic DNA with a number of restriction endonucleases, followed by hyridization with a specific genomic or cDNA probe. Segregation pattern of markers is studied within a panel of reference families. and recombination frequencies are calculated form this finite and limited sample size. By 1997 whole of chromosome 21 was sequenced. Complete human genome is expected to be mapped by 1999.

Somatic Cell Hybridization

The technique of somatic cell hybridization permits hybridization between different lines of somatic cells raised in tissue culture. Cells of different strains of same mammalian species or even cells of different mammalian species fuse followed by the subsequent loss of some of the parental chromosomes. Carlson *et al.* (1972) have shown that cells of different plant species can be made to hybridize by removing cell walls from leaf tissue, and growing them under conditions that permit only fused hybrid cells to survive. These hybrid cells (e.g., *Nicotiana glauca* X *N. langsdorfii*) can be raised to full-grown plants and shown to bear characteristics different from either plant. The method have been used to produce allopolyploids in *Nicotiana*.

In those cases involving human cell lines crossed with the somatic cells of mice or hamsters, the chromosomes most rapidly lost in dividing hybrid cells are human ones. Localization of a particular gene to a particular chromosome is then accomplished by noting whether the retention or loss of a particular gene-determined cellular function or enzyme is associated with the presence or absence of a recognizable human chromosome.

HAT Medium

Weiss and Green (1967) first used this technique to localize the gene involved in the production of the enzyme thymidine kinase, which catalyzes transfer of a phosphate group to the nucleoside thymidine, thereby transforming into nucleotide thymidine monophosphate (dTMP). First, they are mixed together in a common culture medium a TK^- mouse cell strain and a strain of TK^+ human cells. This mixture was placed in a medium containing the drug aminopterin, purine hypoxanthine, and pyrimidine thymidine ("**HAT**" **medium**). Aminopterin prevents the *de novo* cellular synthesis of purines and pyrimidines, so that such compounds must be provided in the culture medium in order for DNA replication to proceed. Because of exposure of this selective medium, the mouse TK^- cells soon die, since they are unable to convert the thymidine in the medium into necessary dTMP nucleotides where the human TK^+ as well as hybrid cells being complements of mouse and human chromosomes survive.

Selection Procedure

Hypoxanthine in HAT medium is converted into hypoxanthine phosphoribosyl purines. If transferase (HPRT) cell is $HPRT^-$, then aminopterin will cause death of such cells since they too, like TK^-, are unable to synthesize nucleic acids. Thus, when a line of $TK^+ HPRT^-$ cells is hybridized with $TK^- HPRT^+$ line, exposure to HAT medium will cause any remaining cells of each parental line to die, and only hybrid cells containing $TK^+ HPRT^+$ will survive. *De novo* synthesis pathway is inhibited by aminopterin. The salvage pathway which utilizes nucleosides resulting from nucleic acid breakdown requires the TK and HGPRT enzymes. Point to be noted is that when aminopterin is present, both TK and HGPRT enzymes are needed for DNA synthesis. Such selective methods ensure that fusion and hybridization have taken place (**Fig. 24.7**).

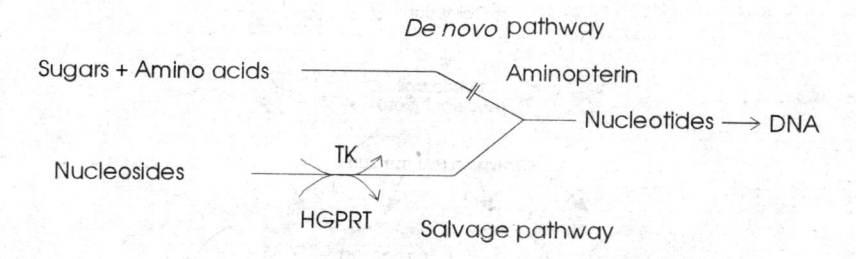

```
                        De novo pathway
Sugars + Amino acids ─────────────╲   Aminopterin
                                   ╲╱
                                   ╱╲──── Nucleotides ──→ DNA
                           TK  ╱
Nucleosides       ─────────╲╱
                           ╱╲
                        HGPRT    Salvage pathway
```

Fig. 24.7. *De novo* and salvage pathways for DNA synthesis

The hybrid mouse-man cells, that continue to grow in HAT medium must obviously retain the human chromosomes bearing the gene TK^+. This chromosome can be identified by the fact that it must be absent from the hybrid. Cells which grow in a medium containing thymidine analogue 5-bromodeoxyuridine (BUDR), since TK permits incorporation of the lethal BUDR nucleotide into DNA. Using both methods of identification, Migeon *et al.* (1968) have shown that *TK* gene is located in human chromosome 17. The scheme is summarized in **Fig. 24.8**.

Techniques involving the use of chromosomal aberrations now enable detection of the particular section of the chromosome carrying a specific gene. For example, strains of somatic cells bearing a "deletion" of a particular section of a chromosome may be associated with absence of a gene, whereas its presence in the other cell lines may be associated with the "translocation of that particular section to another chromosome. Thus the TK genes has been localized to the long arm of chromosome 17 since only when a certain section of this arm is present in mouse-human hybrid cells is the human TK enzyme produced.

Cellular hybridization technique has been applied to a variety of primates. Comparisons show that similar chromosomal location of genes have been preserved throughout the order. The *TK* gene, for example, is found on chromosome 17 in man, chimpanzee, gorilla, and orangutan. De

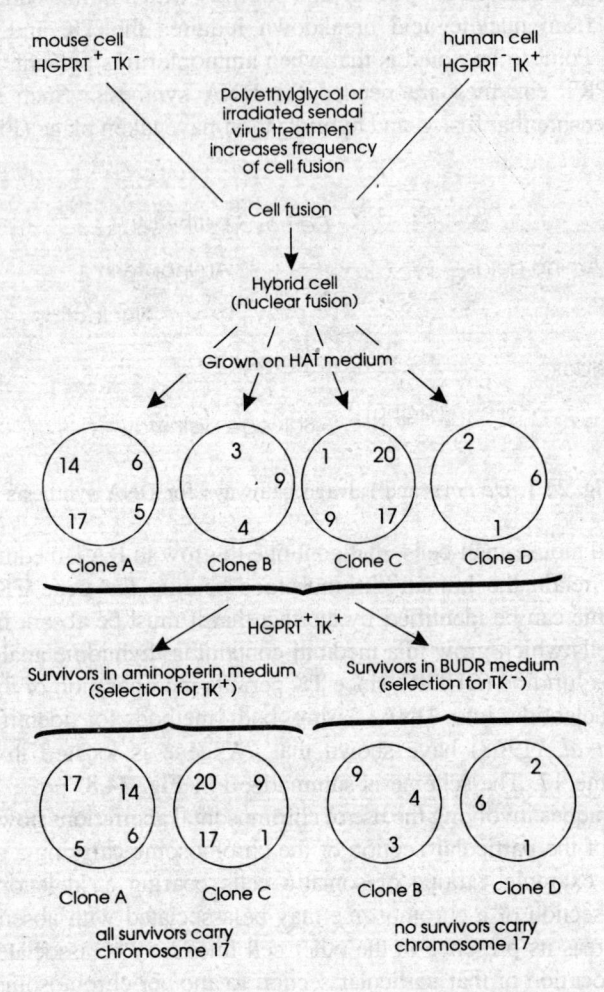

Fig. 24.8. Somatic cell hybridization technique

Grouchy *et al.* (1978) reported about 40 such similar gene locations among primates. Such studies have been extended to mammalian species (Stallings and Siciliano, 1982).

Degree of cell fusion between different cell lines is increased 100 to 1,000-fold through use of irradiated sendai (parainfluenza) virus or the chemical polyethylene glycol. A large number of genes, mostly recognized by their association with specific enzyme products, are now assigned to specific human chromosome. Specific human chromosomes are identified by peculiarities in size, shape, banding pattern, location of centromeres, satellites, etc.

Synteny

In some instances, linkages between two or more different genes on the same chromosome have been established on their concordant association with a particular human chromosome. This is known to synteny. Synteny may be defined as the occurrence of two or more genetic loci on the same chromosome. Depending upon intergene distances involved, they may or may not exhibit nonrandom assortment at meiosis.

Treatment for Inherited Disease

It involves three steps: (a) Diagnosis, (b) prognosis and (c) gene therapy. Cures are not available for inherited defects, but management by diet or drugs can improve the symptoms. This is not correct from human genetics point of view as this is in contradiction to the objectives of eugenics where our aim is to reduce the frequency of undesirable genes. By treating the symptoms of genetic disorders, we are simply perpetuating the undesirable gene as after reproduction it may be passed on to the next generation.

Consanguineous Marriages

Marriages between closely related individuals lead to homozygosity of defective genes and thus are cause for high frequency of many inherited disorders. Such disorders are: albinism, phenylketonuria, craniostenosis, haemophilia B, limb girdle muscular dystrophy. Phenotypes controlled by recessive genes have occurred more frequently in those populations where close marriages are common.

Genetic Counselling

This needs proper diagnosis. For proper diagnosis, detection of carriers of hereditary diseases is essential. Aim of genetic counselling is improvement of human race by decreasing the frequency of undesirable genes. This objective can be met with by decreasing the frequency of defective genes. Hitler tried to achieve this objective by eliminating those human beings that suffered from some serious genetic defects. Muller suggested establishment of sperm banks. Still another way is genetic counselling. This is telling the parents of a child to be born to them whether he/she will also have that defect. It is equally valuable to tell parents or parents-to-be that there is essentially no chance of transmitting some particular genetic disorder. A large number of people have urgent genetic questions. All genetic questions deserve possible answers which get better as knowledge of genetics increases.

Gene Therapy

Cure to genetic disorders will come from replacing the defecting genes. Recombinant DNA technology and other techniques are being used to find methods for safe and sure gene therapy. The hopes of gene therapy lie in the effort to get new or corrected genetic information into the somatic cells of the specific tissue where the critical gene function occurs. This type of gene therapy affects only the somatic cells and will not change transmission of abnormal alleles to the next generation.

Gene therapy is defined as the insertion into an organism of a normal gene which then corrects a genetic defects. Gene therapy has been carried out in *Drosophila melanogaster* and mice. Human diseases candidates for gene therapy are simply inherited conditions, such as, hemoglobin abnormalities (specifically β-thalassemia). Gene therapy should be primarily beneficial for the replacement of a defective or missing enzyme or protein. Three defective genes that produce defective enzymes are initial prime candidates: hypoxanthine guanine phosphoribosyl transferase (HGPRT) absence of which results in Lesch-Nyhan disease; purine nucleoside phosphorylase (PNP), the absence of which results in severe immunodeficiency; and adenosine deaminase (ADA), the absence of which results in severe combined immunodeficiency disease. Three general requirements for a successful gene therapy programme are delivery, expression, and safety.

Delivery of a gene means the new gene is put into the correct target cells and should remain three long enough to be effective. At present, the only human tissues that can be used for gene transfer are bone marrow and skin cells. Techniques for transferring cloned genes into cells include viral, chemical and physical methods. Viral techniques include both RNA (retroviruses) and DNA viruses (adenoviruses). Chemical method is based on calcium phosphate-mediated uptake of DNA (gene). Physical methods of gene transfer are electroporation, microinjection, gene gun, etc. The retroviral-based vectors appear to be most promising approach at present for use in humans. Three should be appropriate expression of the new gene in the target cells. Normal **expression** of exogenous genes is the exception rather than the rule. Appropriate active promoters and enhances are used to increase expression. Some enhances may even be tissue-specific. The new gene should not harm the cell, or by extension, the animal. Although retroviruses have many advantages for gene transfer they also have disadvantages. (1) They can rearrange their own structure as well as exchange sequences with other retroviruses. They may lead to production of infectious recombinant virus. **Safety** of the gene therapy procedures remains the major issue. Gene therapy should not be attempted on human beings until there are good animal data available.

Gene Therapy in Lower Species

In *Drosophila*, the transposable genetic element, the P factor, has been used to transfer a normal gene coding for the enzyme that produces wild-type red eye colour in *Drosophila* that have defective gene. Transposable elements have not so far been identified in vertebrates. A strain of mice called little possess a mutation that results in reduced serum levels of growth hormone, and the mice are therefore dwarfs. A rat growth hormone gene has successfully been inserted into the cells of little mice in such a way that the gene is expressed at a high level. The deficiency in growth hormone was corrected and the animals grew rapidly. However, expression of the new gene was not controlled appropriately. The equivalent human disease is pituitary dwarfism.

Gene Therapy in Humans

Pituitary dwarfism, β-thalassemia, hemophilia, Lesch-Nyhan disease, severe immune-deficiency disease, acquired immunodeficiency syndrome (AIDS), and severe combined immune deficiency disease are the human diseases that are short listed for gene therapy. In HGPRT, PNP and AA deficient patients, defect is found in bone marrow. Production of small level of enzyme in these three cases should be beneficial. Defective bone marrow can be removed from a patient and normal HGPRT/PNP/ADA gene inserted into a therapy. In one of the gene therapy experiments, a tumor necrosis factor was injected into the people with cancer. The necrosis factor often shrinks tumors. The gene is in lymphocytes that have been engineered. In an other study, a child was infused with cells to replace a gene for the enzyme adenosine deaminase, an enzyme whose absence results in dysfunctional immune system.

It would be ethical to insert genetic material into a human being for the sole purpose of medically correcting a severe genetic defect in the patient somatic cell gene therapy. Germline gene therapy deals with correcting a gene in germ line cells. Attempts to correct germ cells do not have societal acceptance at this time. The techniques that are now being developed are for somatic cell, not germ line, gene therapy. Those diseases that are controlled by polygenes are not easy to be subjected to gene therapy.

Mitochondria and Diseases

Some human disorders have been associated with mutations in the mitochondrial DNA. They show typical maternal pattern of inheritance. Gene therapy in mitochondrial genes have not yet been taken up.

Targeted Gene Replacement

Targeted gene replacement is one of the most powerful methods of discovering what a gene does. A gene is knocked out and its effect is observed on the organism. In this way, researchers can create an organism bearing chosen mutations in any known gene (Capecchi, 1994).

DNA Fingerprinting

DNA fingerprinting technique is useful in establishing near-perfect identity of an unidentified body. It is based on the fact that every organism is conferred with DNA variations in form of a **variable number of tandem repeats** (VNTRs). The DNA profile of an individual is also known as **genetic signatures**. Accuracy of this method in 1-in-75 billion error probability ratio. DNA fingerprinting using **microsatellites** also called **simple sequence repeats** (SSRs) has also been utilized both in plants and animals (Jeffreys *et al.*, 1985).

LOD Score

Log_{10} ratio of two probabilities for recombination ratio, i.e., observed recombination and free recombination, is known as LOD score. The higher the LOD value, the stronger the linkage. An LOD score of 3 means that the probability of a linkage value is 10^3 compared to the probability of free recombination.

Human population geneticists can increase their accuracy of linkage analysis by using this probability technique. Logarithms for ease of calculation and the parameter is called z, the *lod* score. Using this method different crossover frequencies can be tried until the one giving the highest *lod* score is found.

Some Useful Reading Hints

Newman *et al.* (1937) used twins for studying heredity and environment. Muller (1949) discusses our load of mutations. Neel (1949) dealt with the detection of the genetic carriers of hereditary disease. Pauling *et al.* (1949) described sickle-cell anemia as a molecular disease. Osborn (1951) provided a preface to eugenics. Morton *et al.* (1956) gives an estimate of the mutational damage in man from data on consanguineous marriages. Tjio and Levan (1956) resolved the dispute on the chromosome number of man. Levine (1958) discussed the influence of the ABO system of Rh hemolytic disease Morton (1962) used segregation and linkage as a methodology in human genetics. Burdette (1962) described methodology in human genetics. Haller (1963) deals with subject of eugenics. Muller (1963) suggested genetic progress by voluntary conducted germinal choice. Race and Sanger (1968) delat with blood groups in man.

Hamerton (1971) dealt with cytogenetics of some human disorders. McKusick (1972)gave classical reading in form of his book *Human Genetics*. Hilton *et al.* (1973) discussed ethical issues in human genetics involving genetic counselling and the use of genetic knowledge. Vogel and Rathenberg (1975) estimated frequency of spontaneous mutation in man. Bajema (1976) dealt with various aspects of eugenics. Witkin *et al.* (1976) discussed criminality in XYY and XXY men. Epstein and Golbus (1977) dealt with prenatal diagnosis of genetic diseases. Text book *Principles of Human Genetics* by Stern (1977) is still a very useful reference source. Brewster and Gerald (1978) dealt with chromosome disorders associated with mental retardation. Therman (1980) dealt with the human chromosomes. Harris (1980) provided an excellent work on the principles of human biochemical genetics. Kolata (1984) expressed happiness that gene therapy method showed promise. Hassold and Jacobs (1984) describes cases of trisomy in man. Patterson (1987) dealt with the cause of Down syndrome. Caskey (1987) discussed diagnosis of genetic diseases by recombinant DNA methods. Nakamura *et al.* (1987) described the use of variable number of tandem repeat (VNTR) markers for human gene mapping. Epstein (1988) dealt with mechanisms of the effects of aneuploidy in mammals. Watkins (1988) described Restriction fragment length polymorphism (RFLP) and its application in human chromosome mapping and genetic disease research. Culliton (1989) gave general comments on gene tests. Kirby (1990) gave an introduction to DNA fingerprinting. Warburton (1990) compileed chromosomal anomalies and prenatal development in form of an atlas. Hartl (1991) in his book *Basic Genetics* focussed on human genetics. Debenham (1992) dealt with probing identity through DNA fingerprinting. Richards *et al.* (1992) provoded evidence of founder chromosomes in fragile X syndrome. Borgaonkar (1994) compiled chromosomal variations in man in form of *A Catalogue of Chromosomal Variants and Anomalies*. McKusick (1994) compiled a catalogue of human genes and genetic disorders. Krontiris (1995) found connection between minisatellites and human diseases.

Carlson (1970) gave new method for induction and isolation of auxotrophic mutants in somatic cell cultures of *Nicotiana tobaccum*. Puck and Kao (1982) discussed technique of somatic cell genetic and its application to medicine. Walters (1986) dealt with the ethical problems of human gene therapy. Waldman (1991) discussed use of monoclonal antibodies in diagnosis and therapy. Miller (1991) reviewed progress made in human gene therapy. Morgan and Anderson (1993) dealt with the subject of human gene therapy. Capecchi (1994) discussed methods of targeted gene replacement and its application. Hoffman (1994) discussed gene therapy and is not sure where to draw the line.

Questions

1. Define relationship between genetics and man. What is eugenics?
2. What approaches are used in understanding genetic basis of genetic disorders?
3. Describe various genetic disorders in man due to changes in chromosome number.
4. Describe various genetic disorders in man due to changes in chromosome structure.
5. Genetic disorders arising due to point mutations can be classified into various types on the basis of type of inheritance exhibited. Comment.
6. Describe the criteria used to conclude in pedigree analysis whether a trait is controlled by an autosomal recessive mutation, an autosomal dominant mutation, a sex-linked recessive mutation, a sex-linked dominant mutations, a Y-linked recessive mutation, and a Y-linked dominant mutation.
7. What is genetic counselling? What is the importance of disease diagnosis and carrier detection in genetic counselling?
8. What is the technique of amniocentesis? Describe its use.
9. What is the use of postnatal diagnosis in genetic counselling?
10. Describe the use of restriction fragment length polymorphisms (RFLPs).
11. What is somatic cell hybridization technique? In what cases is this technique used?
12. Define HAT medium. What are various ingredients of this medium? Describe functions of these ingridients. What is the use of this medium in somatic cell hybridization?
13. Describe the selection procedure used to select hybrid celllls?
14. What is synteny?
15. Treatment for inherited disease provides temporary relief to the patient. Is this approach in line with the principles of eugenics? Give reasons for your answer.
16. What is genetic counselling? What are consanguineous marriages? Why will a genetic counsellor not recommend such marriages?
17. Define gene therapy? What are aims of gene therapy?
18. What is somatic cell gene therapy? Why most of the experiments on human beings use somatic cells for gene therapy?
19. What is germline gene therapy?
20. Describe various success stories of gene therapy in lower species.
21. Describe various examples of gene therapy in humans.
22. What type of genetic disorders are on the top agenda for gene therapy and why?
23. What is relationship between mitochondria and diseases.
24. What is targeted gene replacement? Describe its applications.
25. What is DNA fingerprinting? Describe the use of this technique.
26. What is LOD score. Describe the significance of this parameter.

25

GENETIC ENGINEERING

Genetic engineering deals with isolation, synthesis, adding, removing or replacing genes in order to achieve permanent and heritable changes in plants, microbes, animals or even man. The main advantage of techniques of genetic engineering is that it bypasses all reproductive barriers.

Restriction Endonucleases

Restriction endonucleases are the enzymes which recognize and cut specific nucleotide sequences in DNA, which are four or six nucleotide long. Type II restriction endonucleases recognize, bind and cleave at palindromic sequences. Palindromic sequence is a sequence of letters (DAD, MADAM, MALAYALAM, REDIVIDER), words (AND MADAM DNA), phrases, nucleotides, or nucleotide pairs (GAATTC/CTTAAG) that reads the same regardless of which direction one starts from. **Type II restriction endonucleases** have found applicability in construction of recombinant DNA molecules. Characteristics of some type II restriction endonucleases are given in **Table 25.1**. These restriction endonucleases may produce cohesive (sticky) or blunt ends (**Table 25.2**). **Cohesive ends** are used in construction of recombinant DNA molecules. **Blunt ends** are used to produce sticky ends of interest.

With the use of restriction endonuclease enzymes the **recombinant DNA molecules** can be generated combining DNA segments from any two species. Over 90 restriction enzymes each with different site specificity have so far been isolated. These enzymes provide powerful tools for analysis of DNA organization, gene structure and gene regulation. Recombinant DNA molecules produced using these enzymes are playing important role in genetic engineering and transfer of genes from one organism to another bypassing the limitations imposed by

Table 25.1. Characteristics of some type II restriction endonucleases

Enzyme	Organism from which enzyme isolates	Recognition sequence and position of cut	λ	Ad2	SV40
BamHI	*Bacillus amyloliquefaciens* H	↓ 5'G GATCC3'	5	3	1
BglII	*Bacillus globigii*	↓ A GATCT	5	12	0
EcoRI	*E. coli* RY13	↓ G AATTC	5	5	1
HaeIII	*Haemophilus aegyptius*	↓ GG CC	50	50	18
HhaI	*Haemophilus haemolyticus*	↓ GCG C	50	50	2
HindIII	*Haemophilus infleenzae* Rd	↓ A AGCTT	6	11	6
HpaI	*Haemophilus parainfluenzae*	↓ GTT AAC	11	6	5
PstI	*Providencia stuartii*	↓ CTGCA G	18	25	3
SmaI	*Serratia marcescens*	↓ CCC GGG	3	12	0
SalI	*Streptomyces albus* G	↓ G TCGAC	2	3	0

Number of cleavage sites in DNA from

Table 25.2. Production of cohesive (sticky) or blunt ends by type II restriction endonucleases

Restriction endonuclease	Ends
Cohesive or sticky ends	
EcoRI	-G AATTC- -CTTAA G-
HindIII	-A AGCTT- -TTCGA A-
Blunt ends	
HaeIII	-GG CC- -CC GG-

Site for EcoR I
endonuclease

5'- C - C - A -|G - A - A - T - T - C |- G - C - G - 3'
3'- G - G - T -|C - T - T - A - A - G |- C - G - C - 5'

↓ EcoR I endonuclease

.... 5'- C - C - A -|G   A - A - T - T - C |- G - C - G - 3'
.... 3'- G - G - T -|C - T - T - A - A   G |- C - G - C - 5' ....(1)

Sticky ends

A - A - T - T - C - .... - G
G - .... - C - T - T - A - A   ....(2)

Fragment to be inserted possesses sticky ends
produced by the same enzyme (EcoRI)

↓ Incubation and ligation of
fragments (1) and (2)

.... 5'- C - C - A -|G A - A - T - T - C - .... - G| A - A - T - T - C |- G - C - G - 3'
.... 3'- G - G - T -|C - T - T - A - A G - .... - C - T - T - A - A| G |- C - G - C - 5'

Recombinant DNA molecule

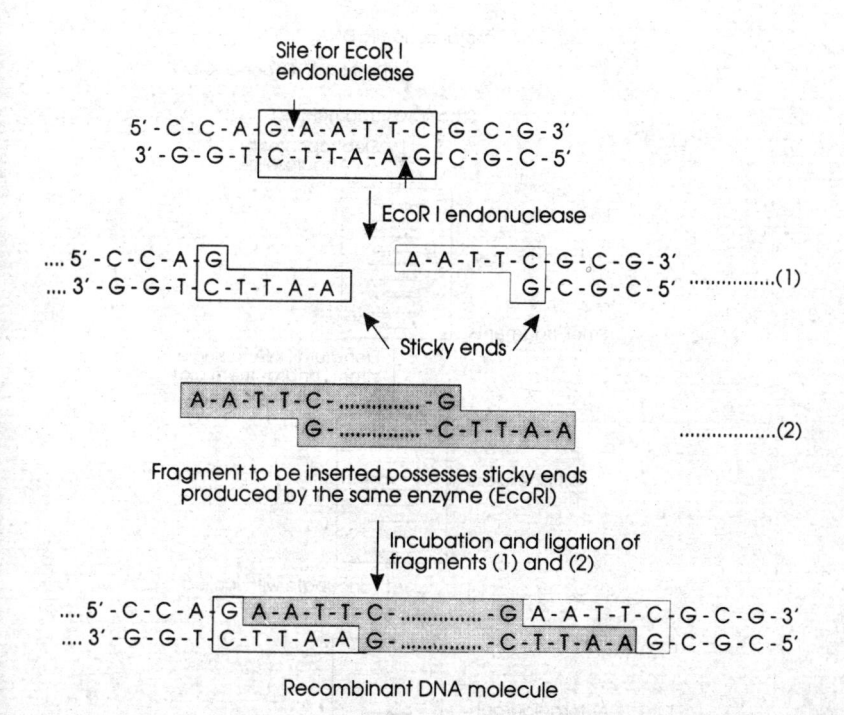

Fig. 25.1. Construction of a recombinant DNA molecule

biological methods (sexual cycle, lack of pollination, fertilization etc.). Procedure for construction of a recombinant DNA molecule has been illustrated in **Fig. 25.1**.

Gene/DNA Cloning

Bacteriophage (**Charon phages, cosmids**) or bacterial plasmid are used as a vector of prokaryotic origin for molecular cloning. Among **eukaryotic vectors** are yeast vectors (yeast artificial chromosome), animal vectors (Simian virus 40, a mammalian virus) and plant vectors (Ti plasmid of *Agrobacterium tumefaciens*). Procedure used for cloning of a DNA piece or a gene of interest is shown in **Fig. 25.2**. Recently use of pure preparations of individual genes, combinations of plasmid and viral vectors and DNA precipitated with calcium phosphate for increased

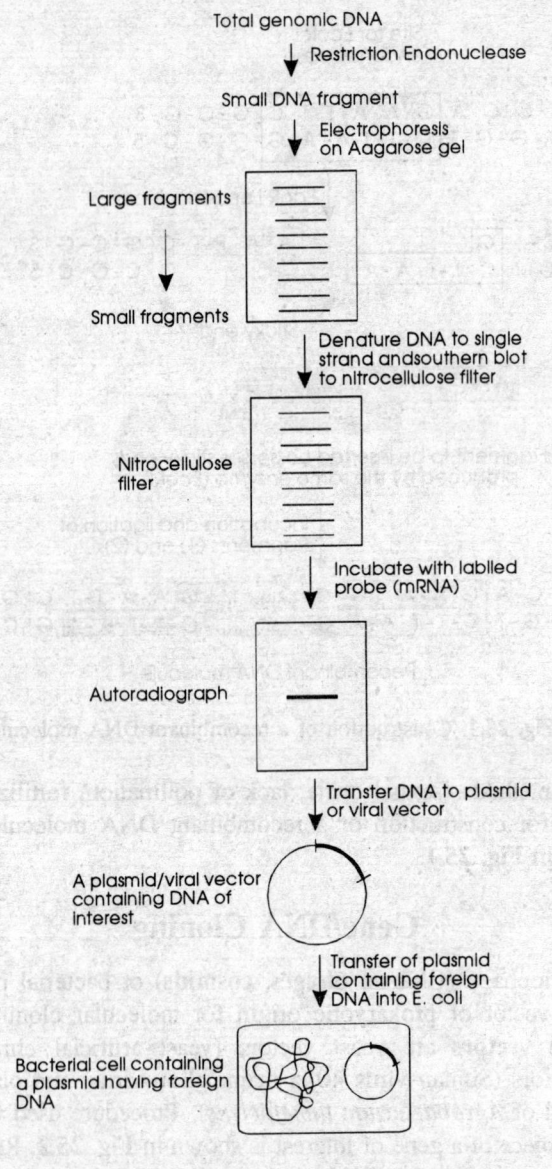

Fig. 25.2. Cloning of foreign DNA into a bacterial cell through its insertion into a plasmid with the help of Southern blotting techhnique

uptake have been introduced. One of the new approaches is to insert the desired gene into an animal virus and then infect target animal cells with the virus. For example, the rabbit globin gene was inserted into DNA of simian virus 40. A line of cultured monkey cells was infected with the recombinant virus, the new gene was shown to be expressed.

Conjugative plasmids in bacteria can be transferred to bacteria outside their own species. For example, F-factor of *E. coli* can be transferred during conjugation to bacteria of other genera within the family Enterobacteriaceae. Similarly, plasmid from *Pseudomonas aeruginosa* (R1822) has an extraordinary broad host range, as it can be transferred to many different families of Gram negative bacteria. Bacteriophage and retroviruses can be used as vector for gene transfer.

Transfer of genes from donor DNA to a recipient cell requires the ability of recipient DNA to replicate autonomously within a cell. For replication of foreign DNA segment in recipient DNA there are two requirements. One, the presence of a region on the DNA which serves as point of origin of DNA replication. Two, controlling or regulating element required for expression of the inserted gene. This piece of DNA in which a foreign DNA is cloned is called **vector**.

Gene Transfer

Once a gene has been isolated and cloned, several methods of gene transfer are available.

Chemical Method

Protoplasts are mixed with vector plasmids containing the gene of interest in presence of **polyethyl glycol (PEG)**.

Microprojectiles

Small tungsten or gold particles are coated with DNA to be transferred. These are bombarded into cells with the help of **microgun** or **particle gun**.

Agrobacterium-mediated Gene Transfer

Bacterium *Agrobacterium tumifaciens* causes tumors in dicots. This bacterium carries a tumor inducing (Ti) plasmid. In **Ti plasmid** a piece of DNA called T-DNA can integrate into host chromosome. By inserting a foreign gene into T-DNA, foreign gene can be inserted into plants.

Microinjection

Direct microinjection of a purified gene into the nucleus of an animal cells is done through a micropipette. The cell could be propagated into a large clone of identical cells having the inserted gene. Raddle and Gordon injected foreign genes into the nuclei of fertilized mouse eggs. The idea was to have the inserted gene in all the body cells resulting from division of the zygote. The new gene was actually found to be present in 3 out of 180 animals that developed from injected cells. This and other experiments showed that foreign DNA is functional in animals that develop form injected eggs.

Co-transformation

Co-transformation method by Mainatis *et al.* (1978) was used to transfer rabbit β-globin gene into mouse cells. The rabbit β-globin gene was isolated from a cloned library of rabbit chromosomal DNA. This gene was carried on the DNA of phage λ as the cloning vector. DNA from the clone was mixed with viral DNA carrying the **Tk gene** and incubated with mouse cells. The transformed cells having the Tk gene were isolated and analyzed for the presence of rabbit β-globin sequence. One to several copies of the β-globin gene were found to be inserted in the mouse genome. Furthermore, the rabbit globin sequences were transcribed in the transformed cells of the mouse.

Transfection

Studies with *Bacillus subtilis* have shown that when DNA from an animal virus or bacteriophage is used in transformation, intact virus particles are formed inside the recipient bacterial cells. This process is called transfection. In this case there is no need for donor DNA to become integrated into the host chromosome. When such a bacteria come in contact with the animal host which the virus is able to infect, it releases the contained virus particles causing the infection of the host animal.

Electroporation

Electroporation is electron-mediated gene transfer technique. A gene can be transferred into a cell and its expression analyzed within hours. Method is as follows: **Protoplasts** are isolated from rapidly growing suspension cells. Protoplasts are resuspended in phosphate-buffered saline, containing mannitol with or without plasmid DNA. Electrical pulse is supplied at 2000 V. The solution containing the protoplasts was held at 0°C for 11 minutes after the electron pulse. Protoplasts were subjected to electroporation and were then assayed in duplicate for chloramphenicol acetyltransferase (CAT) activity resulting from expression of introduced plasmid. Gene transfer efficiency increased with DNA concentration.

Chromosome-mediated Gene Transfer

Chromosome-mediated gene transfer has been done in mammalian cells. It is different from DNA-mediated gene transfer.

Liposome-mediated Gene Transfer

Stable transformation of mouse L cells deficient in thymidine kinase (TK) was achieved by liposome-mediated transfer of a recombinant plasmid carrying the Tk gene. Ten per cent of the recipient cells expressed TK activity. The transformed phenotype (e.g., 200 out of 10^6 cells) was stable under selective and non-selective condition.

DNA-mediated Gene Transfer

DNA-mediated transfer is another method of gene transfer. A purified DNA fragment carrying the gene of interest is mixed with a carrier DNA and is precipitated out of solution with calcium phosphate. When the target cells are incubated with the precipitate, there is increased uptake of DNA and some of the cells are transformed by the desired gene. Axel *et al.* utilized this method in mouse cells. They took a purified DNA fragment from herpes simple virus that carried the gene for thymidine kinase (Tk). A small amount of viral DNA was mixed with several mg of Salmon sperm DNA as a carrier and precipitated by calcium phosphate. The precipitate was incubated with cultured mouse L cells which lacked the Tk gene (Tk⁻) cells. A few of the mouse L cells incorporated the Tk gene and became transformed.

A generalized scheme of gene transfer from *Drosophila* to bacteria is given in **Fig. 25.3**.

Chromosome Walking

Despite the limited use of any one inserted piece of DNA, it is possible to learn about longer stretches of DNA by a using a technique of **overlapping clones** called chromosome walking. For example, let us

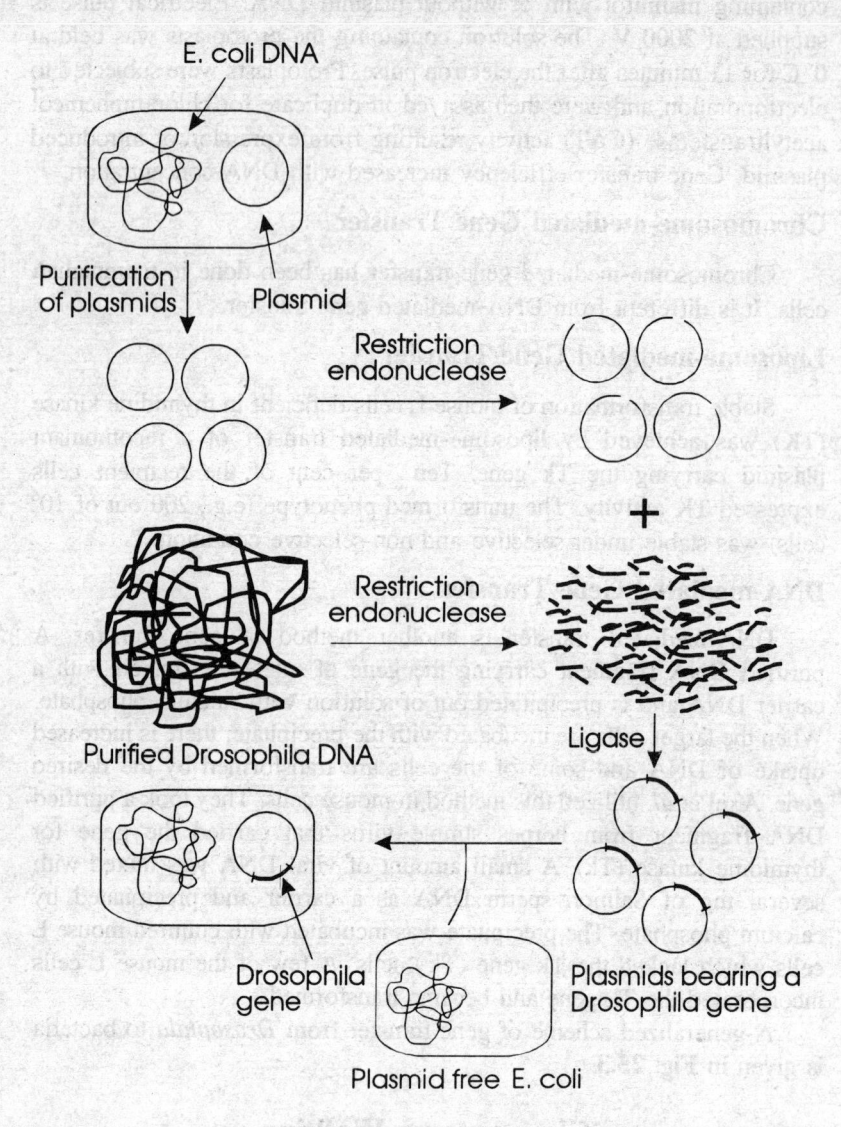

Fig. 25.3. Transferring a *Drosophila* gene into *E. coli*

suppose that a particular gene in region A is cloned in clone 1, as discovered through cloning. The cloned insert can be removed, using the same restriction enzymes used to insert it in the vector initially, and broken into small pieces that are used as probes themselves. The idea is to locate another clone with an inserted region that overlaps the first one (**Fig. 25.4**). The second one is now treated the same way to probe for yet another overlap farther down the chromosome. In this way relatively long segments of chromosomes can be available in overlapping clones.

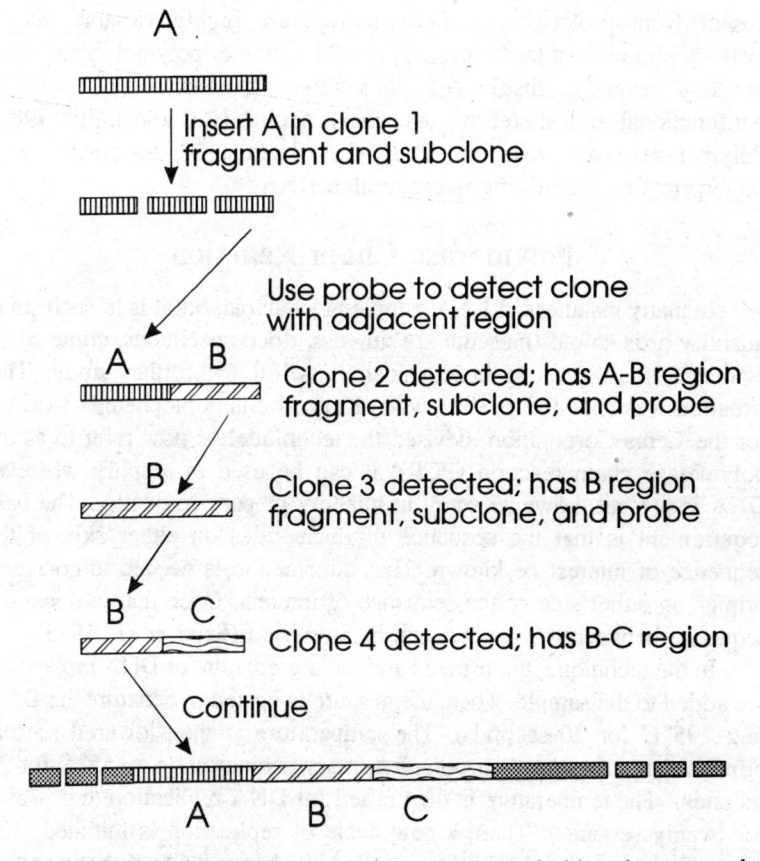

Fig. 25.4. Chromosome walking technique used to study long chromosomal regions by locating overlapping cloned inserts that cover the region

Microsatellites

Microsatellite is a class of hypervariable loci consisting of tandemly repeated, short (less than 10 bp), simple sequences of 100 to 200 bp length distributed in eukaryotic genomes. These sequences serve as molecular markers for selection of agronomically useful traits and for preparation of advanced, high-density, genetic linkage maps. Microsatellites, also called simple **sequence repeats** (SSRs), **simple tandem repeats** (STRs), or simply **simple sequences** (SSs), consist of head-to-tail tandem arrays of short DNA motifs (usually 1 to 5 bases). They are a common component of eukaryotic genomes but are almost absent from prokaryotes. Microsatellites are highly variable due to avariable number of tandem repeat (VNTR)-type of polymorphism, more or less evently distributed throughout genome, and probably nonfunctional and, therefore, selectively neutral (Litt and Lutty, 1989). Polymerase chain reaction (PCR) has been used to create DNA fingerprints by amplifying microsatellite DNA.

Polymerase Chain Reaction

In many instances, a DNA sample is available, but it is in such small quantity or is so old (museum specimenss, dried specimens, crime scene evidence, fossils) as to be considered uselful for further study. That situation was changed in 1983 when Kary Mullis, a biochemist working for the Cetus Corporation, devised the technique we now refer to as the polymerase chain reaction (PCR). It can be used to amplify whatever DNA is present, however small in quantity or poor in quality. The only requirement is that the sequence of nucleotides on either side of the sequence of interest be known. That information is needed to construct primer on either side of the sequence of interest. Once that is done, the sequence between the primers can be amplified (Saiki et al., 1985).

In the technique, the primers and the ingredients for DNA replication are added to the sample. Then, the mixture is heated to denature the DNA (e.g., 95°C for 20 seconds). The temperature is then lowered so that primers can anneal to their complementary sequences (e.g., 55°C for 20 seconds). The temperature is then raised for DNA replication (e.g., 72°C for twenty seconds). Then, a new cycle of replication is initiated. The various stages in the cycle are controlled by changes in temperature since the temperatures for denaturation, primer annealing, and DNA replication are different. In about 20 cycles, a million copies of the DNA are made; in thirty cycles, a billion copies are made. The technique is aided by using DNA polymerase from a hotsprings bacterium, *Thermus aquaticus*, that can withstand the denaturing temperatures. Thus after each cycle of replication new components do not have to be added to the reaction mixture. Rather, the cycling can be continued without interruption in PCR machines that are simply programmable water baths that accurately and rapidly can change the water temperature that surrounds the reaction mixture. Some machines can process ninety-six samples at a time.

Transposons

Transposon is a segment of DNA, which generally consists of more than 2,000 nucleotides, that is capable of moving into and out of a chromosome or plasmid and/or from place to place within the chromosome in both prokaryotes and eukaryotes. It is capable of turning genes "on" and "off". Transposons are also termed as **Transposable genetic elements**. These elements can insert into nonhomologous DNA, exit, and relocate in a reaction that is independent of the general recombination function of the host. Prokaryotic transposable elements are of three types: (a) insertion sequences (Is elements) that are of size 0.8 to 1.5 kb that encode only transposition determinants; (b) transposons (TN) that are greater than 5 kb long and encode additional transposition determinants (such as antibiotic resistance) and are often bracketed by IS elements; and (c) transposable phages that carry out site-specific recombination reactions which neither require DNA replication nor degradation or resynthesis of DNA. Eukaryotic genetic elements fall into three major classes: (a) classical transposons, and (b) retrovirus or retroviral-like retroposons, and (c) retroposons (McClintock, 1951; Cohen, 1976).

Transposase is the enzyme that catalyzes transposition of a transposable genetic element (Berg, 1977). Transposition is the process whereby a transposon or insertion sequence insert into a new site on the same or another DNA molecule. The exact mechanism is not fully understood, and different transposons may transpose by different mechanisms. Transposition in bacteria does not require extensive DNA homology between the transposon and the target DNA. The phenomenon is therefore described as **illegitimate recombination**. Some transposons have the ability to prevent others of the same type from transposing to the same DNA molecule. This phenomenon is known as **transposition immunity**.

Activity of functional genes can be blocked by inserting a foreign sequence with the help of a transposon. This phenomenon is known as

transposon tagging. Genes can be transferred from one species to another by use of a transposon. This method of gene transfer is termed as **transposon-mediated gene transfer**.

High Density Molecular Maps

Morphological markers useful for constructing linkage maps are only in limited numbers. The molecular markers on the other hand are numerous and they provide capacity for complete coverage of nuclear mitochondrial and chloroplast genomes. The **saturated maps** prepared with the help of molecular markers are known as high density molecular maps. Molecular markers used for this purpose are of several kinds: **amplified fragment length polymorphism (AFLP), amplicon length polymorphism (ALP), arbitrarily premied PCR (Ap-PCR), allele-specific PCR (AS-PCR), DNA amplification fingerprinting (DAF), random amplifed polymorphic DNA (RAPD), restriction fragment length polymorphism (RFLP), specific amplicon polymorphism (SAP), sequence-characterized amplified region (SCAR), single strand conformation polymorphism (SSCP), microsatellite simple sequence length polymorphism (MSSLP), simple sequence repeats (SSR)**, and **sequenced tagged sites (STS)**. Of all these markers, RFLP and RAPD are the principle classes of markers (Tanksley *et al.*, 1992). High density molecular maps are helpful in demonstrating **synteny**.

Variable Number of Tandem Repeat Loci

Variable number of tandom repeat (VNTR) loci are **hypervariable loci** because of **tandem repeats**. Such loci are presumably generated by **unequal crossing over**. Since VNTR loci are highly variable, probing for one of these VNTR loci in a population reveals many alleles. The Southern blots create a DNA fingerprint of extreme value in forensics. This technique has a greater power to positively identify individuals than using their fingerprints.

Applications of Genetic Engineering

Genetic engineering using recombinant DNA technology has enhanced the research capabilities in several areas. It has become possible to obtain large number of copies of a particular gene (gene cloning). Study of DNA function and regulation and role of nuclear and cytoplasmic DNA could be intensified. Extensive and accurate mapping of genomes particularly for viruses such as SV40, yeast and human are becoming possible. Now it has become possible to synthesize large quantities of antibiotics, enzymes and hormones for human health. Genetically engineered bacteria such as *E. coli* carrying genes for hormones like insulin or anti-viral proteins like **'interferon'** have been used to produce large quantities of these substances at reduced costs of production. To transfer gene from prokaryotes to plants and animals has become a routine work.

These achievements have numerous applications in industry, agriculture and human health. Genetically altered plants carrying genes from bacteria or other micro-organisms have been obtained. First example of this has been reported in tobacco. Genetically engineered plant carrying a gene from bacteria *Bacillus thuringiensis* has been added to its structure. This *Bt* gene produces a protein which is toxic to caterpillars and others insects feeding on plant leaves, thus making the plants highly resistant to a broad spectrum of insects. Many genetically altered species of plants and animals can be obtained by these techniques. Genes for nitrogen fixation ('*nif*' genes) can be transferred to cereals like wheat, maize and rice which may enable them to fix atmospheric nitrogen and genes for resistance to diseases and tolerance to adverse environmental conditions (salinity, alkality, drought, etc.) can be transferred from wild relatives to cultivated species. Attempts are also being made to produce bacterial strains which can perform many otherwise impossible tasks, e.g., clean the environment from the residues of 2,4-D, DDT and other industrial wastes, plastic degradation, reduce viscosity of crude oils facilitating easy pumping, bacterial fermentation to produce methane and other energy forms, direct convertion of sun light into electricity, degrade sulphur in coal and oil.

It also seems possible that foreign DNA segments could be integrated into the pollen chromosomes. The DNA segments could be of structural gene or regulatory gene and may affect gene expression causing phenotypic changes in plants. Sorghum and rice are genetically distant but DNA segment could be exchanged between them. In China, conventional crossing between these two plants has been attempted and some hybrids which look like rice having bigger spikes, more grain and higher yield have been obtained. These hybrids have also been found to be drought tolerant.

Exogenous DNA segments could be transferred into growing plant cells. Such DNA is integrated into the host cell and phenotypic changes brought about in the parent from different species or genera could be introduced through this technique into cotton plants. Human intestinal bacteria could be made to digest cellulose. It may be possible to add to cereal crop plants the genes needed to fix atmospheric nitrogen. Genetic engineering accelerated researches in the field of vaccines. A day is not far when we will be eating vaccines.

Hazards of Genetic Engineering

Recombinant DNA technology poses some hazards also. Antibiotic-resistance genes might be introduced into pathogenic bacteria. Genes for toxin production might be introduced into bacteria that were not formerly pathogenic. The introduction of DNA from tumor production viruses into viruses or plasmids as those carried by *E. coli* in the intestine of animals and humans might increase the incidence of cancer. Keeping possible hazards of recombination DNA researches in view, the National Institute of Health has proposed restrictions on some of the recombinant DNA researches. The U.S. Government has provided laws that such researches in industry were following safety guidelines.

Some Useful Reading Hints

McClintock (1941) described the stability of broken ends of chromosomes in *Zea mays*. McClintock (1951) discussed chromosome organization and genic expression. Shapiro (1983) discusseds structure, properties and uses of mobile genetic elements. Holzman (1991) related the story of a "jumping gene" caught in the act.

Berg *et al.* (1974) pointed out potential biohazards of recombinant DNA molecules. Fox *et al.* (1975) proformed gene transfer in *Drosophila melanogaster* by genetic transformation induced by the DNA of transformed stocks. Garmeraad (1976) conducted genetic transformation in *Drosophila* by microinjection of DNA. Hohn and Murray (1977) packed recombinant DNA molecules into bacteriophage particles *in vitro*. Bolivar *et al.* (1977) achieved the construction and characterization of new cloning vehicles which serve as a multipurpose cloning system. Sutcliffe (1979) reported complete nucleotide sequences of the *Escherichia coli* plasmid pBR322. Broda (1979) dealt with properties and uses of plasmids. Williams (1981) discussed the preparation and screening of a cDNA clone bank. Denniston and Enquist (1981) described in detail the technique and advantages of recombinant DNA.

Anderson and Diecumakos (1981) discussed scope of genetic engineering in mammalian cells. Cocking *et al.* (1981) describes various aspects of plant genetic manipulation. Old and Primrose (1981) gave principles of gene manipulation as an introduction to genetic engineering. Maniatis *et al.* (1982) gave methods of molecular cloning. Murai et al. (1983) observed that *phaseolin* gene from bean is expresed after transfer to sunflower via tumor-inducing plasmid vectors. Chilton (1983) discussed about a vector for introducing new genes into plants. Marx (1988) reviewed the progress on foreign gene transferred into maize. Tanksley et al. (1992) dealt with consturction and usefulness of high density molecular linkage maps. Gasser and Fraley (1992) discussed production of transgenic crops. Gheysen et al. (1992) described transgenic plants by *Agrobacterium tumefaciens*-mediated transformation and its use for crop improvement. Gupta et al. (1996) discussed microsatellites in plants and considers them a new class of molecular markers.

Andreason and Evans (1988) worked on introduction and expression of DNA molecules in eukaryotic cells by electroporation. Kilbane et al. (1991) achieved instantaneous gene transfer from donor to recipient microorganisms via electroporation. Mannino and Gould-Fogerite (1988) achieved liposome-mediated gene transfer. Lasic (1992) described use of liposomes in gene transfer. Arnheim and Erich (1992) described polymerase chain reaction strategy. Mullis et al. (1994) discussed the polymerase chain reaction technique.

Watson and Tooze (1981) while relating the DNA story, described a documentary history of gene cloning. Motulsky (1983) gave his views on the impact of genetic manipulation on society and medicine. Johnson (1983) produced human insulin with the use of recombinant DNA technology. Wu et al. (1989) discussed the details of recombinant DNA methodology. Watson et al. (1992) dealt in length the recombinant DNA technology. Cooper (1994) reviewed the current status of the human genome project. Cohen and Hogan (1994) dealt with the new genetic medicines. Cole-Turner (1995) discussed religion and gene patenting.

McBridge and Peterson (1980) obtained chromosome-mediated gene transfer in mammalian cells. Herbers et al. (1981) microinjected cloned retroviral genomes into mouse zygotes, integration and expression in the animal. Marx (1982) commented on the building of bigger mice through gene transfer. Southern (1984) discussed DNA sequences in relation to chromosome structure. Smithies et al. (1985) achieved insertion of DNA sequences into the human chromosomal β-globin locus by homologous recombination. Ou-Lee et al. (1986) observed expression of a foreign gene linked to either a plant virus or a *Drosophila* promoter, after electroporation of protoplasts of rice, wheat and sorghum. DePamphilis (1988) used microinjecting DNA into mouse ova to study DNA replication and gene expression and to produce transgenic animals. McCoubrey et al. (1988) observed that microinjected DNA from the X chromosome affected sex determination in *Caenorhabditis elegans*. Shigekawa and Dower (1988) considered electroporation of eukaryotes and prokaryotes as a general approach to the introduction of macromolecules into cells. Williams et al. (1991) dealt with introduction of foreign genes into tissues of living mice by DNA-coated microprojectiles. Wilson (1995) described the extraction, PCR amplification and sequencing of motochondrial DNA from human hair shafts. Pennisi (1997) described production of transgenic lambs. Valander et al. (1997) discussed the production of transgenic livestock as drug factories.

Questions

1. Define of genetic engineering. What is the scope of genetic engineering?
2. Define recombinant DNA technology? What is the role of restriction endonucleases in construction of recombinant DNA molecules?

480 Basic Genetics

3. What is difference between cohesive (sticky) and blunt ends. Show by using cohesive ends created by a restriction endonuclease of your choice their involvement in construction of a recombinant DNA molecule.
4. What is a vector? What different types of vectors have found application in genetic engineering?
5. Define and describe the technique of gene/DNA cloning. What avenues have this technique opened for geneticists?
6. What do you understand by gene transfer? Describe the following methods of gene transfer: chemical method, microprojectiles, *Agrobacterium tumefaciens*-mediated gene transfer, DNA-mediated gene transfer, microinjection, co-transformation, transfection, electroporation, chromosome-mediated gene transfer and liposome-mediated transfer.
7. What is chromosome walking? Describe applications of this technique?
8. What are microsatellites? What place microsatellites have in modern genetics?
9. What is polymerase chain reaction technique? In what different ways is this technique is helpful?
10. Define transposons, transposable genetic element, transposase, transposition, transposition immunity and transposone tagging. What are various properties of transposons? For what purpose transposons have been used in molecular genetics?
11. What do you understand by high density molecular maps? How was such maps constructed?
12. What are variable number of tandem repeat loci? What is the significance of these loci?
13. Describe various applications of genetic engineering.
14. Describe various hazards of genetic engineering.

EVOLUTION

Various theories have been put forward from time to time to explain the process of evolution. These theories will be dealt with briefly.

Lamarck's Theory of Evolution

This theory was given by Lamarck (1744-1829). He was first to draw a phylogenetic tree to illustrate the relationships between the principle groups of animals, with the simplest forms being placed at the bottom and man at the top. He was convinced that evolutionary changes occur at a slow rate. He thus believed that species were not static but were derived from pre-existing species. In 1802 he published the first truly mechanistic theory of evolution called theory of **"use and disuse"** or the **"inheritance of acquired characteristics"**. To explain mechanism of evolution he believed that in response to changes in ecological conditions, certain new needs are caused. These needs are met by modifications of old organs. By continuous use of these structures, they increase in size and functional capacity and by disuse they may degenerate and become lost. Changes brought about by the environment during life time of an individual become heritable and thus can be transmitted to next generation.

Lamarck explained long neck of modern giraffe by his theory. During times when food was scarce, the short-necked ancestors needed to stretch their necks to get at leaves higher up in the trees. This stretching modified their heredity and allowed the production of offspring with slightly longer necks. In each generation, the exercising of the neck in this fashion caused an increment of hereditary change to the added to that contributed by previous generations. This theory may sound plausible, but it is unsupported by any scientific evidence. Cleland (1972) explained cytogenetic and evolutionary aspects of *Oenothera* which do not support Lamarck's theory.

Steele (1976) took two strains of mice CBA and A/J. First he immunized CBA males by infecting with 100 million spleen/bone marrow cells of A/J strain. He crossed the immunized batch of CBA males with normal CBA females. He claimed that as many as 50 to 60 per cent progeny inherited the immunity to A/J cells! He discussed his results with many scientists. He published his work in the *Proceedings of the National Academy of Sciences, USA* in January 1980 issue Paper was communicated by Howard Temin, a Nobel laureate. Medawar from British scientific arena repeated experiments of Steele but found negative results which were published in British journal *Nature* in 1981. Editorial published in *Nature* (19 February. 1981) dismissed Steele's data as "incomplete". The British scientists are content saying "Lamarckism is dead, long live Jean Baptiste Lamarck".

Theory of Natural Selection

Charles Robert Darwin was born in Shrewsbury on February 12, 1809. A great naturalist of nineteenth century, as a child he did not enjoy regular schooling. He went to Cambridge and studied geology and botany. He undertook a voyage in H.M.S. Beagle in (1831-1836) which sailed from England. In Brazil, whenever ashore, he made collections of plants and animals. When aboard ship, he took samples of marine life. He took notes on everything, including geological formations. When Beagle arrived at Galapagos Islands, about 600 miles off the coast of Equador, Darwin made some of his most important observations. He noted that each Island had its own kind of finch, found no where else in the world. Some had beaks adapted to eating large seeds, others to eating small seeds; some had parrot-like beaks adapted to feeding on buds and fruits; still others had slender beaks for feeding on small insects. All the Galapagos **finches** closely resembled certain species of finches on nearby mainlands. Darwin concluded that the closely related species of finches evolved from an original by colonization of the islands by similar members of mainland populations. The only mechanism he saw could drive **organic evolution** was the process of natural selection. He stated that most young animals die, and only those survive which adapt themselves well to their environment. From 1838 to 1841, Darwin was Secretary to the Geological Society and was in frequent contact with geologist Sir Charles Lyell, whose book *Principles of Geology* influenced Darwin's work.

Darwin noted that most traits are not uniformly expressed in natural populations and that variations are common. He, however, was not aware about the question of origin of **hereditary variations**. Some of the hereditary variations may aid an individual in finding or capturing food, or escaping from predators or help in some way exploiting its environment which would make a particular individual more fit than other in surviving under specific circumstances. In **struggle for existence**, individuals possessing useful variations would be selected for and those lacking useful variation would be selected against. Thus there will be **survival of the fittest**.

Each generation would be subject to selective process where individuals with more favourable variations would tend to leave more highly adapted progeny than those with less favourable variations. The idea of natural selection operating over long periods of geological time, seemed to provide an explanation for the origin of different species. After completion of voyage in 1836, for more than 20 years he collected data supporting his theory. Components of theory of **natural selection** are shown in **Fig. 26.1**.

Fact 1 Potential exponential increase of populations (Superfecundity)	Inference 1 Struggle by existence among individuals	
Fact 2 Observed steady-state stability of populations	Fact 4 Uniqueness of the individual	Inference 2 Differential survival i.e., Natural Selection
Fact 3 Limitation of resources	Fact 5 Heritability of much of the individual variation	Inference 3 Through many generations : Evolution

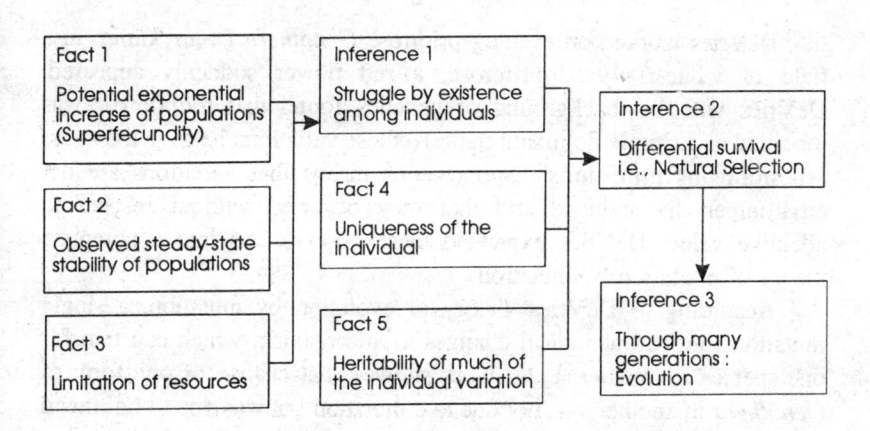

Fig. 26.1. Components of theory of natural selection followed by a certain order of Darwin's thinking that lead him to the formulation of his theory

A.R. Wallace (1823-1913), a naturalist, working in Indonesia, arrived independently at the conclusion that species were mutable, and that natural selection was the force principally responsible. In 1858, Darwin's and Wallace's papers were read before Linnean Society in London. Darwin published book *The Origin of Species* in 1859. He extended his general thesis to man in 1871. Charles married in 1839. Later years of his life were spent in the demonstration of his theories. He received the Prussian order Pour le Merite in 1871. He became a member of French Academy in 1878. He died on April 18, 1882.

There are some conditions for operation of natural selection. Before natural selection plays a role there has to be a radical change in the environment. The organisms must be pre-adapted to new environment.

There were many landmark discoveries after Darwin published his work. Principles of inheritance (Mendel's work and its rediscovery) explained mechanics of inheritance, Weismann's **germ plasm theory** gave a blow to theory of inheritance of acquired traits, chromosomal theory of inheritance pointed out chromosome as bearers of hereditary units, genes. Question that Darwin and Mendel left unanswered was where the inherited differences come from? DeVries answered this question in his theory.

Mutation Theory of Evolution

DeVries worked on evening primrose *Oenothera lamarckiana*. In a field of white-flowered primrose, a red flower suddenly appeared. DeVries with his background of genetics found that red flower was inherited as a simple dominant trait. To these variations he gave the name "**spontaneous mutations**". Spontaneous means that variations are not environmentally induced and that they occurred without respect to adaptive value. DeVries expanded these observations into a complete theory of evolution by mutation.

According to DeVries' theory of evolution by mutation, a single mutation can lead to radical changes in an organism which can transfer one species to another. Later work showed that change of one form of *Oenothera* to another was not due to a mutation but was due to **balanced system** in the organism in form of translocations. Today we know that much evolution can occur without new mutations. DeVries work could not thus give generally acceptable theory of evolution but he certainly contributed significantly in explaining the origin of heritable differences due to spontaneous mutations.

Modern Synthetic Theory of Evolution

Dobzhansky published *Genetics and Origin of Species* in 1937. As none of the existing theories could explain the process of evolution satisfactorily, he emphasized the need of experimentally testing the existing theories and the experimentally proven components of these theories should be put together to synthesize a new theory of evolution. Thus, modern concept of evolution combines various theories into a unified system. It combines the idea of Mendelian genetics, the phenomenon of mutations, the concept of Darwin's natural selection and knowledge gained from molecular cytogenetics and molecular genetics. Several other branches of biology, viz., biochemistry, paleontology, ecology, have also contributed in understanding and solving evolutionary problems. The modern synthetic theory recognizes the following basic processes.

Mutations

Both gene mutations and chromosomal mutations have played role in evolution. **Gene mutations** include nucleotide substitutions or very small additions or deletions. These add to the number of alleles at a locus. Changes in chromosome structure include gross **chromosomal rearrangements** like additions, inversions and translocations. Euploidy, aneuploidy, fusion and fission lead to changes in chromosome number. Gross chromosomal rearrangements change linkage position between genes. They may also alter regulation of gene expression. The classification of mutations relevant to evolution has been given in Chapter 21. Gene and chromosome mutations are ultimate source of all genetic variability. Mutation is a very slow process that, by itself, changes the genetic constitution of populations at a very low rate.

Assume there are two alleles, A_1 and A_2, at a locus, A_1 mutates to A_2 at a rate u per gamete per generation. Assume also that at a given time the frequency of A_1 is p_0. In the next generation, a fraction u of all A_1 alleles become A_2 by mutation. The frequency (p_1) of A_1 after mutation will be the initial frequency (p_0) minus the frequency of mutated allele (up_0), or $p_1 = p_0 - up_0 = p_0(1-u)$. In the following generation, a fraction u of the remaining A_1 allele (p_1) will mutant to A_2, the frequency of A_1 will become $p_2 = p_1 - up_1 = p_1(1-u)$. Substituting the value of p_1 obtained above we have $p_2 = p_1(1-u) = p_0(1-u)(1-u) = p_0(1-u)^2$. After t = generations, the frequency of A1 allele will be $p_t = p_0(1-u)^t$. Because 1-u is less than 1, as it increases, pt becomes even smaller. If this process continues indefinitely, the frequency of A_1 allele will eventually decrease the zero.

Mutations, however, are often reversible: allele A_2 may mutate back to A_1, If A_2 mutants to A_1 at a rate v and as before A_1 mutates to A_2 at a rate u, and frequencies of A_1 and A_2 at a given time are p_0 and q_0, respectively, then after one generation, frequency of A_1 will be $p_1 = p_0 - up_0 + vq_0$. Change in the frequency of A_1, represented by ΔP will be $p_1 - p_0$. Substituting value of p_1, $\Delta P = (p_0 - up_0 + vq_0) - p_0 = vq_0 - up_0$. An equilibrium will exist when $\Delta P = 0$. If we represent equilibrium frequencies by p and q, up = vq. At equilibrium p + q = 1. up = v (1-p) and

$$p = \frac{v}{u + v}$$

Similarly,

$$q = \frac{v}{u + v}$$

Allelic frequencies are usually not in mutational equilibrium, because other process affect them. Natural selection may favour one allele over the other. Allelic frequencies change at a slower rate when there is forward and backward mutation than when mutation occurs in only one direction. Thus mutation by itself takes a very long time in order to effect any substantial change in allele frequencies.

Amplification of Genetic Variation

Amplification at Gene Level. Though mutation is primary source for creating genetic variability, it is not enough, for evolution. One type of amplification is at gene level and the other type is at genomic level. After induction of mutation, genetic variability is amplified at different levels. A mutation produces a new allele (m_1) in certain site of wild-type (+). Another mutation produces another allele (m_2). Assume, there was a heterozygote possessing two mutations m_1 and m_2 in heterozygous form crossing over takes place within two mutant sites, we get one chromatid that is non-mutant and another is mutant at two sites. Thus a new mutant allele has been added to a population by **intragenic recombination**, without a new mutation. Intragenic recombination was first described by Lewis and Oliver (1940). Role of intragenic recombination in generating genetic variability was emphasized much later by Ohno (1970) in which he called it as a **mutational-like event**, since origin of third allele ($m_{1,2}$) involves mutations. If there by a third **unisite mutation** (m_3) the allele $m_{1,2}$ will, upon intragenic recombination give rise to a new allele $m_{1,2,3}$.

Table 26.1 Amplification of genetic variation

No. of unisite mutant alleles	Expected number of alleles mutant at sites				No. of multiple site alleles	Total No. of alleles (n)	Total No. of genotypes
	2	3	4	5			
0	0	0	0	0	0	1	1
1	0	0	0	0	0	2	3
2	1	0	0	0	1	4	10
3	3	1	0	0	4	8	36
4	6	4	1	0	11	16	136
5	10	10	5	1	26	32	528
n	$^nC_2$	$^nC_3$	$^nC_4$	$^nC_5$	$2^n-(n+1)$	2^n	$n+nC_2$

Table 26.1 shows extent of amplification of genetic variation due to intragenic recombination. Some generalization can be drawn from intragenic recombination. Intragenic recombination adds new alleles only when there are at least two mutant alleles, affecting different amino acid residues. Alleles produced by intragenic recombination are **multisite mutant alleles**. When there are at least three mutant alleles, affecting three different amino acid residues, the number of alleles contributed by intragenic recombination is more than the number of unisite mutant alleles. Corresponding to an arithmetic increase in the number of unisite mutant alleles, there is a geometric increase in the number of multisite mutant alleles.

Intragenic recombination occurs at a frequency of 1.4×10^{-2} at the *Esterase-2* locus in *Zaprionus paravittiger* (Kumar, 1978); 2.10^{-3} for the locus controlling blood group B system in cattle (Stormont, 1965); 2.6-2.9 X 10^{-2} for one of the two loci controlling LDH in brook trait (Wright and Antherton, 1968) and 0.6×10^{-2} for the dehydrogenase in quail (Ohno et al., 1969). Coyne (1976) uncovered 23 alleles of *xanthine dehydrogenase* gene in *Drosophila persimillis*. Singh et al. (1976)

uncovered 37 alleles for the same enzyme in *D. pseudoobscura*. Johansson and Rendel (1968) uncovered more than 300 alleles for the blood group B locus in cattle.

If two mutation are very near, recombination frequency is very low. So frequency of intragenic recombination is naturally quite low. But it is not really that low.

Are Multisite Alleles Present in Population? Intragenic recombination results in multisite mutants. Clear-cut evidence that multisite mutants are present comes from human haemoglobin. Out of 200, 182 mutation are single site mutations. If Table 26.1 is correct large number of multisite mutations alleles may appear. But only three 2 site mutations alleles have been, observed in human haemoglobin. One multisite allele Haemoglobin J-Singapore mutation of ∝-polypeptide chain at 68th position where asparagine is replaced by Aspartic acid and at 79th position alanine is replaced by glutamic acid. Haemoglobin-C-Harben mutation in β-polypeptide chain at 6th position replaces glutamic acid with valine, and at 73rd position Aspartic acid is replaced with asparagine. Another multiple site allele haemoglobin-Arlington Park in β polypeptide at 6th position replaces glutamic acid with lysine and at 95th position replaces lysine with glutamic acid. If frequency of mutations is low, probability that we will get heterozygote frequency also low. For intragenic recombination to make contribution, the alleles may have to be present in considerable frequency.

Electrophoretic bands also provide evidence for existence of multiple alleles in populations. When an acidic amino acid is replaced by a neutral amino acid, or when a neutral amino acid is replaced by an acidic amino acid , or when a basic amino acid is replaced by a neutral amino acid, or when a neutral amino acid is replaced by a basic amino acid, difference in electric charge between the wild type and mutant bands will be of one unit. But when a Acidic amino acid is replaced by a basic amino acid or vice versa, difference will be of two units. Single site mutations along with the ancestral allele can explain the existence of 5 such electrophoretic bands. How? In butterfly, *Colias eurytheme*, 13 different bands were observed. They can not be explained as single-site substitutions. To have three unit of difference, there must be five amino acid substitutions. Why? Since single site mutations can explain only five bands. Thus electrophoretic studies demonstrate that multisite mutations in certain cases must be present in a population. Electrophoretic studies provide evidence for existence of multisite mutation in a population, but does not say anything about their origin. One most likely mechanism discussed earlier is intragenic recombination. **Successive mutational** events is another possibility. Although there is greater probability for intragenic recombination, there is no way of excluding the second one.

Amplification at Genome Level. Through independent assortment of unlinked genes and crossing-over between linked genes and within genes enormous new genotypic combinations are generated **(Table 26.2)**. Genetic Recombination depends upon crossing-over frequency of which depends upon the distance between two genes in question. Thus, independent segregation and crossing-over lead to recombination of genes.

Table 26.2 The number of diploid genotypes than can be produced by recombination among various numbers of separate genes, each of which possesses various numbers of allele

Number of alleles of each ene	Number of genes					
	2	3	4	5	n
2	9	27	81	243	3^n
3	36	216	1296	7776	
4	100	1000	10000	100000	
5	225	3375	50625	759375	
6	441	9261	194481	4084104	
7	784	21952	614656	17210368	
8	1296	46656	1679616	60466176	
9	2025	91125	4100625	184528125	
10	3025	166375	9150625	503284375	
y r	$[r(r+1)/2]^2/2$	$[r(r+1)/2]^3/2$	$[r(r+1)/2]^4/2$	$[r(r+1)/2]^5/2$	$[r(r+1)/2]^n/2$

Source: Grant (1963)

Natural selection acts at the phenotype. Total phenotype of individual results from genotype x environment. So genotype is important for natural selection. Genotype of an individual is reflected by genome. If variation at gene level is expressed in genome, it gets amplified. Different types of genomes are formed in different frequencies. In case of non-linked genes, considering only two pairs for sake of simplicity, the four types of gametes are formed in the same frequency whereas in case of linked genes, the four types of gametes are not formed in same frequency unless the two gene pairs are greater than 50 map units apart. Random union of gametes generates further variability by process of random union of gametes. If an individual is heterozygous at n loci, the number of different types of gametes in 2^n and number of different types of diploid genotypes is 3^n.

Migration and Hybridization. Migration includes **emigration** and **immigration**. Migration, or **gene flow**, occurs when individuals move from one population to another and interbreed with the latter. Assume that individuals from surrounding populations migrate at a certain rate into a local population and they interbreed with the residents. The proportion of migrants is m, so that is next generation (1-m) of the genes are descendants of residents, and m are descendants of migrants. Assume that in a surrounding population, a certain allele A_1 has an average frequency P, while in the local population it has the frequency p_0. In next generation, the frequency of A_1 in the local population will be:

$$p_1 = (1-m) p_0 + mP$$
$$= p_0 - m (p_0-P)$$

Change Δp in allele frequency is

$$\Delta p = p_1 - p_0$$

Substituting value of p_1

$$\Delta P = p_0 - m (p_0-P) - p_0$$
$$= -m (p_0 - P)$$

ΔP will be zero only when either m or p_0-P is zero

When m = 0, there is no migration

when $p_0 - P = 0$, then $p_0 = P$

Effect of t generations of migration

$$p_t = (1-m)^t (p_0-P) + P$$

Next step is hybridization between local population and immigrants. Both these accessory processes increase amount of genetic variability available to a population.

Random Genetic Drift. This refers to changes in gene frequencies due to sampling variation from generation to generation. Random genetic drift is a process of pure chance and represents a special case of sampling errors. Magnitude of the errors due to sampling is inversely related to the sample size. Smaller the sample size, larger the effects; smaller the number of breeding individuals in a population, larger the allele frequency changes due to genetic drift are likely to be. Amount of variation that would be found among different samples

$$s^2 = \frac{pq}{2N}$$

Where, p and q are the frequencies of two alleles; N is number of parents and s^2 is variance. The equation shows inverse relationship between sample size, 2N, and the expected variation. The cumulative effect of random genetic drift over indefinite number of generations, all populations become fixed, approximately half for each allele.

Founder Effect and Bottlenecks. Extreme cases of random genetic drift occur when a new population is established by only very few individuals. This has been called by Mayr the **founder effect**. Populations of many species on oceanic islands, although they exist of millions of individuals, are descendants of one or very few colonizers arrived long ago by accidental dispersal. Situation is similar in lakes, isolated forest and other ecological isolates. Chance variation in allelic frequencies similar to those due to the founder effect, occur when populations go through **bottle necks**. When climatic and other conditions are unfavourable, populations may be drastically reduced in numbers and run the risk of **extinction**. Such population may later recover their typical size, but the allelic frequencies may be considerably changed.

Assortative Mating. This type of mating occurs when choices of mates are affected by the genotypes. For example, marriages in the U.S. are assortative with respect to racial features. Assortative matings also occur in organisms other than man. An extreme from of assortative mating is **self-fertilization**, which is the most common form of reproduction in many plants. Thus assortative mating prevails where matings do not occur at random. Assortative mating, but itself does not change gene frequencies, but it does not change genotypes frequencies. If the probability of mating between like genotypes is greater than expected from randomness, the frequency of homozygotes will increase; if the probability is smaller, the frequency of homozygotes will decrease.

Natural Selection

Selection can be defined as **differential reproduction** of alternative genetic variants. Natural selection may act on one or more **components of fitness**: differential mortality in early and juvenile stages, differential mortality in adult stage, Differential viability, differential mating drive and mating success, differential fertility, and differential fecundity. **Adaptedness** of an individual rarely if ever depends upon the independent action of individually genes. Products of different genes interact. Different genes are retained in a population depending upon their ability to form favourable combinations with other genes. Evolution takes place through alterations of the frequencies of genes in a population.

Selection acts directly on phenotypes and only indirectly on the underlying genes. Between the direct act of selection and the resulting change in gene frequency lies whole complex train of events by which gene action is translated into phenotypic characteristics. In order to understand the working of natural selection, one must understand phenotypic expression of genes from various angles. The relativity of selective value, phenotypic modifiability, pleiotropy, expressivity modifiers and dominance modifiers are some such aspects.

Genetic variability generated by gene mutations, changes in chromosomal structure and number and genetic recombination is acted upon by natural selection. Without genetic variability change can not take place. Natural selection cleanses the population of unfavourable genes, and builds up adaptive complexes of genes. Genetic variability is used as a raw material to be guided by natural selection into various adaptive channels (**Fig. 26.2**).

Directional Selection. Selection resulting in a shift in the population mean for the character considered. Selection when an allele is being replaced by a superior one.

Disruptive Selection. Centrifugal selection in which two or more different genotypes are an advantage and intermediate types are at a disadvantage. Selection for phenotypic extremes occurs until a discontinuity is achieved.

Diversifying Selection. Selection in which two or several classes of genotypes have optimal adaptiveness in different subenvironments.

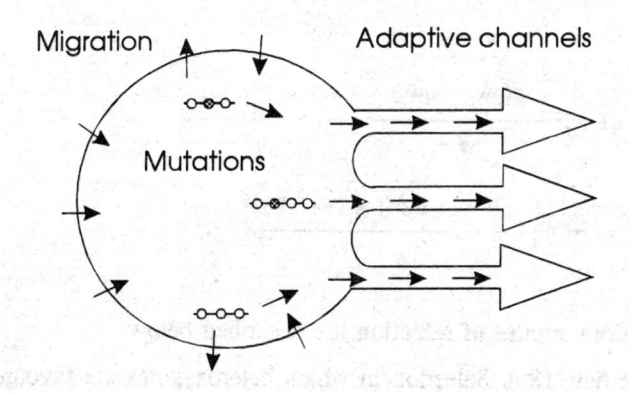

Fig. 26.2. Natural selection guides genetic variation into various adaptive channels

The generalized model of selection at a single gene locus is given in Table 26.3.

Table 26.3. Generalized model of selection at a single locus

	Genotype			Total	Frequency of A_2
	A_1A_1	A_1A_2	A_2A_2		
Initial zygote frequency	p^2	$2pq$	q^2	1	q
Fitness	w_1	w_2	w_3		
Contribution of each genotype to next generation	p^2w_1	$2pqw_2$	q^2w_3	w	
Normalized frequency	$\dfrac{p^2w_1}{w}$	$\dfrac{2pqw_2}{w}$	$\dfrac{q^2w_3}{w}$	1	q_1
Change in allele frequency					Δq

Where,

$$w = p^2w_1 + 2pqw_2 + q^2w_3$$

$$q1 = \frac{q(pw_2 + qw_3)}{w}$$

$$\Delta q = pq\frac{p(w_2 - w_1) + q(w_3 - w_2)}{w}$$

Various **modes of selection** are described below:

Balance Selection. Selection in which heterozygotes are favoured over homozygotes.

Frequency-dependent Selection. This type of selection is also known as **density-dependent selection** in which the adaptive value of a genotype is inversely related to its frequency.

Normalizing Selection. The removal by selection of all genes that produce deviations from the normal (= average) phenotype of a population.

Stabilizing Selection. Stabilizing selective is also known as **centripetal selection**. The elimination by selection of all phenotypic deviation from the population mean, and hence also of genes producing such deviating phenotypes. This is also a type of normalizing selection.

Effects of various modes of selection on some statistical parameters are summed up in **Table 26.4**.

Table 26.4. Effects of various modes of selection on some statistical parameters

Parameter	Directional	Stabilizing	Disruptive
Mean	Changes	No change	No change
Range (interval)	Changes	No change	No change
Peak	No change	Increases	Decreases

Reproductive Isolation

Mutation, recombination and natural selection can transform one species into another when extended over millions of years. But this does not lead to an increase in the number of species. For increasing the number of species there is requirement of **geographical isolation** followed by reproductive isolation. **Reproductive isolation** includes all the barriers to gene exchange between populations. Populations that are reproductively isolated from each other are almost certain to evolve in different directions. Populations that are not reproductively isolated, because of gene exchange they will evolve in the same direction. Various reproductive isolation mechanisms operate at premating or postmating levels.

Premating Mechanisms. These mechanisms are effective before mating; no wastage is involved in this case. Premating isolation mechanisms may be ecological, seasonal, ethological, mechanical or genetic.

Ecological Isolation. Ecological isolation is also known as **habitat isolation**. The two species of scarlet oak (*Quercus coccinea*) and black oak (*O. velutina*) are habitually separated. Scarlet oak inhabits swamps or poorly drained bottom lands having acid soils, while the black oak is found on drier, well-drained soils of the uplands in Eastern United States. Two closely-related species of dragonflies overlap in north Central Florida. The northern species *Progomphus obscurus* is restricted to rivers and streams, the southern species *P. alachuensis* inhabits lakes. The two species occupy different habitats in the same general territory.

Seasonal Isolation. Buffo amaricanus breeds early in the spring while *Buffo fowleri* breeds late. Two speices of pine *Pinus radiata* and *P. muricata* grow together on the Monterey Peninsula in California. The former species sheds its pollen in early February, the latter in April. Two species reach sexual maturity at different times of the year.

Ethological Isolation. In chambers with equal numbers of females and males of two species of *Drosophila bifasciata* and *D. imaii*, the freuency of matings observed were: *bifasciata* ♀ X *bifasciata* ♂, 229; *bifasciata* ♀ X *imaii* ♂, 13; *imaii* ♀ X *bifasciata* ♂, 9; *imaii* ♀ X *imaii* ♂, 375. It was noted that attraction between males and females of different species is either absent or very weak.

Mechanical Isolation Two species of *Salvia, S. apiaha* and *S. mellifera* overlap wisely in California. Flowers of *mallifera* are so constructed that at least 12 species of small and medium sized bees and one species of fly effectively pollinate them. Flowers of *apiana* require visits of larger carpenter or bumble bees to pollinate them. Flower parts in plants and genitalia in animals do not allow cross-pollination or cross-fertilization.

Gametic Isolation. This type of reproductive isolation is found in aquatic forms, e.g., in frogs, sea urchin, fertilization takes place in water. No attraction between gametes of different species was observed. Release of gametes in water may be stimulated by chemical substances produced by **conspecific** individuals.

Postmating mechanisms. These mechanisms are effective after mating, considerable wastage is involved in postmating isolation barriers. Postmating isolation mechanisms include gametic and zygotic mortality and hybrid inferiority, sterility, inviability and breakdown.

Gametic Mortality. The spermatozoa of foreign species may be incapacitated in the sexual ducts of alien females. This occurs, for example, in some species of *Drosophila* that have the so-called **insemination reaction**, i.e., swelling of the vagina by secretion of a fluid. At least superficially similar phenomena in the plant kingdom are failure of the pollen grains of one species to germinate on the stigma of another or slowness of growth of the pollen tube in styles of different species.

Zygote Mortality. Hybrids produced by crossing sheep and goat appear to be normal as early embryos but die much before birth. In the cross *Linum perenee X L. austriacum,* hybrid seeds fail to germinate. Hybrids, between *Rana sylvetica* and *R. pipiens* are completely inviable.

Hybrid Inferiority. This type of reproductive isolation been studied very little.

Hybrid Sterility. Hybrid male produced by crossing *D. pseudoobscura* and *D. persimilis* are sterile. Is the classical example of a vigorous but completely sterile hybrid.

Hybrid Inviability. Hybrids either do not survive or they do not reach sexual maturity.

Hybrid Breakdown. Production of weak or sterile F_2 progeny by vigorous fertile F_1 hybrids.

Molecular Evolution

Considerable work has been done to understand the genomic changes in different prokaryotic and eukaryotic organisms. In general there is no correspondence of **genomic evolution** with **morphological evolution**. For instance, bacteria with limited morphological variations might have experienced more genomic evolution. Conversely, higher forms with limited morphological repertoire might have undergone major genomic evolution.

Most studies of evolution at the molecular level have been conducted on structural genes as these are the ones that are transcribed to produce RNAs that either function as mRNA and are translated to produce proteins, or have functional roles, like rRNAs or tRNAs. The distinction between structural and regulatory genes is more or less arbitrary, because the proteins such as lac-repressor of *E. coli* is the product of a structural gene; however, its function is regulatory in that it directly controls the expression of a set of structural genes coding for enzymes.

Structural Gene Changes

Through electrophoretic analysis of structural genes it has been estimated that about twenty electrophoretically detectable allelic substitutions per hundred loci accumulate before reproductive isolation is completed. Immediately, question arises: Are these changes in structural genes solely important in development of isolating mechanisms? A more critical analysis of such studies has indicated that as a general rule only those structural genes whose products have some important regulatory function contribute significantly to the process of speciation. Otherwise, the **organismal evolution** and **structural gene evolution** go on at virtually independent rates. This view is supported by the indirect evidences which show that speciation can occur even without alteration of structural genes. And, therefore, for the evolution of new species, the evolution of new control mechanisms seems to be more important than the evolution of new proteins.

Regulatory Gene Changes

Regulatory genes code for RNAs or protein products whose function is to control the expression of other genes. Any change in these sequences would be classified under regulatory change. That the chief molecular barrier to development of hybrid organisms are regulatory has got enough support from the extensive studies of A.C. Wilson and his associates. It seems that evolution at organismic level depends predominantly on regulatory changes and, as mentioned earlier, structural gene changes may have secondary role in organismal evolution. The changes in regulatory systems may be caused as a result of a number of gene rearrangements, like, inversions, translocations, duplications, fusions or fissions, thereby changing the order of genes or the chromosomes. This may alter the pattern(s) of gene expression. Enough evidences are available to convince that the evolutionary changes of regulatory genes are primarily responsible for development of reproductive isolation and, thus, for the origin of new species. The noncoding sequences, transposable elements, repetitive DNAs and other such structural features of the various genomes have also been proposed to carry out important functions in the regulation of gene expression and thus contribute effectively to the process of speciation. Although the exact nature of **molecular barriers** to the development of interspefific hybrids is not known, the phenomenon of **allelic repression** occurring in hybrids shows that regulatory barriers may be very important. Comparison of serum albumins of each of the members of 31 pairs of mammalian species and 50 pairs of frogs that produce viable cross-species hybrids suggests that evolutionary loss of the ability of two species to hybridize probably results from the accumulation of incompatibilities between genes during embryonic development. Therefore, mammals appear to have undergone both rapid regulatory evolution and rapid rearrangement of genes. Thus, the **gene rearrangement** provides an important means of achieving new pattern of regulation.

A Model of Speciation

There are a variety of modes of speciation and mechanisms by which reproductive isolation can be achieved in animals and plants (**Fig. 26.3**). Analogous to the complex process of gene regulation, in the genetic control of reproductive isolation several mechanisms work in coordination with each other to prevent hybridization. Each isolating mechanism may be under the control of various genic or chromosomal factors, thereby, suggesting that species do not arise at a single step, as suggested by Hugo de Vries in 1909, but through accumulation of many different genetic changes over time. From this, however, it cannot be generalized that all the genic changes promote reproductive isolation as speciation occurs in diverse ways giving rise to diverse kinds of species.

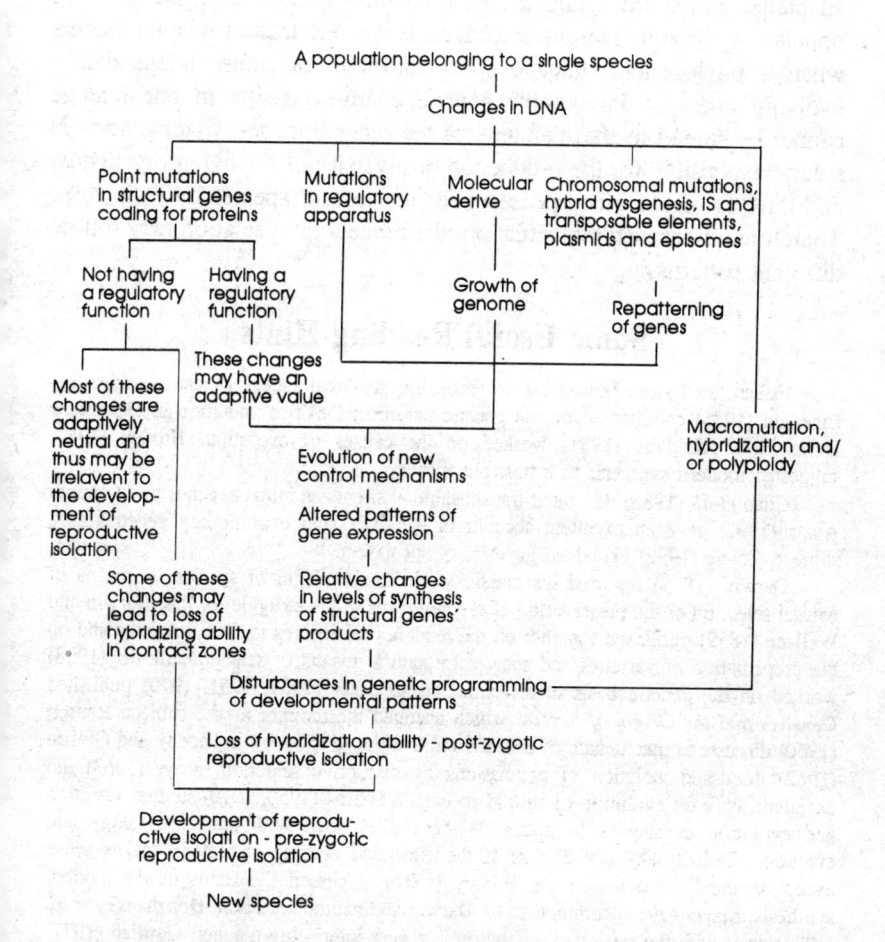

```
                   A population belonging to a single species
                                        |
                                 Changes in DNA
 ┌──────────────────┬──────────────────┬──────────────┬──────────────────────────┐
 Point mutations    Mutations           Molecular      Chromosomal mutations,
 in structural genes in regulatory      derive         hybrid dysgenesis, IS and
 coding for proteins apparatus                         transposable elements,
                                                       plasmids and episomes
 ┌────────┬─────────┐                          |                  |
 Not having  Having a                    Growth of           Repatterning
 a regulatory regulatory                 genome              of genes
 function    function
             |
             These changes
 Most of these may have an
 changes are  adaptive value
 adaptively
 neutral and                                                Macromutation,
 thus may be              Evolution of new                  hybridization and/
 irrelevant to            control mechanisms                or polyploidy
 the develop-                  |
 ment of                  Altered patterns of
 reproductive             gene expression
 isolation                     |
      |                   Relative changes
 Some of these            in levels of synthesis
 changes may              of structural genes'
 lead to loss of          products
 hybridizing ability           |
 in contact zones
             Disturbances in genetic programming
             of developmental patterns
                           |
             Loss of hybridization ability - post-zygotic
             reproductive isolation
                           |
 Development of reprodu-
 ctive isolati on - pre-zygotic
 reproductive isolation
             |
        New species
```

Fig. 26.3. Flow diagram depicting mechanisms responsible for origin of new species

During the passage of species evolution, two processes are recognized: (i) a species may change and become radically different through the process of mutation and natural selection. This is called **phyletic speciation** or **anagenesis**; and (ii) one species may split into two or more daughter species. This is called **true speciation** or cladogenesis. The physical evolution occurs within a given **phylogenetic lineage** as time proceeds. Such changes are gradual and result from increased adaptation to the environment. The favourable genetic changes spread in population through natural selection. However, true speciation occurs when a phylogenetic lineage splits into two or more independently evolving lineages. Favourable genetic changes arising in one lineage cannot be spread to the members of the other lineages. Cladogenesis is solely responsible for the production of diversity of the living organisms, and thus the process compensates for the loss of species by extinction. Therefore, under different situation, the process of speciation may follow different patterns.

Some Useful Reading Hints

Fisher (1930) did poineering work on the genetical theory of natural selection. Emerson (1935) commented on the genetic nature of DeVries' mutation in *Oenothera lamarckiana*. Haldane (1931) worked on the causes of evolution. Huxley (1942) suggested modern synthesis as a possible answer.

Chao et al. (1983) described transposable elements as mutator genes in evolution. Ayala (1992) gave an excellent account of evolution and evolutionary genetics as a subject. Coyne (1992) explained genetics of speciation.

Darwin (1859) reported his classic work on the origin of species by means of natural selection or the preservation of favored races in the struggle for life. Darwin and Wallace (1859) published together on the tendency of species to form varieties and on the prepetuation of vareiites and species by natural means of selection. Lerner (1958) worked on the genetic basis of selection. Dobzhansky (1937, 1951, 1970) published *Genetics and the Origin of Species* which changed the thought on the subject. Rensch (1960) discusseds mechanism of evolution above the species level. Thoday and Gibson (1962) discussed isolation of populations by disruptive selection. Mayr (1963) did excellent work on evolution of animal species. Stebbins (1950, 1970) studied variation and evolution of species in plants. White (1973) dealt with animal cytology and evolution. Dobzhansky (1973) was of the view that nothing in biology makes sense except in the light of evolution. Huxley (1974) discussed the status of the modern synthesis. Mayer (1977) commented on Darwin and natural selection. Dobzhansky et al. (1977) dealt with the subject of evolution in a very interesting manner. Gottlieb (1977) provided evidence for duplication and divergence of the structural gene for phosphoglucoisomerase in diploid species of *Clarkia*. White (1978) described various modes of speciation. Freeling (1978) discussed allelic variation at the level of intragenic recombination. Grant (1977, 1981, 1991) did basic work on the understanding of organismic evolution. Kumar (1978) observed intragenic recombination while studying genetics of esterase enzyme in *Zaprionus* paravittiger.

Satyanarayana (1981) gaves a fresh look at Lamarckism. Landman (1991) commented on the inheritance of acquired characteristics. Minkoff (1983) described various evolutionary processes in most readable manner. Gallagher (1989) described evolution of mitochondria and Gallagher (1991) described role of DNA repair in evolution. Gillespie (1991) dealt with the cause of molecular evolution. Hurst (1995) dealt with origin of introns and their role in molecular evolution.

Rose and Doolite (1983) described molecular biological mechanisms of speciation. Kumar and Miglani (1986) reviewed molecular mechanisms of biological speciation. Li and Graur (1990) described fundamentals of molecualr evolution. Selander *et al.* (1991) discussed evolution at the molecular level. Nowark (1995) discussed a new approach of proteome in evolution by looking at cell's proteins. Wilson *et al.* (1974) discussed two types of molecular evolution obtaining evidence from the studies of interspecific hybridization. Wilson (1975, 1978) described evolutionary importance of gene regulation. Wilson *et al.* (1977) dealt with biochemical aspect of evolution.

Questions

1. What was the first consistent theory of evolution? Describe and comment on various components of this theory.
2. What are the difficulties in accepting the concept of inheritance of acquired traits? Is Lamarckism dead for ever?
3. At one state in the history of evolution it was announced that Darwinism is dead? What impact did this development had on the theory of natural selection.
4. Write a brief account on the life of Sir Charles Darwin.
5. Depict the sequence in Darwin's thinking that led him to the formation of his theory.
6. Write a short note on natural selection.
7. What do you understand by the terms: struggle for existence and survival for the fittest.
8. Describe mutation theory of evolution. What are the weak points of this theory.
9. What is deVries' contribution in development of modern theory of evolution.
10. Describe briefly basic process recognized in the modern synthetic theory of evolution.
11. What is contribution of Th. Dobzhansky in idea formation, arrival and establishment of synthetic theory of evolution?
12. Discuss the role of intragenic recombination in the amplification of genic variation.
13. Discuss the following statements:
    (a) Intragenic recombination is a mutation-like event.
    (b) Polymorphism breeds more polymorphism.
14. Write down generalizations drawn from the Table 26.1 concerning amplification of genic variation.
15. Discuss the evidence for the existence of multisite mutations.
16. Discuss the role of independent segregation of non-homologous chromosomes, intergenic recombination of linked genes, and random union of gametes in amplifying genomic variation.
17. Definite: mutation, recombination, migration, random, drift, founder effect, bottlenecks, assortative meting, selection.
18. What is the ultimate source of all genetic variability? What role does process of mutation play by itself, in changing genetic constitution of populations?
19. How can genetic recombination convert a small initial stock of multiple gene variation into a much greater amount of genotypic variation.

20. Does linkage prevent formation of different types of gametes?
21. Discuss the following statements:
    (a) Increase in genotypic variation takes place, due to recombination, with increase in the number of alleles at each of the two separate loci.
    (b) Astronomically large number of genotypes can be produced by recombination between a moderate number of genes containing a few alleles each.
    (c) Recombination is a mechanism for generating tremendous amount of individual genotypic variation.
22. What is the effect of migration on the allele frequency in a population?
23. How do sampling errors change gene frequencies?
24. Comment: Smaller the sample size larger the effect on gene frequency.
25. How do "Founder effect" and "bottlenecks" affect gene frequencies?
26. Comment: Assortative mating, by itself, does not change gene frequencies but it does change genotypic frequencies.
27. Describe a generalized model of selection at a single gene locus.
28. What is the Darwinian concept of natural selection?
29. Name numerous components of fitness.
30. Does natural selection act on genes or phenotypes?
31. Understanding the working of natural selection requires understanding the phenotypic expression of genes from various angles, comment.
32. Name and describe various modes of selection.
33. Define the following: selection, fitness, directional selection, disruptive selection, stabilizing selection, balancing selection, sexual selection, undertime selection, group selection, population selection, kin selection.
34. What is the significance of reproductive isolation mechanisms in formation of species. At what levels in reproduction do the isolation mechanisms operate?
35. Describe various premating isolation mechanisms with examples.
36. Describe various postmating isolation mechanisms with examples.
37. What do you understand by the term molecular evolution?
38. Is there any correspondence of genomic evolution with morphological evolution?
39. What is the role of structural gene changes in evolution of species?
40. What role regulatory changes in genes play in evolution of species?
41. What types of mutations cause changes in regulatory systems of genes?
42. How important are mutations in introns compared to those in exons in evolution?
43. How does Britten-Davidson model of eukaryotic gene regulation fit in the present theory of evolution?

27

PROBABILITY AND CHI-SQUARE

Probability

Statistically, the possibility of happening of an event is known as probability. An event is certain to happen if all the cases are favourable to it and the probability is = 1 (unity). The probability of happening of that event goes on decreasing as the number of unfavourable cases goes on increasing, whereas its probability is zero, if it is certain that it will not happen. Thus, for all the events when the expected probability is less than one, it can fluctuate owing to **chance**. This can be verified by drawing small samples, from population, knowing to consist of different units in a particular ratio. These samples will never come out in the actually **expected number** of each class is never realized. The principles of probability based on certain assumptions or established principles, are used to calculate the expected number of different classes in a segregating population and X^2 test is employed to examine the goodness of fit between the observed and expected values.

Probability and Genetic Ratios

The relationship of probability to simple **genetic ratios** may be demonstrated by using seeds of two different colours. The equal numbers of brown and white beans are put in a container and samples are drawn at random. The probability of the first bean to be brown or white is 0.5. Similarly, the probability of the second bean to be brown or white is also 0.5. The probability that both beans will be brown is 0.5 X 0.5 = 0.25. But when a sample of say, 40 beans and the white ones are rarely found in equal ratio. This thing can be observed by drawing samples of 40 beans each, from the lot provided to you and see how much deviation is observed in each sample.

Deviations in Segregating Progeny

The exact number of individuals in each class (as expected) are never obtained. In order to test these **deviations** between the observed and expected values, the use of Chi-square is made.

Level of significance

Actually **observed ratios** depart to a greater or lesser extent from the expected. How great must be this departure in order to discard our hypothesis? To decide whether to accept or reject our hypothesis, we must, therefore, evaluate the size of discrepancy between the observed and expected ratios. Discrepancies can be small or large. If we want to assign values to these two kinds of discrepancies, we might say that large discrepancies are the largest 5 per cent, and small discrepancies are the remaining 95 per cent. We can, therefore, propose to reject the hypothesis if the discepancy falls into the large class. This 5 per cent frequency value that enables us to reject a hypothesis is called the 5 per cent level of significance. We may of course change the level of significance so that we will less easily reject a hypothesis.

Degrees of Freedom

The concept of degrees of freedom arises because in statistical analysis of many experimental results, such as genetic ratios, it is desirable to regard the total number of observed individuals in the experiment as fixed or given quantity. Composing this given quantity are the contributions of component classes in the experiment, some of which are variable or free, in respect to their numbers. For instance, if there are only two scored classes, such as tall and short plants, as soon as number of variable classes is set, the size of the other class is automarically determined. It does not matter which particular class is considered to be variable. The rule for biological experiments is simply that the degrees of freedom is equal to the number of classes less one.

Type I and Type II Error

There are two type of correct decisions in rejecting the hypothesis. If our **null hypothesis** (H_o) is correct then the decision to accept it will be correct. If we reject a true null hypothesis, we commit an error called **type I error**. If our null hypothesis is not true then rejecting it will be a correct decision. If we accept null hypothesis that is not true we commit another error called **type II error**.

Chi-square

This is used to test the goodness of fit the genetic ratios. The **expected number** of individuals in each class is worked out on the basis of a ratio. For example, in F_2 progeny of a monohybrid cross, the two phenotypes are expected to appear in 3:1 ratio when one allele is completely dominant over the other. The deviations observed are then tested for the probability of their occurrence. Chi-square test is applied on whole numbers, not on percentages or frequencies.

Procedure for Calculating Chi-square

When number of either of the expected classes is large, calculate Chi-square, using the following formula:

$$\chi^2 = \sum \frac{|Obs. - Exp.|^2}{Exp.}$$

Where obs. = the observed number in a class, Exp. = the expected number in a class, and $\sum$ = the summation of all k classes. Vertical lines on either side of values mean the absolute value of this difference. Look up the chi-square value from **Table 27.1** along the line for appropriate degrees of freedom. Determine the degrees of freedom (df) by subtracting

Table 27.1. Chi-square values

d.f.	Probabilities	
	0.05	0.01
1	3.84	6.64
2	5.99	9.21
3	7.82	11.34
4	9.49	13.28
5	11.07	15.00
6	12.59	16.81
7	14.07	18.48
8	15.51	20.09
9	16.92	21.07
10	18.31	23.21

one from the total number of classes (k). Thus df = k - 1. If the calculated value of Chi-square is less than the tabulated value at a given level of significance, we accept the **null hypothesis** (H_0), i.e., observed = expected, which leads to the conclusion that the segregation is according to the expected ratio and the deviation observed is due to chance only. But if the calculated value of Chi-square is more than the tabulated value at a given level significance, we reject the null hypothesis (H_0) and accept the **alternate hypothesis** (H_1), i.e., observed ≠ expected, which leads to the conclusion that the segregation is not according to the expected ratio and the deviation observed is not due to chance only. In such cases the investigator should explain the reasons for this failure. The investigator may put an other hypothesis and design an experiment to test the same.

Yates Correction Term

When number of either of the expected classes is small, the following formula is used for calculating chi-square.

$$\chi^2 = \sum \frac{[|Obs. - Exp.| - \frac{1}{2}]^2}{Exp.}$$

The reduction of ½ from the absolute value of observed - expected deviation is known as Yates correction term which adds to the accuracy of chi-square determinations. Vertical lines on either side of values mean the absolute value of this difference.

Contingency Chi-square

Some times it is desirable to compare one set of observations under particular conditions to those of a similar nature taken under different conditions. In this case there are no expected values. The question to be answered is whether the results are dependent (contingent upon) or independent of the conditions under which they were observed. The test is, therefore, called a **test for independence** or **contingency test**. To calculate a contingency chi-square, both the observed numbers and the marginal totals must be used If we call total number of observations N, the individual numerical contributions to this value a, b, c, and d, repectively, then the chi-square calculations are as follows:

	Categories of observations		Totals
	1	2	
A	a	b	a + b
B	c	d	c + d
Total	a + c	b + d	a + b + c + d = N

$$\chi^2 = \frac{[|ad - bc| - (\tfrac{1}{2})N]^2 N}{(a + b)(a + c)(c + d)(b + d)}$$

In case of contingency test also the conclusions are drawn as explained for simple chi-square test.

Some Useful Reading Hints

Steel and Torrie (1960) described principles and procedures of statistics. Fisher and Yates (1963) provided statistical tables for biological, agricultural and medical research. Goldstein (1964) provided a useful compilation of statistical methods. Snedecor and Cochran (1967) provided details of statistical methods. Dixon and Massey (1983), in their book *Introduction to Statistical Analysis*, described methods useful to biologists. Moore (1985) dealt with concepts and controversies in statistics.

Questions

1. Define probability. How does this concept relate to biological experiments?
2. How does probability help in expecting genetic ratios in a segregating progeny?
3. Why do the deviations occur in a segregating progeny?
4. Define Chi-square test. What is the formula used for computing the chi-square value?
5. What do the terms degree of freedom and level of significance mean?
6. Describe Type I and Type II errors.
7. What does a chi-square value given in Table 27.1 at a particular degree of freedom and level of significance mean?
8. What are the limitations of chi-square test?
9. What is Yates correction term?
10. What is contingency chi-square test? For what purposes is this test applied?

GLOSSARY

A

ABO blood group. Major blood antigen system containing types A, B, AB, and O, depending upon the presence or absence of the A and/or B antigens.

Activation reaction. A reaction using ATP as energy and conducted by an amino acyl synthetase to activate a specific amino acid. This reaction forms amino acid∼AMP complex.

Activator-Dissociator (Ac-Ds) system. Originally reported by Mc Clinctock in 1956. Responsible for turning the expression of genes on or off. Ds gene located on chromosome 9 causes breakage at the site of its location. It functions in the presence of an other gene Ac located on any other chromosome. Both Ac and Ds elements are transposable and regulate the activity of other genes. Ac-Ds system has been used for gene transfer in eukaryotes.

Activator RNA. A component of eukaryotic gene regulation which is product of an integrator gene. It complexes with a specific receptor gene.

Additive genes. Non-allelic genes affecting the same trait such that each adds to the effect of others in the phenotype.

Additive portion of genetic variance (V_A). Genetic variance due to the difference between homozygotes (for any locus). Heritability in the narrow sense is equal to V_A divided by the total phenotype variancee (V_P).

Adenine. A purine, one of the nitrogenous bases found in DNA and RNA.

Allele. One of two or more alternative forms of a gene that can exist at a single gene locus. Distinguished by their differing effects on the phenotype.

Allele frequency. A measure of the commonness of an allele in a population; the proportion alleles of that gene in the population that are of this specific type.

Allopolyploid. A polyploid having whole chromosome sets from different species.

Allotetraploid. A polyploid formed by doubling of chromosome number in a dihybrid between two organisms with different genomes or by fusion of diploid gametes of such organisms.

Allozymes. Forms of an enzyme, controlled by alleles of the same locus, that differ in electrophoretic mobility. See *Isozymes*.

Alternative RNA splicing. Different functional mRNA molecules are produced from the same primary transcript by differential removal of introns.

Amber codon. The nonsense codon UAG.

Glossary 509

Ames Test. A method which relies on reverse mutations of histidine-requiring (*his⁻*) auxotrophs in *Salmonella typhimurium* to wild type (*his⁺*) auxotophs. This method is widely used to test mutagenicity of environmental chemicals.

Aminoacyl or A-site. The site on the ribosome to which the aminoacyl-tRNA attaches during translation.

Aminoacyl-tRNA. One of the divers family of enzymes responsible for amicoacylation.

Amphidiploid. An organism produced by hybridization of two species followed by somatic doubling. It is an allotetraploid that appears to be a normal diploid.

Amplification at gene level. A mutation at some site produces a new allele. Another mutation at a different site produces another new allele. Intragenic recombination produces another new allele which carries both the mutations. Thus, several **unisite mutant alleles** as a result of intragenic recombination can produce two-site, and **multisite mutant alleles**. Multisite mutant alleles have been reported to be actually present in natural populations. This is thus a mechanism for amplifying genetic variation at gene level.

Amplification at genomic level. Variation at gene level gets further amplified at genomic level. Independent assortment, crossing-over and random union of gametes are the mechanisms for amplification of genetic variation at genomic level.

Anagenesis. The evolutionary process whereby one species evolves into another without any splitting of the phylogenetic tree. See **cladogenesis**.

Anaphase. The stage of nuclear division characterized by movement of chromosomes from spindle equator to spindle poles. It begins with longitudinal division of the centromeres and closes with the end of pole ward movement of chromosomes.

Anaphase I. The stage of meiosis I characterized by separation of chromosomes of the bivalents from spindle equator to spindle pole. Division of the centromeres does not take place. It closes with the end of poleward movement of chromosomes.

Anaphase II. The stage of meiosis II characterized by division of the centromeres and separation of sister chromosomes from spindle equator to spindle pole. It closes with the end of pole ward movement of sister chromosomes.

Ancillary sites. The sequence present between -30 and-40 position in prokaryotic or eukaryotic gene. In bacterial genes, it is known as cAMP-activated protein (CAP) site.

Aneuploidy. Variation in chromosomes number by whole chromosomes, but less than an entire; e.g., 2n + 1 (trisomy), 2n - 1 (monosomy).

Angstrom unit (Å). A measurement of length or distance, often used in describing intra- or intermolecular dimensions; equal to 1×10^{-10} meter and 1×10^{-1} nanometer.

Anticoding strand. The DNA strand that forms the template for both the transcribed mRNA and the coding strand.

Anticodon. The triplet of nucleotides in a tRNA molecule that is complementary to and base pairs with a codon in an mRNA molecule.

Antileader. A nucleotide sequence in DNA upstream of the initiation codon of a gene which is transcribed but not translated. Antileader generally contains Shine-Dalgarno sequence which helps in binding of mRNA to the ribosome.

Antiparallel strands. A term used to refer to the opposite but parallel arrangement of the two sugar-phosphate strands in double-stranded DNA; the 5'-3' orientation of one such strand is aligned along the 3'-5' orientation of the other strand.

Antisense DNA. According to recent terminology, a DNA strand that is used as a template during transcription.

Antisense RNA. According to recent terminology, RNA transcribed on sense DNA strand as a template.

Antitrailer. A nucleotide sequence in DNA downstream of the termination codon of a gene which is transcribed but not translated. Antileader generally contains signal sequence for attachment of poly(A) tail to RNA polymerase II transcript.

AP endonucleases. Endonucleases that initiate excision repair at apurinic and apyrimidinic sites on DNA.

AP sites. Apurinic or apyrimidinic sites resulting from the loss of a purine or pyrimidine residue from the DNA.

Artificial selection. Choosing by man, as far as possible, the genotypes contributing to the gene pool of succeeding generation.

Assortative mating. The mating of individuals with similar phenotypes.

Asynapsis. Failure of pairing during meiosis between homologous chromosomes.

Autogamy. This is the process that occurs in single unpaired animals through a special kind of nuclear organization. This process involves **internal fertilization** which leads to homozygosity in a single generation.

Autopolyploid. A polyploid all of whose sets of chromosomes are those of the same species.

Autoradiography. An image produced on a photographic film or plate due to radiation emitted by a radioactive substance.

Autosome. Any chromosome that is not a sex chromosome. Autosomes are similarly distributed in both sexes. Segregation of autosomes during meiosis does not normally affect determination of sex.

Autotrophic. An living system which obtains complex organic materials from surroundings and uses it as nutrition.

Auxotroph. An individual that, unable to carry on some particular synthesis, requires supplementing of minimal medium by some growth factor.

B

Backcross. The cross of a progeny individual with one of the parents.

Back mutation. A mutation that reverses the effect of a previous mutation by restoring the original nucleotide sequence.

Bacterial transformation. The transformation of a bacterium from one form to another, or the acquisition of a new characteristic by a bacterium, by the uptake of genetic material.

Bacteroid. A stage found in symbiotic nitrogen fixing bacteria after the bacteria are released from the infection thread.

Bacteriophage. A virus whose host is a bacterium.

Balanced lethal system. An arrangement of recessive lethal alleles that maintains a heterozygous chromosome combination. Homozygotes for any lethal-bearing chromosome perish.

Balance in *Drosophila* sex determination. Sex in *Drosophila* is determined by ratio of number of X chromosomes to number of sets of autosomes. X/A ratio of 1.0 is female, 0.5 is male, while 0.67 and 0.75 are intersexes, >1.0 is superfemale and <0.5 is supermale. Zygotes having X/A ratio <0.33 and >1.5 do not develop beyond embrynoic stage.

Balancing selection. Heterozygotes when superior to the corresponding homozygotes in gene vigour or some specific component of fitness have a selective advantage. Both alleles will be maintained and a state of balanced ploymorphism will be maintained.

Balbiani ring. A large RNA puff on giant chromosomes present during specific times of larval development in the chironomid family of Diptera.

Bands of polytene chromosomes. Bands are visible as dense regions that contain the majority of DNA in tightly coiled form; bands of normal chromosomes are relatively much larger and are generated in the form of regions that retain a stain on certain chemical treatments.

Bank, gene. A collection of recombinant DNA molecules containing inserts which together comprise the entire genome of an organism.

Bar eye. A phenotype of *Drosophila* characterized by a narrow, oblong, bar-shaped eye with fewer facets then wild type.

Barr body. The inactive, densely staining, condensed X chromosome, generally found next to the nuclear membrane, in nuclei of somatic cells of XX females. The number of Barr bodies in such nuclei is one less than the total number of X chromosomes. Barr body is absent in the nuclei of normal males.

Basc. A stock of *Drosophila melanogaster* which carries five mutations - a semidominant marker *Bar*, a recessive marker *white-apricot*, and three inversions covering entire X chromosome. Also known as **Muller-5 stock**, it is used to detect sex-linked recessive lethal mutations.

Base analogue. A chemical compound that is sturcturally similar to one of the bases in DNA and whiich may act as a mutagen.

Base pair. The hydrogen-bonded structure formed between two complementary nucleotides.

Basic number. The gametic chromosome number of a true diploid species. It is represented by X.

B chromosomes. The chromosomes which are found in addition to the normal chromosomes. Also known as **supernumerary** or **accessory** or **extra chromosomes**.

β-galactosidase. The enzyme that splits lactose into glucose and galactose and coded by a gene, z, in the *lac* operon.

β-galactoside acetyl transferase. An enzyme that is involved in lactose metabolism and encoded by a gene, y, in the *lac* operon.

β-galactoside permease. An enzyme that is involved in concentrating lactose in the cell and is coded by a gene, a, in the *lac* operon.

Bidirectional replication. Replication accomplished when two replication forks move away from the same origin in different directions.

Binomial expansion. $(a+b)^n$, where a and b are the probabilities of occurrence and nonoccurrence of an event and must total 1, and n is the number of times the event is tried.

Biogenesis. The axiom that life originates only from pre-existing life.

Biological evolution. Evolution of diverse forms of life from first simple living cell. It refers to change in diversity and adaptation of populations. Population adapts itself to environment. Its rate is slow.

Biometrical genetics. A branch of genetics that deals with the inheritance of quantitative traits using statistical concepts and procedures. Also known as **quantitative genetics, statistical genetics** or **mathematical genetics**.

Biological property/ Character/ Characteristic/ Trait. Some attributes of individuals within species for which various heritable differences can be defined.

Biotechnology. The use of living organisms, often but not always microbes, in industrial processes.

Biparental zygote. The zygote which contains mitochondrial or chloroplast genes from both the parents. Two alleles for some cytoplasmic genes may thus occur in the zygote. For example, a *Chlamydomonas* zygote that contains cpDNA from both parents; such cells generally are rare.

Bivalent. A pair of synapsed homologous chromosomes.

Blood group antigens. Those antigens that are principally, but not exclusively, found on erythrocytes.

Blunt end. An end of a DNA molecule, at which both strands terminate at the same nucleotide position with no single-stranded extension.

Bombay phenotype. A blood group in which individuals lack H-antigen.

Bottleneck effect. A brief reduction in size of a population, which usually leads to random genetic drift.

Branch migration. The process in which a crossover point between two duplexes slides along the duplexes.

Breakage and reunion. The general mode by which recombination occurs. DNA duplexes are broken and reunited in a crosswise fashion according to the Holiday model.

Bridge. Chromosome region between the two centromeres of a dicentric, which at anaphase is being pulled toward both poles of a spindle.

C

CAP site. A nucleotide sequence upstream of some bacterial genes and operons; the attachment point for the **catabolite activator protein**.

CAT box. A conserved sequence found within the promoter region of the protein-encoding genes of many eukaryotic organisms. It has the canonical sequence GGPyCAATCT and is believed to determine the efficiency of transcription from the promoter.

Glossary 513

Catenane. Interlocked circular DNA molecules that are the exclusive products of bacteriophage λ integrative reaction *in vitro* when substrate is supercoiled molecule containing both attachment sites, attP and attB.

cDNA cloning. A method of cloning the coding sequence of a gene starting with its mRNA transcript. It is normally used to clone a DNA copy of a eukaryotic mRNA. The cDNA copy, being a copy of a mature messenger molecule, will not contain any intron sequences and may be readily expressed in any host organism if attached to a suitable promoter sequence within the cloning vector.

cDNA library. A library of clones that have been prepared from mRNA after conversion into double-stranded DNA.

Carrier. A heterozygous individual. Term ordinarily used in case of complete dominance.

Cell. A basic unit of structure and function in all living organisms.

Cell cycle. The period between one cell division and the next. Divided into M (mitotic), G_1 (gap 1), S (synthetic) and G_2 (gap 2) phase.

Cell division. The process of reproduction or formation of new cell form the pre-existing cell in living organisms.

Cell cycle. The period between one cell division and the next.

Cell organelles. Various membrane-bound structures that are found within a cell.

Cell transformation. The alteration in morphological and bio-chemical properties that occurs when an animal cell is infected by an oncogenic virus.

CEN. A cloned eukaryotic centromere. Each CEN is given a number corresponding to the chromosome from which it was derived. This notation is confined to yeast molecular biology at the moment but, in principle, could be widely applied.

Centimorgan (cM). The map unit used to describe the distance between two genes on a chromosome. 1 cM is the distance that corresponds to a 1% probability of recombination in a single meiotic event.

Central Dogma. The key hypothesis in molecular genetics, proposed by Crick in 1958, which states that DNA makes RNA makes protein.

Centromere. The strucural feature of a chromosome that is the point at which the pair fo chromatids of the metaphase chromosome are held together.

Chargaff's rule. Chargaff's observation that in the base composition of DNA the quantity of adenine equaled the quantity of thymine and the quantity of guanine equalled the quantity of cytosine (equal purine and pyrimidine content).

Charged tRNA. A tRNA carrying an amino acid.

Charon phages. A phage which contains two nonsense mutations that prevent phage from growing in *E. coli* strains that do not carry an appropriate nonsense suppressor mutation. This limits phage growth only to special laboratory bacterial strains and thus acting as a safety feature aganst spreading of undesirable strains.

Chaperone. Molecular chaperone is a protein that is needed for the assembly or proper folding of some other protein, but which is not itself a component of the target complex.

Chiasma (Pl. chiasmata). A cross-shaped or X-shaped configuration of the chromosomes in a bivalent in prophase of meiosis I, usually the visible result of prior cytological crossing-over.

Chi form. An intermediate structure in recombination between DNA molecules.

Chimeric gene. A recombinant gene having regulatory sequence from one gene and coding sequence from another gene.

Chimeric plasmid. Hybrid, or genetically mixed, plasmid used in DNA cloning.

Chi-square (X^2) test. A statistical test for testing goodness of fit between expected (on the basis of a gene hypothesis) and observed numbers.

Chloroplast (cp) DNA. DNA present in chloroplasts. It is circular double stranded DNA (as in prokaryotes) and it replicates independently of the nuclear DNA by rolling circle method. Molecular weight = 55×10^6 to 97×10^6; Length = 419 µm; Size = 1.3×10^5 to 1.5×10^5.

Chromatid. A chromosome arm.

Chromatin. Originally, the deeply staining material present in the nuclei of cells and corresponding to the chromosomes. Now used more specifically to refer to the structural association between DNA and protein in chromosomes.

Chromocenter. is an aggregate of heterochromatin from different chromosomes.

Chromosomal mutation. Segments of chromosomes, whole chromosomes, or even entire sets of chromosomes may be involved in genetic change.

Chromosomal theory. The theory, stated in its most convincing form by Sutton in 1903, that genes reside on chromosomes.

Chromosome imprinting. A phenomenon where, in some organisms, paternal X chromosome is inactivated in some or all the cells. Chromosome imprinting is obviously heritable but is reversed in germ line. DNA methylation is involved in this process.

Chromosome puffs. Diffused, uncoiled regions in polytene chromosomes where transcription is actively taking place. See *puffs*.

Chromosomes. The structures, comprising DNA and protein, That are found in eukaryotic nuclei and which carry the bulk of the cell's genes.

Chromosome segregation. Separation of the members of a pair of homologues in a manner that only one member is present in any post-meiotic nucleus.

Chromosome walking. A technique for studying segments of DNA, larger than can be individually cloned, by using overlapping probes.

Circularization. A DNA fragment generated by digestion with a single restriction endonuclease will have complementary 5' and 3' extensions (sticky ends). If these ends are annealed and ligated the DNA fragment will have been converted to a covalently-closed circle or circularized.

***Cis-trans* complementation test.** A genetic analysis that tests whether two mutations lie in the same or different genes, and which can also provide infromation on dominance and recessiveness. The test involves introducing the two mutated genes into a single cell, for example, by introducing an F' plasmid carrying one mutated gene into a recipient bacterium with a chromosomal copy of the second gene.

Cistron. A segment of DNA specifying one polypeptide chain in protein synthesis. Under the concept of a triplet code, one cistron must contain three times as many nucleotide pairs as amino acids in the chain it specifies.

Cladogenesis. The evolutionay process whereby one species splits into two or more species. This is also known as **true speciation**. Also see *Anagenesis*.

Glossary 515

Classical genetics. Mendelian genetics; a study of genetics on an organismal rather than cellular level. Also known as **Mendelian genetics** or **transmission genetics**.

ClB method. A technique devised by Muller to rapidly screen in fruit flies recessive X chromosome lethal mutations. The ClB chromosome carries a recessive lethal (*l*), a dominant marker (B), and an inversion (crossover suppressor, C).

Clone. A population of identical cells, generally those containing identical DNA molecules.

Clone library. A collection of clones that contain a number of different genes. See **genomic library**.

Cloning vector. A DNA mloecule, capable of replication in a host organism, into which a gene is inserted to construct a recombinant DNA base paired.

Cloverleaf. A convenient two-dimensional representation of the strucrure of a tRNA molecule.

Code dictionary. A listing of the 64 possible codons and their translational meanings (the corresponding amino acids).

Coding strand. The DNA strand with the same sequence as the transcribed mRNA (given U in RNA and T in DNA). Compare with **anticoding strand**. According to recent terminology, coding strand is called **antisense strand**.

Codominance. The situation whereby both members of a pair of alleles contribute to the phenotype.

Codon. A triplet of nucleotides that code for a single amino acid.

Codon-anticodon relationship. This is alignment on the ribosome of an anticodon of amino acid charged-tRNA molecule against its complementary codon on mRNA.

Codon bias (preference). The idea that for amino acids with several codons, one or a few are preferred and are used disproportionately. They would correspond with tRNAs that are abundant.

Cognate tRNAs. tRNAs that are recognized by a particular aminoacyl-tRNA synthetase.

Cohesive end. An end of a double-stranded DNA molecule where there is a single-stranded extension.

Coincidence coefficient. The observed frequency of double cross overs, divided by their calculated or expected frequency. Expressed as a pure number; a measure of interference. In positive interference the coincidence is > 1.

Colchicine. An alkaloid that produces plyploids by dissolving spindle apparatus.

Colinearity. Refers to the fact that a gene and the polypeptide for which it codes are related in a direct linear fashion, with the 3'- end of the template strand of the gene corresponding to the amino-terminus of the polypeptide.

Colour blindness. Inability to distinguish certain colours. Red-green colour blindness is inherited as a sex-linked recessive gene in the X chromosome in human beings.

Col plasmids. Plasmids that produce antibiotics (colicinogens) used by the host to kill other strains of bacteria.

Combining ability, general. The average or over-all performance of a genetic strain in a series of crosses.

Combining ability, specific. The performance of specific combinations of genetic strains in crosses in relation to the average performance of all combinations.

Commaless code. Said of a genetic code in which successive codons are contiguous and not separated by noncoding bases or groups of bases.
Commensalism. An association in which one symbiont benefits and the other is neither harmed nor benefited.
Competitive exclusion. The principle that no two species can coexist in the same place if there ecological requirements are identical. This is also known as **Gause principle**.
Complementary genes. Genes which interact to produce an effect distinct from the effects of an individual gene. Complementary genes yield same mutant phenotype when present separately but when present together they interact to produce a wild type phenotype. Two heterogyzotes in F_2 progeny produce ratio 9 wild type : 7 mutant.
Complementary DNA (cDNA). DNA produced on an RNA template.
Complementary RNA. Synthetic RNA produced by transcription from a specific DNA single-stranded template.
Complementary sequence. A nucleotide sequence that base pairs with another nucleotide sequence.
Complementation map. A map developed from the complementation relationships between alleles, normally in a small segment of the chromosome.
Complementation matrix. A tabular representation of complementation tests involving a number of phenotypically similar mutants.
Complete linkage. A condition where no cross overs are observed in a test cross or F_2 progeny. Closely linked genes may show complete linkage. In some species crossing over is absent in one or the other sex. This is also an example of complete linkage.
Complete medium. A culture medium that is enriched to contain all the growth requirements of a strain of organisms.
Complete dominance. Resemblance of F_1 with one of its parents.
Complex gene. The gene or its protein product undergoes various rearrangements, cleavages or modifications before a functional product is obtained.
Compound gene. The genes in which coding sequences are separated by non-coding sequences. Also known as **split gene**.
Concept of dominance. The concept that dominance is not a universal property of a gene. It is relative property of a gene to produce a particular phenotype in a particular genetic and environmental background.
Confidence limits. A statistical term for a pair of numbers that predict the range of values within which a particular parameter lies.
Conjugation. Physical contact between two bacteria, usually associated with transfer of DNA from one cell to the other.
Conjugation mapping. The technique that allows the relative positions of bacterial genes to be mapped by determining the time it takes for each gene to be transferred during conjugation.
Cojugative plasmids. Bacterial plasmids which can be used to transfer genes to bacteria outside their own species.
Conjugative transposons. Those transposons which are transferred between different bacterial cells through conjugation.

Consanguineous mating. Mating between blood relatives. See **inbreeding**.
Consensus sequence. A sequence of nucleotides most often present in a DNA segment of interest.
Constitutive gene. A gene whose expression is not regulated. Its product is continuously synthesized by the cell whether or not substrate is present in the cell.
Constitutive heterochromatin. Heterochromatin that surrounds the centromere.
Contingency table. A table of frequencies showing two classifications simultaneously, and used in testing their independence.
Continuity of life. Germ plasm is continuous between all descendent generations. It occurs through reproduction.
Continuous replication. In DNA uninterrupted replication in the 5' to 3' direction using a 3' to 5' template.
Continuous trait. Quantitative trait; variation that can not be represented by discrete classes and requires measurement data. Multigenic or polygenic inheritance represent this type of variation.
Copy-choice hypothesis. An incorrect hypothesis that stated that recombination resulted from the switching of the DNA-replicating enzyme from one homologue to the other.
Corepressor. A small molecule that must bind to a repressor protein before the latter is able to attach to its operator site.
Co-transduction. The transfer of two or more genes on a single DNA molecule during transformation of a bacterium.
Co-transfection. The simultaneous transfection of two markers.
Co-transformation. The uptake of two or more genes on a single DNA molecules during transforming of a bacterium.
Coupling linkage. Tendence of two diminant characters, inherited by F_1 from one parent stay together in F_2.
Criss-cross inheritance. Transmission of a trait or an X- chromosome from heterogametic sex to homogametic sex and then back to heterogametic sex.
Cross-fertilization. Fertilization of a female gamete with a male gamete derived from a different individual.
Crossing-over. A process inferred genetically from recombination of linked genes in the progeny of heterozygotes. A process evidenced cytologically from the formation of chiasmata between homologous chromosomes during diplotene stage. Exchange of corresponding segments between chromatids of homologous chromosoms, by breakage and reunion, following pairing.
Crossover suppression. The apparent lack of crossing-over within an inversion loop in heterozygotes. Due to mortality of zygotes carrying defective crossover chromosomes rather than actual suppression.
Crossover unit. The per cent of recombination of linked genes. One per cent of recombination equals one crossover unit in a linkage map.
Crown gall. A tumor formed, usually, on the stems of broad-leaved plants when infected with *Agrobacterium tumefaciens* containing a Ti-plasmid. The bacterium is only necessary for the initiation of the tumor. The genome of the affected plant cells

contains several copies of a segment of the Ti-plasmid (the T-DNA). Crown galls can be of two types, **octopine** or **nopaline**, depending on the type of Ti-plasmid which initiated the tumor. Whole plants can be regenerated from crown-gall tissue and some of these will contain the T-DNA.

Cruciform structure. The cross-shaped structure that can arise by intramolecule pairing within a double-stranded DNA molecule that contains an inverted repeat.

Cryptic plasmids. Plasmids to which phenotypic traits have yet not been ascribed.

Cryptomorphic gene. The gene has a cryptic structure in that the ultimate active product or products are carried within the precursorial molecule. The active protein is released after enzymatic breakdown of the precursor followed by processing.

Cumulative effect. The action of two alleles of a gene giving a more pronounced effect than one in the heterozygous conditions. Probably one allele supplies insufficient enzyme for the reaction conditioned by two. The hybrids are distinguishable from parents. The effect is sometimes erroneously referred to as **incomplete dominance**.

Curly-Lobe-Plum Drosophila. A balanced lethal system in chromosome 2 in *Drosophila.* Used to detect autosomal recessive lethals of chromosome 2.

C-value paradox. The estimated number of genes in eukaryotes is much less than the amount of DNA present, e.g. haploid human genome has 2.8×10^7 bp and should contain 3×10^6 genes but the number estimated is 50,000. This anomalous situation is referred to as C-value paradox.

Cyclic AMP. A modified version of AMP in which an intramolecular phosphodiester within a double-stranded DNA molecule that contains an inverted repeat. It is synthesized from ATP by adenolate cyclase.

Cytogenetics. Combined study of cytology and genetics.

Cytogenic male sterility. Male sterility controlled by the interaction of cytoplasmic and nuclear genes.

Cytohet. A genetic condition where a zygote contains in its cytoplasm genetically different mitochondria contributed by two parents. Thus the individual is cytoplasmically heterozygous.

Cytokinesis. The division of the cytoplasm during cell division in mitotic cells.

Cytokinesis I. The division of the cytoplasm during meiosis I division giving rise to two haploid cells.

Cytokinesis II. The division of the cytoplasm during meiosis II division giving rise to four haploid cells.

Cytological map. The assignment of genes to particular chromosomes by observation of chromosomal morphology.

Cytology. The study of the structure and function of cells.

Cytoplasm. The protoplasm of a cell other than the nucleus.

Cytolasmic inheritance. Inheritance of characters whose governing genes are located in the cytoplasmic organelles like mitochondria, chloroplasts than in nucleus.

Cytoplasmic segregation and recombination (CSAR). A process suggested to explain assortment and recombination of organelle-based genes.

Cytosine. A pyrimidine, one of the bases found in DNA and RNA.

D

Darwinian evolution. A theory of evolution proposed by Darwin which explains evolution in terms of natural selection.

Degeneracy. Refers to the genetic code and the fact that most amino acids are coded for by more than one triplet codon.

Daughter chromosomes. The two chromosomes produced by the replication of a single parental chromosome.

Degree of dominance. Square root of ratio of H/D, where $H=\Sigma(d)^2$ and $D=\Sigma(h)^2$. d represents difference in performance between homozygous parents and their mid-parent and h represents difference in performance between F_1 and mid-parent. If this ratio is = 1, the trait shows complete dominance, if the ratio is > 1, the trait shows overdominance, and if the ratio is < 1, the trait shows incomplete dominance.

Degrees of freedom. An estimate of the number of independent categories in a particular statistical test or experiment.

Delayed segregation in Snails. Segregation occurs in F_3, rather than in F_2 since the phenotype of the individual is determined by the genotype of the mother. Example of delayed segregation is direction of coiling snails.

Deletion (deficiency). The loss of a part of a chromosome, usually involving one or more genes (rarely a portion of one gene).

Deletion mapping. Mapping mutation by use of overlapping deletion mutants to determine whether or not a mutation includes the site of a mutant gene.

Deoxyribonuclease. An enzyme that breaks a DNA polynucleotide by cleaving phosphodiester bonds.

Deoxyribonucleic acid (acid). A usually double stranded, helically coiled, nucleic acid molecule composed of deoxyribose-phosphate "backbones" connected by paired bases attached to the deoxyribose sugar; the genetic material of all living organisms and many viruses.

Desynapsis. Falling apart of chromosomes during diplotene or diakinesis which paired during zygotene and remained paired at pachytene.

Detection of linkage. Statistical analysis in which total X^2 is partitioned into various components to determine presence of linkage in F_2 or test cross progeny.

Developmental genetics. A study of the operation of genes during development.

Developmental homeostasis. The capacity of the developmental pathway to produce a normal phenotype in spite of developmental or environmental disturbances.

de Vriesism. The hypothesis that evolution in general and speciation in particular are the results of drastic mutations (saltations). Also see **mutation theory of evolution**.

Diakinesis. A substage of prophase I of meiosis I during which chromosomes have separated except at the ends. The homologous chromosomes are present as bivalents.

Dicentric chromosome. A chromosome or chromatid with two centromeres.

Dideoxynucleotide. A modified nucleotide that lacks the 3' hydroxyl group and so prevents further chain elongation when incorporated into a growing polynucleotide.

Differential segments. Portions of chromosomes that do not pair in meiosis.

Digenic inheritance. Pattern of inheritance observed in a cross between two individuals identically heterozygous at two gene pairs. It gives 9:3:3:1 ratio in F_2.

Dihybrid cross. A sexual cross in which the inhertance of two pairs of alleles is followed (Tall, Yellow X Dwarf, Green).

Dimorphic gene. A gene whose protein product exists in two forms - inactive and active; active form is produced by cleaving a part of polypeptide chain of the inactive protein.

Diploid. Having two copies of each chromosome.

Diplotene. A substage of prophase I of meiosis I during which chromosomes start separating and chiasmata are visible at the sites where crossing over had occurred.

Directional selection. A type of selection that removes individuals from one end of a phenotypic distribution and thus causes a shift in the distribution.0

Discontinuous gene. A gene in which the biological information is divided between two or more exons, separated by introns.

Discontinuous replication. In DNA the replication in short 5' to 3' segments using the 5' to 3' strand as a template while going backwards, away from the replication fork.

Discontinuous trait. Variation in which discrete classes are easily recognized, e.g., tall vs. dwarf plants.

Discontinuous variation. Variation that falls into discrete categories (e.g., the color of garden peas).

Disjunction. The separation of homologous chromosomes during anaphase-I of meiosis.

Disruptive selection. A type of selection that removes individuals from the center of a phenotypic distribution and thus causes the distribution to become bimodal.

Diversifying selection. Selection in which two or more genotypes have optimal adaptiveness in different subenvironments.

Dizygotic twins. Twins formed from two eggs fertilized at the same time.

DNA. Deoxyribonucleic acid, the genetic material in living systems except some plant viruses.

DNA-binding protein. Any protein that attaches to DNA as a part of its normal function, e.g., histone, RNA polymerase, *lac* repressor.

DNA clone. A section of DNA that has been inserted into a vector molecule, such as a plasmid or a phage chromosome, and then replicated to form many copies.

DNA crosslinking. Interstrand thymines of DNA form dimers thus blocking replication.

DNA G primase. An enzyme which synthesizes primer required for λ phage and *E. coli* DNA replication; this enzyme is localized in these systems by proteins λP and Dna C, respectively..

DNA A helicase. An *E. coli* protein which binds to replication origin site Ori C. Twenty to thirty molecules are required for the formation of initiation complex.

DNA helix A. Right-handed helix; rotation/base pair 33.6°; mean base pairs/turn 10.7; inclination of base to helix axis +19°; rise/base pair along helix axis 2.3Å; pitch/turn of helix 24.6Å, mean propeller twist +18°; glycosyl angle confirmation anti; sugar pucker conformaion c3'-endo. This form of DNA occurs at high humidity. Also see **alpha DNA**.

DNA helix B. Right-handed helix; rotation/base pair 38.0°; mean base pairs/turn 10.0; inclination of base to helix axis -1.2°; rise/base pair along helix axis 0.32Å; pitch/turn of helix 33.2Å; mean propeller twist +16°; glycosyl angle confirmation anti; sugar pucker conformaion c1'-endo to C2'-endo. The most commonly found right-handed form of DNA.

DNA helix Z. Left-handed helix; rotation/base pair -60°/2; mean base pairs/turn 12.0; inclination of base to helix axis -9°; rise/base pair along helix axis 3.8Å; pitch/turn of helix 45.6Å; mean propeller twist 0° glcosyl angle confirmation anti at C, syn at G', sugar pucker conformaion G2'-endo at C, G1' exo to C1'-exo at G.

Dnase C. An *E. coli* protein required to localize replication proteins.

DNA fingerprinting. A technique developed by Jeffreys *et al.* in 1985 useful in establishing near perfect identity of an unidentified body. The technique is based on the fact that every human being is conferred with DNA variations in form of variable number of tandem repeats (VNTRs). DNA profile of an individual is also known as **genetic signatures.** Accuracy of this method is 1-in-75 billion error probability ratio.

DNA ligase. An enzyme that repairs single-stranded discontinuities in double-stranded DNA molecules. In the cell, ligases are involved in DNA replication. Purified ligases are used in construction of recombinant DNA molecules.

DNA-mediated gene transfer. The method of gene transfer in which a purified DNA fragment carrying gene of interest is mixed with a carrier DNA and is precipitated out of solution with calcium phosphate. When target cells are incubated with the precipitate, there is increased uptake of DNA and some of the cells are transformed by the desired gene.

DNA polymerase l. The *E. coli* enzyme that completes synthesis of individual Okazaki fragments during DNA replication by degrading the primer and filling in the gaps. May also be helpful in termination of DNA replication.

DNA polymerase II. The main DNA replicating enzyme of *E. coli*.

DNA polymerase III. The enzyme required for *in vivo* DNA replication.

DNA puff. Cross band in a polytene chromosome swollen due to the relaxation of chromatin of banded region at the time of high activity.

DNA replication. The process by which a DNA molecule makes its identical copies.

DNA sequencing. Determination of the order of nucleotides in a DNA molecule. Methods - (1) Chemical cleavage method of Maxam and Gilbert (1977), (2) Plus and minus method of Sanger *et al.* (1978), and Wandering Spot Analysis.

DNA strand. One of the two DNA strands in a double-helical DNA molecule.

DNA synthesis *in vitro*. Synthesis of DNA outside a cell (in a test tube) involving four types of triphosphate nucleosides. A template is used in artificial DNA synthesis.

DNA topoisomerase. An enzyme that introduces or removes turns from the double helix by transient breakage of one or both polynucleotides.

Dominance hypothesis. The theory that heterosis is caused by the masking of harmful recessive alleles by dominant alleles. See also **overdominance hypothesis.**

Dominance modifiers. Those genes which through relative strength of action of the two alleles at a locus or action of other genes (epistasis) affect dominance-recessiveness relationship.

Dominance portion of genetic variance (V_D). That portion of the genetic variance for a given trait that results from the fact that heterozygotes do not always score exactly midway between the homozygotes.

Dominant. Refers to the allele whose phenotypic effect is expressd in a heterozygote.

Dominant epistasis. A condition where a dominant gene masks the expression of non-allelic member(s).

Dominant lethal. Action of a gene lethal in both the homozygous and heterozygous conditions; a gene which frequently kills in a single dose.

Dominant phenotype. The phenotype of a genotype containing the dominant allele; the parental phenotype that is expressed in a heterozygote.

Dosage compensation. Any mechanism by which the effective dosages of sex-linked genes in organisms with an XX-XY or XX-XO mechanism of sex determination are made equal.

Dosage effect. A type of inheritance in which two like dominant alleles have a stronger effect than a single one of the same type.

Double *Bar* Drosophila. A more extreme bar phenotype, manifest as a narrower bar with reduced facet number, brought about by the duplication of a segment of the X chromosome at the *bar* locus.

Double crossover. Two crossover occurring in a chromosomal region under study.

Double fertilization. In angiosperms, the fusion of one male gamete with the egg nucleus (which forms an embryo) at the same time as a second male gamete fuses with a second female gamete to beget nutritive tissue (endosperm).

Down's syndrome. A manifestation of trisomy for chromosome 21 in man. Characterized by mental retardation, oriental eyes, short stature, abnormal palm prints, and malformation of the heart, ears, hands, and feet. Also known as **monogolism**.

Downstream. Towards the 3'-end of a polynucleotide.

***Drosophila*.** A dipteran fruit fly which was a central experimental organism in the development of classical genetics. Experiments on *Drosophila* led to the chromosome theory of inheritance by Morgan, Bridges, Sturtevant and Muller. *Drosophila* continues to be an important experimental organism in genetics.

Duplicate dominant epistasis. Dominant alleles at either of two loci can mask the expression of recessive alleles at the two loci resulting in 15:1 ratio. Also called **duplicate epistasis**.

Duplicate recessive epistasis. Recessive alleles at either of the two loci can mask the expression of dominant alleles at the two loci resulting 9.7 ratio. Also called **complimentary epistasis**.

Duplicate genes. Either dominant or both dominant genes together produce the same phenotype to give 15:1 ratio in F_2 progeny.

Duplication. A chromosomal aberration in which a segment of the chromosome bearing specific loci is repeated.

Dyad. Two sister chromatids attached to the same centromere.

Dysgenic. Any effect or situation that is or tends to be harmful to the genetics of future generations.

E

Effector gene. The gene which drives transcription of an other gene with the help of a promoter from another gene.

Effector molecule. A molecule (a sugar, an amino acid or a nucleotide) that can bind to a regulator protein and thereby change the ability of the regulator molecule to interact with the operator.

Element. The particulate hereditary determiner of Mendel. Gene.

Electroporation. A technique for transfecting cells by the application of a high-voltage electric pulse.

Elongation factors. The proteins that play an ancillary role in the elongation step during translation process molecule (EF-Ts, EF-Tu, EF-G).

Endomitosis. An increase in somatic DNA content which takes place within an intact nuclear envelop and gives rise to **endopolyploidy**.

Endonuclease. An enzyme that breaks phosphodiester bonds within a nucleic acid molecule.

Endoplasmic reticulum. A double membrane system in the cytoplasm, continuous with the nuclear membrane and bearing numerous ribosomes.

Endopolyploidy. An increase in the number of chromosome sets caused by replication without cell division.

Endoreplication. Replication without separation of chromatids.

Endosperm. A polyploid (in many species, triploid) food storage tissue in many angiosperm seeds formed by fusion of two (or more) gametes and a sperm.

End-product inhibition. Describes the ability of a product of a metabolic pathway to inhibit the activity of an enzyme that catalyzes an early step in the pathway.

Enhancer. A special type of eukaryotic regulatory sequence that can increase the rate of transcription of a gene located some distance away in either direction. They are mostly *cis*-acting but *trans*-acting ones are also known. They are effective whether lying upstream or downstream. They are active whether they lie in same or opposite polarity as the gene. They are equally effective regardless of the organism or the gene from which they are derived when attached to foreign DNA.

Environmental mutagenesis. See **genetic toxicology**.

Environmental variance. The variance due to environmental variation.

Environmental variation. When individuals with the same genotypes develop in different environments their phenotypes may be quite different, e.g., identical twins raised in different environments show different phenotypes due to environmental variation.

Episome. A plasmid capable of integration into the host cell's chromosome. These are those particles that are added to a genome through an external source - not by mutation or rearrangement of the existing genome.

Epistasis. The ability of one gene to mask the phenotype derived from a second gene.

Epistatic genes. The genes that suppress expression of other genes.

Equational division. The second meiotic division is an equational division because it does not reduce chromosome numbers.

Equatorial plate. The figure formed at the spindle equator in nuclear division.

Equilibrium. A state of dynamical systems in which there is no net change.

Equilibrium frequency. In a population, gene frequency that varies nondirectionally about a mean by an amount described by the standard deviation (which see) under conditions of unchanging selection pressure and mutation rate, with no intermixing from other populations.

Erythroblastosis fetalis. A pathological condition in which red blood cells are destroyed, with resultant anemia.

Escherichia coli. A species of human intestinal bacteria genetically well understood and often used in biological research.

Estimation of linkage. Calculating distance between genes form F_2 or TC_1 progeny on the basis of crossing-over frequency.

Euchromatin. The regions of a eukaryotic chromosome that appear less condensed and stain less deeply with DNA-specific dyes.

Eugenics. Means and methods of social control to improve the hereditary qualities of future generations.

Eukaryote. An organism whose cells are characterized by the presence of membrane-bound nuclei.

Euploid. An individual having changes by complete sets or exact multiples of the monoploid, e.g., triploid, tetraploid and so on.

Evolution. The transformation of an organism in a way that descendants differ from their predecessors.

Exon. One of the coding regions of a discontinuous gene. The sequence present in a gene which is complementary to that present in its mRNA.

Exon theory of genes. Split genes arise not by insertion of introns into unsplit genes but from combinations of primordial "minigenes" separated by spacers.

Exonuclease. An enzyme that sequentially removes nucleotides from the ends of a nucleic acid molecule.

Extrachromosomal gene. Any gene that is not carried by the cell's chromosome(s). For example, gene present on mitochondrial or chloroplast genomes, genes carried by plasmids.

Explant. The plant part which is used for regeneration in tissue culture.

Expression vector. A hybrid vector (plasmid) that expresses its cloned genes.

Expressivity. The degree of phenotypic expression within one phenotype under a variety of environmental conditions.

Expressivity modifiers. Those genes which are responsible for variable expressivity of a phenotype.

Extended anticodon hypothesis. The structure of anticodon loop and the proximal anticodon stem are related to the sequence of anticodon. Thus anticodon is extended to (i) two nucleosides at the 5' end of the anticodon, (ii) three nucleosides of the anticodon and (iii) five pairs of nucleosides in the anticodon stem. Extended anticodons are involved in translation efficiency.

Extrachromosomal gene. Any gene that is not carried by the cell's chromosome(s). For example, gene present on mitochondrial or chloroplast genomes, genes carried by plasmids.

Extranuclear genes. The genes which reside in organelles such as mitochondria and chloroplasts outside the nucleus.

Extranuclear inheritance. It is characterized by differences in reciprocal crosses, progeny shows characters of the female parent, no linkage with the nuclear genes. See **cytoplasmic inheritance**.

Eyes. Referring to the configuration of replicating DNA in eukaryotic chromosomes.

F

F_1. Refers to first filial generation.

F_2. Refers to second filial generation.

Factor. Mendelian unit of inheritance.

F^+ cell. A becterium that carries an F plasmid.

F^- cell. A becterium that does not carry an F plasmid.

F' cell. A becterium that carries a modified F plasmid, one which carries a small piece of DNA derived from the host bacterial DNA molecule.

F-duction. The transfer of bacterial genes from donor to recipient bacterium via an F' plasmid, used as the basis to a gene mapping technique.

F-plasmid. A fertilty plasmid carrying genes that direct conjugal transfer of DNA between bacteria. See **fertility factor**.

F-duction. The transfer of bacterial genes from donor to recipient bacterium via an F' plasmid, used as the basis to a gene mapping technique.

Feedback inhibition. A post-translational control mechanism in which the end product of a biochemical pathway inhibits the activity of the first enzyme of its pathway.

Fertility (F) factor. The fertility factor in the bacterium *Escherichia coli*; it is composed of DNA and must be present for a cell to function as a donor in conjugation. Its presence confers donor ability (maleness). Also see **sex factor**.

Fertility restorer. Cytoplasmic male sterile lines that will produce viable pollen in certain genotypes. Restorers have genes that restore fertility to a cytoplasmic male sterile line.

First division segregation (FDS). Segregation of two alleles at a locus at meiosis I. FDS gives 1:1 ratio during tetrad analysis. Compare with **second division segregation (SDS)**.

f^{Met}. N-formylmethionine, the modified amino acid carried by the $tRNA_f^{met}$ that initiates translation in bacteria.

Footprinting. A technique to determine the length of nucleic acid in contact with a protein. While in contact, the free DNA is digested. The remaining DNA is then isolated and characterized.

Forces of evolution. Mutation, recombination, migration, random genetic drift, Founder effect, bottlenecks, assortative mating, selection, etc. are regarded as forces of evolution.

Forward mutation. Any change away from the standard (wild type).

Frequency-dependent selection. A selection whereby a genotype is at an advantage when rare and at a disadvantage when common.

Frequency distribution. A synthetic presentation of a series of observations, obtained by specifying the number of observations falling in each "class" i.e., the frequencies of individuals with values of x between x_1 and x_2, between x_2, or x_3, and so on, defines a class. May be graphically presented as frequency diagrams, histograms, or line graphs.

F-test. A test of statistical significance which is used to compare the differences among several mean sum of square values.

Functional alleles. Alleles determined on the basis of complementation test.

Furrowing. A mechanism through which animal cells undergo cytokinesis.

G

G1 phase. The first gap period of the cell cycle.

G2 phase. The second gap period of the cell cycle.

Gamete. A reproductive cell, usually carrrrying the haploid chromosome complement, that can fuse with a second gamete to produce a new cell during sexual reproduction.

Gametophyte. The haploid stage of a plant life cycle that produces gametes (by mitosis). It alternates with a diploid, sporophyte generation.

Gap period. One of two intermediate periods within the cell cycle. Gap 1 occurs between mitosis and the DNA synthesis phase, gap 2 occurs between DNA synthesis and next mitosis.

Gemmule. An old and no longer appropriate term used to describe hypothetical units in the body that carried the essence of specific body traits to the gonads, from which they were transmitted to the offspring.

Gene. A segment of DNA that contains biological ingromation and codes for an RNA and/or polypeptide molecule. Recognised through its variant forms which transmit specification(s) from one generation to the next.

Gene amplification. A process or processes by which the cell increases the number of a particular gene within the genome.

Gene cloning. Insertion of a fragment of DNA, containing a gene, into a colning vector, and subsequent propagation of the recombinant DNA molecule in a host organism.

Gene conversion. In Ascomycete fungi a 2:2 ratio of alleles is expected after meiosis, yet a 3:1 ratio is sometimes observed. The mechanism of gene conversion is explained by repair of heteroduplex DNA produced by the Holliday model of recombination.

Gene delivery. In gene therapy, this term refers to putting the new gene into correct target cells.

Gene dosage. The number of times an allele is present in a particular genotype.

Gene-environment interaction. A gene produces a particular effect only in the presence of a particular environment.

Gene expression. The process by which the biological information carried by a gene is released and made available to the cell, through transcription possibly followed by translation.

Glossary 527

Gene frequency. The frequency of a given gene in a population, relative to its alleles.

Gene-gene interaction. Interaction between products of different genes leads to modification in phenotype or phenotypic ratios.

Gene library. A large collection of cloning vectors containing a complete set of fragments of the genome of an organism.

Gene manipulation. The formation of new combination of heritable material by the insertion of nucleic acid molecules, produced by whatever means outside the cell, into any virus, bacterial plasmid, or any other vector system so as to allow their incorporation into a host organism in which they do not naturally occur, but in which they are capable of continued propagation.

Gene mutation. These changes occur at or within a single gene, so they some times are called point mutations. As a result a gene can mutate from one allelic form to another.

Gene patenting. It is a sort of protection provided by a government of a country to discoverer of a new gene, genotype, a genetic strain, a gene test or a genetic procedure so that the detailed information can be declared publically.

Gene pool. The total of all genes in a population.

Generalized recombination. Recombination between two double-stranded DNA molecule which share extensive nucleotide sequence similarity.

Generalized transduction. Form of transduction in which any region of the host genome can be transduced. See **specialized transduction**.

Gene regulation theory of ageing. Sequential changes in the expression of genes occur from the beginning of the life span of an organism. until the attainment of reproductive maturity and then ageing results.

Gene regulatory mechanism in eukaryotes. Transcription of a **producer gene** occurs only if at least one of its **receptors** was activated by forming a sequence-specific complex with **activator RNA**. The activator RNA is synthesized by **integrator genes** in response to signals by sensor genes that are sensitive to external/development signals.

Gene regulatory mechanism in prokaryotes. They are based on presence or absence of substrate (inducer), end product (co-repressor), or product of a regulatory gene (repressor). Based on whether control is inducible or repressible, or whether control is positive or negative, several models of gene regulation in prokaryotes are conceived. Basically, gene regulation involves interaction between RNA polymerase and promoter to initiate transcription, repressor and operator thus blocking passage of RNA polymerase, and repressor and inducer to inactivate the repressor.

Gene synthesis. Chemical synthesis of a gene involving blocking of certain reactive groups of nucleotides and joining the same one by one. No template is used. For using this method, nuleotide sequence of the gene to be synthesized needs to be known.

Gene therapy. It includes development of methods for curing genetic disorders by replacing a defective gene with a normal gene.

Genetic analysis. Deals with analysis of a gene in term of its (a) nature, i.e., nuclear or cytoplasmic, dominant or recessive, autosomal or sex linked (if applicable), (b) location on a particular chromosome, (c) position on the chromosome, and (d) nucleotide sequence.

Genetic block. In biochemical genetics, a block in a step in the synthesis or segregation of a biochemical brought about by a mutant gene.

Genetic code. The rules that determine which triplet of nucleotides codes for which amino acid during translation.

Genetic engineering. The use of experimental techniques to produce DNA molecules containing new genes or new combination of genes. It deals with isolation, synthesis, adding, removing or replacing genes in order to achieve permanent and heritable changes in diverse forms of life bypassing all reproductive barriers.

Genetic code dictionary. A table of all code words or codons that specify amino acids.

Genetic counselling. This is telling the parents of a child who suffers from a genetic defect the probability that an other child born to them will also have that defect.

Genetic dissection. The use of recombination and mutation to piece together the various components of a given biological function.

Genetic distance. Average number of electrophoretically detectable codon substitutions per gene that have accumulated in the population studied since they diverged from a common ancestor.

Genetic drift. Random change (in any direction) in gene frequencies due to sampling error.

Genetic equilibrium. The condition of a population in which successive generations consist of the same genotypes with the same frequencies, in respect of particular genes or arrangements of genes.

Genetic fine structure. The structure of the gene analyzed at the level of the smallest units of recombination and mutation (nucleotides). Benzer used *cis-trans* complementation test to study fine structure of the game.

Genetic homeostasis. The tendency of populations under selection to regress toward the original mean.

Genetic isolation. Isolation of a variety or species by genetic means, e.g., cross sterility in corn and cross incompatibility of amphidiploid species with parents or other species.

Genetic load. The relative decrease in the mean fitness of a population due to the presence of genotypes (usually comprising of lethal genes) that have less than the highest fitness.

Genetic male sterility. Male sterility controlled by nuclear genes.

Genetic mapping. The process of determining the location of genes and distance between genes on a chromosome.

Genetic marker. An allele whose phenotype is easily recognizable and which can, therefore, be used to follow the inheritance of its gene during a genetic cross.

Genetic material. The chamical material of which genes are made, now known to be DNA in most organisms, RNA in a few.

Genetic medicines. It is use of snippets of genetic material which can block the expression of defective genes. These blocks may be at transcription level (triplex approach) or at translation level (antisense approach). These techniques are yet to be perfected.

Genetic mutagen. A gene or a DNA sequence that causes other genes to mutate, e.g., the *AC-Ds* system in corn, or the *Dt* gene in corn.

Genetic prognosis. It illustrates some basic principles underlying genetic predictions in terms of probability of a particular genetic disorder to appear in an individual.

Genetic risk assessment. Quantitative estimate of probable impact on the gene pool of subsequent generations from a specific mutagenic exposure.

Genetics. A branch of life sciences that precisely understands the nature, structure, organization, function, regulation, manipulation of the hereditary units called genes that are responsible for carrying out different life processes and for transmission of biological properties from parents to offspring. Genetics also clvers finding out ways and means to use knowledge of genetics for welfare of mankind.

Genetic screening. The testing of individuals for a particular genetic trait.

Genetic toxicology. A subdiscipline of toxicology which identifies and analyzes the action of agents with toxicity directed towards the genetic material of the living systems.

Genetic variance. The phenotypic variance due to the presence of different genotypes in the population.

Gene transfer. The passage of a gene or group of genes from a donor to a recipient organism.

Genic balance theory. The theory of Bridges that the sex of a fruit fly is determined by the relative number of X chromosomes and autosomal sets (X/A).

Genic explanation of hybrid vigour. Hybrid vigor is widely explained as the result of a dominance of linked genes. Many dominant genes are involved. Different inbreds contribute different dominant genes, so that the F_1 has more than either parents.

Genic and plasmagenic interaction. Interaction between a plasma gene and a gene in the chromosome to produce an effect. For example in *Paramecium*, plasma gene *kappa* and nuclear dominant gene *K* interact to produce a killer phenotype.

Genome imprinting. A process that temporarily and erasably marks the genes passed on by females and males in different ways. Offspring that receives marked genes from their mothers are consequently different from those that receive the genes from their fathers.

Genomic library. A collection of clones sufficing in number to include all the genes of a particular organism.

Genome. The entire genetic complement of a cell.

Genotype. A description of the genetic composition of genes, cell or an organism. Genotype is relatively stable throughout the life of an organism.

Genotype-environmental interaction. The interplay of a specific genotype and a specific environment, affecting the phenotype. The extent and nature of this interaction varies with each genotype and environment. The variance due to this interaction (V_1) is part of the total phenotypic variance of a trait.

Genotypic ratio. Ratio of different genotypes in a segregating generation. Genotypic ratio of F_2 of a monohybrid cross is $1AA : 2Aa : 1aa$.

Genotypic variation. When two or more individuals develop in the same environment and come to possess different phenotypes implies that these individuals have different genotypes. Genotypic variation arises due to changes in the germplasm.

530 Basic Genetics

Germplasm. (1) The genetic makeup of an organism - the sum total of its genes, both dominant and recessive; (2) The potential hereditary materials within a species, taken collectively.

Germplasm theory. A theory stating that multicellular organisms give rise to two types of tissues, **somatoplasm** and **germ plasm**. Somatoplasm is essential for functioning of the organism but lacked the property of entering into sexual reproduction. Changes in somatic tissues were not heritable. Germ plasm is set aside for reproductive purposes. A change in germ plasm could be heritable.

Group I introns. Self-splicing introns that do not require an external nucleotide for splicing. The intron is released in a lariat form.

Group II introns. Introns that are not self-splicing. They need small guide RNAs or U RNAs for their splicing.

GT-AG rule. Refers to the fact that with introns present in nuclear protein-coding genes, the first two nucleotides of the intron are GT and the last two AG.

Guanine. 2-amino, 6-oxypurine, one of the nucleotides found in DNA and RNA.

Gyrase. An enzyme that relaxes supercoiling caused by wnwinding of double helix of DNA.

H

Haploid. Refers to a cell that contains a single copy of each chromosome.

Hairpin loop. A double-helical region formed by base pairing between adjacent (inverted) complementary sequences in a single strand of RNA or DNA.

Haploid. Refers to a cell that contains a single copy of each chromosome.

Haplotype. It is the particular combination of alleles in a defined region of some chromosome, in effect the genotype in miniature. Originally used to described combinations of MHC alleles, it now may be used to describe particular combinations of RFLPs.

Hardy-Weinberg equilibrium . Occurrence in a population of three genotypes, two homozygous and one heterozygous (e.g., AA, Aa, and aa), where the frequencies of the three genotypes are p^2 for homozygous dominants, $2pq$ for heterozygotes, and q^2 for homozygous recessive and p = frequency of A and q = frequency of a. At these frequencies there is no change in genotypic frequencies as long as isolation mechanisms, selection mutation, and drift are absent.

HAT medium. A medium containing hypoxanthine, aminopterin and thymidine which is used to select somatic hybrid cells.

Helicase. The enzyme responsible, during DNA replication, for breaking the hydrogen bonds that hold the double helix together.

Hemizygous. Refers to a gene that is present in only one copy in a diploid cell, such as those genes on the differential regions of the X and the Y chromosomes.

Heredity. The passage of characteristics from parents to offspring.

Heritability. A statistic measuring degree to which the total phenotypic variation is result of genetic factors.

Heritability in broad sense (H_b). An estimate of heritability measuring contribution of al genetic components contributing towards the phenotype.

Heritability in narrow sense (H_n). An estimate of heritability measuring contribution of only additive genetic effects contributing towards the phenotype.

Hermaphrodite. An individual with both male and female genitalia.

Heteroallele. The allele which arise due to base pair replacement in different mutons; those alleles between which recombination is theoretically possible.

Heterochromatic. Entire chromosomes or portions of the chromosomes that do not manifest the usual prophase-telophase transformations and appear to lack genes with major phenotypic effects.

Heterochromatin. The regions of a chromosome that appear relatively condensed and stain deeply with DNA-specific stains.

Heterodimer. A dimeric protein having two different polypeptides.

Heteroduplex. A base-paired structure formed between two polynucleotides that are not entirely complementary.

Heteroduplex DNA model. A model that explains both crossing-over and gene conversion by assuming the production of a short stretch of heteroduplex DNA (formed from both parental DNAs) in the vicinity of a chiasma.

Heteroduplex modified DNA. DNA having no methyl group on a particular adenine or cytosine on only one strand included in a host specificity site.

Heterochromatin. The regions of a chromosome that appear relatively condensed and stain deeply eith DNA-specific stains.

Heteroduplex. A base-paired structure formed between two polynucleotides that are not entirely complementary.

Heterogametic sex. Sex producing gametes of two kinds with regard to sex determination.

Heterogeneous nuclear RNA (hnRNA). The nuclear RNA fraction that comprises the unprocessed trasncripts synthesized by RNA polymerase II.

Heterokaryon. Existence of genetically different nuclei in the same cytoplasm.

Heterosis. (1) The increased vigour, growth, size, yield or function of a hybrid progeny over the parents that results from crossing genetically unlike organisms; (2) The increase in vigour or growth of a hybrid progeny in relation to the average of the parents.

Heterozygous. Refers to a diploid cell or organism that contains two different alleles for a particular gene.

Heterozygote advantage. A selection model in which heterozygotes have the highest fitness.

Heterozygous. A gene pair having non-identical alleles in two homologous chromosomes. (e.g., *Aa* is a heterozygous genotype.)

Hexaploid. (Usually) a diploid with duplicate sets of three different genomes. Common bread wheats (42 chromosomes) are hexaploids in contrast to primitive wheat with 14 chromosomes (a diploid).

High frequency recombination (Hfr) cell. A bectrium whose DNA molecule contains an intergrated copy of the F plasmid.

Highly repetitive sequences. When number of copies of a sequence in eukaryotic DNA is 10^6 to 10^7. These repeats are usually found in centromeric regions.

Histocompatibility. This refers to tissue compatibilty. Tissues can be transplanted between genetically identical individuals without concern for immunological rejection whereas transplants between genetically non-identical individuals are usually rejected with time.

Histocompatibility antigens. Antigens that determine the acceptance or rejection of a tissue graft.

Histocompatibility genes. The genes that code for the histocompatibility antigens.

Histone. One of the basic proteins that make up nuclosomes and have a fundamental role in chromosome structure.

HLA complex. Human leukocyte antigens complex comprises of genes that control the synthesis of HLAs that are located on the surfaces of the leukocytes and which affect tissue compatibility in organ transplants and skin grafting. These antigens occur in quite a large number of types and are produced by at least four closely linked loci (in the sequence D B C A) on autosome 6.

Hogness box. A nucleotide sequence that makes up part up of the eukaryotic promoter. Same as **TATA box**.

Holandric gene. A gene located on the Y-chromosome. It is transmitted form father to son.

Holliday structure. An intermediate structure believed to be formed during recombination between two DNA molecules.

Holoenzyme. The version of the *E. coli* RNA polymerase that has the subunit composition $a_2 \beta \beta' \sigma$ and is involved in efficient recognition of promoter sequences.

Homoallele. The allele which arise due to base pair replacement in the same muton; recombination is theoretically not possible between homoalleles.

Homodimer. A dimeric protein having two identical polypoptides.

Homogametic sex. Sex producing gametes of only one kind with regard to sex determination.

Homologous chromsomes. Two or more identical chromosomes (one received from father and the other received from mother) which are able to undergo synapsis during prophase of meiosis-I.

Homozygous. A gene pair having identical alleles in two homologous chromosomes. An individual having a homozygous gene pair is known as **homozygote**. (*AA* and *aa* are homozygous genotypes.)

Host-mediated assay. A method in which microorganisms are introduced into a host organism (such as rat). Then the host is treated with the chemical whose mutagenicity is to be tested. After a period of time, the microorganisms are recovered and checked for mutations

Hotspot. It is a site at which the frequency of mutation (or recombination) is very much increased.

House keeping genes. The genes whose products are required by the cell at all time, their activity is controlled by constitutive factors. Also see **constitutive genes**.

H-Y antigen. A histocompatibility factor determined by the Y chromosome. It is thought to be the major male-determining factor in mammals. In humans, the gene for the H-Y antigen is on the short arm of the Y chromosome.

Hybrid. An individual that results from a cross between two genetically unlike parents.

Hybrid-arrested translation (HART). A method used to identify the polypeptide coded by a cloned gene.

Hybrid breakdown. Production of weak or sterile F_2 progeny by vigorous fertile F_1 hybrids.

Hybrid DNA. DNA whose two strands have different origins.

Hybrid inferiority. Hybrid is inferior to either of the parents.

Hybridization *in situ*. Finding the location of a gene by adding specific radioactive probes for the gene and detecting the location of the radioactivity on the chromosome after hybridization.

Hybridization probing. A method that uses a labelled nucleic acid molecule to identify complementary or homologous molecules through the formation of stable base-paired hybrids.

Hybridize. (1) To form a hybrid performing a cross. (2) To anneal nucleic acid strands from different sources.

Hybrid plasmid. A plasmid that contains an inserted piece of foreign DNA.

Hybrid vigour. Superiority of hybrid over the better parent in one or more traits. This is also known as **positive heterosis**.

Hypermorph. An allele having an effect similar to but greater than that of the wild form.

Hypostatic genes. The genes whose expression is suppressed by other genes.

I

Illegitimate recombination. Recombination between two double-stranded DNA molecules which have little orders of structure of many biomolecules.

Inborn errors of metabolism. Genetically caused metabolic dysfunctions.

Incestuous mating. A mating between blood relatives who are more closely related than the law of the land allows.

Incomplete dominance. The situation where neither of a pair of alleles displays dominance and the phenotype of the heterozygote is intermediate between the phenotypes of the two alternative homozygotes.

Induced mutations. Genetic changes produced by some physical or chemical agent or under changed growth conditions.

Inducer. A molecule that induces expression of a gene or operon by binding to a **repressor** protein and thereby preventing the repressor from attaching to its **operator** site. Effectors in inducible operons are known as **inducers**.

Inducible control. In this case substrate acts to induce the production of the enzymes. Repression occurs in the absence of the substrate.

Induction. A chemical or physical treatment which results in excision from the host genome of the integrated form of a lysogenic phage, followed by the switch to the lytic mode of infection.

Inhibitory epistasis. One gene when dominant is epistatic to other but other when recessive is epistatic to the first. These two genes give 13:3 ratio in F_2 progeny.

Initiation codon. The codon, usually but not exclusively 5'-AUG-3', which indicates the point at which translation of an mRNA should begin.

Initiation complex. The complex which comprises mRNA, a small ribosomal subunit, aminoacylated initiator-tRNA and initiation factors and which forms during the initiation stage of translation.

Initiation factors. Protein molecules that play an ancillary role in the initiation stage of translation (IF1, IF2, IF3).

Initiation site. The site on the DNA molecule where the synthesis of RNA begins.

Inosine. A newly discovered nucleotide which is found in third position in an anti-codon (on tRNA) and can pair with A, U and C resulting in wobble base pairing. It a deamination product of adenosine.

Insertion sequences (IS elements). Small, simple transposons. See **transposable element**.

In situ hybridization. Hybridization performed by denaturing the DNA of cells squashed on a microscope slide so that reaction is possible with an added single-stranded RNA or DNA; the added preparation is radioactively labelled and its hybridization is followed by autoradiography.

Integrator gene. A component of eukaryotic gene regulation which responds to the signals provided by **sensor gene**. These genes are transcribed to produce **activator RNAs**.

Interallelic complementation. The change in the properties of a heteromultimeric protein brought about by the interaction of subunits coded by two different mutant alleles; the mixed protein may be more or less active than the protein consisting of subunits only of one or the other type.

Intercalate. Certain drugs or dyes, such as ethidium bromide, are able to insert into DNA or double-stranded RNA between adjacent base pairs. Molecules with this property are called **intercalating agents**. The binding of such dyes reduces the buoyant density of the DNA. The DNA duplex increases in length and, if the DNA is supercoiled, increasing concentrations of the dye first unwind the superciols and then wind the molecule up again in the opposite sense.

Interference. The increase (negative interference) or decrease (positive interference) in likelihood of a second crossover closely adjacent to another. In most organisms interference increase with decreased distance between crossover. See **coincidence**.

Interphase. The period between cell division. Misnamed as **resting stage**. Metabolically very active stage as DNA synthesis occurs during this phase.

Interrupted mating. A mapping technique in which bacterial conjugation is disrupted after specified time intervals.

Intragenic mutations. Mutations which occur within the same gene.

Intron. A segment of DNA that is transcribed, but removed from within the transcript by splicing together the sequences (exons) on either side of it.

Inversion mutation. Alteration of the sequence of a DNA molecule by removal of a segment followed by its reinsertion in reverse orientation.

Inverted repeats. The repeats that comprise two copies of the same sequence of DNA repeated in opposite orientation on the same molecule. Adjacent inverted repeats constitute a **palindrome**.

Isoalleles. Alleles with similar phenotypic effect.

Isochromosome. A chromosome with two genetically and morphologically identical arms.

Isolating mechanisms. Any structural, physiological, behavioural, or other features of an individual, or any geographical or geological barrier, that prevents individuals of one population from successfully interbreeding with those of other populations.

Isozymes. Different forms of the same enzyme which catalyze particular biochemical reactions during metabolism.

J

Jacob and Monod model. The operon model of gene regulation in *lac* region of *E. coli*. This model explains mechanism of gene expression in prokaryotes.

Jumping gene. The gene which keeps on changing its position in a chromosome and also between the chromosomes in a genome.

K

Kappa particle. A particle present in the cytoplasm of some Paramecia (killers) that secretes a substance that kills Paramecia not possessing the kappa particle. They contain their own DNA of length 0.4 microns.

Karyokinesis. The division of the nucleus during cell division.

Karyotype. The entire chromosome complement of a cell, with each chromosome described of terms of its appearance at metaphase.

Kinetochore. The part of the centromere to which microtubules of the spindle apparatus attach.

Kinetoplast. These are modified mitochondria located near to the base of flagella. A kinetoplast may be larger than a nucleus.

Kinetoplast DNA. DNA present in kinetoplasts. This DNA exists in form of a large number of interlocked "minicircles".

Kin selection. The mode of natural selection that acts on an individual's inclusive fitness.

Klinefelter's Syndrome. A genetic disease in man due to the XXY karotype. It produces sterile males with some mental retardation.

L

***Lac* operon.** The cluster of three structural gene that code for enzumes involved in utilization of lactose by *E. coli*.

Lactose repressor. The regulatory protein that controls transcription of the *lac* operon in reponse to the levels of lactose in the environment.

Lagging strand. The strand of the double helix which, during DNA replication, is synthesized in a discontinuous fashion in form of **Okazaki fragments**.

Lamarckism. Theory of evolution proposed by Lamarck which espouses the inheritance of acquired characteristics. Also see **theory of inheritance of acquired traits**.

Lampbrush chromosome. A chromosome that has paired loops extending laterally, and occurs in primary oocyte nuclei; they represent sites of active RNA synthesis.

Leader transcript. The untranslated segment of mRNA that lies upstream of the initiation codon on an mRNA molecule.

Leader segment. The untranslated segment that lies upstream of the initiation codon on an mRNA molecule.

Leading strand. The strand of the double helix which, during DNA replication, is copied in a continuous fashion.

Leptotene. A substage of prophase I of meiosis I during which the chromosomes appear thread-like.

Leucine zipper. A structural motif found in several DNA-binding proteins.

Level of significance. The probability value in statistics used to reject the null hypothesis.

Life. Any entity that is capable of making a reasonable accurate reproduction of itself and duplicate being able to produce the same task and subjected to low rate of alteration and these changes being heritable.

Life cycle. The entire series of development stages undergone by an individual from zygote to maturity and death.

Linkage. The physical association between two genes that results from them being present on the same chromosome. Tendancy of the parental combination of genes to stay together in F_2.

Linkage detection. Deals with partitioning of total χ^2 into different components to find out presence of linkage.

Linkage group. A group of genes that display linkage. Eukaryotes, a single linkage group usually corresponds to a single chromosome.

Linkage map. Arrangements of genes in a linkage group in such a way that the distance between any two of them reflects per cent crossing-over between them.

Linker DNA. The DNA that links nucleosomes together and which makes up the 'string' in the beads-on-a-string model for chromatin structure.

Locus. The specific place on chromosome where a gene is located.

LOD score method. A technique *(logarithmic odds)* for determining the most likely recombination frequency between two loci from pedigree data.

Long-term gene regulation. Gene regulation recognized in eukaryotes which operates during determination, differentiation, or more generally development.

Luxury genes. The genes coding for specialized functions; their products are synthesized (usually) in large amounts in particular cell types.

Lyon hypothesis. The hypothesis which states that in any given cell of a female one X chromosome is active and the other is inactive.

Lysis. Disintegration or dissolution; usually, the destruction of a bacterial host cell by infecting phage particles.

Lysogeny. The pattern of bacteriophage infection that involves integration of the phage genome into the host DNA molecule.

M

Major gene. A gene that may cause sufficiently large variation in the trait studied to be easily detected.

Major groove. The larger of the two grooves that spiral around the surface of the double helix model for DNA.

Major histocompatibility antigens. These antigens act with high intensity and reject a graft fast.

Major histocompatibility complex. A group of highly polymorphic genes whose products appear on the surface of cells imparting to them the property of "self" (belonging to that organism). Some other functions are also involved.

Male sterility. A condition in which pollen is absent or nonfunctional in flowering plants.

Map unit. A unit used to describe the distance between two genes on a chromosome, now superceeded by **centiMorgan**.

Marker. A locus or allele whose phenotype provides information about a chromosome or chromosomal segment during genetic analysis.

Masked mRNA. mRNA complexed with protein so that it is not translated or enzymatically degraded.

Maternal effect. A nuclear gene product in the cytoplasm determines the phenotype of the organism.

Maternal inheritance. Phenotypic differences due to factors such as those in chloroplasts and mitochondria transmitted by the female gamete.

Maximum likelihood method. The method of estimation depending on the maximization of the log likelihood function. Always leads to an efficient statistic.

Meiosis. The series of events, involving two cell divisions by which diploid cells are converted to haploid cells.

Meiosis I. First division of meiosis during which chromosome number is reduced to one-half. This is thus a reductional division.

Meiosis II. Second division of meiosis during which chromosome number remains constant. This is thus an equational division.

Mendelian genetics. A branch of genetics which deals with the inheritance of qualitative

Mendelian inheritance. The mechanism of the inheritance of chromosomal genes. Mendel's law of segregation states that genes occur in pairs and segregate from each other during meiosis. Mendel's law of independent assortment states that member of one pair of genes do not influence the way in which other pair of genes are distributed.

Mendelian population. A random mating population with equal survival of all genotypes.

Mendel's first law. The two members of gene pair segregate from each other during meiosis; each gamete has an equal probability of obtaining either member of the gene pair. Now known as **principle of segregation**.

Mendel's second law. The law of independent assortment; unlinked or distantly linked segregating gene pairs behave independently. Now known as **principle of independent assortment (segregation)**.

Merodiploidy. Temporary partial diploidy in bacteria due to the presence of transferred genes.

Messenger RNA (mRNA). A transcript of a protein-encoding gene.

Messenger RNA (m RNA) caps. Addition of m7G group(s) at 5′end of the most of the eukaryotic primary transcript. These caps are of three types - (i) **Cap O**. Cap with a single methyl group, found in 100% cases, (ii) **Cap 1**. methyl group may be present on penultimate base at 2′ O position of the sugar moiety, present in most cases, and (iii) **Cap 2**. methyl group may be present in the third base also at the 2′ O position of sugar, present in 10-15% cases.

Messenger RNA (mRNA) decay. A process responsible for differential survival of mRNAs in the cytoplasm. Site-specific endonucleases seem to control this process.

Metacentric chromosome. A chromosome with a centrally located centromere.

Metafemale (superfemale). Abnormal females in *Drosophila*, usually sterile and week, with an overbalance of X chromosomes with respect to automsomes; X/A ratio greater than 1.0.

Metamale (supermale). Abnormal males in *Drosophila* with an overabundance of autosomes to X chromosomes; X/A ratio less than 0.5.

Metaphase. A stage of mitosis at which the nuclear membrane disappears, spindle apparatus is completely formed and the chromosomes are most condensed and are arranged on the equatorial plate of the spindle apparatus.

Metaphase I. A stage of meiosis I at which the nuclear membrane disappears, spindle apparatus is completely formed and the homologous chromosomes, in form of bivalents, are arranged on the equatorial plate of the spindle apparatus.

Metaphase II. A stage of meiosis II at which the nuclear membranes in the two haploid cells seen at prophase II disappear, spindle apparatus is completely formed in each haploid cell and the chromosomes are arranged on the equatorial plates of the spindle apparatus.

Metaphase chromosome. A chromosome at the metaphase stage of cell division, when the structure is in its most organized state and features such as the banding pattern can be visualized.

Microbial Genetics. A branch of genetics which deals with microorganisms like bacteria, viruses and fungi.

Microinjection. Purified gene is directly injected into the nucleus of an animal cell through a micropipette.

Minichromosome. Chromosome of SV40 or polyoma is the nucleosomal form of the viral circular DNA.

Minor gene. A gene whose effect on a given trait is so small that it is not easily detected.

Glossary 539

Minor groove. The smaller of the two grooves that spiral around the surface of the double helix model for DNA.

Minor histocompatibility antigens. These antigens act with low intensity and reject the graft slowly.

-25 box. A component of the nucleotide sequence that makes up the prokaryotic promoter.

Missense mutation. An alteration in a nucleotide sequence that converts a codon specifying one amino acid into a codon for a second amino acid.

Mitochondrial (mt) DNA. In most eukaryotic cells mt DNA is double stranded and circular (as in bacteria) in structure. It replicates independently of nuclear DNA in semi-conservative manner. Molecular weight = 9×10^6 to 10×10^6; length = 5 μm; size = 14,000.

Mitochondrial DNA diseases. Refers to human diseases associated with mutations in mitochondrial DNA. Diseases due to single deletions or duplications, multiple deletions or point mutations in mitochondrial DNA are known.

Mitosis. The series of events that result in division of a single cell into two daughter cells.

Moderately repetitive sequences. When number of copies of a sequence in eukaryotes is 10^3 to 10^5. These sequences are interspersed with unique sequences, and are also called **dispersed repeats**.

Modification of DNA/RNA. Includes all changes made to the nucleotides after their initial incorporation into the polynucleotide chain.

Modified bases. All those bases except the usual four from which DNA (T, C, A, G) or RNA (U, C, A, G) are synthesized; they result from post-synthetic changes in the nucleic acid.

Modifier (modifying) gene. A gene that affects or modifies the expression of an other gene.

Molecular genetics. A branch of genetics that involves the study not only of the moleclar nature of genes and gene expression.

Molecular zippers. These are the regulatory proteins which reveal the motifs which are composed primarily of amino acids that may not make direct contact with DNA. These motifs form three-dimensional scaffolds which steer the particular amino acid side chain of a regulatory protein into the grooves of double helical DNA where they can interact directly with DNA bases. These zippers my be of several types - steroid receptors, helix-turn-helix motifs, acid blobs, amphipathetic, helix-loop-helix motifs, zinc finger motifs, leucine zippers.

Monogenic inheritance. Pattern of inheritance observation a cross between two individual identically heterozygous at one gene pair. It gives 3:1 ratio in F_2.

Monintron gene. The genes which contain only one intron. Transcripts of such genes are not used for translation.

Monocistronic mRNA. An RNA molecule, mostly in eukaryotes, that contains information from only one cistron.

Monohybrid cross. A sexual cross in which the inheritance of just a single pair of alleles is followed (Tall X Dwarf).

Monosomic. An individual lacking one chromosome of a set (2n-1).

Monozygotic twins. Twins formed from a single fertilized egg that at some time in early cleavage divided into two embryos.

Morphogen. A factor that induces development of particular cell types in a manner that depends on its concentration.

M phase. The period of the cell cycle when mitosis or meiosis occurs.

Multifactorial (polygenic) trait. A trait whose phenotypic expression is influenced by the cumulative effects of many genes.

Multigene family. A group of genes, possibly although not always clustered, that are related either in nucleotide sequence or in terms of function.

Multigenic inheritance. Inheritance determined by several genes with cumulative effect, e.g., ear length in corn.

Multintron gene. The gene that contains more than one intron. Transcripts of such genes are translated into proteins.

Multiple alleles. The different alternative states of a gene that has more than two alleles.

Multiple factor hypothesis. A hypothesis that explains quantitative inheritance on the basis of action and segregation of a number of allelic pairs having duplicate and accumulative effects without complete dominance.

Multiple genes. Two or more independent pairs of genes produce complementary or cumulative effects upon a single character of the phenotype.

Multisite mutant allele. A mutant differing from its wild type form at two or more sites.

Mutagen. A chemical or physical agent able to cause a mutation in a DNA molecule.

Mutagenesis. Experimental treatment of a group of cells or organisms with a mutagen in order to induce mutations.

Mutant. A cell or organism with an abnormal genetic constitution.

Mutation. An alteration in the nucleotide sequence of a DNA molecule.

Mutation theory of evolution. deVries advocated that a single mutation can lead to radical changes in an organism which can transform one species to another. Also known as **deVriesism**.

Muton. The smallest segment of DNA or subunit of a cistron that can be changed and thereby bring about a mutation; can be as small as one nucleotide pair.

N

n, 2n. The gametic (haploid) and zygotic (diploid) chromosome numbers, respectively.

Natural selection. The process in nature whereby one genotype leaves more offspring than another because of superior life history attributes such as survival or fecundity. Darwin gave natural selection as mechanism of evolution.

Negative control. Repression of an operator site by a regulatory protein that is produced by a regulator site.

Negative eugenics. Prevention of the deterioration of the human race through a reduction in the birth rate among defectives; discouraging reproduction among those likely to perpetuate deleterious genes. Compare with **positive eugenics**.

Negative interference. The phenomenon whereby a crossover in a particular region enhances the occurrence of other apparent crossovers. Compare with **positive interference**.

Neo-Darwinism. Term that refers to merger of classical Darwinian evolution with population genetics and thus leading to the synthesis of modern theory of evolution. Also see **synthetic theory of evolution**.

Neurospora. A pink mold, commonly found growing on old food. Being haploid it is very useful in genetic studies. Working on biochemical mutants of *Neurospora*, Beadle and Tatum put forward one gene-one enzyme hypothesis.

Neutral alleles. The alleles whose differential contribution to fitness is so small that their frequencies change more due to generations than to natural selection.

Neutral gene hypothesis. The hypothesis that most genetic variation in natural populations is not maintained by selection.

Neutrality theory of protein evolution. Rates of amino acid replacements in proteins and nucleotide substitutions in DNA during evolution may be approximately constant because the vast majority of such changes are selectively neutral.

Nitrogenous base. One of the purine or pyrimidine compounds that form part of the molecular structure of a nucleotide.

Nonambiguous code. Nature of genetic code where one codon codes for only one amino acid.

Non-Darwinian evolution. A theory of evolution which considers natural selection as incompetent to account for arrival of the fittest. Saltation hypothesis, punctuation equilibrium model and evolution by random walk are in contrast to the concept of **Darwinian evolution**.

Non-disjunction. The failure of homologous chromosomes to separate at anaphase-I of meiosis. **Primary non-disjunction** may occur in an XX female, and lead to production of XX or O eggs (in addition to the normal X eggs); or it may occur (first division) in an XY male, and result in XY and O sperms or (second division) in XX and O or YY and O sperm (in addition to normal Y or X sperms). **Secondary non-disjunction** may occur in an XXY female, giving rise to eggs with XX, XY, X, or Y chromosomal combinations.

Non-histone proteins. The proteins remaining in chromatin after the histones are removed. The scaffold structure is made of nonhistone proteins.

Non-Mendelian ratio. An unusual ratio of progeny phenotypes that does not reflect the simple operation of Mendel's laws; for example, mutant: wild ratio of 3:5, 5:3, 6:2, or 2:6 in tetrads indicate that gene conversion has occurred.

Non-parental ditype (NPD). A spore arrangement in Ascomycetes that contains only the two recombinant-type ascospores (assuming two segregating loci).

Non-reciprocal recombination. Recombination in which homologous chromosomes do not undergo reciprocal exchange of genetic material. Only one chromosome may be recombinant, the order never forming.

Non-overlapping codon. Nature of genetic code where one nucleotide is part of only one codon.

Non-repetitive DNA. The DNA which shows reassociation kinetics expected of unique sequences.

Nonsense codon. One of the mRNA sequences (UAA, UAG, UGA) that signals the termination of translation.

Nonsense mutation. An altertion in a nucleotide sequence that converts triplet coding for an amino acid into a termination codon.

Non-sister chromatids. The chromatids which are derived from partner chromosomes at pachytene, or from any distinct pair of chromosomes at mitosis.

Nontranscribed spacer. The region between transcription units in a tandem gene cluster.

Normal distribution. Any of a family of bell-shaped frequency curves whose relative position and shape is defined on the basis of the mean and standard deviation.

Naomalizing selection. The removal by selection of all genes that produce deviations from the normal (= average) phenotype of a population.

Northern blotting. A gel transfer technique used for RNA.

Nucleic acid. Originally, the acidic chemical compound isolated from the nuclei of eukaryotic cells. Now, the polymeric molecules comprising nucleotide monomers: DNA and RNA.

Nucleo-cytoplasmic interaction. Activity of nuclear genes during development is limited by properties of the cytoplasm. Phenotype is regulated but not determined by cytoplasm. Genes determine a cell's potential; the cytoplasm determines whether or not that potential will be reached.

Nucleoid. The DNA-containing region within a prokaryotic cell.

Nucleolar organizer. The chromosomal region around which the nucleolus forms; site of tandem repeats of the major rRNA gene.

Nucleolus. The region of the nucleus in which rRNA transcription occurs.

Nucleoside. A chemical compound comprising a purine or pyrimidine base attached to a five-carbon sugar.

Nucleosome. The subunits of chromatin produced during chromosome coiling in eukaryotes with roughly spherical shape, which is composed of a core octamer of histones and approximately 140 nucleotide pairs of DNA. The structure comprising histon proteins and DNA that is the basic organizational unit in chromatin structure.

Nucleotide. A chemical compound comprising a purine or pyrimidine base attached to five-carbon sugar, to which a monodi-, or tiphosphate is also attached. The monomeric unit of DNA and RNA.

Nucleotide pair substitution. The replacement of a specific nucleotide pair by a different pair; often mutagenic. More commonly known as **bae pair substitution**.

Nucleus. The membrane-bound structure of a eukaryotic cell within which the chromosome are contained.

Null allele. An allele whose protein product shows no histochemically detectable activity.

Null hypothesis. The hypothesis from which the expectations are predicted for the purpose of testing significance.

Nullisomic. An aneuploid cell in which both members of one particualr pair of homologous chromosomes are missing from the chromosome complement.

O

Ochre codon. The nonsense codon UAA. One of the three nonsense codons that cause termination of protein synthesis.

Okazaki fragment. One of the short segments of RNA-primed DNA that are synthesized during replication of the lagging strand of the double helix.

Oligogenes. One or few genes governing the same qualitative character.

Oligonucleotide. A linear sequence of a few (generally not over 10) nucleotides.

Oligonucleotide-directed mutagenesis. An *in vitro* mutagenesis technique that involves the use of a synthetic oligonucleotide to introduce a predetermined nucleotide alteration into the gene to be mutated.

One gene-one enzyme hypothesis. One gene controls the synthesis of one enzyme. This hypothesis is valid for only those enzymes or proteins that are made up of only one type of polypeptide.

One gene-one polypeptide hypothesis. One gene controls the synthesis of only one polypeptide. This hypothesis is valid for all the protein-encoding genes.

One gene-one primary function hypothesis. One gene performs one specific primary cellular function.

Opal codon. The nonsense codon UGA. One of the three nonsense codons that cause termination of protein synthesis.

Open promoter complex. The complex formed between *E. coli* RNA polymerase and a promoter, in which the double helix is partially unwound in readiness for the start of RNA synthesis.

Open reading frame (ORF). A series of codons with an initiation codon at the 5'-end but no termination codon. Often considered synonymous with 'gene' but more properly used to describe a DNA sequence which looks like a gene but to which no function has been assigned.

Operator. A nucleotide sequence element to which a repressor protein attaches in order to prevent transcription of a gene or operon.

Operator-constitutive mutation (O^c). Mutation in the operator which leads to constitutive synthesis of all the gene products of the operon regardless of presence or absence of the substrate. The mutation affects activity of all the genes of the operon in *cis* arrangement not in *trans*-arrangement.

Operon. A system of cistrons, operator and promoter sites, by which a given genetically controlled metabolic activity is regulated.

Opine. The general name give to rare amino acid and sugar derivatives found in crown-gall tumors.

Ordered tetrads. In fungi refers to those ascii that are so narrow that ascospores can not change their position during meiotic divisions and subsequent mitosis. This is in contrast to unordered tetrads in which case ascospores can change their position.

Organelle. A structure within a cell that carries out a specific function; an analogue of organs of multicellular organisms.

Origin of species. The production of new types of organisms in descent from extant types. Natural selection is the process by which changes are favoured leading to the production of new types. The term was Introduced by Charles Darwin.

Overdominance. Superiority of F_1 for one or more characters over both the parents.

Overdominance hypothesis. The theory that heterosis is caused by the Aa genotype being superior to either the *AA* or the *aa* parent.

Overlapping genes. Genes whose coding regions are overlapping such that a single nucleotide sequence produces more than one polypeptide.

P

p. Relative frequency of a dominant gene.

P_1. Parental generation.

Pachytene. A substage of prophase I of meiosis I during which homologous chromosomes are completely synapsed and undergo crossing over.

Palindrome. A sequence of DNA base pairs that reads the same on the completely strands. For example, 5'GAATTC3' on one strand, and 5'CTTAAG3' on the other strand. Palindrome sequence are recognized by restriction endonucleases.

Pangenesis. A belief that very small exact but invisible copies of each body organ and component, called **gemmules** were transported by blood stream to the sex organs which were assembled into gametes.

Paracentric inversion. A rotation of a segment of chromosome a full 180°, with the centromere beyond inversion, which is all within one arm of chromosome Para means "alongside of", "beyond".

Parental ditype (PD). A spore arrangement in Ascomycetes that contains only the two nonrecombinant-type ascospores.

Partial diploids. Bacteria containing two alleles for a particular locus owing to gene transfer by transduction or sexduction.

Partial dominance. The characteristic of a selection regime in which fitness of the heterozygote is anywhere between the fitness of the two homozygotes.

Particulate inheritance. The model proposing that genetic information is transmitted from one generation to the next in discrete units ("particles"), so that the character of the offspring is not a smooth blend of essences from the parents. Compare **blending inheritance.** Also see **Mendel's law.**

Paternal inheritance. Inheritance of the cytoplasmic organelles from male parent to the offspring. First discovered by Russel in 1980 in an angiosperm *Plumbago zeylamica*.

pBR322. A particular type of artificial plasmid, frequently used as a vector for cloning genes in *E. coli*.

PCR (Polymerase chain reaction). A technique in which cycles of denaturation, annealing with primer, and extension with DNA polymerase, are used to amplify the number of copies of a target DNA sequences by more than 10^6 times.

Pedigree. The ancestral history of an individual; a chart showing such history.

Penetrance. The extent to which a phenotype is expressed, measured as the proportion of individual with a particular genotype that actually display the associated phenotype.

Peptidyl-or P-site. The site on the ribosome to which the tRNA attanhed to the growing polypeptide is bound during translation.

Peptidyl transferase. The enzyme activity responsible for peptide bond synthesis during translation.

Pericentric inversion. Refers to an inversion that does include the centromere, hence involves both arms of a chromosome.

Permanent heterosis. A hybrid condition maintained by a balanced lethal system so that *Oenothera lamarckiana* is a permanent hybrid of *gaudensvelans* complex.

Petite. A slow-growing strain of yeast (*Saccharomyces*) that lacks certain respiratory enzymes and forms unusually small colonies on agar. Segregational petites bear mutant nuclear gene(s), whereas neutual petites (sometimes called vegetative petites) bear mutant mitochondrial DNA.

Phage "ghost". The phage protein shell left behind after the phage has injected its DNA.

Phasmid. A hybrid molecule formed *in vivo* between a plasmid containing multiple att sites of λ and a bacteriophage λ derivative. The formation of a plasmid or its breadown to release the plasmid and phage is controlled by the site-specific recombination system of λ. Phasmids can replicate as a plasmid (non-lytically) or as a phage (lytically). A number of sophisticated *in vivo* genetic manipulation can be acomplished using phasmids.

Phenocopy. An evironmentally induced phenotype that resembles the phenotype produced by a mutation.

Phenotype. The observable characteristics displayed by a cell or an organism. Phenotype changes continuously during developmental processes of an individual. Phenotype is a result of specific interaction of a genotype and an environment.

Phenotypic ratio. Ratio of different phenotypes in a segregating generation, e.g., phenotypic ratio in F_2 of a monohybrid cross is 3 tall : 1 dwarf.

Phenotypic varition. When individuals with the same genotype develop in different environments, their phenotypes may be quite different.

Phenylketonuria. An autosomal recessive metabolic disorder due to accumulation of phenylalanine because of defective enzyme phenylalanine hydroxylase. It is characterized by a serious mental and physical retardation.

Phenylthiocarbamide (PTC). A substance which is tasteless to some persons and extremely bitter to others. Ability to taste is inherited as a single gene difference, nontasting being recessive.

ϕX174. A bacteriophage which attacks *E.coli*. The nucleic acid within the viral particle is a single-stranded DNA circle of 5375 bases. The replicative form is a double-stranded circle and was the first complete DNA molecule to be sequenced.

Phosphodiester bond. The chemical bond that links adjacent nucleotides in a polynucleotide.

Phyletic speciation. The process of speciation caused by the gradual change in the genetic constitution of a population without the population splitting into demes and without any increase in the number of species produced by that population at any one time.

Picornavirus. A small RNA virus.
Plasmagene. A self-replicating, cytoplasmically located gene.
Plasmid. A usually circular piece of DNA, primarily independent of the host chromosome, often found in bacterial and some other types of cell.
Plastid inheritance. Inheritance which is governed by chloroplast genes.
Pleiotropy. The ability of a single gene to produce a complex phenotype that consists of two or more distinct characteristics.
Ploidy. A variation in the number of chromosomes sets per cell.
Point mutation. A mutation that results from a single nucleotide alteration in a DNA molecule.
Poisson distribution. A statistical equation that descrives the process of sampling in situations where the number of events per sample is potentially very large but in practice is very small.
Pollen mother cell (PMC). The microsporocyte (2n) in plants immediately before the reduction division. It is the place hwere meiotic chromosomes are readily studied.
Polyadenylation. The post-transcriptional addition of a series of adenine residuces to 3'-end of a eukaryotic mRNA molecule.
Poly(A) polymerase. The enzyme responsible for polydenylation of a eukaryotic mRNA molecule.
Poly(A) tail. The initially long sequence of adenine nucleotides at the 3' end of mRNA; added after transcription.
Polyacrylamide gel electrophoresis (PAGE). A method for separating nucleic acid or protein molecules according to their molecule size. The molecules migrate through the inert gel matrix under the influence of an electric field. In the case of protein PAGE, detergents such as sodium dodecyl sulphate are often added to ensure that all molecules have a uniform charge.
Polycentric. A chromosome or chromatid having several centromeres.
Polycistronic mRNA. An RNA molecule, mostly in prokaryotes, that contains information from more than one cistrons.
Polydactyly. The occurrence of more than the usual number of fingers or toes.
Polygenes. Two to more different pairs of alleles, with a presumed cumulative effect that governs such quantitative traits as size, pigmenation, intelligence, among others. Those contributing to the traits are termed as **contributing (effective)** alleles; those appearing not to do so are referred to as **noncontributing** or **noneffective alleles**.
Polymeric genes. Two dominant alleles having similar effect when they are separate but produce enhanced effect when they come together resulting in 9:6:1 ratio.
Polyploidy. A condition in which individuals have more than two sets of chromosomes.
Polyribosome (polysome). An mRNA molecule in the process of being translated by several ribosomes at once.
Polytene chromosome. Many-stranded gaint chromosomes produced by repeated replication during synapsis in certain dipteran larval tissues. Synonym, **giant chromosome**.
Population genetics. Study of the frequencies of genes and genotypes in a Mendelian population.

Position effect. A phenotypic effect dependent on a change in position on the chromosome of a gene or group of genes. It may produce variegation, chimera, or a mosaic phenotype.

Positive control. Activation of an operator site by a regulatory protein that is produced by a regulator site. Compare with **negative control**.

Positive eugenics. Enhancement of desirable characteristics through identification of 'good' genes and encouraging facourable mating by restricting child-bearing to those best fitted to perpetuate the species. Compare with **negative eugenics**.

Positive interference. When the occurrence of one crossover reduces the probability that a second will occur in the same region. Compare with **negative interference**.

Post-zygotic isolating mechanism. Any one of several mechanisms that keep populations reoroductively isolated from each other even through fertilization and hybrid zygotes may form. These hybrids are either sterile, nonviable, or so weak that they do not survive.

Pre-mRNA. The primary, unprocessed transcript of a protein-encoding gene.

Pre-rRNA. The primary, unprocessed transcript of a gene or group of genes specifying rRNA molecules.

Pre-tRNA. The primary, unprocessed transcript of a gene or group of genes specifying tRNA molecules.

Pre-zygotic isolating mechanism. Any one of several mechanisms that keep populations reproductively isolated from each other by preventing fertilization and zygote formation.

Pribnow box. Relatively invariable six nucleotide DNA sequence TATAAT, located in eukaryotic promoters upstream (at position -10) from the start codon.

Primary cellular function. Synthesis of a transcript - mRNA, tRNA or rRNA.

Primary constriction. A constriction which is determined by and associated with the centromere region.

Primary oocytes. The cells that undergo meiosis in female animals.

Primary nondisjunction. Failure of homologous chromosomes to separate at anaphase in normal *Drosophila* female.

Primary spermatocytes. The cells that undergo meiosis in male animals.

Primary transcript. The immediate product of transcription of a gene or group of genes, which will subsequently be processed to give the mature transcript(s).

Primary trisomics. The trisomics in which the additional chromosome is a normal one.

Primase. The RNA polymerase enzyme that synthesizes the primer needed to initiate replication of a DNA polynucleotide.

Primer. A short oligonucleotide that is attached to a single-stranded DNA molecule in order to provide a site at which DNA replication of synthesis of Okazaki fragments.

Primosome. A complex of two proteins, a primase and helicase, that initiates RNA primers on the lagging DNA strand during DNA replication.

Principle of independent assortment (segregation). Different segregation gene pairs behave independently during meiosis.

548 Basic Genetics

Principle of segregation. The separation of homologous chromosomes, or members of allele pairs, into different gametes during meiosis.

Prion. An unussual infectious agent that appears to consist purely of protein with no nucleic acid.

Probability. The likelihood of a given event. Usually expressed as a number between 0 (complete certainty that the event will not occur) and 1 (complete certainty that the event will occur).

Probe. In recombinant DNA work, a redioactive nucleic acid complmentary to a region being searched for in a restriction digestor genome library.

Processed pseudogene. A pseudogene whose sequence resembles the mRNA copy of a parent gene, and which probably arose by integration into the genome of a reverse transcribed version of the mRNA.

Producer gene. A eukarytic structural gene that produces a pre-mRNA molecule which after going through some processing steps becomes mRNA. A producer gene may be under the control of several **receptor genes**.

Prokaryote. An organism whose cells are characterezed by the abesnce of a distinct nucleus and a general lack of membranous architecture.

Promiscuous DNA. The occurrence of the some DNA sequences in more than one cellular compartment. It suggests that DNA may have been exchanged between organelles, or between organelles and the nucleus.

Promoter. The nucleotide sequence, upstream of a gene, to which RNA polymerase binds in order to initiate transcription.

Proofreading. The ability of a DNA polymerase to correct misincorporated nucleotides as a result of its 3' to 5' exonuclease activity.

Prophage. The integrated form of the DNA molecule of a lysogenic phage.

Prophase. The first stage of nuclear division, including all events up to (but not including) arrival of the chromosomes at the equator of the spindle.

Prophase I. Longest period of meiosis I during which chromosomes condense, homologous chromosomes pair, undergo crosing over and then separate out but stil close at the ends. It includes substages leptotene, zygotene, pachytene, diplotene and diakinesis.

Propositus. A member of a family who first comes to the attention of a geneticist and through whom pedigree is discovered.

Protoplast. A cell from which the cell wall has been completely removed.

Protoplast fusion. A technique for producing hybrids between two cells which would not normally mate. The two cells may belong to the same or different species. The cell walls are removed from the two parent cells to create protoplasts and then fusion of the two cell membranes is promoted, usually by the addition of polyethylene glycol and Ca^{2+} ions. These fusogenic agents cause proteins to migrate from certain regions of the two cell membranes and allow areas of naked phospholipid to fuse. Subsequent regeneration of the cell wall allows the propagation of the hybrid organism. If nuclear fusion does not follow cell fusion then **heterokaryons**, rather than diploid organsims, are produced. In some crosses the genetic contribution of the two parents to the stable hybrid can be markedly unequal.

Prototroph. An organism that has no nutritional requirements beyond those of the wild type and is, therefore, able to grow on minimal medium.

Provirus. The DNA copy of retroviral genome that is integrated into host chromosomal DNA.

pSC101. A small, non-conjugative plasmid which encodes tetracycline resistance. Its origin is not clear as it is now thought to have come from a contaminant in a transformation experiment. It was one of the first vectors used in genetic engineering. The pSC101 tetracycline-resistance gene was used in the construction of pBR322.

Pseudoalleles. Nonalleles so closly linked that they are often inherited as one gene, but shown to be separable by crossover studies. Peudoalleles are found to be allelic in complementation test but nonallelic in recombination test. Now this term is discarded.

Pseudoautosomal inheritance. Pattern of inheritance shown by genes that are located on the pairing region of the X and Y chromosomes in individuals having XY mechanism of sex determination.

Pseudodominance. The expression (apparent dominance) of a recessive gene at a locus opposite a deletion.

Pseudogene. A nucleotide sequence that has similarity to a functional gene but within which the biological information has become scrambled so that the pseudogene is not itself functional.

Puff. A localized synthesis of RNA occurring at specific sites on giant chromosomes of Diptera. Also see **Balbiani ring**.

Punctuation codon. A codon that designates either the start or the end of a gene.

Punnett square. A tabular treatment used to predict the genotypes of the progeny resulting from a genetic cross in which a number of alleles are followed.

Purine. One of the two types of nitrogenous base compounds that are components of nucleotides. It has a double-ring structure.

Pure line. A population that breeds true for a particular character or a set of characters (Tall → Tall; Dwarf → Dwarf).

Pyrimidine. One of the two types of nitrogenous base compounds that are components of nucleotides. It has a single-ring structure.

Q

q. The relative frequency of a gene; when used together with p, indicates the relative fequency of the recessive alleles.

Qualitative trait. A trait for which there are relatively a fewer number of discrete phenotypes that can be distinguished by visual observation.

Qualitative inheritance. Mendelian inheritance, where discrete phenotypes give expected ratios. Also see **oligogenic inheritance**.

Quantitative genetics. A branch of genetics which deals with the inheritance of quantitative characters. Also called **biometrical genetics** or **statistical genetics**.

Quantitative inheritance. The mechanism of genetic control of traits showing continuous variation.

Quantiatative trait. A trait, such as height, which has a continuous distribution pattern within a population, and which is typically determined by the combined effects of a number of genes. There is a range of phenotypes differing in degree.

R

R_1, R_2, R_3, etc. The first, second, third, etc., generations following any type of irradiation to induce mutations.

R plasmid. A plasmid containing one or several transposons that bear resistance genes.

Radiation genetics. A branch of genetics which deals with the effects of various types of radiations on chromosomes and genes.

Random genetic drift. Changes in allelic frequency due to sampling error.

Randomized block design. An experimental design which controls fertility variation in one direction only.

Random mating. When an individual of one sex has the equal probability of mating with any individual of the opposite sex.

Random sample. A finite series of observations or individuals taken at random from the hypothetical infinitely large population of potential observations or individuals.

Rapid lysis (r) mutants. These mutants display a change in the pattern of lysis of *E. coli* at the end of an infection by a T-even phage.

Rare base pairing. Pairing between two purines or pyrimidines in DNA.

Reading frame. One of the three overlapping sequences of triplet codons that are contained in any DNA sequence.

Reassociation of DNA. Pairing of complementary single strands to form a double helix.

recA. The product of the *recA* locus of *E. coli* dual activities, activating proteases and also able to exchange single strands of DNA molecules. The protease-activating activity controls the **SOS response**; the nucleic acid handling facility is involved in recombination-repair pathways.

recBC. The genetic notation of an *E. coli* gene whose product mediates recombination. The enzyme encoded by the *recBC* gene is an ATP-dependent nuclease specific for double-stranded DNA. This enzyme will degrade linear DNA that has been taken up by transformation.

Receptor element. A controlling element that can insert into a gene (making it a mutant) and can also exit (thus making the mutation unstable); both of these functions are nonautonomous, being under the influence of the regulator element.

Receptor gene. A component of eukaryotic gene regulation with which a specific **activator RNA** molecule complexes. This interaction activates the receptor gene.

Recessive allele. The allele whose phenotype is not expressed in a heterozygote. Recessive allele expresses its effect in homozygous or hemizygous state.

Recessive epistasis. A condition where a homozygous recessive gene pair masks the effect of another gene.

Recessive lethal. An allele that is lethal when the cell is homozygous for it.
Recipient cell. The cell that receives DNA during gene transfer between bacteria.
Reciprocal cross. A second cross of the same individuals in which the sexes of the parental generation are reversed. Tall X Dwarf and Dwarf X Tall is a set of reciprocal crosses.
Reciprocal recombination. Production of new genotypes with the reverse arrangements of alleles according to maternal and paternal origin.
Reciprocal translocation. The exchange of segments between two nonhomologous chromosomes.
Recognition site. A sequence of bases within the promoter region that serves to recognize RNA polymerase molecule. See also **promoter** and **initiation site**.
Recombinant. A cell, derived from a genetic cross, that displays neither of the parental combinations for the alleles under study.
Recombinant DNA. A DNA molecule created in the test-tube by ligating together pieces of DNA that are not normally contiguous.
Recombinant DNA technology. All the techniques involved in the construction, study and use of recombinant DNA molecules.
Recombinant joint. The point at which two recombining molecules of duplex DNA are connected (the edge of the heteroduplex region).
Recombinant RNA. A term used to describe RNA molecules joined *in vitro* by T4 RNA ligase. This technique may be used to join an RNA sequence to a phage Qβ replicase template. The recombinant RNA molecule can be autocatalytically replicated by Qβ replicase to produce large amounts of an RNA sequence of interest.
Recombination. A physical process that can lead to exchange of segments of polynucleotides between two DNA molecules and which can result in the progeny of a genetic cross possessing combinations of alleles not displayed by either parent.
Recombination frequency. The proportion of recombinant progeny in the total progeny from a genetic cross.
Recon. The smallest segment of DNA or subunit of a cistron that is capable of recombination; may be as small as one deoxyribonucleotide pair.
Reductional division. The first meiotic division. It reduces the number of chromosomes and centomeres to half that of the original cell.
Regulated gene. A gene whose expression is regulated. It product is synthesized in the cell type, at a time and in the amount at which it is required.
Regulator(y) gene. A gene that codes for a protein, such as a **repressor**, involved in regulation of expression of other genes.
Release factors. Proteins (RF1, RF2) in prokaryotes responsible for termination of translation and release of the newly synthesized polypeptide when a nonsense codon appears in the A site of the ribosome. Replaced by eRF in eukaryotes. Also known as **termination factors**.
Renner complexes. Specific gametic chromosome combinations in *Oenothera*.
Repair. The correction of alterations in DNA structure before they are inherited as mutations.

Repeated genes. Some genes may be present in a haploid genome in many copies. These copies may be identical or only similar.

Repetitive and unique sequence hypothesis. Only small proportion of DNA is unique and is meant for genes carrying the genetic information and rest of DNA is Repetitive and has some other functions like control of gene activity and serving as raw material for evolution.

Repetitive DNA. DNA made up of copies of the same nucleotide sequence.

Replica-plating. A technique used to make replicas of bacterial colonies from a master plate by placing a velvet covering wooden block over the master plate and making an imprint of the same on a fresh plate. This method is used to select particular mutants from a plate.

Replicase. An enzyme that unwinds double helical DNA.

Replication. (1) The process of copying; replication of DNA. (2) The equal incorporation of all combinations two or more times in a n experimental design, which is then said to be replicated.

Replication fork. The region of a double-stranded DNA molecule that is beings unwound to enable DNA replication to occur.

Replication origin. A site on a DNA molecule where unwinding begins in order for replication to occur.

Replication-defective virus. The virus which has lost one or more genes essential for completing the infective cycle.

Replication eye. A region in which DNA has been replicated within a longer, unreplicated region.

Replicator. A DNA segment which contains an origin of replication and is able to promote the replication of a plasmid DNA molecule in a host cell.

Replicon. A sequentially replicating segment of a nucleic acid, controlled by a sebsegment known as a **replicator**. A single replicator is present in the bacterial "chromosome", whereas the chromosomes of eukaryotes bear large numbers of replicons in series.

Replisome. The DNA-replicating structure at the Y-junction consisting of DNA polymerase III enzymes and a primosome (primase and DNA helicase).

Repressible control. End product (co-repressor) acts to repress reproduction of the enzymes. Transcription is initiated in the absence of the end product.

Repressible enzyme system. A coordinated group of enzymes, involved in a synthetic pathway (anabolic), is repressible if excessquantities of the end product of the pathway lead to the termination of transcription of the genes for the enzymes. These systems are primarily prokaryotic operons.

Repression. The ability of a bacteria to prevent synthesis of certain enzymes when their products are present; more generally, refers to inhibition of transcription (or translation) by binding of repressor protein to a specific site on DNA (or mRNA).

Repressor. A regulator molecule (protein) produced by a fegulator gene that can combine with and repress action of an associated operator. sample.

Repulsion. Allelic arrangement in which each homologous chromosome has mutant and wild-type alleles.

Repulsion linkage. Tendency of two dominant characters, in an F_1 inherited one from one and the other from the other parent, to stay apart in the F_2 generation.

Resistance transfer factor (RTF). Infectious transfer part of R plasmids.

Resistant. Characteristic of a host plant such that it is capable of suppressing or retarding the development of a pathogen or other injurious factor.

Restriction digest. The results of the action of a restriction endonuclease on a DNA

Restriction endonuclease. Any of a group of enzymes that break internal bonds of DNA at highly specific points.

Restriction fragment length polymorphism (RFLP). Variations in banding patterns of electrophoresed restriction digests. RFLPs have certain traits - lack of dominance, multiple allelic forms, lack of pleiotropic effects on agronomic traits, co-dominance, no measurable effect on phenotype, no effect of environment.

Restriction map. A physical map of a piece of DNA showing recognition sites of specific restriction endonuclease separated by lengths marked in number of bases. Also known as **cleavage map**.

Restriction site. The base sequence at which a restriction endonuclease cuts the DNA molecule, usually a point of symmetry within a palindrome sequence.

Retroposon. A transposon that mobilize via an RNA form; the DNA element is transcribed into RNA, and then reverse-transcribed into DNA, which is inserted at a new site in the genome. Such as **retrotransposon**.

Retrovirus. A viral retroelement whose encapsidated genome is made of RNA.

Reversal of central dogma. A phenomenon mostly observed in RNA viruses where reverse transcriptase uses RNA as a template to produce complementary DNA.

Reverse mutation. Any change towards the standard (wild type) by way of a second mutational event. Also known as **reversion** or **back mutation**.

Reverse transcriptase. An enzyme that synthesizes a DNA copy on an RNA template.

Reverse transcription. Synthesis of a DNA copy on an RNA template.

Reversion. Reversion of mutation is a change in DNA that either reverse the original alteration (**true reversion**) or compensates for it (**second site reversion** in the same gene).

Revertants. Revertants are derived by reversion of a mutant cell or organism.

R factors. Plasmids that carry genes that control resistance to various drugs. Also known as **R plasmids**.

Rh factor. An antigen first detected in the red blood cells or Rhesus monkeys.

Rho factor. A protein that is required for termination of transcription.

Ribonuclease. An enzyme that degrades RNA.

Ribonucleic acid (RNA). A single-stranded nucleic acid molecule, synthesized principally in the nucleus from deoxyribonucleic acid, composed of a ribose-phosphate backbone with purines (adenine and guanine) and pyrimidines (uracil and cytosine) attached to the sugar ribose. RNA is of several kinds of functions to carry the "genetic message" from nuclear DNA to the ribosomes.

Ribonucleoside. Protion of an RNA molecule composed of one ribose molecule plus either a purine or a pyrimidine.

Ribonucleotide. Portion of an RNA molecule composed of one ribose-phosphate unit plus a purine or a pyrimidine.

Ribosomal RNA (rRNA). The RNA molecules that act as structural components of ribosomes.

Ribosome. One of the protein-RNA structures on which translation occurs.

Ribosome binding site. The nucleotide sequence that acts as the site for attachment of a ribosome to an mRNA molecule.

Ribozyme. An RNA molecule that possesses catalytic activity.

Right splicing junction. The boundary between the right end of an intron and the left end of the adjacent exon.

Ring chromosomes. A physically circular chromosome. Usually found in bacteria.

RNA-dependent DNA polymerase. A group of enzymes that catalyze formation of DNA molecule from RNA templets. These occur in some viruses (e.g. those that produce tumors). See also **reverse transcriptase**.

RNA editing. A process by which nucleotides not coded by the gene are introduced at specific positions in an mRNA molecule after transcription.

RNA phages. Phages whose genetic material is RNA.

RNA polymerase. An enzyme capable of synthesizing an RNA copy of a DNA template.

RNA polymerase core enzyme. In bacteria it consists of 5 subunits ($\alpha^2 \beta \beta' \omega$) and is capable of elongating already initiated transcription by holoenzyme.

RNA polymerase holoenzyme. The version of the *E. coli* RNA polymerase that has the subunit composition $\alpha_2 \beta \beta' \omega \sigma$ and is involved in efficient recognition of promoter sequences and is capable of initiation of transcription.

RNase. An enzyme that hydrolyzes RNA.

RNase D. The enzyme responsible for processing pre-tRNA by cleaving at the 3'-termini of the mature tRNA sequences.

RNase P. The enzyme responsible for processing pre-tRNA by cleaving at the 5'-termini of the mature tRNA sequences.

RNA processing. Involves the steps which heterogenous nuclear RNA (pre-mRNA) transcribed by RNA polymerase II undergoes to produce a finished mRNA molecule. These steps involve (a) addition of a "cap" (5-m^{7}Gppp) at the 5' end, (b) addition of a poly(A) tail at the 3' end, (c) splicing of noncoding sequences (introns), and methylation of one out of every 400 adenines present.

RNA replicase. A polymerase enzyme that catalyzes the self replication of single-stranded RNA.

RNA splicing. The removal of large noncoding sequences (introns) from the primary RNA transcript followed by rejoining of the nonadjacent coding sequences (exons) to produce the functional mRNA.

RNA transcript. An RNA copy of a gene.

Rolling circle replication. A model of DNA replication that accounts for a circular DNA molecule producing linear daughter double helices.

Rough endoplasmic reticulum (RER). It consists of endoplasmic reticulum associated with ribosomes.

S

s. The symbol representing the selection coefficient in a mathematical treatments of the effects of natural selection.

Salivary chromosomes. The giant chromosomes in the cells of the salivary glands of larval flies and other dipterans such as mosquitoes and *Drosophila* characteristically each shows a highly specific pattern of contrasting bands. See **grant chromosomes**.

Satellite. A portion of the chromosome separated from the main body of the chromosome by secondary constrictions.

Satellite chromosome. Chromosomes that seem to be additions to the normal genome.

Satellite DNA. DNA comprising of clustered repetitive sequences, so-called because it forms a satellite band in a density gradient.

Scaffold. The eukaryotic chromosome structure remaining when DNA and histones have been removed; made from nonhistone proteins.

Second-division segregation (SDS). Segregation of two alleles at a locus at meiosis I. SDS gives 1:1:1:1 ratio during tetrad analysis.

Second messenger. A molecule (cAMP) that mediates response of eukaryotic cells to altered extracellular environment.

Second site reversion. A second mutation that reverses the effect of a previous mutation in the same gene although without restoring the original nucleotide sequence.

Sedimentation coefficient. A value used to express the velocity with which a molecule or structure sediments when centrifuged in a dense solution.

Segmental allopolyploid. An allopolyploid in which the combined genomes are homologous in many small segments throughout complement; crossing-over may recombine material from different genomes.

Segregation. The separation of homologous alleles at meiosis.

Segregation distorter (SD). A factor that alters the segregation ratio in heterozygous *Drosophila*; SD + cells are eliminated during spermiogenesis in SD + /SD males.

Segregational load. Genetic load caused when population is segregating less fit homozygotes because of heterozygote advantage.

Segregational petite. Petite in yeast that arises due to mutation in nuclear gene.

Segregation ratio. The expected (Mendelian) or observed ratio between genotypes or phenotypes in the progeny of a cross.

Selection. Non-random reproduction of different genotypes that induce evolutionary changes.

Selector gene. A gene that influences the development of specific body segments in *Drosophila*; a homeotic gene.

Self-fertilization. Fertilization of a female gamete with a male gamete derived from the same individual.

Self-incomparability alleles. These are alleles of a multiple allelic series which determine compatibility of perfectly fertile male and female flowers. If a pollen grain bears a self-incomaptibility (S) allele that is also present in the maternal parent, then it will not germinate but if that allele is not present in the maternal tissue, then the pollen grain produces a pollen tube containing the male nucleus, and this tube affects fertilization.

Selfish DNA. DNA that appears to have no function and apparently contribution nothing to the cell in which it is found. Also known as **ignorant** or **junk** DNA.

Self-pollination. Pollinating a plant with its own pollen; selfing.

Self-sterility. Incapability of producing seed when self-pollinated. Several alleles, S_1, S_2, S_3, etc., are responsible for this phenomenon.

Semi-conservative replication. The mode of DNA replication in which each daughter double helix comprises one polynucleotide from the parent and one newly synthesized polynucleotide.

Sense codon (sense word). A codon specifying a particular amino acid in protein synthesis.

Sense DNA strand. The strand of DNA which is complementary to the one used as a template during transcription.

Sensor gene. A eukaryotic regulatory gene that is sensitive to external/developmental signals.

Sex cell. A cell that divides by meiosis.

Sex chromatin. Barr body; facultative heterochromatin of the X-chromosomes of normal mammalian females but not males.

Sex chromosome. A chromosome which is involved in sex determination. These chromosomes are not similarly distributed in both sexes. Distribution of a sex chromosome to one but not to other of the products of meiosis determines differences in sex of the offspring.

Sex determination. Mechanisms responsible for determining sex of a zygote during development.

Sexduction. Incorporation of bacterial chromosomal genes in the fertility plasmid, with subsequent transfer to a recipient cell in conjugation.

Sex factor (F). An episome of bacteria which, when integrated into the chromosome, causes chromosome breakage at conjugation and facilitates genetic exchange; also exists autonomously in F^+ cell. Also known as **fertility** factor or **sex-particle**.

Sex-influenced trait. A trait in which the dominance-recessiveness relationship of two alleles in a heterozygote is influenced by sex of the individual and thus the same genotype expresses differently in the two sexes. Aso known as **sex-controlled trait**.

Sex-lethal. A gene in *Drosophila*, located on the X chromosome, that is a sex switch, directing development regulated by numerator and the denominator elements that act to influence the genic balance equation (X/A).

Sex-limited gene. The gene that is differently expressed in the two sexes because of the presence or absence of sex hormones.

Sex-linked gene. Refers to a gene that is located on a X-chromosome.

Sex-linked inheritance. Inheritance of sex-linked genes. Same as **criss cross inheritance**.

Sex pilus. A tube structure present on the exterior of F^+ bacteria, thought to be involved in transfer of DNA during conjugation.

Sex plasmid. Actually an episome, it is able to initiate the process of conjugation, by which chromosomal material is transferred from one bacterium to another.

Sex reversal. Transformation of one sex into another.

Shine-Dalgarno sequence. The prokaryotic ribosome-binding site AGGAGG located on mRNA just prior to AUG initiation codon which has complementarity with the 3' end of 16S rRNA.

Short-term gene regulation. Gene regulation recognized in eukaryotes which operates in response to fluctuations in environment, changes in activities or concentrations of substrates, end product or hormone levels. It is a feature of both developing and fully differentiated cells, or operates even when the cell is undergoing differentiation

Shotgun experiment. Cloning of an entire genome in the form of randomly generated fragments.

Shuttle vector. A plasmid constructed to have origins for replication for two hosts (for example, *E. coli* and *S. cerevisiae*) so that it can be used to carry a foreign sequence in either prokaryotes or eukaryotes.

Sibling species. Morphologically similar or identical populations that are reproductively isolated. These species may be sympatric or allopatric. These species may exit at the same time or at different time.

Sickle cell trait. Trait shown by individuals characterized as heterozygote, h^B/h^S. In the homozygous condition, h^S/h^S, red blood cell become sickle-shaped under reduced oxygen tension.

Sigma factor. A proteinous component of RNA polymerase which is must for initiation of transcription. It is not required for elongation of the transcript.

Simple gene. A continuous sequence in a nucleic acid that specifies a particular polypeptide or functional RNA.

Single-strand binding (ssb) protein. One of the proteins that attaches to single-stranded DAN in the region of the replication fork, preventing reannealing of unreplicated DNA.

Sister chromatids. Copies of a chromosome produced by its replication and joined together by a centromere.

Site-directed mutagenesis. It is construction of mutations at the predetermined site of a cloned DNA, precisely defining the nature of mutational change and then test functional effect of that mutation *in vivo* or *in vitro*.

Small nuclear ribonucleoproteins (snRNPs). Small nuclear ribonucleoproteins; components of the spliceosome, the intron-removing apparatus in eukaryotic nuclei.

Small nuclear RNA (snRNA). The RNA component of the nucleus that comprises relatively small molecules thought to be involved in splicing and other transcript processing events.

Smooth Endoplasmic Reticulum (SER). It consists of a regions of endoplasmic reticulum devoid of ribosomes.

S1 nuclease. An enzyme that degrades specifically single-stranded molecules or single-stranded regions in predominantly double-stranded nucleic acid molecules.

Solenoid. A higher order structure in a eukaryotic chromosome in which 100 Å fibres of nucleosomes are stacked with 5 or 6 nucleosomes/helix. H1 histone helps in solenoid formation which gives a package ratio of 1:50.

Somaclonal variation. Variation produced in plants regenerated from tissue culture involving callus formation.

Somatic cell hybridizatyon. A technique that permits hybridization between somatic cells of same or different species in tissue culture.

Somatic cell mutation and recombination (SMART). A test which detects induction of mutations and recombination in somatic cells (like eyes and wings of *Drosophila*). it is used to test mutagenicity of individual chemical compounds and mixtures.

SOS box. The region of the promoters of various genes that is recognized by the LexA repressor. Release of repression results in the induction of the SOS response. SOS stands for "**save our souls**".

SOS repair system. Repair systems (recA, uvr) induced by the presence of single-stranded DNA that usually occurs from post-replicative gaps caused by various types of DNA damage. The RecA protein, stimulated by single-stranded DNA, is involved in the inactivation of the LexA repressor, thereby inducing the response. This is also known as **SOS response**.

Southern blotting. A method, first devised by E.M. Southern, used to transfer DNA fragments from an agarose gel to a nitrocellulose gel for the purpose of DNA-DNA or DNA-RNA hybridization during recombinant DNA work.

Spacer DNA. Regions of nontranscribed DNA between transcribed segments. Term generally applies to immunoglobulin genes.

Specialized gene. A gene whose activity is under the control of constitutive as well as regulatory factors.

Specialized nucleoriboprotein structures (snurps). These are small ribonucleoprotein particles that help to remove meaningless introns from the massage issued by a cell's genes.

Specialized transduction. A type of transduction where only a few bacterial genes are transferred because the phage has only specific sites of integration on the host chromosome.

Speciation. A process whereby, over time, one species evolves into a different species (**anagenesis** or **phyletic speciation**) or whereby one species diverges to become two or more species (**cladogenesis** or **true speciation**).

Species. (1) A group of actually or potentially interbreeding natural populations which are reproductively isolated from other such groups (Mayrs, 1910). (2) A largest and most conclusive reproductive community of sexual and cross-fertilizing individuals which share a common gene pool (Dobzhansky, 1937). (3) A lineage evolving separately from others and with its own evolutionary role and tendencies (Simpson, 1949).

S phase. See **syntheitc phase**.

Spindle. The microtubule apparatus that controls chromosome movement during mitosis and meiosis.

Spliceosome. The protein-RNA structure believed to be responsible for splicing.

Splicing. The removal of introns from the primary transcript of a discontinuous gene.

Split gene. A discontinuous gene. A gene in which coding sequences are separated by noncoding sequences.

Spontaneous mutations. Genetic changes produced under normal growth conditions.

Stabilizing selection. A type of selection that removes individuals from both ends of a phenotypic distribution divided by deviation of a sample of means.

Stacking forces. Weak forces between two adjacent base pairs of double helical DNA.
Standard deviation. A measure of the variation in a sample. It is square root of the variance (σ). Symbolized by s.
Standard error of difference in means. Measure of the significance of the difference in two sample means. Symbolized by S_d.
Standard error of sample mean. An estimate of the standard deviation of a series of hypothetical sample means that serves as a measure of the closeness with which a given sample mean approximates the population mean. Symbolized by s_x.
Start codon. A codon which codes for initiation of protein synthesis, i.e., AUG.
Start point. The position on DNA corresponding to the first base incorporated into RNA. Also known as **start site**.
Stasipatric speciation. Instantaneous speciation caused by polyploidy.
Statistic. Actual value of some quantitative character for a sample from which estimates of parameters may be made.
Statistics. Measurements of attributes of a sample from a population; denoted by Roman letters. See **parameters**.
Stem-loop structure. A lollipop-shaped structure formed when a single-stranded nucleic acid molecule loops back on itself to form a complimentary double helix (stem), topped by a loop.
Sticky end. An end of a double-stranded DNA molecule where there is a single-stranded extension. Also known as **adhesive end**.
Stop codon. The codons which provide signal for termination of polypeptide chain, viz., UAA, UAG and UGA.
Strain. A group of individuals from a common origin. Generally, a more narrowly defined group than a variety.
Strand displacement. It is a mode of replication of some viruses in which a new DNA strand grows by displacing the previous (homologous) strand of the duplex.
Strong promoter. A promoter whose copy number is under strict control and, therefore, has only one or two copies per chromosome, i.e., a low copy number plasmid, e.g., the F plasmid.
Structural alleles. Alleles determined on the basis of recombination test.
Structural gene. A gene that codes for an RNA molecule or protein other than a regulatory protein.
Submetacentric chromosome. A chromosome where the centromere is nearer one end than the other, resulting in the arms not being of equal length.
Super female. An abnormal type, almost completely sterile, in *Drosophila* with an X chromosome/autosome ratio greater than 1, e.g., 3X/2A=1.5.
Super male. An abnormal type, almost completely sterile, in Drosophila with the X chromosome/autosome ratio less than 0.5, e.g. 1X/3A=0.33.
Super gene. Advantageous grouping of genes within an inversion; the entire gene complex is inherited as a whole.
Supernumerary chromosomes. chromosome present, often in varying numbers, in addition to the characteristic relatively invariable complement.

Supplementary genes. Two dominant genes when present together produce a novel phenotype. These genes yield 9:3:4 ratio in F_2 progeny.

Suppression. A mutation in a gene that reverses the effect of a previous mutation in a different gene. This term also refers to **rescue** of an allele by means of another gene.

Suppressive petite. Petite in yeast that occurs due to mutation of DNA in mitochondria.

Suppressor, Extragenic. Usually a gene coding a mutant tRNA that reads the mutated codon either in the sense of the original codon or to give an acceptable substitute for the original meaning.

Suppressor gene. A gene that, when mutated, apparently restores the wild-type phenotype to a mutant of another locus.

Surrogate genetics. A branch of genetics which deals with introduction of a manipulated DNA into a living nucleus where its expression can be monitored. Effects of directed DNA changes, changes in RNA transcript and changes in protein encoded by it can all be studied.

Survival of the fittest. In evolutionary theory survival of only those organisms best able to obtain and utilize resources (fittest). This phenomenon is the cornerstone of **Darwin's theory.**

S value. The unit of measurement of a sedimentation coefficient. Also known as **Svedberg unit.**

Swivel. The place in DNA at which replication starts.

Synapsis. The pairing of homologous chromosomes in prophase-I of meiosis.

Synaptonemal complex. A structure that forms between the paired homologous chromosomes at the pachytene stage, consisting of a tripartite ribbon of parallel, dense, and lateral elements surrounding a medial complex.

Syndrome. A group of symptoms that occur together and characterize a disease.

Syngamy. The union of the nuclei of sex cells (gametes) in reproduction.

Synonymous codons. Different codons that specify the same amino acid.

Synteny. The occurrence of two or more genetic loci on the same chromosome. Depending on intergene distance(s) involved, they may or may not exhibit nonrandom assortment at meiosis.

Synthetic theory of evolution. The current extension of Darwin's theory of evolution, in which mutation, natural selection, and reproductive isolation play major roles.

Synthetic phase. A stage in interphase during which DNA synthesis occurs. At the end of this phase the amount of DNA doubles.

T

t. The ratio of an observed deviation to its estimated standard deviation.

Tachytelic evolution. Evolution at a much faster rate than horoletic evolution.

TACTAAC box. A consensus sequence surrounding the lariat branch point of eukaryotic mRNA.

TATA box. A component of the nucleotide sequence located at position -25 that makes the prokaryotic promoter.

Tautomeric shift. Reversible shift of proton position in a molecule. Based on nucleic acids shift between keto and enol forms or between amino and imino forms.

Tay-Sachs disease. A degenerative brain disorder of infancy due to an autosomal recessive allele in the gene controlling the enzyme hexosaminidase A. The age of onset of the disease is 4 to 6 months. Affected children show progressive mental deterioration, paralysis, deafness, blindness, and convulsions, leading to death usually between the ages of 3 and 5 years.

T-DNA. Complete Ti plasmid is not found in plant tumor cells, only 20kb DNA of Ti plasmid called T-DNA is found integrated in plant nuclear DNA. Genes of T-DNA are in eukaryotic in origin and have been captured by Ti plasmid during evolution.

Telocentric chromosome. A chromosome with a terminal centromere.

Telomerase. The enzyme that maintains the ends of chromosomes by synthesizing telomeric repeat sequences.

Telomere. The end of a chromosome.

Telophase. The concluding stage of nuclear division, characterized by the reorganization of interphase nuclei.

Telophase I. The concluding stage of meiosis I which is characterized by the reappearance of the two nuclei which carry haploid number of chromosomes.

Telophase II. The concluding stage of meiosis II which is characterized by the reappearance of the nuclei, four in number, which carry haploid number of chromosomes.

Temperate phage. A bacteriophage that is able to follow a lysogenic mode of infection.

Template. A pattern serving as a mechanical guide. In DNA replication each strand of the duplex acts as a template for the synthesis of new double helix.

Termination codon. One of the three codons (5'-UAA-3', 5'-UAG-3' and 5'-UGA-3' in the standard genetic code) that mark the position where translation of an mRNA should stop.

Termination factors (TF). Proteins required to obtain release of newly synthesized polypeptide chain from tRNA. Also known as **release factors**.

Terminator stem. A configuration of the leader transcript that signals transcription termination in attenuator-controlled amino acid operons.

Test cross. A genetic cross between an individual (generally of dominant phenotype) and the one having recessive phenotype. It is use to determine genotype (homozygous or heterozygous) of an individual of dominant phenotype. It is also used to construct linkage maps.

Testing of hypothesis. The determination of whether to accept or reject a proposed null hypothesis based on the likelihood of the experimental results.

Test systems. The organisms and methodology used to test mutagenicity and/or carcinogenicity of environmental chemical and physical agents.

Tetrad. The four haploid cells that result from a single meiotic event in a eukaryotic germ cells.

Tetrad analysis. The genetic analysis of the products of a single meiosis division. Term generally applied to Ascomycetes.

Tetraploid. A polyploid cell, tissue, or organism with four sets of chromosomes (4n).

Tetrasomic. Having two extra chromosomes of a given kind, making four of the kind in question (2n + 2).

Tetratype (TT). A spore arrangement in Ascomycetes that consists of two parental and two recombinant spores, indicating a single crossover between two linked loci.

Thalassemias. Group of diseases of red blood cells resulting from lack of either α or β globin.

Theory of inheritance of acquired traits. Acquired characters are inherited and adaptive, and thus account for evolutionary change. Also see **Lamarckism**.

Theta-structure. A θ-like structure seen as an intermediate in replication of a circular DNA molecule.

3'-OH terminus. The end of a polynucleotide which terminates with a hydroxyl group attached to the 3'-carbon of the sugar.

Thermoregulation of a gene. Regulation of expression is controlled by temperature. Histone-like bacterial proteins seem to play role in this process.

Three-point test cross. A trihybrid testcross (e.g., ABC/abc X abc/abc, etc.) used primarily in chromosome mapping.

Thymidine. The deoxyribonucleoside that contains the pyrimidine thymine.

Thymine. 5-mehtyl uracil, also written as 2,6-dioxy, 5-methyl pyrimidine, one of the four nitrogenous bases found in DNA.

Thymine dimer. Hydrogen bonding of two molecules of thymine by action of ultraviolet light.

Ti plasmid. The large plasmid found in *Agrobacterium tumefaciens* cells and used as the basis for a series of cloning vectors for higher plants.

Tissue culture. A technique by which a full fledged individual can be developed, on a well defined medium, from a somatic cell or an individual sex cell.

Tolerable mutations. The mutations which affect function of the gene to such a degree that they do not have a drastic effect on the individual and are tolerable. Such mutations accompany the process of evolution and play important role in speciation. Tolerable mutations are of two types - neutral and favourable.

Topoisomerase. An enzyme that can relieve (or create) supercoiling in DNA by creating transitory breaks in one (type I) or both (type II) strands of the helical backbone.

Totipotency. The property of a cell (or cells) whereby it develops into a complete and differentiated organism.

Trailer segment. The untranslated segment that lies downstream of the nonsense codon of a mRNA molecule.

Trans-**acting.** Referring to mutations of, for example, a repressor gene, that act through a diffusible protein product; the normal mode of action of most recessive mutations.

Transcribed spacer. Part of an rRNA transcription unit that is transcribed but discarded during maturation; that is, it does not give rise to part of rRNA.

Transcription. The synthesis of an RNA copy of a gene or a DNA template.

Transcription factors. Eukaryotic proteins that aid RNA polymerase to recognize promoters. Analogous to procaryotic sigma factors.

Transcription unit. The distance between sites of initiation and termination by RNA polymerase; may include more than one gene.

Transduction. Recombination in bacteria whereby DNA is transferred by a phage from one cell to another. Generalized transduction involves phages that have incorporated a segment of bacterial chromosome during packaging of the phage; specialized transduction involves temperate phages that are always inserted into the bacterial chromosome at a site specific for that phage.

Transduction mapping. The use of transduction to determine the relative positions of two or more genes on a bacterial DNA molecule.

Transfection. The introduction of foreign DNA into eukaryotic cells.

Transfer operon (*tra*). Sequence of loci that impart the male (F-pili-producing) phenotype on a bacterium. The male cell can transfer the F plasmid to an F⁻ cell.

Transfer reaction. A reaction that attaches an amino acid with its specific tRNA. his reaction is conducted by a specific amino acyl synthetase.

Transfer RNA (tRNA) One of the small RNA molecules that act as adaptors during translation and are responsible for decoding the genetic code.

Transformant. A cell that has been become transformed by the uptake of naked DNA.

Transformation. (1) The acquisition by a bacterial cell of new genes by the uptake of naked DNA. (2) Also refers to **cancerous growth**.

Transformation-competent viruses. These viruses are able to transform the infected cells and may replicate via DNA or RNA intermediate and can transfer genetic information horizontally or vertically.

Transformation, eukaryotic. Conversion of eukaryotic cells to a state of unrestrained growth in culture, resembling or identical with the turmoigenic condition.

Transformation-incompetent viruses. They are able to replicate but not transform the infected cells. They may replicate via DNA or RNA intermediate and can transfer genetic information horizontally or vertically.

Transformation mapping. A method of gene mapping in bacteria through conjugation.

Transformed cells. Cells that have become capable of sustained proliferation *in vitro*.

Transformer. An allele in fruit flies that converts chromosomal females into sterile males.

Transforming principle. The chemical substance responsible for transforming *Streptococcus* bacteria from one serotype to another, shown by Avery and colleagues to be DNA.

Transgenic. Eukaryotic organisms that have taken up foreign DNA.

Transgressive segregation. The phenomenon through which we get variation in F_2 or later generation outside the range of both the parents.

***Trans*-heterozygote.** A heterozygote of two linked genes with a mutant and wild type allele linked, a+/+b. This is more appropriate term than "repulsion phase" of linkage, since like alleles do not repel.

Transient polymorphism. Some populations contain detrimental mutations which are gradually eliminated by selection. Selection in such cases acts to decrease polymorphism.

Translation. The synthesis of a polypeptide, the amino acid sequence of which is determined by the nucleotide sequence of an mRNA in accordance with the rules of the genetic code.

Translocase (EF-G). Elongation factor in prokaryotes necessary for proper translocation at the ribosome during the translation process. Replaced by eEF2 in eukaryotes.

Translocation of chromosome. A rearrangement in which part of a chromosome is detached by breakage and then becomes attached to some other chromosome.

Translocation of gene. Appearance of a new copy of a gene at location in the genome elsewhere from the original copy.

Translocation of protein. The movement of a protein from (the site of synthesis) to the (site of action).

Translocation of ribosome. It is movement a ribosome from one codon to the next along mRNA after the addition of each amino acid to the polypeptide chain.

Transposon (Transposable genetic element). A segment of DNA, which generally consists of more than 2,000 nucleotides, and is capable of moving into and out of a chromosome or plasmid and/or from place to place within the chromosome in both prokaryotes and eukaryotes. It is capable of turning genes "on" and "off".

Transmission genetics. The study of the mechanisms involved in the passage of a gene from one generation to the next. Alos known as **classical genetics** or **Mendellian genetics**.

Transposase. The enzyme that catalyses transposition of a transposable genetic element.

Transposition. The process whereby a transposon or insertion sequence inserts into a new site on the same or another DNA molecule. The exact mechanism is not fully understood and different transposons may transpose by different mechanisms. Transposition in bacteria does not require extensive DNA homology between the transposon and the target DNA. The phenomenon is, therefore, described as illegitimate recombination.

Transposon tagging. Blocking activity of a functional gene by inserting a foreign DNA sequence.

Transposon-mediated gene transfer. Transferring a gene from one species to another by use of a transposon.

Trans-sexual. A person who has adopted the appearance of the opposite sex with the aid of surgical procedures and hormone therapy.

Trans-specific evolution. The evolution of groups above the species level. Also known as **macroevolution**.

Transvection. The ability of a locus to influence activity of an allele on the other homologue only when two chromosomes are synapsed.

Transversion. A point mutation that results in a purine being replaced by a pyrimidine, or vice versa.

Trihybrid. An individual heterozygous for three pairs of alleles. For example, +++/abc.

Triple X. Trisomy for X chromosome.

Triplet code. A group of three successive nucleotide in RNA or DNA that, in the genetic code, specifies a particular amino acid in the synthesis of polypeptide chains.

Triplet binding assay. An experimental technique that enables the coding specificity of a triplet of nucleotide to be determined.

Triplicate genes. Three genes, any one of which produces the same phenotypic effect, e.g., red seed coat colour in wheat may be produced by R_1, R_2, R_3.

Triploid. A polyploid cell, tissue, or organism with three sets of chromosomes (3n).
Trisomic. An individual with one extra chromosome of a set (2n + 1).
Trisomy 21. A genetic disease of man in which chromosome 21 is represented three times per cell; **mongolism**. Also known as **Down's Syndrome**.
Trivalent. Three synapsed homologous chromosomes.
tRNA deacylase. The enzyme responsible for cleaving the bond between a tRNA molecule and the growing polypeptide during translation.
tRNA nucleotidal transferase. The enzyme responsible for the post-transcriptional attachment of the nucleotide sequence 5'-CCA-3' to the 3'-end of a tRNA molecule.
t-test. A test of statistical significance which is used for comparing two means when sample size is small (less than 30). It is of two types. Student's t-test used is with paired observations, and Fisher's t-test used is when observations are not paired.
Tumor-inducing (Ti) plasmid. A 140-235 kb plasmid found in *Agrobacterium tumefaciens* having virulence trait.
Tumor suppressor genes. Those genes which code for proteins that inhibit cell replication.
Turner's syndrome. A series of abnormalities in humans due to monosomy for the X chromosome. Individuals are phenotypically female, but are sterile.
Twisting number. In double helical DNA, this is the number of base pairs divided by the number of base pairs per turn of the double helix.
Type I error. In statistics the rejection of a true hypothesis.
Type II error. In statistics the accepting of a false hypothesis.
Tyrosinosis. A human disease caused by metabolic defect and characterized by excretion of p-hydroxy phenyl pyruvic acid and tyrosine into urine.

U

U1-U7 RNA. A family of stable, small nuclear RNA molecules originally found in rat Novikoff hepatoma cells but subsequently discovered in a range of mammals and other higher eukaryotes. The U RNAs are used in pre-mRNA processing.
μm (micrometer). 1×10^{-6} meter.
Ubiquitin. A heat-shock protein present in yeast and chicken cells, although apparently not present in *Drosophila*.
Underdominance. A phenotypic relation in which the phenotypic expression of the heterozygote is less than that of either homozygote.
Underwinding. Underwinding of DNA is produced by negative supercoiling (because the double helix is itself coiled in the opposite sense from the intertwining of the strands).
Unequal crossing-over. Non-reciprocal crossing over caused by mismatching of homologous chromosomes. Usually occurs in regions of tandem repeats.
Unidentified reading frame (URF). An open reading-frame recognized from a DNA sequence for which no genetic function is known.
Unidirectional replication. Movement of a single replication fork from a given origin.

Uninemic chromosome. A chromosome consisting of one double helix of DNA.

Uniparental inheritance. Inheritance pattern where the offspring have received certain phenotypes from only one parent. This inheritance pattern is due to transmission of DNA containing cytoplasmic particles.

Unique DNA. A length of DNA with no repetitive nucleotide sequences.

Unisite mutant allele. A mutant allele differing from its wild type form at only one site.

Unit factor. The term used by Mendel for the gene.

Univalent. A single chromosome seen segregating in meiosis.

Universal codon. A triplet codon codes for same amino acid in diverse forms of life, viz., bacteria, wheat, *Drosophila*, man.

Universal donor. A person with group O blood, whose erythrocytes, therefore, bear neither A nor B antigens, and whose blood can be donated to members of groups O, A, B, and AB, if necessary.

Universal recipient. A person with group AB blood, who can receive blood from members of groups O, A, B, and AB, if necessary.

Unstable gene. A gene which mutates frequently.

Unstable mutation. A mutation that has a light frequency of reversion; a mutation caused by the insertion of a controlling element, whose subsequent exit produces a reversion.

Unusual bases. Other bases, in addition to adenine, cytosine, guanine, and uracil, found primarily in tRNAs.

Unusual chromosomes. The chromosomes which show adaptational forms of the normal chromosomes (e.g. lamp brush chromosomes, polytene chromosomes) or these may be permanently specialized structures (e.g. B chromosomes, sex chromosomes).

Unwinding proteins. Proteins that bind to, and unwind, the DNA helix at the replicating fork.

Up promoter mutations. Mutations in promoter that increase the frequency of initiation of transcription.

Upstream. Towards the 5'-end of a polynucleotide represents sense strand of DNA.

Upstream activating sequence (UAS). This is a DNA segment required for activating transcription of yeast genes.

Uracil. 2,6-dioxypyrimidine, one of the nitrogenous bases found in RNA.

Uridine. The ribonucleoside that contains the pyrimidine uracil.

U-RNP. A nuclear particle, consisting of one or two U-RNAs and several proteins, involved in splicing and other transcript processing events.

Use and disuse doctrine. The discredited idea that the more a structure is used, the more prominent it becomes in future generations, and that the less it is used, the less prominence it assumes in later generations. Also known as the **doctrine of acquired characteristics**.

U snRNPs. U-class small ribonucleoproteins that regulate sex-specific alternative splicing of transcripts required for sexual differentiation.

V

Vacuole. A cytoplasmic empty organelle surrounded by a membrane called tonoplast and contains sugar, organic acids, inorganic salts, proteins and pigments.

Variable-number-of-tandem repeats (VNTR) loci. Loci that are hypervariable because of tandem repeats. Presumably, variability is generated by unequal crossing over.

Variable region. An immunoglobulin chain is coded by the V gene and varies extensively when different chains are compared, as the result of multiple (different) genomic copies and changes introduced during construction of an active immunoglobulin.

Variance (σ). A statistic providing an unbiased estimate of population variability and is calculated as the average squared deviation about the mean of a set of data. It is the square of the standard deviation.

Variance, environmental. The variance resulting from environmental or non-genetic causes.

Variance, genetic. The variance resulting from genetic causes.

Variance, phenotypic. The total variance, the sum of the environmental and the genetic variance.

Vector. A DNA molecule capable of replication into which a gene is inserted by recombinant DNA techniques.

Vehicle. The host organism used for the replication or expression of a cloned gene or other sequence. (Compare with **vector**, the DNA molecule which contains the cloned gene.) The term is little used and is often confused with **vector**.

Vehicle plasmid. A plasmid containing a piece of passenger DNA; used in recombinant DNA work.

Virion. The genome of a virus.

Viroids. Bare RNA particles that are plant pathogens.

Virulent phage. Refers to a bacteriophage that follows the lytic mode of infection and destroys its host bacterial cell.

Virus. An infective particle, composed of protein and nucleic acid, that must parasitize a host cell in order to replicate.

W

Watson-Crick base pairing. The normal base pairing, viz., A with T and G with C in DNA and A with U and G with C in RNA.

W chromosome. Used for X chromosome where female is the heterogametic sex.

Western blotting. A technique for probing for a particular protein using antibodies. See **Southern blotting**.

Wild-type. A gene, cell or organism that displays the typical phenotype and/or genotype for the species and is, therefore, adopted as a standard.

Wobble base pairing. The pairing of mRNA codon with tRNA anticodon in which first two bases of a codon have normal pairing and the third base has abnormal base pairing.

Wobble hypothesis. The partial or total lack of specificity in the third base of some triplet codons whereby two, three, or four codons differing only in the third base may code for the same amino acid.

X

x. Basic number.

X_1, X_2, X_3, etc. Different generations obtained from an individual that has been irradiated.

X/A ratio. Ratio between number of X chromosomes to number of sets of autosomes that determines sex in *Drosophila*.

X chromosome. The sex chromosome that is represented twice in the homogametic sex but only once in the heterogametic sex.

Xenia effect. Direct effect of male gamete on tissues other than embryonic ones.

Xeroderma pigmentosum. A genetic disorder in which the skin is extremely sensitive to sunlight and death usually occurs from skin cancer. It is inherited as an autosomal recessive condition.

X inactivation. The genetic inactivation of all X chromosomes in excess of one, taking place on a random basis in each cell in an early stage in embryogenesis.

X-linkage. It refers to genes located on the X chromosome. Also see **sex-linkage**.

XO condition. Having an X chromosome but no Y chromosome, as the female in poultry and some other birds.

X syndrome. Turner's syndrome, 45 chromosomes, chromatin negative.

XXY syndrome. Klinefelter syndrome, 47 chromosomes, chromatin positive.

Y

Yates's correction. A correction applied in the calculation of normal deviates or X^2's to allow for the discrepancy arising by the observations being discontinuous while tables of the normal deviate and X^2 are calculated on the supposition of continuity in the variate.

Y chromosome. The chromosome that is not represented in the homogametic sex but only once in the heterogametic sex.

Yeast artificial chromosome (YAC). A cloning vector comprising the structural components of a yeast chromosome and able to clone very large piece of DNA.

Y-linkage. It refers to genes located on the Y chromosome. Also see **holandric genes**.

Z

Z chromosome. Used for the Y chromosome where female is the heterogametic sex.

Zinc finger. A structural motif found in several DNA-binding proteins.

Zygote. The cell that results from fusion of gametes during meiosis.

Zygotene. A substage of prophase I of meiosis I during which homologous chromosomes begin to synapse.

Z-DNA. The DNA in which sugar and phosphate linkage follow a zig zag pattern. Such DNA has left handed double helical model.

Z-test. A test of significance which is used to compare two means when sample size is large (more than 30).

BIBLIOGRAPHY

Adams, R.L.P., J.T. Knowler, and D.P. Leader. 1986. The Biochemistry of Nucleic Acids London: Chapman and Hall.
Adhya, S. and M. Gottesman 1978. Control of transcription termination. Ann. Rev. Biochem. 47:967-996.
Aggarwal, S. 1992. Antisense oligonucleotides as antiviral agents, TIBTECH 10: 152-158.
Allard, R.W. 1960. *Principles of Plant Breeding*. New York: Wiley.
Altenburg, E. 1934. The artificial production of mutations by ultraviolet light. American Naturalist 68:491-507.
Altman, S. (ed.). 1978. *Transfer RNA*. Cambridge: MIT Press.
American Association for the Advancement of Science. 1995. Frontiers in biotechnology: Emerging plant science. Science 268:654-691.
Ames, B. 1979. Identifying environmental chemicals causing mutations and cancer. Science 204:587-593.
Ames, B.N., J. McCann and E. Yamasaki. 1975. Methods for detecting carcinogens and mutagens with *Salmonella*/mammalian -microsome mutagenicity test. Mut. Res. 31:347-364.
Andreason, G.L. and G.A. Evans. 1988. Introduction and expression of DNA molecules in eukaryotic cells by electroporation. BioTechniques 6:650-660.
Anderson, W.F. and E.G. Diecumakos. 1981. Genetic Engineering in mammalian cells. Scient. American 245:106.
Arnheim, N. and H. Erich. 1992. Polymerase chain reaction strategy. Ann. Rev. Biochem. 61:131-56.
Ashburner, M. and T.R. Wright (eds.). 1978-1980. *The Genetic and Biology of Drosophila*. Volumes 2a-2d. New York: Academic Press.
Auerbach, C. 1949. Chemical Mutagenesis. Biol. Rev. 24:355-391.
Auerbach, C. 1961. *The Science of Genetics*. New York: Harper & Bros.
Auerbach, C. 1976. *Mutation Research: Problems, Results and Perspectives*. London: Chapman and Hall.
Auerbach, C. and J.M. Robson. 1946. Chemical production of mutations. Nature 157:302.
Avery, O.T., C.M. MacLeod, and M. McCarty. 1944. Studies on the Chemical Nature of Substance Inducing Transforming of Pneumococcal Types. Inducation of Transformation by a Desoxyribonucleic Acid Fraction Isolated from Pneumococcus Type III. J. Exptl. Med. 79:137-158.

Bibliography 571

Ayala, F.J. 1992. *Evolution and Evolutionary Genetics - A Primer*. Menlo Park: Benjamin/Cummings Publ. Co.

Bachmann, B.J. and K.B. Low. 1980. Linkage map of *Escherchia coli* K-12, edition 6. Microbiol. Rev. 44:1-56.

Baglioni, C. 1963. Correlation between genetics and chemistry of human hemoglobins. In: *Molecular Genetics*. Part I, J.H. Taylor (ed.). New York: Academic Press. pp. 405-475.

Bejema, C.J. (ed.). 1976. *Eugenics: Then and Now*. Stroudsburg, Pa: Dowden, Hutchinson and Ross.

Bak, A.L., J. Zeuthen, and F.H.C. Crick. 1977. High order structure of human mitotic chromosomes. Proc. Natl. Acad. Sci. USA 74:1595-1599.

Baker, B.S., M. Gorman, and I. Marin. 1991. Dosage compensation in *Drosophila*. Ann. Rev. Genet. 28:491-521.

Baltimore, D. 1970. RNA dependent DNA Polymerase in virions of RNA tumor viruses. Nature 226:1209-1211.

Barkley, M.D. and S. Bourgeosis. 1978. Repressor recognition of operators and effectors. In: *The Operon*. J.H. Miller and W.S. Reznikoff (eds). Cold Spring Harbor Laboratory. New York: Cold Spring Harbor. pp. 177-200.

Barratt, R.W., D. Newmeyer, D.D. Perkins, and L. Garniobst. 1954. "Map Construction in *Neurospora crassa*". Adv. Genet. 6:1-93.

Barrell, B.G., G.M. Air, and C.A. Hutchison, III 1976. Overlapping genes in bacteriophage φX174. Nature 264:33-41.

Bateman, A.J. 1947. Number of S-alleles in a population. Nature 160:337.

Baylor, M.B., A.Y. Hessler and J.P. Baird. 1965. The circular linkage map of bacteriophage T2. Genetics 51:351-361.

Beadle, G.W. 1945. Genetics and Metabolism in *Neurospora*. Physiol. Revs 25:643-663.

Beadle, G.W. and B. Ephrussi. 1973. Development of eye colors in *Drosophila*:Diffusable substances and their interrelations. Genetics 22:76-86.

Beadle, G.W. and E.L. Tatum. 1941. Genetic control of biochemical reactions in *Neurospora*. Proc. Nat. Acad. Sci. 27:499-506.

Beadle, G.W. and E.L. Tatum. 1945. *Neurospora*. II. Methods of producing and detecting mutations concerned with nutritional requirements. American J. Bot. 32:678-686.

Beale, G.H. 1954. *The Genetics of Paramecium aurelia*. Cambridge: Cambridge Univ. Press.

Beale, G. and J. Knowles. 1978. *Extranuclear Genetics*. London: Edward Arnold.

Beckwith, J.R. and D. Zipser (eds.). 1970. *The Lactose Operon*. New York: Cold Spring Harbor Laboratory.

Beisson, J., and T.M. Sonneborn. 1965. Cytoplasmic inheritance of the organization of cell cortex in *Paramecium aurelia*. Proc. Nat. Acad. Sci. USA 53:275-282.

Bell, P.B. and B. Stillman. 1992. ATP-dependent recognition of eukaryotic origins of DNA replication by a multiprotein complex. Nature 357:128-134.

Benzer, S. 1961. Genetic Fine Structure. In: *Harvey Lectures 56*. New York: Academic Press.

Benzer, S. 1962. The fine structure of the gene. Scient. Am. 206(1):70-64.
Berg, P., D. Baltimore, H.W. Boyer, S.N. Cohen, R.W. Davis, D.S. Hogness, D. Nathans, R. Roblin, J.D. Watson, S. Weissman and N.D. Zinder. 1974. Potential biohazards of recombinant DNA molecules. Proc. Nat. Acad. Sci. USA 71:2593-2594.
Berlyn, M.B. and S. Letovsky. 1992. COTRANS:A program for cotransduction analysis. Genetics 131:235-41.
Beyreuther, K. 1978. Chemical structure and functional organization of *lac* represor from *Escherichia coli*. In: *The Operon*, J.H. Miller and W.S. Reznikoff (eds.). Cold Spring Harbor Laboratory. New York: Cold Spring Harbor. pp. 123-154.
Biemont, C. 1983. Homeostasis, enzymatic heterozygosity and inbreeding depression in natural populations of *Drosophila melanogaster*. *Genetica* 61:179-189.
Birchler, J.A. 1996. X chromosome dosage compensation in *Drosophila*. Science 272:1190.
Birge, E.A. 1981. *Bacterial and Bacteriophage Genetics*. New York: Springer-Verlag.
Bolivar, F., R.L. Rodriguez, P.J. Greene, M.C. Betlach, H.L. Heyneker, H.W. Boyer, J.H. Crosa, and S. Falkow. 1977. Construction and characterization of new cloning vehicles. II. A multipurpose cloning system. Gene 2:95-113.
Bonner, D.M. 1965. Gene-enzyme relationships. In: *Genetics Today*. Vol. II, Proc. XI Intern. Congr. Genet., S.J. Geerts (ed.). Oxford: Pergamon. pp. 141-148.
Borgaonkar, D.S. 1994. *Chromosomal Variation in Man: A Catalogue of Chromosomal Variants and Anomalies*. 7th ed. New York: Alan R. Liss, Inc.
Boveri, T. 1904. Erebnnisse über die Konstitution der chromatischen substanz des Zellkerns. G. Fischer, Jena.
Boycott, A.E. and C. Diver. 1923. On the inheritance of sinistrality in *Limnaea peregra*. Proc. Roy. Soc. Lond. (B), 95:207-213.
Brennan, C.A., A.J. Dombroski, and T. Platt. 1987. Transcription termination factor rho is an RNA-DNA helicase. Cell 48:945-52.
Brenner, S. 1955. Tryptophan biosynthesis in *Salmonella typhimurium*. Proc. Natl. Acad. Sci. USA 41:862-863.
Brenner, S., F. Jacob, and M. Meselson. 1961. An unstable intermediate carrying information from genes to ribosomes for protein synthesis. Nature 190:576-581.
Brenner, S., L. Barnett, F.H.C. Crick, and A. Orgel. 1961. The theory of mutagenesis. J. Mol. Biol. 3:121-124.
Brenner, S., A.O.W. Stretton and S. Kaplan. 1965. Genetic code:The "nonsense" triplets for chain termination and their suppression. Nature 206:994-998.
Brewbaker, J.L. 1964. *Agricultural Genetics*. Englewood Cliffs: Prentice-Hall.
Brewster, T. and P. Gerald. 1978. Chromosome disorders associated with mental retardation. Pediatric Annals 7:82-89.
Bridges, C.B. 1916. Nondisjunction as proof of the chromosome theory of heredity. Genetics 1: 1-52, 107-163.
Bridges, C.B. 1925. Sex in Relation to Chromosomes and Genes. Am. Naturalist 59:127-137.

Bridges, C.B. 1936. The bar "gene", a duplication. Science 83:210-211.
Britten, R. and D. Kohne. 1968. Repeated sequences in DNA. Science 161:529-540.
Britten, R.J. and E.H. Davidson. 1969. Gene regulation for higher cells:a theory. Science 165:349-357.
Broda, P. 1979. *Plasmids*. San Francisco: W.H. Freeman.
Brock, T.D. 1990. *The Emergence of Bacterial Genetics*. Cold Spring Harbor, New York: Cold Spring Harbor Laboratory Press.
Brown, D.D. 1981. Gene expression in eukaryotes. Science 211:667-674.
Brown, T.A. 1992. Genetics: A Molecular Approach. London: Van Nostrand Reinhold Int.
Brown, W.V. 1972. *Textbook of Cytogenetics*. St. Louis: C.V. Mosby.
Brusick, D. 1987. *Principles of Genetic Toxicity*. 2nd ed. New York: Plenum.
Buc, H. 1986. Mechanism of activation of transcription by the complex formed between cyclic AMP and its receptor in *E. coli*. Biochem. Soc. Trans. 14:196-199.
Bull, J.J. and R.C. Vogt. 1979. Temperature-dependent sex determination in turtles. Science 206:1186-1189.
Burdette, W.J. 1962. *Methodology in Human Genetics*. San Francisco: Holden-Day.
Burdette, W.J. 1963. *Methodology in Basic Genetics*. San Francisco: Holden-Day.
Burgoyne, P.S. 1982. Genetic homology and crossing over in the X and Y chromosomes in mammals. Human Genet. 61:85-90.
Burnham, C.R. 1962. *Discussions in Cytogenetics*. Minneapolis: Burgess.
Burns, G.W. and P.J. Bottino. 1989. *The Science of Genetics*. New York: Maxwell Mac Millan Intern.
Burton, G.W. 1951. Quantitative Inheritance in Pearl Millet. Agr. J. 43:409-417.
Busby, S. and R.H. Ebright. 1994. Promoter structure, promoter recognition, and transcription activation in prokaryotes. Cell 79:43-46.
Byers, R.E., L.R. Baker, H.M. Sell, R.C. Herner, and D.R. Dilley. 1972. Ethylene, a natural regulator of sex expression of *Cucumis melo*. Proc. Nat. Acad. Sci. USA 69:717-720.
Cairns, J. 1963. The chromosome of *Escherichia coli*. Cold Sp. Harb. Symp. 28:43-45.
Cairns, J., G.S. Stent, and J.D. Watson (eds.). 1966. *Phage and the Origin of Molecular Biology*. Cold Spring Harbor Laboratory Quantitative Biology, Cold Spring Harbor, New York.
Callan, H.G. 1963. The nature of lampbrush chromosomes. Intern. Rev. Cytol. 15:1-34.
Callan, H.G. 1986. *Lampbrush Chromosomes*, Vol. 36. *Molecular Biology, Biochemistry, and Biophysics*. New York: Springer-Verlag.
Campbell, A.M. 1962. Episomes. Adv. Genet. 11:101-145.
Campbell, A.M. 1969. *Episomes*. New York: Harper & Row.
Capecchi, M.R. 1994. Targeted gene replacement. Scient. Am, 270(3): 34-41.
Carlson, E.A. 1959. Comparative genetics of complex loci. Quart. Rev. Biol. 34:33-67.
Carlson, P.S. 1970. Induction and isolation of Auxotrophic mutants in somatic cell cultures of *Nicotiana tobaccum*. Science 168:487-489.
Carr, K. 1996. Gene expression: RNA bound to silence. Nature 379:676.

Caruthers, M.H. 1985. Gene synthesis machines. DNA chemistry and its use. Science 230:281-85.

Caskey, C.T. 1987. Disease diagnosis by recombinant DNA methods. Science 236:1223-1229.

Castle, W.E. and C.C. Little. 1910. On a modified Mendelian ratio among yellow mice. Science 32:868-870.

Cazenave, C. and C. Hélène. 1991. Antisense oligonucleotides, In *Antisense Nucleic Acids and Proteins*, Mol, J.N.M. and vander Krol, R.R. (eds). New York: Marcel Dekker. Inc. pp. 47-03.

Cech, T.R. 1990. Self-splicing of group 1 introns. Ann. Rev. Biochem. 59:543-568.

Chambon, P. 1981. Split genes. Scient. Am. 244(5): 48-59.

Chao, L., C. Vargas, B. B. Spear, and E.C. Cox 1983. Transposable elements as mutator genes in evolution. Nature 303:633-635.

Chapeville, F., F. Lipmann, G. von Ehrenstein, B. Weisblum, W.J. Roy, Jr., and S. Benzer. 1962. On the rule of soluble ribonucleic acid in coding for amino acids. Proc. Nat. Acad. Sci. USA 48:1086-1092.

Chargaff, E. and J.N. Davidson. 1955. *The Nucleic Acids*. New York: Academic Press.

Charles, D.R. and H.H. Smith. 1939. Distinguishing between two types of gene action in quantitative inheritance. Genetics 24:34-48.

Cheverud, J.M. and E. Routman. 1995. Epistasis and its contribution to genetic variacne components. Genetics 139:1455-1461.

Chilton, M.D. 1983. A vector for introducing new genes into plants. Scient. Ame. 248(6):51-59.

Chovnick, A. 1961. The garnet locus in *Drosophila melanogaster*. I. Pseudoallelism. Genetics 46:493-507.

Clayton, D.A. 1982. Replication of animal mitochondrial DNA. Cell 28:693-705.

Cleland, R.E. 1972. *Oenothera: Cytogenetics and Evolution*. London: Academic Press.

Clewell, D.B. 1993. *Bacterial Conjugation*. New York: Plenum Press.

Cocking, E.C., M.R. Davey, D. Pental, and J.B. Power. 1981. Aspects of plant genetic manipulation. Nature 293:265-270.

Corcos, A.F. 1983. Pattern baldness: Its genetics revisited. Am. Biol. Teacher 45:371-375.

Corcos, A. and F. Monaghan. 1985. Role of deVries in the recovery of Mendel's work. I. Was deVries an indepenent disoverer of Mendel? J. Hered. 76:187-90.

Corcos, A. and F. Monaghan. 1987. Correns, an independent descoverer of Mendelism? I. An historical/critical note. J. Hered. 78:330.

Correns, C. 1909. Verebungsversuche mit blass (gleb) grünen und bunblättrigen bei *Mirabilis jalapa*, *Urtica pilulifera* und *Lunaria annua*. Z. Ind. Abst. Vererb.-lehre. 1:291-329.

Coyne, J.A. 1992. Genetics and speciation. Nature 355:511-515.

Creighton, H.B. and B. McClintock. 1931. A correlation of cytological and genetical crossing over in *Zea mays*. Proc. Nat. Acad. Sci. USA 17:492-497.

Crick, F.H.C. 1957. *Nucleic Acids*. Scient. Am. Available as Offprint 54. San Francisco: W.H. Freeman and Co.

Crick, F.H.C. 1958. On protein synthesis. Symposia, Society for Experimental Biology 12:138-163.
Crick, F.H.C. 1966. Codon-anticodon pairing: The wobble hypothesis. J. Mol. Biol. 19:548-555.
Crick, F.H.C. 1966. The genetic code III. Scient. Am 215(4):55-62.
Crick, F.H.C. 1968. The origin of the genetic code. J. Mol. Biol. 38:367-379.
Crick, F.H.C. 1970. Central dogma of molecular biology. Nature 227:561-563.
Crick, F.H.C. 1979. Split genes and RNA splicing. Science 204:264-267.
Crick, F.H.C., L. Barnett, S. Brenner and R.J. Watts-Tobin. 1961. General nature of the genetic code for proteins. Nature 192:1227-1232.
Crow, J.F. 1963. *Genetics Notes*. Minneapolis: Burgess Publishing Co., 5th ed.
Crow, J.F. 1986. *Basic Concepts in Population, Quantitative, and Evolutionary Genetics*. New York: Freeman.
Cuenot, L. 1980. Sur quelques anomalies apparentes des proportions mendeliennes. Arch. Zool. Exp. et Genet. 9:7-15.
Daniel, A., ed. 1988. *The Cytogenetics of Mammalian Autosomal Rearrangements*. New York: Alan R. Liss, Inc.
Darlington, C.D. and Ammal, E.K.J. 1945. *Chromosome Atlas of Cultivated Plants*. London: G. Allen & Unwin, Ltd.
Darlington, C.D. 1958. *The Evolution of Genetic Systems*. 2nd ed. New York: Basic Books.
Darlington, C.D. 1965. *Cytology*. London: Churchill.
Darwin, C. 1859. *The Origin of Species by Means of Natural Selection or the Preservation of Favored Races in the Struggle for Life*. London: John Murray.
Darwin, C. and A.R. Wallace, 1859. On the tendency of species to form varieties; and on the prepetuation of varieites and species by natural means of selection. J. Linn. Soc. 3:45-62.
Das, A. 1993. Control of transcription termination by RNA binding proteins. Ann. Rev. Biochem. 62:893-930.
Davidson, J.N. 1957. *The Biochemistry of Nucleic Acids*. London: Methuen and Co.
Davidson, E.H. and R.J. Britten 1979. Regulation of gene expression: possbile role of repetitive sequences. Science 204:1052-1059.
Debenham, P.G. 1992. Probing identity: The changing face of DNA fingerprinting. Trends in Biotech. 10:96-102.
DeCrombrugghe, B., S. Busby, and H. Buc 1984. Cyclic AMP receptor protein: Role in transcription activation. Science 224:831-838.
Demerec, M. 1937. Frequency of sontaneous mutation in certain stocks of *Drosophila melanogaster*. Genetics 22:469-478.
Demerec, M. (ed.). 1950. *The Biology of Drosophila*. New York: John Wiley.
Demerec, M. and B.P. Kaufman. 1950. *Drosophila Guide*. Washington, DC: Carneige Institution of Washington.
Denniston, K.J. and I.W. Enquist (eds.). 1981. *Recombinant DNA*. Benchmark Papers in Microbiology. New York: Academic Press.

DePamphilis, M.L. 1988. Microinjecting DNA into mouse ova to study DNA replication and gene expression and to produce transgenic animals. BioTechniques 6:662-680.
DeVries, H. 1901. *Die Mutationtheorie.* Veit, Leipzig.
Dickerson, R., H.R. Drew, B.N. Conner, R.M. Wing, A.W. Fratini, and L.M. Kopka. 1982. The anatomy of A-, B-, and Z-DNA. Science 216:475-85.
Dickson, R.C., J. Abelson, W.M. Barnes, and W.S. Roznikoff. 1975. Genetic regulation: The lac control region. Science 187:27-35.
Dillon, L.S. 1987. *The Gene - Its structure, Function and Evolution.* New York: Plenum Press.
Dixon, W.J. and F.J. Massey, Jr. 1983. *Introduction to Statistical Analysis.* 4th ed. New York: McGrew-Hill.
Dobzhansky, T.H. 1951. *Genetics and the Origin of Species.* 3rd ed. New York: Columbia Univ. Press.
Dobzhansky, T.H. 1962. *Mankind Evolving.* New Haven, Conn.: Yale Univ. Press.
Dobzhansky, T.H. 1970. *Genetic of the Evolutionary Process.* New York: Columbia Univ. Press.
Dobzhansky, Th. 1973. Nothing in biology makes sense except in the light of evolution. Am. Biol. Teacher, 35:125-129.
Dobhansky, Th., F.J. Ayala, G.L. Steebins, and J.W. Valentine. 1977. *Evolution.* San Francisco: W.H. Freeman & Co.
Doermann, A.H. and M.B. Hill. 1953. Genetic structure of bacteriophage T4 as described by recombination studies of factors influencing plaque morphology. Genetics 38:79-90.
Douglas, L. and E. Novitski. 1977. What chance did Mendel's experiments give him of noticing linkage? Heredity 38:253-57.
Dover, G.S. and R.B. Flavell (eds.). 1982. *Genome Evolution.* London: Academic Press.
Drake, J.W. 1969. Comparative rates of spontaneous mutation. Nature 221:1132.
Drake, J.W. 1970. *The Molecular Basis of Mutation.* San Francisco: Holden-Day.
Dressler, D. and H. Potter. 1982. Molecular mechanisms in genetic recombination. Ann. Rev. Biochem. 51:727-761.
Dreyfuss, G., M.J. Matunis, S. Pinol-Roma, and C.G. Burd. 1993. hnRNP proteins and the biogenesis of mRNA. Ann. Rev. Biochem. 62:289-321.
Drlica, K. and M. Riley, eds. 1990. *The Bacterial Chromosome.* Washington, DC: American Society of Microbiology.
Dunn, L.C. 1962. Cross currents in the history of human genetics, Am. J. Hum. Genet. 14:1-13.
Eanes, W.F. 1978. Morphological variance of enzyme heterozygosity in the monarch butterfly. *Nature* 276:263-264.
East, E.M. 1916. A Mendelian interpretation of variation that is apparently continuous. Am. Naturalist 44:65.
East, E.M. 1936. Heterosis. Genetics 21:375-397.
Eaton, G.J. and M.M. Green. 1962. Implantation and lethality of the yellow mouse. Genetica 22:14-42.
Echols, H. and H. Murialdo. 1978. Genetic map of bacteriophage lambda. Microbiol. Rev. 42:577-591.
Editorial. 1990. The secret of sex. The Lancet 336:348-349.

Edwards, J.W. and G.M. Curuzzi. 1990. Cell-specific gene expression in plants. Ann. Rev. Genet. 24:275-304.
Eguchi, Y., T. Itoh, and J.-I. Tomazawa. 1991. Antisense RNA. Ann. Rev. Biochem. 60:631-652.
Ehling, U.H. 1991. Genetic risk assessment. Ann. Rev. Genet. 25:255-280.
Ehrenstein, G. von, B. Weisblum, and S. Benzer. 1963. The function of sRNA as amino acid adaptor in the synthesis of hemoglobin. Proc. Nat. Acad. Sci. USA 49:669-675.
Ellis, J. 1982. Promiscuous DNA-chloroplast genes inside plant mitochondria. Nature 299:678-679.
Ellis, R.J. 1991. Molecular chaperones. Ann. Rev. Biochem. 60:321-347.
Emerson, S. 1963. Meiotic recombination in fungi with special reference to tetrad analysis. In: *Methodology in Basic Genetics*. W.J. Burdette (ed.). San Francisco: Holden-Day. pp. 167-206.
Emerson, S.H. 1935. The genetic nature of DeVries' mutation in *Oenothera lamarckiana*. Am. Natural. 69:545-559.
Emerson, S.H. 1969. Linkage and recombination at the chromosome level. In: *Genetic Organization*. Vol, I. E.W. Caspari and A.W. Ravin (eds.). New York: Academic Press. pp. 267-360.
Englesberg, E. and G. Wilcox 1974. Regulation: Positive control. Ann. Rev. Genet. 8:219-242.
Ephrussi, B. 1953. *Nucleo-Cytoplasmic Relations in Micro-Organisms*. London: Oxford Univ. Press.
Epstein, C.J. and M.S. Golbus. 1977. Prenatal diagnosis of genetic diseases. Am. Scient. 65:703-711.
Erskine, A.G. and W.W. Socha. 1978. *Principles and Practice of Blood Grouping*. 2nd ed. St. Louis: Mosby.
Fahring, R. 1974. Development of host-mediated mutagenicity tests. Mut. Res. 26:29-36.
Falconer, D.S. 1960. *Introduction to Quantitative Genetics*. New York: Ronald Press.
Farnsworth, W.W. 1988. *Genetics*. New York: Harper & Row.
Fincham, J.R.S. 1966. *Genetic Complementation*. New York: W.A. Benjamin.
Fincham, J.R.S. and R.R. Day. 1963. *Fungal Genetics*. Philadelphia: F.A. Davis & Co.
Fisher, R.A. 1930. *The Genetical Theory of Natural Selection*. Oxford: Clarendon Press.
Fisher, R.A. 1936. Has Mendel's work been rediscovered? Annals of Science 1:115-37.
Fisher, R.A. and F. Yates 1963. *Statistical Tables for Biological, Agricultural and Medical Research*. 6th ed. Edinburgh: Oliver and Boyd.
Flanders, P.H., S.C. West, and A. Stasiale. 1984. Role of RecA protein spiral filaments in genetic recombination. *Nature* 309:215-220.
Fogel, S. and R.K. Mortimer. 1969. Informational transfer by meiotic gene conversion. Proc. Nat. Acad. Sci. USA 62:96-103.
Fox, T.D. 1987. Natural variation in the genetic code. Ann. Rev. Genet. 21:67-91.
Fraenkel-Conrat, H. and B. Singer. 1957. Virus reconstitution: combination of protein and nucleic acid from different strains. Biochem. et Biophys. Acta 24, 540-548.
Freese, E. 1963. Molecular mechanism of mutation. In: *Molecular Genetics*. Part I (Taylor, J.H. ed.), London: Academic Press. p. 207-269.

Frederick, J.E. (ed.). 1981. *Origins and Evolution of Eukaryotic Intracellular Organelles.* New York: Ann. N.Y. Acad. Sci.

Freeling, M. 1978. Allelic variation at the level of intragenic recombination. Genetics 80:211-224.

Freese, E. 1971. Molecualr mechanisms of mutations. In: *Chemical Mutagens.* Volume 1. A. Hollaender (ed.). New York: Plenum Press. pp. 1-56.

Freifelder, D. 1987. *Microbial Genetics.* Boston: Science Books International.

Frelbery, C., R. Fellay, A. Bairoch, W.J. Brougton, A. Rosenthal, and X. Perret. 1997. Molecular basis of symbiosis between Rhizobium and legumes. Nature 387: 394-401.

Garmeraad, S. 1976. Genetic transformation in *Drosophila* by microinjection of DNA. Nature 252: 229-231.

Garrod, A.E. 1902. The incidence of alkaptonuria: A study in chemical individually. The Lancet 2: 1616-1620.

Garrod, A.E. 1909. *Inborn Errors of Metabolism.* Oxford: Oxford Univ. Press.

Garza, D., J.W. Ajioka, D.T. Burke and D.L. Hartl. 1989. Mapping the *Drosophila* genome with yeast artificial chromosomes. Science 246:641-646.

Gasser, C.S. and R.T. Fraley. 1992. Transgenic crops. Scient. Am. 266(6): 34-39.

Gheysen, G., G. Angenon and M. Van Montagu. 1992. Transgenic plants: *Agrobacterium tumefaciens*-mediated transformation and its use for crop improvement. In: *Transgenesis,* ed. J.A.H. Murray 187-232. West Sussex, England: John Wiley.

Gilbert, W. and D. Dressler. 1968. DNA replication: the rolling circle model. Cold Sp. Harb. Symp. 33:473-484.

Gilbert, W. and B. Muller-Hill. 1966. Isolation of the lac repressor. Proc. Natl. Acad. Sci. USA 56:1891-1898.

Gilbert, W. and B. Müller-Hill 1967. The lac operator is DNA. Proc. Nat. Acad. Sci. USA 58:2415-2421.

Gill, K.S. 1981. Developmental homeostasis in self and cross pollinated crops. *Summer Institute on "Recent advances in Genetics & Unconventional Tech. in Crop. Imp.,* P.A.U., Ludhiana (June-July, 1981).

Gillespie, J.H. 1991. *The Cause of Molecular Evolution.* Oxford: Oxford University Press.

Gillham, N.W. 1965. Linkage and recombination between non-chromosomal mutations in *Chlamydomonas reinhardi.* Proc. Nat. Acad. Sci. USA 54:1560-1567.

Gillham, N.W. 1978. *Organella Heredity.* New York: Raven Press.

Gillis, A.M. 1994. Turning off the X chromosome. BioScience 44:128-132.

Glover, J. 1984. *What Sort of People Should There Be?* Middlesex, England: Penguin Books.

Gold, L. 1988. Posttranscriptional regulatory mechanisms in *Escherichia coli.* Ann. Rev. Biochem. 57:199-233.

Goldberger, R.F. 1974. Autogenous regulation of gene expression. Science 183:810-816.

Goldstein, A. 1964. *Biostatistics, An Introductory Text.* New York: Macmillan, Inc.

Goodfellow, P.J., S.M. Darling, N.S. Thomas, and P.N. Goodfellow. 1986. A pseudoautosomal gene in man. Science 234:740-743.

Gorman, M. and B.S. Baker. 1994. How flies make one equal two: Dosage compensation in *Drosophila*. Trends Genet. 10:376-380.

Gowen, J.W. (ed). 1952. *Heterosis*. Ames: Iowa State Coll. Press.

Gowen, M.S. and J.W. Gowen. 1922. Complete linkage in *Drosophila melanogaster*. Am. Natural. 56:286-288.

Grant, V. 1977. *Organismic Evolution*. San Francisco: W.H. Freeman & Co.

Grant, V. 1981. *Plant Speciation*. 2nd ed. New York: Columbia Univ. Press.

Grant, V. 1991. *The Evolutionary Process: A Critical Review of Evolutionary Theory*. 2nd ed. New York: Columbia University Press.

Green, M.C. 1963. Methods for testing linkage. In: *Biology of the Laboratory Mouse*. 2nd ed., E.L. Green (ed.). New York: McGraw-Hill. pp. 87-150.

Green, M.M. 1961. Phenogenetics of the lozenge loci in *Drosophila melanogaster*. II. Genetics of *lozenge-Krivshenko* (lz^k). Genetics 46:1169-1176.

Green, M.M. and K.C. Green. 1949. Corssing over between alleles at the lozenge locus in *Drosophila melanogaster*. Proc. Natl. Acad. Sci. USA 35:586-591.

Griffith, F. 1928. The significance of pneumococcal types. J. Hygiene, 27:113-159.

Grivell, L.A. 1983. Mitochondrial DNA. Scient. Am. 248(3): 60-73.

Grun, P. 1976. *Cytoplasmic Genetics and Evolution*. New York: Columbia University Press.

Gupta, P.K. 1994. *Elements of Biotechnology*. Meerut: Rastogi & Co.

Gupta, P.K. 1994. *Genetics*. Meerut: Rastogi Publications.

Gyllensten, U., D. Wharton, A. Josefsson, and A.C. Wilson. 1991. Paternal inheritance of mitochondrial DNA in mice. Nature 352:255-257.

Haldane, J.B.S. 1931. *The Causes of Evolution*. New York: Harper.

Hammerling, J. 1953. Nucleo-cytoplasmic relationships in the development of *Acetabularia*. Intern. Rev. Cytol. 2:475-498.

Haqq, C.M., C.-Y. King, E. Uklyama, S. Falsaff, T.N. Haqq, P.K. Donachoe, and M.A. Weiss. 1994. Molecular mechanism of sexular determinaion. Activation of Mullarian inhibiting substance gene expression by SRY. Science 266:1494-1500.

Hardy, G.H. 1908. Mendelian Proportions in a Mixed Population. Science 28:49-50.

Hardy, K. 1986. *Bacterial Plasmids*. 2d ed. Washington, DC: American Society of Microbiology.

Harris, H. 1980. *The Principles of Human Biochemical Genetics*. 3rd ed. New York: Elsevier.

Hartl, D.L. 1991. *Basic Genetics*. Boston: Jones & Bartlott.

Hassold, T.J. and P.A. Jacobs. 1984. Trisomy in man. Ann. Rev. Genet. 18:69-97.

Hayes, W. 1968. *The Genetics of Bacteria and Their Viruses*, 2nd ed. New York: John Wiley.

Heard, E., Clerl, P. and Anvel, P. 1997. X chromosome inactivation in mammals. Ann. Rev. Genet. 31: 571-610.

Herbers, K., D. Jähner, and R. Jaenisch. 1981. Microinjection of cloned retroviral genomes into mouse zygotes, integration and expression in the animal. Nature 293:540.

Hersh, A.H. 1929. The effect of different sections of the X-chromosome upon bar eye in *Drosophila melanogaster*. Anat. Rec. 68:378-382.

Hershey, A.D. 1958. The production of recombinants in phage crosses. Cold Sp. Harb. Symp. 23:19-46.

Hershey, A.D. and M. Chase. 1951. Genetic recombination and heterozygosis in bacteriophage. Cold Sp. Harb. Symp. 16:471-479.

Hershey, A.D. and M. Chase. 1952. Independent functions of viral protein and nucleic acid in growth of bacteriophage. J. Gen. Physiol. 36:39-56.

Hershey, A.D. and R. Rotman. 1949. Genetic recombination between host-range and plaque-type mutants of bacteriophage in single bacterial cells. Genetics 34:44-71.

Hershey, J.W.B. 1980. The translational machinery: Components and mechanism. In: *Cell Biology, A Comprehensive Treatise*. Volume 4. D.M. Prescott and L. Goldstein (eds.). New York: Academic Press. pp. 1-68.

Herskowitz, I. and D. Hagan 1980. The lysis-lysogeny decision of phage λ: Explicit programming and responsiveness. Ann. Rev. Biochem. 14:399-445.

Heslot, H. 1965. The nature of mutations. In: *The Use of Induced Mutations in Plant Breeding Programmes*. New York: Pergamon Press. p. 3-45.

Hoagland, M.B. 1959. The present status of the adaptor hypothesis. Brookhaven Symp. Biol. 12:40.

Hoagland, M.B. and M.L. Stephenson, J.F. Scott, L.I. Hecht and P.C. Zamecnik. 1958. A soluble ribonucleic acid intermediate in protein synthesis. J. Biol. Chem. 231:241-257.

Hodgkin, J. 1989. Drosophila sex determination: a cascade of regulated splicing. Cell, 56:905-906.

Hodgkin, J. 1990. Sex determination compared in *Drosophila* and *Coenorhabditis*. Nature 344:721-728.

Hodgkin, J. 1992. Genetic sex determination mechanisms and evolution. BioEssays 14:253-261.

Hoffman, M. 1994. Gene Therapy, Where to draw the line? Am. Scient. 82(4): 322-23.

Hollaender, A. 1954. *Radiation Biology*. New York: McGraw-Hill.

Hollaender, A. and F.J. DeSerres (eds.). 1971-1982. *Chemical Mutagens, Principles and Methods for Their Detection*. Volume. 1-7. New York: Plenum Press.

Holley, R.W., J. Apgar, G.A. Everett, J.T. Madison, M. Marquisee, S.H. Merrill, J.R. Penwick, and A. Zamir. 1965. Structure of a ribonucleic acid. Science 147:1462-1465.

Holliday, R. 1964. A mechanism of gene conversion in fungi. Genet. Res. 5:282-304.

Holliday, R. 1974. Molecular aspects of genetic exchange and gene conversion. Genetics 78:273-287.

Holliday, R. 1987. The inheritance of epigenetic changes. Science 238:163-170.

Holzman, D. 1991. A "jumping gene" caught in the act. Science 254:1728-1729.

Hood, L. 1972. Two genes, one polypeptide chain - fact or fiction? Fed. Proc. 31:177-187.

Horne, R.W. 1963. The structure of viruses. Scient. Am. 208(1):48-56.

Hotchkiss, R.D. and E. Weiss. 1956. *Transformed Bacteria*. Scient. Am, November. Available as Offprint 18. San Francisco: W.H. Freeman and Co.

Hurst, L.D. 1995. Molecular evolution: The uncertain origin of introns. Nature 371:381-382.
Huxley, J. 1942. *The Modern Synthesis.* New York: Harper.
Huxley, J. 1974. *Evolution: The Modern Synthesis.* New York: Allen and Unwin.
Ilan, J. ed. 1993. *Translational Regulation of Gene Expression 2.* New York: Plenum Press.
Immer, F.R. 1930. Formulae and Tables for Calculating Linkage Intensities. Genetics 15:81-98.
Ingram, V.M. 1957. Gene mutation in human hemoglobin: The chemical difference between normal and sickle-cell hemoglobin. Nature 180:326-328.
Ingram, V.M. 1958. *How Do Genes Act?* Scient. Am. Available as Offpring 104. San Francisco: W.H. Freeman and Co.
Ippen-Ihler, K.A. and E.G. Minkley, Jr. 1986. The conjugation system of F, the fertility factor of *Escherichia coli.* Ann. Rev. Genet. 20:593-624.
Jackson, I.J. 1994. Molecular and developmental genetics of mouse coat color. Ann. Rev. Genet. 28:189-217.
Jacob, F. and J. Monod. 1961. Genetic regulatory mechanisms in the synthesis of proteins. J. Mol. Biol. 3:318-356.
Jacob, F. and E.L. Wollman. 1961. *The Sexually and the Genetics of Bacteria.* New York: Academic Press.
Jenkins, J.B. 1979. *Genetics.* Boston: Houghton Mifflin Co.
Jinks, J.L. 1963. Cytoplasmic inheritance in fungi. In: *Methodology in Basic Genetics.* ed. W.J. Burdette. San Francisco: Holden-Day. pp. 325-354.
Jinks, J.L. 1964. *Extrachromosomal Inheritance.* Englewood Cliffs, New Jersey: Prentice-Hall, Inc.
Johnson, I.S. 1983. Human insulin from recombinant DNA technology. Science 219:632-637.
Jones, R.N. and H. Pees. 1982. *B Chromosomes.* Academic Press, London.
Judd, B.H., M.W. Shen, and T.C. Kaufman. 1972. The anatomy and function of a segment of the X chromosome of *Drosophila melanogaster.* Genetics 71:139-156.
Kainz, M. and J. Roberts. 1992. Structures of transcription elongation complexes *in vivo.* Science 255:838-841.
Khorana, H.G., H. Büchi, H. Ghosh, N. Gupta, T.M. Jacob, H. Kössel, R. Morgan, S.A. Narang, E. Ohtsuka, and R.D. Wells 1967. Polynucleotide synthesis and the genetic code. Cold Sp. Harb. Symp. 31:39-49.
Kilbane, J.J., II and B.A. Bielaga. 1991. Instantaneous gene transfer from donor to recipient microorganisms via electroporation. BioTechniques 10:354-365.
Kilbey, B.J., M. Legator, W. Nichols, and C. Ramel. 1977. *Handbook of Mutagenicity Test Procedures.* Amesterdum: Elsevier-Scientific.
Kim, Y., J.H. Geiger, S. Hahn, and P.B. Sigler. 1993. Crystal structure of a yeast TBP/TATA-box complex. Nature 365:512-520.
Kimura, M. 1983. *The Neutral Theory of Molecular Evolution.* New York: Cambridge University Press.
Kimura, M. 1987. Molecular evolutionary clock and the neutral theory. J. Mol. Evol. 26:24-33.

King, M.C. and A.C. WIlson. 1975. Evolution at two levels in humans and chimpanzees. Science 188:107-116.
King, R.C. (ed.). 1975. *A Handbook of Genetics.* Vols. 2-4. New York: Plenum Press.
King, R.W., P.K. Jackson, and M.W. Kirschner. 1994. Mitosis in transition. Cell 79:563-571.
King, R.C. and W.D. Stansfield. 1990. *Dictionary of Genetics.* 4th ed. New York: Oxford University Press.
Kirby, L.T. 1990. *DNA Fingerprinting: An Introduction.* New York: Stockton Press.
Kornberg, A. 1980. *DNA Replication.* San Francisco: Freeman.
Kornberg, A. and T.A. Baker. 1992. *DNA Replication.* New York: Freeman.
Kornberg, R.D. and A. Klug. 1981. The nucleosome. Scient. Am. 244(2): 52-64.
Kornberg, R.D. and J.O. Thomas. 1974. Chromatin structure: oligomers of the histones. Science 184:865-868.
Koshland,D. 1994. Mitosis: Back to the basics. Cell 77:951-964.
Krafka, J., Jr., 1920. The effect of temperature upon facet number in the bar-eyed mutant of *Drosophila.* J. Gen. Physiol. 2:409-464.
Krontiris, T.G. 1995. Minisatellites and human disease. Science 269:1682-1683.
Kumar, Ashok. 1985. Molecular basis of heterosis and homesotasis in *Drosophila malerkotliana.* Ph.D. dissertation. Punjab Agricultural University, Ludhiana.
Kumar, Ashwni. 1978. Genetics of esterase enzyme in *Zaprionus paravittiger.* M.Sc. Thesis, Punjab Agricultural University, Ludhiana.
Kumar, R. and G.S. Miglani. 1986. Molecular Mechanisms of Biological Speciation. Everyman's Science 21(5):132-140.
Kumaria, R., R. Verma, and M.V. Rajam. 1998. Potential application of antisense RNA technology in plants. Curr. Scientific 74: 341.
Landauer, W. 1948. Hereditary abnormalities and their chemically-induced phenocopies. Growth Symposium 12:171-200.
Landis, W. 1987. Factors determining hte frequency of the killer trait within populations of the *Paramecium aurelia* complex. Genetics 115:197-205.
Landman, O.E. 1991. The inheritance of acquired characteristics. Ann. Rev. Genet. 25:1-20.
Landschulz, W.H., P.F. Johnson and S.L. McKnight. 1988. The Leucine Zipper: A hypothetical structure common to a new class of DNA binding proteins. Science 240:1759-1764.
Lang, W.H. B.E. Morrow, Q. Ju, J.R. Warner, and H.R. Reeder. 1994. A model for transcription termination by RNA polymerase I. Cell 79:527-534.
Lasic, D. 1992. Liposomes. Am. Scient. 80:20-31.
Lawn, R.M. and A. Vehar. 1986. The molecular genetics of hemophilia. Scient. Am. 254(3):40-55.
Lawrence, W.J.C. and V.C. Sturgess. 1957. Studies on Streptocarpus. III. Genetics and chemistry of flower color in the garden forms, species, and hybrids. Heredity 11:303-336.
Lea, D.E. 1955. *Actions of Radiations on Living Cells.* 2nd ed. Cambridge: Cambridge Univ. Press.

Leary, R.F., F.W. Allendorp and K.L. Knudsen. 1983. Developmental stability and enzyme heterozygosity in rainbow trout. Nature 301:71-72.

Lederberg, J. 1947. Gene recombination and linked segregation in *Escherichia coli*. Genetics, 32:505-525.

Lederberg, J. 1987. Genetic recombination in bacteria: A discovery account. Ann. Rev. Genet. 21:23-46.

Lederberg, J. 1989. Replica-plating and indirect selection of bacterial mutants in bacteria by sib selection. Genetics 121:395-99.

Lederberg, J. and E.M. Lederberg. 1952. Replica plating and indirect selection of bacterial mutants. J. Bact. 63:399-406.

Ledergberg, J., E.M., Lederberg, N.D., Zinder, and E.R. Lively. 1951. Recombination analysis of bacterial heredity. Cold Sp. Harb. Symp. Quant. Biol. 16:413-43.

Lennox, E.S. 1955. Transduction of linked genetic characters of the host by bacteriophage P1. Virol. 1:190-206,

Lepag, T., M.C. Stephen, F.J.D. Benjumea, and S.M. Parkhurst. 1995. Signal transduction by cAMP-dependant protein kinase A in *Drosophila* limb patterning. Nature 373:711-715.

Lerner, I.M. 1954. *Genetic Homeostasis*. Edinburgh: Oliver and Boyd.

Lerner, I.M. 1958. *The Genetic Basis of Selection*. New York: Wiley.

Levene, P.A. and L.W. Bass 1931. *Nucleic Acids*. New York: Chemical Catalog Co.

Levine, R.P. 1955. Chromosome structure and the mechanism of crossing over. Proc. Nat. Acad. Sci. USA 41:727-730.

Levine, P. 1958. The influence of the ABO system of Rh hemolytic disease. Hum Biol. 30:14-28.

Levine, R.P. and U.W. Goodenough. 1970. The genetics of photosynthesis and of the chloroplast in *Chlamydomonas reinhardi*. Ann. Rev. Genet. 4:397-408.

Levinthal, C. 1959. Bacteriophage genetics. In: *The Viruses*. Vol. II, F.M. Burnet and W.M. Stanley (eds.). New York: Academic Press. pp. 281-318.

Lewin, B. 1990. *Genes IV*. New Delhi: Oxford Univ. Press.

Lewin, B. 1994. Gene V. Oxford: Oxford Univ. Press.

Lewis, E.B. 1950. The phenomenon of position effect. Adv. Genet. 3:73-116.

Lewis, E.B. 1951. Pseudoallelism and gene evolution. Cold Sp. Harb. Symp. 16:159-172.

Li, C.C. 1955. *Population Genetics*. Chicago: University of Chicago Press.

Li, W. and D. Graur. 1990. *Fundamentals of Molecualr Evolution*. Sunderland, Mass: Sinauer Associates.

Lindergren. C.C. 1933. The genetics of *Neurospora*. III. Pure bred stocks and crossing-over in *N. crassa*. Bull. Torrey Bot. Club 60:133-154.

Lindsley, D.L. and G.G. Zimm. 1992. *The Genome of Drosophila melanogaster*. San Diego: Academic Press.

Lloyd, R.G. and C. Buckman. 1995. Conjugational recombination in *Escherichia coli*. Genetic analysis of recombinant formation in Hfr XF-crosses. Genetics 139:1123-48.

Logadon, J.M. and J.D. Palmer. 1994. Origin of introns - early or late. Nature 369:526-528.

Long, E. and I. Dawid. 1980. Repeated genes in eukaryotes. Ann. Rev. Biochem. 49:727-764.

Luce, W.M. 1926. The effect of temperature on Infrabar, an allelomorph of Bar Eye in *Drosophila*. J. Exp. Zool. 46:301-316.

Lyon, M.F. 1962. Sex chromatin and gene action in the mammalian X chromosome. Am. J. Hum. Genet. 14:135-148.

Lyon, M.F. 1992. Some milestones in the history of X-chromosomes inactivation. Ann. Rev. Genet. 26:17-28.

Lyon, M.F. 1996. X-chromosome inactivation: Pinpointing the centre. Nature 379:116.

L'Heritier, Ph. 1958. The hereditary virus of *Drosophila*. Adv. Virus Res. 5:195-245.

Makino, Sajiro. 1951. *An Atlas of the Chromosome Numbers in Animals*. Ames:Iowa State College Press.

Maniatis, T., E.F. Firtsch, and J. Sambrook. 1982. *Molecular Cloning: A Laboratory Manual*. Cold Spring Harbor Laboratory. New York: Cold Sp. Harb.

Mannino, R.J. and S. Gould-Fogerite. 1988. Liposome-mediated gene transfer. BioTechniques 6:682-90.

Manuelidis, L. 1990. A view of interphase chromosomes. Science 250:1533-1540.

Marx, J. 1982. Building bigger mice through gene transfer. Science 218:1298.

Marx, J.L. 1988. Foreign gene transferred into maize. Science 240:145-146.

Marx, J. 1991. The Cell Cycle: Spinning further afield. Science 252:1490-1492.

Mather, K. 1951. *The Measurement of Linkage in Heredity*. London: Methuen.

Mather, K. 1943. Polygenic Inheritance and Natural Selection. Biol. Rev. 18:32-64.

Mather, K. 1973. *Genetical Structure of Populations*. London: Chapman and Hall.

Matton, D.P. N. Nass, A.E. Clarke, and E. Newbigin. 1994. Self-incompatibility: How plants avoid illegitimate offspring. Proc. Natl. Acad. Sci. USA 91:1992-97.

Mayer, W. 1977. Darwin and Natural Selection. Am. Scient. 65:321-327.

Mayloy, S.R. 1989. *Experimental Techniques in Bacterial Genetics*. Boston: Jones and Bartlett.

Mayr, E. 1963. *Animal Species and Evolution*. Cambridge, Mass.: Harvard University Press.

Mays, L.L. 1981. *Genetics, A molecular Appraoch*. New York: McMillan Publishing Co.

McBridge, O.A. and J.L. Peterson. 1980. Chromosome mediated gene transfer in mammalian cells. Ann. Rev. Genet. 14:321-345.

McCarthy, J.E.G., and R. Brimacombe. 1994. Prokaryotic translation: The interactive pathway leading to initiation. Trends in Genetics 10:402-407.

McClintock, B. 1951. Chromosome organization and genic expression. Cold Sp. Harb. Symp. 16:13-47.

McClung, C.E. 1902. The accessory chromosomes-sex determinant. Bio. Bull. 3:43-84.

McCoubrey, W.K., K.D. Nordstrom, and P.L. Meneely. 1988. Microinjected DNA from the X chromosome affects sex determination in *Caenorhabditis elegans*. Science 242:1146-1151.

McKeown, M. 1992. Alternative mRNA splicing. Ann. Rev. Cell Biol. 8:133-155.

McKnight, S.L. 1991. Molecular zippers in gene regulation. Scient. Am. 26(4): 32-39.

McKusick, V.A. 1972. *Human Genetics*. New Delhi: Prentice-Hall.

McKusick, V.A. 1994. Mendelian inheritance in Man: A Catalog of *Human Genes and Genetic Disorders*. 11th ed. with the assistance of C.A. Francomano, S.E. Antonarakis, and P.L. Pearson. Baltimore: Johns Hopkin University Press.

McLaren, A. 1981. *Germ Cells and Soma*. A New Look at an Old Problem. New Haven: Yale University Press.

McPeek, M.S. and T.P. Speed. 1995. Modeling interference in genetic recombination. Genetics 139:1031-44.

Meselson, M. 1964. On the mechanism of genetic recombination between DNA molecules. J. Mol. Biol. 9:734-745.

Meselson, M.S. and C.M. Radding. 1975. A general model for genetic recombination. Proc. Nat. Acad. Sci. USA 72:358-361.

Meselson, M. and J.J. Weigle. 1961. Chromosome breakage accompanying genetic recombination in bacteriophage. Proc. Nat. Acad. Sci. 47:857-868.

Miescher, F. 1871. On the chemical composition of pus cells. Hoppe-Seyler's Med.-Chem. Untersuch. 4:441-460.

Miglani, G.S. 1986. Methodologies for genetic risk evaluation of environmental chemicals. XI Annual Conference of Environmental Mutagen Society of India, University of Madras, Madras.

Miglani, G.S. 1998. *Dictionary of Plant Genetics and Molecular Biology*. New York: Food Products Press. pp. 348.

Miglani, G.S. and F.R. Ampy. 1981. *Drosophila* alcohol dehydrogenase: developmental studies on gyptic variant lines. Biochem. Genet. 19(9/10):947-954.

Miller, J.H. 1992. *A Short Course in Bacterial Genetics*. Cold Spring Harbor, New York: Cold Spring Harbor Laboratory Press.

Minkoff, E.C. 1983. *Evolutionary Biology*. Reading: Addison-Wesley Publ. Co.

Moazed, D. and H.F. Noller. 1989. Interaction of tRNA with 23S rRNA in the ribosomal A, P, and E sites. Cell 57:585-597.

Moldave, K. 1985. Eukaryotic protein synthesis. Ann. Rev. Biochem. 54:1109-1149.

Moore, D.S. 1985. *Statistics: Concepts and Controversies*. San Francisco: Freeman.

Morgan, T.H. 1910. Sex-limited inheritance in *Drosophila*. Science 32:120-122.

Morgan, T.H. 1911. Random segregation versus coupling in Mendelian inheritance. Science 34:384.

Morgan, T.H. 1911. An attempt to analyse the constitution of the chromosomes on the basis of sex-limited inheritance in *Drosophila*. J. Exp. Zool. 11:365-414.

Morgan, T.H., A.H. Sturtevant, H.J. Muller, and C.B. Bridges. 1922. *The Mechanism of Mendelian Heredity*. 2nd ed. New York: Holt.

Morse, M.L., E.M. Lederberg and J. Lederberg. 1956. Transduction in *Escherichia coli* K12. Genetics 41:142-156.

Morgan, T.H. 1919. *The Physical Basis of Heredity*. Philadelphia: Lippincott.

Morgan, T.H. 1926. *The Theory of the Gene*. New Haven: Yale University Press.

Morton, N.E. 1962. Segregation and linkage. In: *Methodology in Human Genetics*. W.J. Burdette (ed.). New York: Plenum Press. pp. 3-30.

Morton, N.E. 1995. LODs past and present. Genetics 140:7-12.

Motulsky, A. 1983. Impact of genetic manipulation on society and medicine. Science 219:135-140.

Mouro, I., Y. Colin, B. Cherif-Zahar, J.-P. Cartron, and C. Le Van Kim. 1993. Molecular genetics basis of the human Rhesus blood group system. Nature Genetics 5:62-65.

Muller, H.J. 1916. The mechanism of crossing-over. II. Am. Natural. 50:284-305.

Muller, H.J. 1938. The remaking of chromosomes. Collect. Net 8:182-195.

Muller, H.J. 1949. Our Load of Mutations. Am. J. Hum. Genet. 1:1-18.

Muller, H.J. and I.I. Oster. 1963. Some mutational techniques in *Drosophila*. In: *Methodology in Basic Genetics*. W.J. Burdette (ed.). San Francisco: Holden-Day. pp. 249-274.

Mullis, K.B., F. Ferre and R.A. Gibbs, eds. 1994. *The Polymerase Chain Reaction*. Basal: Birkhauser.

Murai, N., D.W. Sutton, M.G. Murray, J.L. Slinghtom, D.J. Merlo, N.A. Reichert, C. Sengupta-Gopalan, C.A. Stock, R.F. Barker, J.D. Kemp, and T.C. Hall. 1983. Phaseolin gene from bean is expresed after transfer to sunflower via tumor-inducing plasmid vectors. Science 222:476-482.

Murray, A.W and J.W. Szostak 1983. Construction of artifical chromosomes in yeast. Nature 305:189-193.

Murray, A.W. and J.W. Szostak. 1987. Artificial chromosomes. Scient. Am. 257(5): 60-70.

Murray, A.W. and M.W. Kirschner. 1991. What controls the cell cycle? Scient. Am. 264(3): 34-41.

Murray, R.K. 1990. *Harper's Biochemistry*. 24th ed. London: Prentice-Hall Int.

Nakamura, Y., M. Leppert, P. O'Connell, R. Wollf, T.H. Holm, M. Culver, C. Martin, Em Fujimoto, M. Hoff, E. Kumlin, and R. White. 1987. Variable number of tandem repeat (VNTR) markers for human gene mapping. Science 235:1616-1622.

Nanninga, N. 1973. Structural aspects of ribosomes. Intern. Rev. Cytol. 35:135.

Nasrallah, J.B., J.C. Stein, M.K. Kandasamy, and M.E. Nasrallah, 1994. Signalling the arrest of poolen tube development in self-incompatible plants. Science 266: 1505-1508.

Neel, J.V. 1949. The Detection of the Genetic Carriers of Hereditary Disease. Am. J. Human Genetics 1:19-36.

Neuffer, M.G. and E.H. Coe, Jr., 1974. Corn (maize). In: *Handbook of Genetics*. Vol. 2. R.C. King (ed.). New York: Plenum Press. pp. 3-30.

Nevins, J.R. 1983. The pathway of eukaryotic mRNA formation. Ann. Rev. Biochem. 52:441-466.

Nicklas, R.B. 1971. Mitosis. Adv. Cell Biol. 2:225-297.

Nilsson-Ehle, H. 1909. Kreuzungsuntersuchungen an Hafer und Weizen. Lnnds. Univ. Aarskr, N.F. Afd. Ser. 2. Vol. 5. No. 2. pp. 1-122.

Nirenberg, M.W. 1963. The genetic code: II. Scient. Am. 190:80-94.

Nirenberg, M.W. and P. Leder. 1964. RNA codewords and protein synthesis:The effect of trinucleotides upon the binding of sRNA to ribosome. Science 145:1399-1407.

Nirenberg, N.W. and J.H. Matthaei. 1961. The dependence of cell-free protein synthesis in *E. coli* upon naturally occurring or synthetic polyribonucleotides. Proc. Natl. Acad. Sci. USA 47:1588-1602.
Noll, M. 1977. DNA folding in the nucleosome. J. Mol. Biol. 116:49-71.
Noller, H.F. 1984. Structure of ribosomal RNA. Ann. Rev. Biochem. 53:119-162.
Noller, H.F. 1991. Ribosomal RNA and translation. Ann. Rev. Biochem. 58:1029-1049.
Nomura, M., A.N. Tissieres, and L.H. Lengyel. 1974. *Ribosomes*. New York: Cold Spring Harbor Laboratory.
Normark, S., S. Bergström, B. Jaurin, F.P. Lindberg, and O. Olsson. 1983. Overlapping genes. Ann. Rev. Genet. 17:499-525.
Notani, N.K. 1984. *Genetics and Molecular Biology of Rhizobia*. Perspective Series Report 10. New Delhi: Indian National Science Academy.
Ochoa, S. 1963. Synthetic polynucleotides and the genetic code. Symposium on Genetic Mechanics. Fed. Proc. 22:62-74.
Ogawa, T. and T. Okazaki. 1980. Discontinous DNA replication. Ann. Rev. Biochem. 59:421-457.
Okazaki, R., T. Okazaki, K. Sakabe, K. Sugimoto, and A. Suino. 1968. Mechanism of DNA chain-growth. I. Possible discontinuity and unusual secondary structure of newly synthesized chains. Proc. Nat. Acad. Sci. USA 59:598-605.
Olby, R. 1974. *The Path to the Double Helix*. London: MacMillan.
Old, R.W. and S.B. Primrose. 1981. *Principles of Gene Manipulation: An Introduction to Genetic Engineering*. Berkeley: Univ. of California Press.
O'Malley, B.W., H.C. Towle, and R.J. Schwartz. 1977. Regulation of gene expression in eukaryotes. Ann. Rev. Genet. 11:239-275.
Ou-Lee, T., R. Turgeon, and R. Wu. 1986. Expression of a foreign gene linked to either a plant virus or a *Drosophila* promoter, after electroporation of protoplasts of rice, Wheat and Sorghum. Proc. Natl. Acad. Sci. USA 83:6815-6819.
Owen, A.R.G. 1950. The Theory of Genetical Recombination. Adv. Genet. 3:117-157.
Parkurst, S.M. and P.M. Meenley. 1994. Sex determination and dosage compensation lessons form flies and worms. Science 264:924-932.
Patterson, D. 1987. The cause of Down syndrome. Scient. Am. 257(2): 42-48.
Pauling, L. and R.B. Corey. 1956. Specific hydrogen-bond formation between pyrimidines and purines in deoxyribonucleic acids. Arch. Biochem. Biophys. 65:164-181.
Pauling, L., H.A. Itano, S.J. Singer and I.C. Wells. 1949. Sickle-cell anemia, a molecular disease. Science 110:543-548.
Peacock, W.J. 1971. Cytogenetic aspects of the mechanism of recombination in higher organisms. Stadler Genet. Symp. 2:123-152.
Pennisi, E. 1997. Transgenic Lambs from Cloning Lab. Science 277: 631.
Perry, P.R. 1976. Processing of RNA. Ann. Rev. Biochem. 45:650-629.
Peters, J.A. 1959. *Classical Papers in Genetics*. Englewood Cliffs, New Jersey: Prentice Hall.
Pilgrim, I. 1986. A solution to the too-good-to-be-true paradox and Gregor Mendel. J. Hered. 77:218-20.

Pontecorvo, G. 1958. *Trends in Genetic Analysis*. New York: Columbia University Press.
Portugal, F.H. and J.S. Cohen. 1978. *A Century of DNA*. Mass., Cambridge: MIT Press.
Preer, J. R., Jr. 1950. Microscopically visible bodies in the cytoplasm of the "killer" strains of *Paramecium aurelia*. Genetics 35:344-362.
Preer, L.B. and J.R. Preer, Jr. 1977. Inheritance of infectious elements. In: *Cell Biology*. Vol. I. L. Goldstein and D.M. Prescott (eds.). New York: Academic Press. pp. 319-373.
Preer, J., J. Preer and A. Jurand. 1974. Kappa and other endosymbionts in *Paramectium aurelia*. Bacteriol. Rev. 38:113-163.
Preer, J.R., Jr. L.B. Preer, B. Rudman and A. Jurand. 1971. Isolation and composition of bacteriophage-like particles from kappa of killer Paramecia. Molec. Gen. Genet. 111:202-208.
Punnett, R.C. 1911. *Mendelism*. New York: Macmillan.
Race, R.R. and R. Sanger. 1968. *Blood Groups in Man*. 5th ed. Philadelphia: F.A. Davis.
Rhoades, M.M. 1946. Plastid mutations. Cold Sp. Harb. Symp. 11:202-207.
Rhoades, M.M. 1955. Interaction of genic and non-genic heredity units and the physiology of nongenic function. In: *Encyclopedia of Plant Physiology*. ed. W. Ruhland, 1:2-57.
Rhoades, M.M. 1961. Meiosis. In: The Cell, Vol. III, J. Brachet and A.E. Mirsky (eds.). New York: Academic Press. pp. 1-75.
Richardson, R.J. 1997. Blocking adenosine with antisense. Nature 385: 684-685.
Rick, C.M. and G.C. Hanna, 1943. Determination of Sex in *Asparagus officinalis* L. Am. J. Botany 30:711-714.
Ris, H. 1957. Chromosome structure. In: *The Chemical Basis of Heredity*, W.D. McElroy and B. Glass (eds.). Baltimore: Johns Hopkins Press. pp. 23-69.
Risch, N. 1992. Genetic linkage: Interpreting LOD scores. Science 255:803-4.
Robertson, M. 1979. Regulatory sequences involved in the promotion and termination of transcription. Ann. Rev. Genet. 13:319-354.
Robertson, M. 1988. RNA processing: The post-RNA world. Nature 335:16-18.
Rodgers, J. 1991. Mechanism Mendel never knew. Mosaic 22:2-11.
Rogers, H.J. 1985. Mechanism of RNA splicing. Int. Rev. Cytol 99:188-235.
Roman, H. and M.M. Ruzinski. 1990. Mechanisms of gene ocnversion in *Saccharomyces cerevisiae*. Genetics 124:7-25.
Rose, M.R. and W.F. Doolite. 1983. Molecular biological mechanisms of speciation. Science 220:157-161.
Rothman, J.L. 1965. Transduction studies on the relation between prophage and host chromosome. J. Mol. Biol. 12:892-912.
Rowen, L. and A. Kornberg. 1978. J Primase, the dnaG protein of *Escherichia coli*:An enzyme which starts DNA synthesis. Biol. Chem. 253:758-764.
Roy, R.P. and P.M Roy. 1971. Mechanism of sex determination in *Coccina indica*. J. Indian Bot. Soc. 50A:391-400.
Russell, L.B. 1963. Mammalian X-chromosome action:Inactivation limited in spread and region of origin. Science 140:976-978.

Ryder, L.P., A. Svejgaard, and J. Dausset. 1981. Genetics of HLA diseases association. Ann. Rev. Genet. 15:169-187.
Saenger, W. 1984. *Principles of Nucleic Acid Structure.* New York: Springer-Verlag.
Sager, R. 1972. *Cytoplasmic Genes and Organelles.* New York: Academic Press.
Saks, M.E., J.R. Sampson, and J.N. Abelson. 1994. The transfer RNA identity problem: A search for rules. Science 263:191-197.
Sandler, L. and E. Novitski. 1957. Meiotic drive as an evolutionary force. Am. Natural. 91:105-110.
Sanger, R. and Z. Ramanis. 1963. The particulate nature of nonchromosomal genes in *Chlamydomonas.* Proc. Nat. Acad. Sci. USA 50:260:268.
Sarabhai, A., A.D.W. Stretton, S. Brenner and A. Bolle. 1964. Colinearity of the gene with polypeptide chain. Nature 201:13-37.
Sarin C. 1985. *Genetics.* New Delhi: Tata McGraw Hill Publishing Co. Ltd.
Savageau, M.A. 1979. Autogenous and classical regulation of gene expression:A general theory and experimental evidence. In: *Biological Regulation and Development.* Volume I: *Gene Expression*, R.F. Goldberger (ed.). New York: Plenum Press. pp. 57-108.
Scaife, J., D. Leach and A. Galizzi. 1985. *Genetics of Bacteria.* London: Academic Press.
Schulz-Schaeffer, J. 1980. *Cytogenetics.* New York: Springer-Verlag.
Searle, A.G. 1968. *Comparative Genetics of Coat Colour in Mammals.* New York: Academic Press.
Selander, R.K., A.G. Clark, and T.S. Whittam, eds. 1991. *Evolution at the Molecular Level.* Sunderland, Mass.: Sinauer Associates.
Shahi, V.K. 1985. Biochemical and physiological basis of heterosis and homeostasis in *Drosophila punjabiensis.* Ph.D. dissertation, Punjab Agricultural University, Ludhiana.
Sharp, P.A. 1987. Splicing of messenger RNA precursors. Science 235:766-771.
Sharp, P.A. 1992. TATA-binding protein is a classless factor. Cell 68:819-821.
Sharp, P.A., M. Hsu, and N. Davidson. 1972. Note on the sturcture of prophage λ. J. Mol. Biol. 71:499-501.
Shekman, R., A. Weiner, and A. Kornberg. 1974. Multienzyme systems of DNA replication. Science 186:987-993.
Shermann, F., M. Jackson., S.W. Liebman, A.M. Schweinyruber, and J.W. Stewart. 1975. Cyc-1 mutants and its correspondance to mutationally altered iso-1-cytochrome c of yeast. Genetics 81:51-73.
Shigekawa, K. and W.J. Dower. 1988. Electroporation of eukaryotes and prokaryotes. A general approach to the introduction of macromolecules into cells. BioTechniques 6:742-751.
Siddiqui, O. and M.S. Fox. 1973. Integration of donor DNA in bacterial conjugation. J. Mol. Biol. 71:499-501.
Silvers, W.K. 1979. *The Coat Colors of Mice.* New York: Springe-Verlag.
Simons, R.W. and N. Kleckner. 1988. Biological regulation by antisense RNA in prokaryotes. Ann. Rev. Genet. 22:567-600.
Singer, B. and J.T. Kusmierak. 1982. Chemical mutagenesis. Ann. Rev. Biochem. 52:655-693.

Singer, B., R. Sager, and Z. Ramanis. 1976. Chloroplast genetics of *Chlamydomonas*. III. Closing the circle. Genetics 83:341-354.

Singer, M. and P. Berg. 1991. *Gene and Genomes: A Cloning Perspective*. California: University Science Books.

Singleton, W.R. 1962. *Elementary Genetics*. New York: Van Nostrand, Reinhold.

Sinnott, E.W., L.C. Dunn and T. Dobzhansky. 1958. *Principle of Genetics*. New York: McGraw Hill.

Sinsheimer, R.L. 1959. A single-stranded deoxyribonucleic acid from bacteriophage ϕX174, J. Mol. Biol. 1:43-53.

Skleuar, V. and J. Feigon. 1990. Formation of a stable triplex from a single DNA strand. Nature 345:836-838.

Smith, H.O., D.B. Danner, and R.O. Deich. 1981. Genetic transformation. Ann. Rev. Biochem. 50:41-68.

Smith, C.W.J., J.G. Patton, and B. Nadal-Ginard. 1989. Alternative splicing in the control of gene expression. Ann. Rev. Genet. 23:527-577.

Smithies, O., R.G. Gregg, S.S. Boggs, M.A. Koralesvski, and R.S. Kucherlapati. 1985. Insertion of DNA sequences into the human chromosomal β-globin locus by homologous recombination. Nature 317:230-234.

Snedecor, G.W. and W.G. Cochran. 1967. *Statistical Methods*. 6th ed. Ames: Iowa State Univ. Press.

Soans, A.B., D. Pimentel and J.S. Soans. 1974. Evolution of reproductive isolation in allopatric and sympatric populations. Am. Natural. 108:117-124.

Sonneborn, T.M. 1947. Recent Advances in the Genetics of Paramecium and Euplotes. Adv. Genet. 1:263-358.

Sonneborn, T.M. 1950. Methods in the general biology and genetics of *Paramecium aurelia*. J. Exp. Zool. 113:87-148.

Sonneborn, T.M. 1959. Kappa and related particles in *Paramecium*. Adv. Virus Res. 6:229-356.

Sonneborn, T.M. 1965. Degeneracy of the genetic code:Extent, nature and genetic implications. In: *Evolving Genes and Proteins*. V. Bryson and H.J. Vogel (eds.). New York: Academic Press. pp. 377-397.

Srb, A.M. 1963. Extrachromosomal factors in the genetic differentiation of *Neurospora*. Symposia Soc. Exptl. Biol. 17:175-187.

Stahl, F.W. 1979. *Genetic Recombination: Thinking About It in Phage and Fungi*. San Francisco: W.H. Freeman.

Stahl, F.W. 1987. Genetic recombination. Scientific American 256(2): 90-101.

Stahl, F.W. and R. Lande. 1995. Estimating interference and linkage map distance from two-factors tetrad data. Genetics 139:1449-54.

Stansfield, W.D. 1977. *The Science of Evolution*. New York: MacMillan Publ. Co.

Stebbins, G.L. 1970. *Processes of organic evolution*. Englewood Cliffs: Prencie-Hall.

Steel, R.G.D. and J.H. Torrie. 1960. *Principles and Procedures of Statistics*. New York: McGraw-Hill.

Steitz, J.A. and K.T. Tycowski. 1995. Small RNA chapterones for ribosome progenies. Science 270:1626-1627.

Stent, G.S. 1963. *Molecular Biology of Bacterial Viruses.* San Francisco: W.H. Freeman and Co.

Stent, G.S. and R. Calendar. 1978. *Molecular Genetics: An Introductory Narrative.* 2nd ed. San Francisco: W.H. Freeman.

Stern, C. 1977. *Principles of Human Genetics.* San Francisco: Freeman.

Streisinger, G. 1956. Phenotypic mixing of host range and serological specificities in bacteriophages T2 and T4. Virology. 2:388-398.

Streisinger, G. and V. Bruce. 1960. Linkage of genetic markers in phages T2 and T4. Genetics 45:1289-1296.

Streisinger, G., R.S. Edgar, and G.H. Denhardt. 1964. Chromosomes structure in phage T4. I. Circularity of the linkag map. Proc. Nat. Acad. Sci. USA 51:775-779.

Strickberger, M.W. 1962. *Experiments in Genetics with Drosophila.* New York: John Wiley.

Strickberger, M.W. 1986. *Genetics.* New York: MacMillan Publishing Co.

Stubbe, H. 1972. *History of Genetics.* Cambride, Mass: MIT Press.

Sturtevant, A.H. 1923. Inheritance of direction of coiling in *Limnaea.* Science 58:269-270.

Sturtevant, A.H. 1925. The effects of unequal crossing over at the *Bar* locus in *Drosophila melanogaster.* Genetics 10:117-147.

Sturtevant, A.H. 1926. A crossover reducer in *Drosophila melanogaster* due to inversion of a section of the third chromosome. Biol. Zentralbl. 46:697-702.

Sturtevant, A.H. 1965. *A History of Genetics.* New York: Harper & Row.

Sturtevant, A.H. and G.W. Beadle. 1936. The relation of inversion in the X chromosome of *Drosophila melanogaster* to crossing over and disjunction. Genetics 21:554-604.

Sutton, W.S. 1903. The chromosomes in heredity. Biol. Bull. 4:213-251.

Suzuki, D.T., A.J.F. Griffiths, and R.C. Lowentin. 1989. *An Introduction to Genetic Analysis.* New York: Freeman.

Swanson, C.P. 1957. *Cytology and Cytogenesis.* Englewood Cliffs, New Jersey: Prentice-Hall.

Tamarin, R.H. 1996. *Principles of Genetics.* Chicago: Win C. Brown Publishers.

Tanksley, S.D. 1993. Mapping polygenes. Ann. Rev. Genet. 27:205-233.

Tanksley, S.D., M.W. Ganal, J.P. Prince, M.C. de Vicente, M.W. Bonierbale, P. Brown, T.M. Fulton, J.J. Giovannoni, S. Grandillo, G.B. Martin, R. Messeguer, J.C. Miller, L. Miller, A.H. Paterson, O. Pineda, M.S. Röder, R.A. Wing, W. Wu, and N.D. Young. 1992. High density molecular linkage maps of the tomato and potato genomes. Genetics 132:1141-1160.

Tartof, K.D. 1975 Redundant genes. Ann. Rev. Genet. 9:355-385.

Taylor, A.L. and M.S. Thoman. 1964. The genetic map of *Escherichia coli* K-12. Genetics 50:659-677.

Taylor, J.H. (ed.). 1965. *Selected Papers on Molecular Genetics.* New York: Academic Press.

Temin, H.M. 1972. RNA-directed DNA synthesis. Scientific American 226(1): 24-33.

Temin, H.M. and S. Mizutani. 1970. RNA-dependent DNA polymerase in virions of Rous sarcoma virus. Nature 226:1211-1213.

Thoday, J.M. and J.B. Gibson. 1962. Isolation by disruptive selection. Nature 193:1164-1166.

Tjian, R. 1995. Molecular machines that control genes. Scientific American. 272(2): 38-45.

Trehan, K.S. and K.S. Gill. 1983. Homeostasis at the molecular level. Abstract XV International Congress of Genetics, New Delhi, Part-I:196.

Trehan, K.S. and K.S. Gill. 1987. Sub-unit interation; a molecular basis of heterosis. Biochem. Genet. 25(11/12): 855-862.

Valander, W.H., H. Lubon, and W.H. Drohan. 1997. Transgenic livestock as drug factories. Scient. Am. 276(1): 55-58.

Visconti, N. and M. Delbruck. 1953. The mechanism of genetic recombination in phage. Genetics 38:5-33.

Vogel, F. and R. Rathenberg. 1975. Spontaneous mutation in man. Adv. Hum. Genet. 5:223-318.

Wagner, R.P., M.P. Maguire, and R.L. Stallings. 1993. *Chromosomes: A Synthesis.* New York: Wiley-Liss.

Wagner, R.P. and H.K. Mitchell. 1964. *Genetics and Metabolism.* New York: Wiley.

Walbot, V. and E.H. Coe, Jr. 1979. Nuclear gene iojap conditions a programmed change to ribosome-less plastids in *Zea mays.* Proc. Nat. Acad. Sci. USA 76:2760-2764.

Wang, T.S.-F. 1991. Eukaryotic DNA polymerase. Ann. Rev. Biochem. 60:513-522.

Warburton, D., J. Byme, and N. Canki. 1990. *Chromosomal Anomalies and Prenatal Development: An Atlas.* New York: Cambridge University Press.

Warmke, H.E. 1946. Sex determination and sex balance in *Melandrium.* Am. J. Bot. 33:648-660.

Watkins, P.C. 1988. Restriction fragment length polymorphism (RFLP): Application in human chromosome mapping and genetic disease research. BioTechniques 6:310-319.

Watson, J.D. 1968. *The Double Helix.* New York: Atheneum.

Watson, J.D. and F.H.C. Crick. 1953. A structure for deoxyribose nucleic acids. Nature 171:737-738.

Watson, J.D. and F.H.C. Crick. 1953. Genetic implications of the structure of deoxyribonucleic acid. Nature 171:964-969.

Watson, J.D., N.H. Hopkins, J.W. Roberts, J.A. Steiz, and A.M. Weiner. 1987. *Molecular Biology of the Gene.* 4th ed. Menlo Park, California: Benjamin/Cummings.

Watson, J.D., M. Gilman, J. Witkowski, and M. Zoller. 1992. *Recombinant DNA.* 2nd ed. New York: Freeman.

Watson, J.D. and J. Tooze. 1981. *The DNA Story: A Documentary History of Gene Cloning.* San Francisco: W.H. Freeman.

Weijland, A. and A. Parmeggiani. 1994. WHy do two Ef-Tu molecules act in the elongation cycle of protein biosynthesis? Trends Biochem. Sci. 19:188-193.

Weiling, F. 1971. Mendel's "too good" data in *Pisum* experiments. Folia Mendeliana 6:75-77.

Weinberg, W. 1980. Über den Nachweis der Verebung beim Menschen. Jahreshefte Verein f. vaterl. Naturk. in Württemberg 64:368-382.

Weinstein, A. 1958. The geometry and mechanics of crossing-over. Cold Sp. Harb. Symp. 23:177-196.

Weintraub, H.M. 1990. Antisense RNA and DNA. Scient. Am. 262(1): 34-40.
West, S.C. 1992. Enzymes and molecular mechanisms of genetic recombination. Ann. Rev. Biochem. 61:603-640.
White, M.J.D. 1978. *Modes of Speciation.* San Francisco: W.H. Freeman.
White, R.J. and S.P. Jackson. 1992. TATA-binding protein: a central role in transcription by RNA polymerase I, II and III. Trends Genet. 8:284-288.
Whitehouse, H.L.K. 1963. A theory of crossing over by means of hybrid DNA. Nature 199:1034-1040.
Whitehouse, H.L.K. 1982. *Genetic Recombination: Understanding the Mechanisms.* New York: John Wiley & Sons.
Whitehouse, H.L.K. and P.J. Hastings. 1965. The analysis of genetic recombination on the polaron hybrid DNA model. Genet. Res. 6:27-92.
Wichterman, R. 1953. *The Biology of Paramecium.* New York: Blakiston.
Wilkie, D. 1964. *The Cytoplasm in Heredity.* London: Methuen and Co.
Wilkins, M.H.F. 1963. The molecular configuration of nucleic acids. Science 140:941-950.
Wilkins, M.H.F., A.R. Stokes, and H.R. Wilson. 1953. Molecular structure of desoxypentose nucleic acids. Nature 171:738-740.
Williams, N. 1995. Dosage compensation - How males and females achieve equality. Science 1826-1827.
Williams, R.S., S.A. Johnston, M. Riedy, M.J. DeVit, S.C. McElligott, and J.C. Sanford. 1991. Introduction of foreign genes into tissues of living mice by DNA-coated microprojectiles. Proc. Natl. Acad. Sci. USA 88:2726-2730.
Wilson, A.C. 1975. Evolutionary importance of gene regulation. Stadler Symp. 7:117-133.
Wilson, A.C. 1978. Gene regulation in evolution. In: *Molecular Evolution.* (ed. F.J. Ayala). Sinaver Associates. Inc., Sunderland. pp. 225-236.
Wilson, A.C., L.R. Maxon, and V.M. Sarich. 1974. Two types of molecular evolution. Evidence from studies of interspecific hybridization. Proc. Natl. Acad. Sci. USA 71:2843-2847.
Wilson, A.C., S.S. Carlson, and T.J. White. 1977. Biochemical evolution. Ann. Rev. Biochem. 46: 573-639.
Wilson, M.R. 1995. Extraction, PCR amplification and sequencing of motochondrial DNA from human hair shafts. BioTechniques 18:662-669.
Woese, C.R. 1965. On the evolution of the genetic code. Proc. Nat. Acad. Sci. USA 54:1546-1552.
Woese, C.R. 1967. *The Genetic Code.* New York: Harper & Row.
Woese, C.R. 1967. *The Genetic Code: The Molecular Basis for Gene Expression.* New York: Harper & Row.
Wollman, E.L. and F. Jacob. 1956. *Sexuality in Bacteria.* Scientific American, July. Available as Offprint 50. San Francisco: W.H. Freeman and Co.
Wollman, E.L., F. Jaboc, and W. Hayes. 1956. Conjugation and genetic recombination in *Escherichia coli* K-12. Cold Sp. Harb. Symp. 21:141-162.
Wood, W.B. and H.R. Revel. 1976. The genome of bacteriophage T4. Bact. Rev. 40:847-868.

Woolf, C. 1968. *Principles of Biometry*. Van Nostrand Reinhold, New York.

Wright, J.E. and Antherton, L. 1968. Genetic control of intra-allelic recombination at the LDH-B locus in brook tront. Genetics 60:240.

Wright, R.E. and J. Lederberg. 1957. Extranuclear transmission in yeast heterokaryons. Proc. Nat. Acad. Sci. USA 43:919-923.

Wu, R., L. Grossman, and K. Moldave, eds. 1989. *Recombinant DNA Methodology*. San Diego: Academic Press.

Xu, L., M. Thali, and W. Schaffner. 1992. Upstream box/TATA box order is the major determinant of the direction of transcription. Nucl. Acids Res. 19:6699-6704.

Yanofsky, C. 1984. Comparison of regulatory and structural regions of genes of tryptophan metabolism. Mol. Biol. Evol. 1:143-161.

Yanofsky, C., B.C. Carlton, J.R. Guest, D.R. Helinski, and U. Henning. 1964. On the colinearity of gene structure and protein structure. Proc. Natl. Acad. Sci. USA 51:266-272.

Yanofsky, C., G.R. Drapeau, J.R. Guest, and B.C. Carlton. 1967. The complete amino-acid sequence of the tryptophan synthetase A protin (α subunit) and its colinear relationship with the genetic map of the A gene. Proc. Nat. Acad. Sci. USA 57:296-298.

Yanofsky, C., V. Horn and D. Thorpe 1964. Protein structure relationships revealed by mutational analysis. Science 146:1593-1594.

Yamamoto, F., H. Clausen, T. White, J. Marken, and S. Hakomori. 1990. Molecular genetic basis of the histo-blood group ABO system. Nature 345:229-33.

Young, R.A. 1991. RNA polymerase II. Ann. Rev. Biochem. 60:689-715.

Yule, G.U. 1906. On the theory of inheritance of quantitative compound characters on the basis of Mendel;s laws:a preliminary note. Rept. 3rd. Intern. Congr. Genet. pp. 140-142.

Zimmerman, A. (ed.). 1981. *Motisis/Cytokinesis*. New York: Academic Press.

Zinder, N.D. 1958. Transduction in bacteria. Scient. Am. 199(5): 38-43.

Zinder, N.D. 1992. Forty years ago: The discovery of bacterial transduction. Genetics 132:291-94.

Zinder, N.D. 1958. *Transduction in Bacteria*. Scientific American, Available as Offpring 106. San Francisco: W.H. Freeman and Co.

Zinder, N.D. and J. Lederberg. 1952. Genetic exchange in *Salmonella*. J. Bact. 64:679-699.

Zubay, G. 1987. *Genetics*. Menlo Park, California: Benjamin/Cummings.

SUBJECT INDEX

ABO blood group, 508
Acetabularia, 60
  grafting experiment, 424
Acid phosphatase, 46
Acidic proteins, 44
Acquired characteristics, 481
Activation reaction, 508
Activator-dissociator (Ac-Ds) system, 508
Activator protein, 327
Acyl (A) site, 298
Adaptedness, 492
Adaptive channels, 493
Adenine (A), 216, 508
Adrenocorticotropin, 334
Adsorption, 128
AIDS, 461
Agglutination, 78
Agouti patterns, 90
Agrobacterium tumifaciens, 469
Akpattonuria, 257
Alanine tRNA, 281
Allele, 508
Allele frequency, 508
Allele-specific PCR (AS-PCR), 476
Allelic repression, 498
Allopolyploid(s), 348, 508
Allosomes, 435
Allopolyploidy, 348
Allosteric interaction, 333
Allotetraploid, 508
Allozymes, 104, 508
Allozyme, heterodimeric, 105
Allozyme, homodimeric, 105
Alternate hypothesis (H_1), 506
Alternate splicing, 443
Amber codon, 508
Ambiguous codons, 287

Ames test, 372, 509
Amino acids structures, 296
Aminoacyl or A-site, 509
Aminoacyl-tRNA, 509
Amniocentesis, 454
Amphibian oocytes, 60
Amphidiploid(s), 348, 509
Amplicon length polymorphism (APL), 476
Amplification at gene level, 486, 509
Amplification at genome (genomic) level, 489, 509
Amplified fragment length polymorphism, (AFLP), 476
Anagenesis, 499, 509
Anaphase, 32, 509
Anaphase I, 509
Anaphase II, 509
Ancillary protein factors, 247
Ancillary sequences, 248
Ancillary site(s), 248, 509
Aneuploids, 347
Aneuploidy, 509
Aneuploidy, human disorders, 449
Angstrom unit (Å), 509
Anthocyanin pigments, 257, 259
Anti-Rh serum, 79
Antibiotic polypeptides, 306
Antibiotics, 125
Antibodies, 78
Anticoding strand, 509
Anticodon, 280, 286, 509
Antigens, 78, 125
Antileader, 509
Antiparallel strands, 510
Antisense strand, 272
Antitermination, 248

Antitermination factors, 249
Antitrailer, 510
AP endonucleases, 510
AP sites, 510
Apomixis, 111
Apurinic acid, 365
Arbitrary primed PCR (Ap-PCR), 476
Artificial selection, 510
Ascomycetes, 118, 169
Ascospore, 121
Ascus, 121
Aspergillus, 355
Assortative mating, 491
Astral rays, 35
Asymmetric characters, 104
Asymmetry, 102-103
Asynapsis, 510
Attenuation, 248, 249-250
Attenuators, 273
Autogamy, 510
Autogenous regulation, 334
Autoimmune disease, 83
Autonomous development, 259
Autonomous genetic element, 428
Autopolyploid(s), 348, 510
Autoradiography, 25, 510
Autosomal recessive lethals, detection of, 375
Autosome(s), 435, 510
Autotrophic, 510
Autotroph(s), 372
Auxotroph, 128, 510
Ayrshir cattle, spots on body of, 73
B-A translocations, 57, 58
Backcross, 14, 510
Back mutation, 510
Bacteria, 190
 calculating gene order in, 202
 conjugation mapping in, 193-194
 cytoplasmic connection, 193
 DNA synthase, 193
 donor strain, 191
 gene mapping in, 192, 193
 intrupted mating technique, 193
 quantitative estimation of, 125
 recipient, 191, 193
 recombination mapping in, 194
 replicating unit, 193
 replication cycle, 193
Bacteria, sexual contact, 193
Bacterial chromosome, directional orientation, 193
Bacterial conjugation, 125
Bacterial fission, 123
Bacterial transformation, 510
Bacteroid, 510
Bacteriophage, 201, 511
Balanced lethal system, 484, 511
Balance in *Drosophila* sex determination, 511
Balancing selection, 511
Balbiani rings, 60, 511
Baldness in man, 73
Bar eye, 511
Bar, semi-dominant mutation, 374
Barr body, 437, 441, 511
Basc, 511
 analogue, 511
 pair, 511
 pairing, complementary, 190
 pairing, forbidden, 352, 353
 pairing, rare, 351
Basic
 number, 511
 proteins, 38
B chromosomes, 511
Beans, seed weight in, 397
Beans, pure lines, 397
β-galactosidase, 315, 511
β-galactoside acetyl transferase, 511
β-galactoside permease, 315, 511
β-glycosydic bonds, 216
β-lipoprotein, 334
Bidirectional replication, 512
Binomial expansion, 512
Biogenesis, 512
Biological evolution, 512
Biological property/trait/character/characteristic, 1, 512
Biotechnology, 512
Biparental percentage, 110

Subject Index 597

Biparental zygote(s), 413, 512
Birpartite intermediate, 277
Bivalent, 35, 512
Blood
  clotting factor, 68
  group antigens, 512
  groups, 77
Blunt ends, 466, 512
Bombay phenotype, 512
Bonellia viridis, 440
Bottleneck effect, 512
Bottlenecks, 491
Branch migration, 512
Breakage and copying hypothesis, 158
Breakage and reunion, 160, 512
  hypothesis, 159
Bridge, 512
Britten-Davidson Model, 327
Caenohabditis elegans, dosage compensation in, 442
cAMP, *see* Cyclic AMP
Cap (m$_7$ GpppN1), 277, 424
CAP site, 512
Capsule shape in bursa, 93
Carrier, 64, 454
Cascade interactions, 443
CAT box, 512
Catenane, 513
cDNA
  cloning, 513
  library, 513
Caedobacter laeniospiralis, 428
Carrier, 513
Cell(s), 513
  animal 32
  bacterial, 125
  colicinogenic, 431
  daughter, 36
  F$^+$, 525
  F, 525
  F$^-$, 525
  ganglia, 59
  HeLa, 47
  Hfr, 428
  ovarian nurse, 59

  ovarian nurse, 59
  pollen mother (PMC), 546
  polytene, 60
  recipient, 193, 551
  salivary 59
  sex, 556
  somatic hybrid, 558
  suspensor, 59
  transformed, 563
Cell cycle,
  biochemistry of, 30
  control of, 28
  decision making points, 30
  duration of, 27
  genes 30
  genetics of, 30
  measurement of, 27
  phase transitions, 30
  post start events, 30
  representative, 29
  start point, 30
  start promotion factor (SPF), 30
  yeasts, 30
Cell division, 31-36, 513
Cell organelles, 513
Cell transformation, 513
Cell plate, 33
Cellular product, 245
CEN, 513
Central dogma, 270
  reversal of, 270
Centimorgan (cM), 139, 513
Central Dogma, 513
Centromere, 35, 48, 513
Centromeric chromatin, 51
Chance, 503
Characters
  Mendel studied, 11
  morphometric, 102
Chargaffs rule, 513
Charged tRNA, 513
Chargon phages, 467
Charon phages, 513
Chaperone, 513
Chiasma (singular), 35, 136, 136, 513

Chiasma interference, 185
Chiasmata (plural), 35, 61, 136, 513
Chi form, 514
Chimeric gene, 514
Chimeric plasmid, 514
Chi-square (X^2), 505-50, 514
  calculating, 505
  contigency, 506
  test, 136
  values, 505
Chlamydomonas,
  cpDNA, Buoyant densities, 413
  chloroplastgene mapping, 413, 414
  detection of linkage, 186
  estimation of linkage, 186
  gene mapping in, 186
  mating types, 123, 124
  minute mutants in, 415
  streptomycin-resistance in, 412
  uniparental inheritance, 412
Chromatid(s), 48, 514
  interference, 185
  sister 32, 35
Chromatin, 38, 45, 48, 514
Chromocenter, 59, 514
Chromomeres, 50, 55, 61
Chromosomal
  mutation, 514
  rearrangements, 485
  theory, 514
Chromosome(s) 32-36, 38, 514
  acentric, 344
  acrocentric, 49
  arms, 49
  bacterial, 44, 193
  behaviour, 39
  changes in, 347, 449
  dicentric, 54, 344
  diploids, 49
  eukaryotic, 44, 48
  haploids, 54
  homologous, 35, 40
  imprinting, 514
  independent assortment/segregation
    of, 39, 40
  lampbrush 60, 61
  maps, 171
  maternal, 38
  metacentric, 49
  monocentric, 54
  morphology, 50
  movement, 54
  mutations, 342
  number, animals,
    52-53
  parts of, 51
  paternal, 38
  plants, 53-54
  pairing, 38, 160 (*also see*
    synapsis)
  polycentric, 54
  polynemic 58
  polytene, 58, 511
  prokaryotic, 44
  puffs, 514
  segregation, 39, 514
  separation, 35
  shape, 49
  size, 48
  somatic, 49
  structure, changes in, 342, 449
  submetacentric, 50
  telocentric, 49
  types, 50
  unusual, A-chromosome, 55
  unusual, B- 55-58
  unusual, B-genes, 56
  unusual, B-effects of, 56
  unusual, B-uses of, 56
  unusual, supernumerary, 55
  viral, 44, 128, 204
  walking, 471, 514
  X, 64
  Y, 68
Citric acid cycle, 103
Circularization, 514
Cis-arrangement, 317
cis-position, 235, 236
Cis-trans complementation test,
  238, 243, 514
Cistron, 243, 514
Cladogenesis, 500, 514

Subject Index 599

Clathrin-coated vesicles, 310
CIB technique, 374, 375, 515
Cleavage enzymes, 282
Cleavage map, 553
Cleft palate, 75
Clone, 515
Clone library, 515
Cloning vector, 515
Cloverleaf, 515
cms factor, 418, 419, 420
Coat colour in horse, 97
Coat colour in mice, 90, 97
Coat colour in rabbits, 79
Code dictionary, 515
Code letter, 286
Coding sectors,
  major, 283
  minor, 283
Coding strand, 515
Codominance, 14, 77, 515
Codon, 286, 515
  -anticodon relationship, 515
  assignments, nuclear and mitochondrial genome, 422
  bias (preference), 515
  families, mixed, 288
  families, unmixed, 288
  length, 286
Coefficient of coincidence, 136
Cognate tRNAs, 515
Cohesive ends, 429, 466, 515 (*see also* sticky ends)
Coincidence coefficient, 515
Colchicine, 348, 515
Colinearity, 272, 251, 515
Colourblindness, 66, 515
Col factors, 431
Col plasmids, 515
Comb shape in fowls, 88-89
Combining ability, general, 515
Combining ability, specific, 515
Commaless code, 287, 516
Commensalism, 516
Compartmentalization, 326
Compensation mechanism, 443
Competitive exclusion, 516

Complementary
  base pairing, 220
  classes, 138
  sequence, 516
Complementation, 236
  map, 242, 516
  mapping, 236, 244
  matrix, 239, 244, 516
Complete
  copy choice hypothesis, 159
  linkage, 516
  medium, 516
Completed proteins, fate of, 306
Complex locus, 236
Components of fitness, 492
Confidence limits, 516
Conjugation, 516
Conjugation mapping, 516
Conjugative
  plasmids, 469, 516
  transposons, 516
Consanguineous
  marriages, 459
  mating, 517
Consensus sequence, 517
Conspecific, 496
Constitutive
  genes, 517
  heterochromatin, 517
  secreton, 309
Contingency
  table, 517
  test, 506
Continuous
  replication, 517
  trait, 517
Controlled matings, 113
Co-ordinated transcription, 319
Copy-choice hypothesis, 517
Corepressor, 312, 313, 517
Corn,
  ear length in, 396
  flinty seeds, 133, 134
  floury seeds, 133, 134
  meiosis in, 132
  segregation in, 132

starchy pollen, 132, 133
sugary pollen, 132, 133
xenia effect, 133
Cosmids, 467
Co-transduction, 517
Co-transformation, 517
Co-translational factors, 307
Criss-cross inheritance, 517
Cross,
  dihybrid, 10, 14-18, 520
  monohybrid, 10-14, 539
  reciprocal, 10, 64, 551
  dihybrid, gene hypothesis, 15, 17
  monohybrid, gene hypothesis, 13
Cross-fertilization, 517
Cross-linking, interstrand, 364
Crossing-over, 35, 36, 134, 135, 138, 184, 517
  double, 152
  four strand, 154-156
  single, 152
  significance of, 170
  unequal, 476, 565
  within a gene, 240
Crossover
  suppression, 517
  suppressor (C), 374
  unit, 517
Crown gall, 517
Cruciform structure, 518
Cryptic plasmids, 518
Cumulative effect, 518
Curve, 360
C-value paradox, 518
Cyclic AMP (cAMP), 518
  -activated protein (CAP) site, 248
  -CAP Complex, 323
Cytogenetic male sterility, 518
Cytogenetics, 518
Cytogenic inheritance, 417
Cytohet, 413, 518
Cytokinesis, 518
Cytokinesis I, 518
Cytokinesis II, 518
Cytological map, 518
Cytology, 518

Cytoplasm, 518
Cytoplasmic
  connection, 193
  gene, 410
  inheritance, 518
  male sterility, 417, 418
  organelles, 33 (*see also* organelles)
  segregation and recombination (CSAR), 411, 412, 518
Cytosine (C), 216
Darwinian evolution, 519
Daughter chromatid, 59
Daughter chromosomes, 519
Davis-U-tube, 213
Degeneracy, 519
Degenerate code, 287
Degree of
  dominance, 519
  freedom (d.f.), 136, 504, 519
Delayed segregation in snails, 519
Deletion(s) 239, 342, 519
  loop, 343
  mapping, 238, 519
  mapping, principle of, 241
Depurination, 365
Deoxyadenosine-5'-phosphate, 216
Deoxycytidine-5'-phosphate, 216
Deoxyguanosine-5'-phosphate, 216
Deoxyribonuclease, 519
Deoxyribonucleic acid, 4, 38, 211, 519 (*See also* DNA)
Deoxythmidine-5'-phosphate, 216
Desynapsis, 519
Detection of linkage, 519
Developmental
  homeostasis, 519
  signals, 327
  stability, 102
Deviations, 504
deVriesism, 519
Diakinesis, 519
Dicentric chromosome, 519
Dideoxynucleotide, 519
Differential segments, 519
Digenic inheritance, 520
Dionne quadruplets, 101

Subject Index 601

Diplococcus pneumonia 211
Diploid, 520
Diplotene, 520
Diplotene, 60
Dipteran spp., 59
   Malpighian tubules, 59
   ovarian nurse cells, 59
   gut epithelial cells, 59
Directional selection, 520
Discontinuous
   replication, 520
   trait, 520
Disjunction, 35, 520
Disomic, 347
Disorders, genetic basis of, 447
Disruptive selection, 520
Dizygote twins, 520
DNA, 4, 5, 38, 211, 520
   A helicase, 520
   amplification fingerprinting (DAF), 476
   antisense, 510
   antisense strand, 246
   -binding protein, 520
   centriole 410
   chloroplast, 409, 514
   clone, 520
   cloning, 467 (*also see* gene cloning)
   coding strand, 246
   complementary (cDNA), 516
   cross-linking, 369, 520
   dangling, 160
   diploid amount of, 28
   diseases, mitochondrial, 539
   doubled helix, 216
   duplex, 246
   fingerprinting, 462, 521
   genetic material, 211-215
   G primase, 520
   helix A, 520
   helix B, 521
   helix Z, 521
   heteroduplex, 161
   hybrid, 161, 226, 533
   kinetoplast 410, 535
   ligase, 44, 521
   linker, 536
   -mediated gene transfer, 521
   mitochondrial, 409, 539
   newly synthesized 226
   non-repetitive, 542
   organelle, 409
   plasmid, 432
   polymerase, 44, 228, 521, 554
   polymerase I, 521
   polymerase II, 521
   polymerase III, 521
   promiscuous, 548
   puff, 521
   recombinant, 465, 467, 551
   repetitive, 552
   satellite, 59, 555
   selfish, 556
   self-replicating, 431
   sense strand, 246, 556
   sequencing, 521
   spacer, 421, 558
   strand, 521
   structure, 221
   synthase, 193
   synthesis, 28
   synthesis, *de novo*, 457
   synthesis *in vitro*, 521
   synthesis, salvage pathways, 457
   synthesis, single strand, 165
   topoisomerase, 274, 521
   2C, 26
   unique, 566
DNA replication, 25, 521
   bidirectional, 228
   enzymes and proteins of, 229
   Escherichia coli, 226
   eukaryotes, 231-232
   origin of, 230
   modes of, 224
   prokaryotes, 226-231
   rolling circle model, 230
   semi-conservative, 227
Docking protein, 308
Doctrine of acquired characteristics, 566

Dominance, 87
  complete, 516
  concept of, 19, 516
  degree of, 405
  hypothesis, 521
  incomlete, 14, 16, 533
  modifers, 521
  partial, 544
  scale of, 22
  sex-influenced, 73
Dominant, 10, 522
  epistasis, 93-94, 522
  lethal, 522
  phenotype, 522
Dosage
  compensation, 435, 441, 444, 522
  effect, 522
Double
  crossover, 522
  fertilization, 522
  monosomic, 347
  trisomic, 347
Down's syndrome, 522, 565
Downstream, 522
Drosophila (D.), 27, 28, 47, 355, 376, 380, 522
  attached X chromosomes, 155, 156
  balanced lethal system, 378
  Bar gene, 234
  bobbed trait in, 68
  Curly-Lobe-Plum, 518
  double *Bar*, 522
  eye disc transplantation experiments in, 259-261
  eye pigment biosynthesis, 261
  integument, 114
  larval instars, 115
  Lobe gene in, 291
  metamorphosis, 115
  moulting, 114
  sex-determination in, 511
  spiracles, 114
  tumor-head (tu-h) phenotype, 98-100
  ultrabar gene, 234
  white eyes in, 64, 65
  white gene in, 21, 64
Drosophila malerkotliana, 104, 105
Drosophila melanogaster (D.m.), 59, 113, 460
  brain ganglia cells, 59
  CO_2 sensitivity in, 415
  dosage compensation in, 442
  eye colour in, 79
  garnet locus of, 236
  lozenge locus of, 236
  salivary cells, bands, 59
  salivary cells, endomitosis, 59
  salivary cells, interbands, 59
  salivary cells, somatic synapsis, 59
  sigma virus (particles), 415
  white locus in, 234
D. persimillis, 487
D. pseudoobscura, 488
D. punjabiensis, 104, 105
D. willistoni, sex-ratio strains in, 416
Duchenne's muscular dystrophy (DMD), 68
Duplicate
  dominant epistasis, 522
  recessive epistasis, 522
Duplication, 344, 522
  loop, 344
Dyad, 522
Dysgenesis, gonadal, 440
Dysgenic, 522
Ecdysone, 332
Eclipse period, viral chromosome, 128
Egg
  fertilization, 439
  production in chicken, 73
Effector molecule, 313, 523
Electronic structure of bases, 350
Electrophoresis, 377
Electroporation, 523
Element, 523
  denominator, 442
  numemator, 442
Elongation factors, 523
Embryogenesis, 62

Subject Index 603

Emigration, 490
Encapsulation, 2
End-product inhibition, 333
Endocytosis, 310
Endonuclease(s), 44, 46, 225, 523
Endomitosis, 523
Endonuclease, 523
Endoplasmic reticulum, 46, 306, 523
Endopolyploidy, 523
Endoreplication, 523
Endosperm, 523
End-product inhibition, 523
Enhancer(s), 248, 523
Environmental mutagenesis, 523
Epigenesis, 3, 100
Epigenetic systems, 3
Epilobium hirsutum, 420
Episome(s), 125, 202, 419, 428, 523
Epistasis, 87, 523
  dominant, 93-94
  dominant-recessive, 95
  recessive, 90
Equational division, 523
Equatorial plate, 524
Equilibrium, 524
Equilibrium frequency, 524
Erythroblastosis fetalis, 79, 524
Escherichia (E.) coli, 130, 333, 524
  fertility factor, 428
  gene symbols of, 195-201
  tryptophan synthetase gene, 252
Estimation of linkage, 524
  F_2, Emerson's method, 143
  F_2, maximum-liklihood method, 143
  F_2, product-ratio method, 144
Eugenics, 446
Exchange reaction, 300
Exconjugant, 118
Exocytosis, 309
Exon, 524
Exon sharing, 251
Exonuclease, 44, 46, 524
Expected number, 503
Explant, 524
Expression vector, 524

Expressivity, 290, 524
Expressivity modifiers, 524
Extended anticodon hypothesis, 524
External stimuli, 327
Extinction, 491
Extranuclear inheritance, 525
Euchromatin, 524
Eugenics, 524
Eukaryote, 524
Eukaryotic vectors, 467
Euploid(s), 347, 524
Evagination, 46
Evolution, 384, 481, 524
Eyes, 525
F_1, 525
F_2, 525
F_2 ratios, modified, 87, 100
F^+ cell, 125, 525
F^- cell, 125, 525
F' cell, 525
Factor(s), 12, 525
Favourable attributes, in bacteria, 123
F-duction, 202, 429, 525
F-element, 125
F-factor, 125, 192
  autonomous state, 428
  excission, 429
  integrated state, 429
  integration, 429
F-mediated
  sex-duction, 202
  transduction, 429
F-plasmid, 525
F' factor, 318
Feather colour in fowls, 95
Feedback inhibition, 333, 525
Fertility (F) factor, 525
Fertility restorer, 525
Finches, 482
First division segregation (FDS), 525
Fissure of upper lip, 75
Flanking regions, 246
Flower colour in sweet peas, 90
f^{Met}, 525
Footprinting, 525

Forces of evolution, 525
Forward mutation, 525
Founder effect, 491
Frameshift mutations (fms), 342
Free-martin, 440
Frequency-dependent selection, 525
Frequency distribution, 526
Fruit colour in summer squash, 94
Fruit shape in summe squash, 95, 97
F-test, 526
Functional alleles, 234, 526
Furrowing, 526
G_1 phase, 526
G_2 phase, 526
Gamete, 39, 40, 41, 526
Gametophyte, 526
Gap period, 526
Garden pea, 10
Gemmules, 2, 526
Gene(s), 12, 526
  additive, 508
  α-globin, 308
  amplification, 48, 526
  as functional entity, 268
  bank, 511
  battery model, 330
  B-chromosome, 56
  Bt, 477
  chimeric, 417, 514
  cloning, 467, 526
  complementary, 90, 516
  complex, 251, 516
  compound, 516
  constitutive, 311, 517
  concept, 234
    classical phase, 234
    modern phase, 236
    transitional phase, 235
  conversion, 168-170, 526
    mismatched strands, 169
  cryptomorphic, 518
  definition, 245-246
  deleterious, 68
  delivery, 460, 526
  dimorphic, 520
  discontinuous, 520
  dosage, 526
  duplicate, 93, 522
  effector, 523
  -environment interaction, 526
  enzyme relationship, 257
  epistatic, 87, 523
  exogenous, 461
  exon theory of, 524
  expression, 271, 526
    antisense inhibition of, 336
    hormonal control of, 330
    selective, 337
  extrachromosomal, 524
  extranuclear, 525
  female determining, 438
  flow, 490
  frequency,
    definition, 527
    determination of, 387
  function, 257
  holandric, 68, 532
  house keeping, 532
  hypostatic, 87
  interaction(s), 87-97, 527
    complementary, 90
    nuclear and cytoplasmic, 97
    supplementary, 90-92
    suppression, 95
    conclusions from, 100
    different types of, 96
  hypostatic, 533
  independent assortment of, 40
  inhibitory, 94, 95
  integrator, 327, 534
  jumping, 535
  K, 427
  library, 527
  lethal, 84
  luxury, 536
  lytic, 429
  major, 537
  mapping in
    bacteria,, 192, 193
    Chlamydomonas, 186
    Neurospora, 176
    viruses, 203-207
    yeast, 186

Subject Index 605

manipulation, 527
mature, 246
minor, 538
mitochondrial, 422
modifier (modifying), 97, 539
monintron, 250, 539
multintron, 250, 540
multiple, 540
mutation, 342, 485, 527
nif, 477
overlapping, 252, 544
patenting, 527
pleiotropic, 101
polymerics, 546
pool, 527
producer, 327, 548
rearrangement, 498
receptor, 327, 550
regulated (regulatory), 311, 551
regulation, 311
  antisense RNA in, 336
  inducible, 312
  in eukaryotes, 324-332, 326
  long-term, 536
  meaning of, 311
  models, 312, 313
  prokaryote, 312
  repressible, 312
  short-term, 557
  theory of ageing, 527
regulatory
  mechanisms in eukaryotes, 527
  mechanisms in prokaryotes, 527
regulators, amino acid sequence of, 338
repeated, 552
replacement, targeted, 462
restorer, 418
rRNA, 271
segregation, 38
selector, 555
sensor, 327, 556
sex-limited, 556
sex-linked, 64, 556
simple, 557
specialized, 558

split, 250-251, 558
structural, 271, 272, 559
structure, 234
supplementary, 90, 560
suppressor, 560
synthesis, 527
therapy, 460, 461, 527
transfer,
  definition, 529
  Agrobacterium-mediated, 469
  chemical method, 469
  chromosome-mediated, 471
  cotransformation, 470
  DNA-mediated, 469
  Drosophila to *E. coli*, 472
  electroporation, 470
  liposome-mediated, 471
  microinjection, 470
  microprojectiles, 469
  transfection, 470
  translocation of, 564
  transposon-mediated, 476
  triplicate, 564
  tRNA, 271
  tumor suppressor, 565
  unstable, 566
Generalized transduction, 527
Genetic analysis,
  bacteria, 190
  definition, 7, 527
  diploid eukaryotes, 132-175
  experimental methods in, 109
  haploid eukaryotes, 176-189
  organisms used in, 7
  systems used in, 109
Genetic
  block, 528
  code, 284, 528
    properties of, 287
    working of, 288, 289
    dictionary, 286, 528
  control, biochemical reactions, 257
  counselling, 459, 528
  crossover, 166
  dissection, 528
  distance, 528

drift, 528
equilibrium, 384
engineering, 7, 465, 528
　applications of, 476-478
　hazards of, 478
equilibrium, 528
　principle of, 384
exchange, 163, 201
factors, 5
fine structure, 528
homeostasis, 102, 528
inferences, 109, 110
information, continuity of, 224
isolation, 528
load, 528
male sterility, 528
mapping, 528
marker, 528
material, 528
　nature of, 211
　properties of, 222
　storage of information, 222
medicines, 528
mutagen, 528
perspective, 5-6
prognosis, 529
ratios, 503
relatedness, 83
risk assessment, 529
screening, 529
system, choice of, 109
toxicology, 529
viruses, 190
Genetics,
　basis of science of, 2
　biometrical, 512
　classical, 6, 515, 564
　definition of, 1, 529
　developmental, 519
　endosperm, 113, 132
　evolutionary, 6
　father of, 9
　human, 446
　medical, 446
　Mendelian, 537, 564
　microbial, 538
　molecular, 539
　population, 6, 546
　quantitative, 549
　radiation, 550
　surrogate, 560
　transmission, 6, 564
Genic
　balance theory, 529
　explanation of hyrid vigour, 529
　and plasma interaction, 529
Genome, 529
Genome imprinting, 529
Genomic
　evolution, 497
　library, 529
Genotype, 5, 529
　-environmental interaction, 5, 529
Genotypic frequencies, binomial
　distribution of, 383
Genotypic ratio, 529
　monohybrid cross, 10
　dihybrid cross, 14-15, 18
Genotypic variability, 170
Geographical isolation, 495
Germline, 443
Germplasm, 3, 530
Germplasm theory, 3, 484, 530
Glucose effect, 322
Glycolysis, 103
Goldberg-Hogness box, 246
Gout, 74
Graft rejection, 83
Green algae, 123
Griffith effect, 211
Group I introns, 530
Group II introns, 530
GT-AG rule, 530
Guanine (G), 216, 530
Gyrase, 530
Haemoglobin, 266
Hairpin loop (structure), 530, 248
Hairy ear rims, 68
Haploid, 49, 46, 530
　genome, human, 48
　genome, yeast, 48
Haplotype, 82, 530

Subject Index 607

Hardy-Weinberg
  equilibrium, 530
  law, 383
HAT medium, 456, 530
HCN production in plants, 90
Helicase, 530
Hemizygous, 64, 152, 391, 530
Hemophilia, 67, 68
Heredity, 1
Heritability, 403, 530
  in broad sense (H_b), 405, 531
  in narrow sense (H_n), 405, 531
  meaning of, 405
  estimation of, 405
Hermaphrodite, 531
Hermaphroditic, 347
Hermaphroditism, 440
Heteroallele, 531
Heterochromatic, 531
Heterochromatin, 45, 531
  constitutive, 55, 517
Heterochromatization, 442
Heterodimer, 531
Heterodimeric allozymes, 105
Heteroduplex, 160, 531
  DNA model, 161, 531
  modified DNA, 531
Heterogametic sex, 64, 435, 531
Heterogeneous nuclear RNA
  (hnRNA), 531
Heterokaryon, 240, 531
Heterologous base sequence, 160
Heterosis, 531
Heterozygosity, 102-103
Heterozygote advantage, 531
Heterozygous, 64, 531
Hexaploid(s), 347, 531
High density molecular maps, 476
High frequency recombination (Hfr)
  cell, 428, 531
Highly repetitive sequence, 532
Histidine biosynthesis, 264
Histocompatibility, 532
  alleles, 80
  antigens, 80, 532
  antigens, major, 537

antigens, minor, 539
complex, major, 537
genes, 80, 532
Histones, 38, 44, 48, 532
HLA
  antigens and diseases, 83-84
  complex, 532
  genes, 83
  system, 81
Hogness box, 532
Holandric Genes, 68, 532
Holliday intermediate, 160
Holliday's model, 160-168
  duplex chromosomes, 160
  bridge migration, 160
  heterologous base sequence, 160
  heteroduplex, 160
  hybrid DNA, 161
  heteroduplex DNA, 161
  branch migration, 161-163
  planar molecule, 163
  D-loop, 163
  repair synthesis, 164
  DNA synthesis, single strand, 165
  recombination intermediates, 166
  figure eight, 166
  chi-structure, 167
  dimeric circle, generation of 167
  monomeric circles, generation of, 168
Holliday structure, 532
Holoenzyme, 247, 272, 532
Homeostasis, 102
  biochemical basis of, 104
  developmental, 102
Homoallele, 532
Homodimer, 532
Homodimeric allozymes, 105
Homogametic, 64
Homogametic sex, 435, 532
Homologous chromosomes, 532
  disjunction, 38
  pairing, 38
Homologues, 49 (*also see* homologous chromosomes)
Homozygous, 532

Homunculus, 2
Hormone-receptor protein complexes, 331
Hormones, 289
Horns in sheep, 74
Host-mediated assay, 377, 532
Hotspot, 532
House keeping genes, 532
Human
  disorders, 347
  hereditary defects, 257
  leukocyte antigen (HLA) complex, 80, 81
  sex-anomalies, 440
H-Y antigen, 533
Hybrid 10, 533
  -arrested translation (HART), 533
  breakdown, 533
  inferiority, 533
  plasmid, 533
  seed production, 417, 420
  vigour, 533
Hybridization, 490
  in situ, 533
  probing, 533
Hybridize, 533
Hydrogen bonds, 216
Hypermorph, 533
Hypervariable repeats, 476
Hypoploidy, 57
i gene mutants, 318
Illegitimate offspring, 85
Immigration, 490
Immune system, 80
Immunological acceptability, 81
Inborn errors of metabolism, 533
Incestuous mating, 533
Induced mutations, 533
  applications of, 379
  history of, 355
Inducer(s), 313, 533
Inducible control, 312, 533, 128, 533, 534
Inheritance,
  blending, 3
  blood, 3

  chromosome theory of, 42, 211
  clover butterfly colour, 73
  cock-feathering, 70
  co-dominant, 388
  criss-cross, 64
  cytoplasmic, 411
  dominant, 387
  extranuclear, 409, 428-432
  extrachromosomal, 409
  holandric, 437
  horns in rambowillet sheep, 73
  incompletely dominant, 388
  quantitative, 395, 397, 401
  maternal, 409, 413
  mechanics of, 9
  plastid, 411
  plumage in birds, 70
  skip-generation, 64
  sex-linked, 64, 391, 437
  X-linked, 437
  X- and Y-linked, 68
  theories of, 2-3
  temperature-sensitive gene, 19-20
  uniparental, 123, 411, 412, 413, 566
  Y-linked, 437
Inherited disease, treatment for, 459
Inhibitory epistasis, 534
Initiation
  codon(s), 286, 276, 288, 534
  complex,
    definition, 534
    complex, 30S, 298
    complex, 70S, 299
    factors, 534
    site, 534
Inosine, 534
Insemination reaction, 496
Insertion sequences (IS elements), 534
In situ hybridization, 534
Integral membrane protein, 307
Interallelic complementation, 534
Intercalate, 534
Intercistron nucleotide sequences, 324
Interference, 136, 153, 185, 534

Subject Index 609

Interferon, 477
Intergenic spacers, 246
Intermediate lesions, 342
Internal control signals, 250
Interphase, 25, 31, 534
  S phase, 25
  G_1 phase, 25
  G_2 phase, 25
Interrupted mating, 534
Intervening sequences (IVS), 250
Intragenic mutations, 534
Intron, 421, 534
Introns, removal of, 277-278, 280
Invariant control, 312
Inversion, 344
  intragenic, 344
  paracentric, 344
  pericentric, 344
  loops, 345
  mutation, 535
Inverted repeats, 535
Irreversible regulation, 326
i[s] mutants, 318
Isoalleles, 535
Isochromosome, 535
Isolating mechanisms, 535
Isozymes, 535
Jacob and Monod model, 535
  confirmation of, 320
Johannsen's
  experiment, 399
  pure line 13, 399
Joint molecule, 161
Jumping gene, 535
Kappa, 425, 535
Karyokinesis, 535
Karyotype, 535
Kin selection, 535
Kinetochore, 49, 51
Kinetoplast, 535
Klinefelter's Syndrome, 535
Knobs, 50
Lac operon, 314-322, 535
  constitutive synthesis of, 318
  controlling region of, 322
  structure of, 314
  superrepressors of, 318
  working of, 315
Lactate dehydrogenase, 266
Lactose repressor, 536
Lagging strand, 536
Lamarckism, 536, 562
Lamarck's theory of evolution, 481
λ phage, 429
  circular form, 429
  integration and excision, 430
  linear genome, 429
  genophore, 429
Lampbrush
  chromosome, 536
  loops, 61
Lariate molecule, 277
Leader, 246-248
  processing of, 277
  transcript, 536
  segment, 536
Leading strand, 536
Leptotene, 536
Lethal genes in mice, 84
Leucine zipper(s), 337, 536
Level of significance, 504, 536
Life, 536
  continuity of, 4, 517
  facts about, 1
Life cycle,
  bacteria, 123-128
  budding yeast, 121-122
  Chlamydomonas, 123
  corn, 112-113
  definition, 536
  Drosophila melanogaster, 113-115
  man, 111
  Neurospora crassa, 118-121
  Paramecium aurelia, 115-118
  Saccharomyces cerevisiae, 121-122
  viruses, 128-129
Life cycles, 110-129
Life sketch of Mendel, 9
Ligase, 225 (*see also* DNA ligase)
Linkage, 133, 135, 536
  complete, *Bombyx mori*, 154
  complete, *Drosophlia*, 154

coupling, 135, 142
detection of, 536
detection of, after partitioning chi-square, 140
detection of, for autosomal genes, 145, 148-149
detection of, for sex-linked genes, 149-152
detection of, without partitioning chi-square, 137
repulsion, 135, 142
estimation of, for autosomal genes, 152-153
estimation of, for sex-linked genes, 153-154
estimation of, in plant F_2 population, 141
estimation of, in test cross progeny, 138
group, 192, 536
in animal systems, 145-154
in animal systems, three-point crosses, 145
in animal systems, two-point crosses, 145
in plant systems, 136-145
incomplete, 137
map(s), 113, 136, 140, 536
Locus, 536
LOD score, 462
score method, 536
Long-term regulation, 326
Loops, 280
Lymphocyte death, 80
Lyon's hypothesis, 442, 536
Lysis, 537
Lysis, spontaneous, 128
Lysogenic bacteria, 128
Lysogeny, 429, 537
Lytic phase, 201
Macrochromosomes, 49
Macrolesions, 342
Macroevolution, 564
Macronucleus, 115
Major groove, 537
Major histocompatibility
antigens, 537
complex, 537
Malandrium album, 436
Male sterility, definition, 537
in plants, 417
genetic, 417, 418
Male sterile lines, maintenance of, 417, 419
Malvidin, 95
Menstruation, onset of, 73
Map unit, 136, 139, 537
Marker, 537
Masked mRNA, 537
Maternal effect(s), 423-424, 537
Maternal inheritance, 537
Maximum likelihood method, 537
Meiosis, 33-36, 537
Ascaris megalocephala, 40
corn, 132
diagrammatic summary, 34
grasshopper, 41
significance of, 36
Meiosis I, 33-36, 537
anaphase I, 35
diakinesis, 35
diplotene, 35
leptotene, 35
metaphase I, 35
pachytene, 35
phases of, 33-36
prophase I, substages, 35
reductional division, 33
telophase I, 35
zygotene, 35
Meiosis II, 36, 537
anaphase II, 36
equational division, 36
metaphase II, 36
prophase II, 36
stages of, 36
telophase II, 36
Membrane-bound organelles, 307
Mendelian
factors, 39
genes, 397
inheritance, 537

Subject Index 611

population, 537
segregation, 138
Mendelism, rediscovery of, 19
Mendel's
  first law, 538
  second law, 538
Merodiploid(y), 317, 429, 538
Merozygote, 203, 317, 429
Meselson-Stahl experiment, 226-227
Messenger RNA (mRNA), 275, 538
  caps, 538
  decay, 538
  -interfering complementary (mic)
    RNA, 337
  stability of, 277
Metacentric chromosome, 538
Metafemale (spermale), 538
Metaphase, 538
  chromosome, 538
  plate, 35
Metaphase I, 538
Metaphase II, 538
Mice, milk factor in, 417
Microchromosomes, 49
Microinjection, 538
Microlesions, 342
μm (micrometer), 565
Micronucleus, 115
Microsatellite simple sequence
  length polymorphism, 476
Microsatellites, 462, 473
Migration, 490
Milk production in mammals, 73
Minochromosome, 428, 538
Minor groove, 539
Minor histocompatibility
  antigens, 539
-10 sequence, 247
-25 box, 539
-35 sequence, 247
Mirabilis jalapa, plastid inheritance, 411
Missense mutation, 539
Mitochondria, 46
Mitochondrial
  codon assignments, 422

DNA diseases, 422, 462, 539
genes, 422
genome, yeast, 420-422
Mitosis, 28, 539
  animal cells, 32
  centriole, 32
  centromere, 32
  cyclins, 30
  cytokinesis, 33
  dephosphorylation, 30
  furrowing, 33
  G_0 state, 28
  G_1 cyclins, 30
  G_2 to M transition, 28
  Golgi apparatus, 31
  karyokinesis, 32
  lamina proteins, 30
  maturation Promotion factor
    (MPF), 30
  metaphase, 32
  nuclear envelope, 32
  nucleoli, 32, 35
  phosphorylation, 30
  positive signal, S phase, 28
  pre-start G_1, 30, 31-33
  prophase, 32
  significance of, 33
  telophase, 33
Mitotic inhibition, 358
Mixed populations, equilibrium in, 386
Moderately repetitive sequences, 539
Modern synthetic theory of
  evolution, 485-496
Modification of DNA/RNA, 539
Modified bases, 539
Molecular
  barriers, 498
  evolution, 497-498
  zippers, 337, 539
Mongolism, 565
Monocistronic mRNA, 539
Monoecious, 113
Monogenic inheritance, 539
Monoploid, 345
Monosomic, 347, 540

612 Basic Genetics

Monozygotic twins, 540
Mosquitoes, incompatibility in, 416
M phase, 540
Morphogen, 540
Morphological evolution, 497
Muller-5 technique, 374, 376
Multifactorial (polygenic) trait, 540
Multigene family, 540
Multigenic inheritance, 540
Multiple
  alleles, 77, 234, 540
  allelism, 77
  allelism, conclusions from, 85
  births, 101
  factor hypothesis, 396, 398, 540
Multisite
  mutant alleles, 487, 488, 540
  mutations, 238
Mutagen, 355, 540
  5-aminopurine, 362
  5-bomouracil, 361
  high temperature, 366
  hydrazine, 365
  hygrogen peroxide, 366, 367, 368
  hydroxylamine, 364
  nitrous acid, 363
  maleic hydrazide, 366
  pH, 365
Mutagenic hazards, environmental, 379-380
Mutagenesis, 341, 540
Mutagens,
  acridine dyes, 362
  acridine orange, 362
  affecting replicating nucleic acids, 360
  affecting resting nucleic acids, 363
  alkylating agents, 368
    mono-functional, 369
    alkylating agents, bifunctional, 369
  antibiotics, 366
  azide, 366
  base analogues, 360, 361
  dyes, 361
  inhibitors of nucleic acid
    precursors, 360
  physical, 356-360
Mutant, 355, 540
Mutants,
  drug resistant, 415
  nutritional, 373
Mutation(s), 225, 485, 540
  back, 348
  forward, 348
  gain-of-function, 444
  induced, 355-371
  mis-sense, 349
  neutral, 349
  non-sense, 349
  Reverse, 348
  same sense, 349
  spontaneous, 350-355
  visible, 377
  -like event, 487
  theory of evolution, 484, 540
Mutational load, 447
Mutations,
  autosomal recessive, 450
  autosomal dominant, 451
  characteristics of, 341
  classification of, 342
  definition, 341
  detection of, 371
  frameshift, 362
  locating position of unknown, 241
  point, 450
  replica-plating technique, 371-372
  sex linked recessive, 451
  sex-linked dominant, 452
  size of, 342
Muton, 245, 540
Mycelium, 121
n, 2n, 540
NADH dehydrogenease, 422
Natural selection, 483, 492-494, 540
Negative
  control, 312, 314, 315, 540
  eugenics, 540
  interference, 541
  supercoiling, 274
Neocentromere, 54

Neo-Darwinism, 541
Neurospora, 266, 541
 arginine biosynthesis in, 265
 detection of mutants, 263
 gene mapping in, 176
 detection of linkage in, 179
 estimation of linkage in, 179
 nutritional mutants, 262
 poky, 413
 nutritional mutants in, induction of, 263
 origin of tetrads in, 176, 178, 181
 tetrad analysis in, 155-156
 three point cross, 183-185
 two point cross, 179-183
 types of tetrads in, 176
Neurospora crassa, 373
 life cycle, 118-121
 mating types, 121
Neutral
 alleles, 541
 gene hypothesis, 541
Neutrality theory of protein evolution, 541
Nicks, 225
Nitrogenous bases, 541
Non-ambiguous code, 541
Non-Darwinian evolution, 541
Non-degenerate code, 287
Non-disjunction, 14, 55, 347, 541
Non-disjunction, primary, 547
Non-genetic factors, 5
Non-histone proteins, 541
Non-ionizing, radiations, 357
Non-Mendelian ratios, 541
Non-overlapping codon, 288, 541
Non-parental ditype (NPD) tetrads, 176, 541
Non-reciprocal recombination, 541
Non-recombination products, 205
Non-sense codon, 287, 542
Non-sense mutation, 349, 542
Non-sister chromatids, 35, 160, 542
Non-transcribed spacer, 542
Non-waxy pollen, 132
Normal distribution, 396, 542

Normalizing selection, 542
Northern blotting, 542
Novel phenotypes, 87
Nuclear
 envelope, 35, 45
 fusions, 116
 membrane, 32, 45
 pore, 45, 46
Nucleic acid, 542
Nucleo-cytoplasmic interactions, 424-428, 542
Nucleoid, 44, 413, 542
Nucleolar organiser, 46, 542
Nucleoli organizing regions (NOR), 48, 54
Nucleolus, 32, 45
Nucleus, 38, 45, 542
 amount of DNA in, 45
 chemical structure of, 45
Nucleolus, 542
 amorphous zone, 46
 -associated chromatin zone, 46
 fibrillar zone, 46
 granular zone, 46
 zones of 46
Nucleoside, 542
 phosphorylase, 46
Nucleosome, 48, 542
Nucleotide, 542
 pair substitution, 542
Null allele, 542
Null hypothesis (H_0), 504, 506, 542
Nullisomic, 347, 542
Observed ratios, 504
Ochre codon, 543
Oenothera, 481
 lamarckiana, 350
Okazaki fragments, 228, 543
Oligogenes, 543
Oligonucleotide, 543
Oligonucleotide-directed mutagenesis, 543
Ommochromes, 260
Ommatidia, 261
One cistron-one polypeptide hypothesis, 267

One gene-one
  enzyme hypothesis, 262-265, 543
  polypeptide hypothesis, 265-267, 543
  primary cellular function hypothesis, 267
  primary function hypothesis, 543
Oogenesis, 61
Opal codon, 543
Open promoter complex, 543
Open reading frame (ORF), 543
Operator, 272, 312, 543
  -constitutive, mutants, 317
  -constitutive mutation (O^c), 543
  mutations in, 317
Operon, 272, 312, 543
Operon concept, 326
Opine, 543
Ordered tetrad, 121, 543
Organelle, 543
Organic evolution, 482
Organismal evolution, 497
Origin of species, 499, 544
Overdominance, 544
Overdominance hypothesis, 544
Overlapping clones, 471, 473
Overlapping code, 287
p, 544
P_1, 544
P_1 bacteriophage, linkage map of, 202
Pachytene, 138, 544
Palindrome, 544
Pangenesis, 2, 3, 544
Paracentric inversion, 544
Parallelism, chromosome behaviour, 39
Parallelism, Mendelian factors, 39
Paramecin, 425
Paramecium,
  gene *K*, 427
  killer trait in, 425
  killer x sensitive strain, 426
Paramecium aurelia,
  autogamy, 118, 119
  binary fission, 115

diploid nuclei, 117
haploid nuclei, 117
internal fertilization, 118
macronucleus, 115
micronucleus, 115
nuclear fusions, 116
structural organization, 115
Parental classes, 16
Parental ditype (PD), 544
Partial diploids, 317, 544
Particulate inheritance, 544
Parthenogenesis, 439
Parthenogenetic, 347
Paternal inheritance, 544
pBR322, 544
Pedigree, 544
Pedigree analysis, 67, 447
  symbols, 448
  autosomal dominant trait, 452
  autosomal recessive trait, 451
  sex-linked dominant trait, 453
  sex-linked recessive trait, 453
Penetrance, 290, 545
Penicillium, 379
Pentaploids, 347
Peptide hormones, 330
Peptidyl (P)-site, 298, 545
Peptidyl transferase, 300, 302, 545
Pericentric inversion, 545
Perinuclear space, 45
Permanent heterosis, 545
Petite, 545
Phage
  genomes, 205
  "ghost", 545
  maturation, 206
Phasmid, 545
Phecomilia, 291
Phenocopy, 545
Phenocopying
  abnormal, 291
  definition, 291
  normal, 292
Phenocopies, examples, 291-292
Phenotype, 5, 259, 545
Phenotypic differences, 395

Phenotypic ratio,
  definition, 545
  dihybrid F_2, 14-15
  monohybrid F_2, 10
Phenotypic variability, 170
Phenylalanine, metabolism of, 258
Phenylketonuria, 257, 545
Phenylthiocarbamide (PTC), 545
Phenylthiocarbamide tasters, 387
φX174, 545
Phosphodiester bond(s), 216, 277, 545
Phyletic speciation, 545
Phylogenetic lineage, 499
Physiological attributes, in bacteria, 123
Piconavirus, 546
Pisum sativum, 10 (*See also* garden pea)
Plasmagene, 546
Plasmid(s), 428, 546
  importance of, 432
Plastid inheritance, 411, 546
Plastids, 46
Pleiotropic genes, 101
Pleiotropy, 101, 546
Ploidy, 546
Point mutations, 238, 342, 546
Poisson distribution, 546
Polar mutations, 321
Pollen mother cell (PMC), 546
Polyadenylation, 546
Poly(A) polymerase, 546
Poly(A) tail, 546
Polyacrylamide gel electrophoresos (PAGE), 546
Polyadenylation, 277
Polyadenylation signal, 277
Polycentric, 546
Polycistronic mRNAs, 276, 316, 546
Polydactyly, 546
Polyethyl glycol, 469
Polygenes, 402, 546
Polymerase chain reaction (PCR) 474-475, 544

Polymeric effect, 95
Polypeptide chain
  elongation, 299, 301
  initiation, 298
  synthesis, 295
  termination, 300, 303
Polyploidy, 546
Polyribosome (polysomes), 301, 304, 546
Polytene
  cells, 60
  chromosomes, 511, 546
Populations,
  genes in, 383
  laboratory, 383
  natural, 383
Position effect, 547
Positive
  control, 312, 314, 322, 547
  eugenics, 547
  interference, 547
  supercoiling, 274
Postmating mechanisms,
  gametic mortality, 496
  hybrid breakdown, 496
  hybrid inferiority, 496
  hybrid sterility, 496
  zygote mortality, 496
Post-natal diagnosis, 455
Post-translational
  control, 334
  steps, 304-305
Post-zygotic isolating mechanism, 547
Precursor rRNA, 47
Preformation, 2
Preformationism, 100
Pre-edited region (PER), 279
Premating mechanisms,
  ecological isolation, 495
  seasonal isolation, 495
  ethological isolation, 495
  mechanical isolation, 495
  gametic isolation, 496
Pre-mRNA, 276, 547
Prenatal diagnosis, 454

Pre-rRNA, 282, 547
Pre-tRNA, 280, 547
Pre-zygotic isolating mechanism, 547
Pribnow box, 246, 547
Primary
  cellular function, 101, 547
  cellular product, 268
  constriction, 49, 547
  oocytes, 547
  nondisjunction, 547
  spermatocytes, 547
  transcript, 547
  trisomics, 547
Primase, 547
Primer, 547
Probability, 503-504
Proband, 447
Promosome, 547
Promoter, 246
Primula, 95
Principle of
  independent assortment 9, 547
  segregation, 9, 12, 14, 548
Prion, 548
Probability, 548
Probability value (p), 136
Probe, 548
Processed pseudogene, 548
Prokaryote, 548
Promoter (promotor), 272, 275, 312, 548
  consensus sequences, 247
  mutations in, 320
  optimal, 247
  transcriptional, 432
Proofreading, 548
Prophage, 128, 429, 548
Prophase, 25, 26, 548
Prophase I, 548
Propositus, 447, 548
Protamines, 38, 44
Protein
  kinase, 331
  secretion, 307
  synthesis, 298

Proton migration, 350
Protoplast(s), 470, 548
Protoplast fusion, 548
Prototrophs, 128, 372, 549
Provirus, 549
Pyrimidine(s), 216
pSC101, 549
Pseudoalleles, 235-236, 549
Pseudoautosomal inheritance, 437, 549
Pseudoautosomal region, 437
Pseudodominance, 343, 549
Pseudogene(s), 344, 422, 549
Pterins, 260
Puff(s), 56, 60, 290, 549
Pulse labelling, 26
Punctuation codon, 549
Punnet square, 549
Pure line, 10, 549
Purine(s), 216, 549
Pyrimidine, 549
q, 549
Qualitative
  trait, 549
  inheritance, 549
Quantitative
  inheritance, 550
  trait, 550
R_1, R_2, R_3, etc., 550
rII locus, fine structure of, 243
R bodies, 428
R plasmid, 550
Radiation absorbed dose (RAD), 360
Radiation(s),
  α-particles, 356
  β-particles, 356
  neutrons, 356
  neutrons, fast 356
  neutrons, slow 356
  particulate, 356
  protons, 356
  units of, 359-360
  ultraviolet light, 357
Random
  amplified polymorphic DNA (RAPD), 476

genetic drift, 384, 491, 550
mating, 550
sample, 550
Randomized block design, 550
Rapid lysis (r) mutants, 550
Rare base pairing, 550
Ratio, X/A, 438
Rat killikrein gene, 334, 335
Rat preproinsulin peptide, 334, 335
Read through, 248
Reading frame, 287, 550
Reassociation, 550
recA, 550
recBC, 550
Receptor element, 550
Recessive, 10
  allele, 550
  epistasis, 550
  lethal, 551
Recipient cell, 551
Reciprocal
  crosses, genetic inferences, 109, 110
  exchanges, 160
  translocation, 551
Recognition site, 551
Recombinant, 551
  classes, 16
  DNA, 465, 467, 551
  DNA technology, 7, 551
  joint, 551
  RNA, 551
Recombinants, 133
Recombination, 1, 225, 551
  bacteria, 190
  generalized, 160, 527
  genetic, 133
  high frequency of (Hfr), 125
  illegitimate, 475, 533
  intragenic, 486
  non-reciprocal, 541
  reciprocal, 551
  distances, 192
  fraction, 142
  frequency, 185, 551
  intermediates, 166

joint, 161
mechanism, 157-159
models, 159-168
Recon, 245, 551
Reductional division, 551
Regulated (regulatory) gene, 551
Regulation, upgrading, 442
Regulator molecule, 313
Regulators, master, 443
Regulatory
  elements, 317
  gene, 551
  gene mutations, 318
  gene changes, 497
  mechanisms, 326
Relative biological efficiency (RBE), 360
Release factors, 300, 551
Rem, 360
Renner, 551
Repair, 551
Repetitive and unique sequence
  hypothesis, 552
Repetitive rDNA cistrons,
  B. subtilis, 47
  E. coli, 47
  Drosophila, 47
  HeLa cells, 47
  Xenopus, 47
Replica-plating, 552
Replicase, 552
Replication, 552
  viral chromosome, 204
  fork, 552
  origin, 552
  -defective virus, 552
  eye, 552
  fork, 230
Replicator, 552
Replicon, 27, 232, 428, 552
Replisome, 552
Repressible
  control, 312, 552
  enzyme system, 552
Repression, 552
Repressor, 552

618 Basic Genetics

Reproduction, 4
  asexual, 120
  cell 29
  differential, 492
  molecular basis of, 4
  sexual, 2, 121
Reproductive isolation, 494-496
  post-mating mechanisms, 496
  pre-mating mechanisms, 495
Repulsion, 552
  linkage, 553
Resistance transfer factor(s) (RTF), 431, 553
Resistant, 553
Restorer, 417
Restriction digest, 553
Restriction endonuclease(s), 465, 553
  cleavage sites of, 465
  type I, 465
  type II, 465
Restriction
  fragment length polymorphisms (RFLPs), 455, 476, 553
  map, 553
  site, 553
Retroposon, 553
Retrovirus, 553
Reversal of central dogma, 553
Reverse
  mutation, 553
  transcriptase, 553
  transcription, 270, 553
Reversible regulation, 326
Reversion(s), 239, 372, 553
  equivalent, 348
  exact, 348
Revertants, 553
Reterotransposon, 553
R factors, 553
Rhesus alleles, 78, 79
Rhesus (Rh) factor, 78, 553
Rh incompatibility, 79
Rhizoid, 424
Rho
  -dependent terminators, 249
  -independent terminators, 249

  factor, 273, 553
rRNA, 554
Ribonuclease, 553
Ribonucleic acid, 5, 211, 553 (*See also* RNA)
Ribonucleoprotein(s)
  fibrils, 46
  granules, 46
  particle (RNP), 277
  small nuclear (snRNPs), 557
Ribonucleoside, 553
Ribonucleotide, 554
Ribonucleotide structure, 223
Ribosomal RNA (rRNA) genes, 271
Ribosome(s), 282, 554
  bacterial, 123
  binding site, 554
  E. coli, 297, 298
  eukaryotic, 47, 305
  translocation, 303, 564
Ribozyme, 554
Right splicing junction, 554
Ring chromosome, 554
RNA(s), 6, 211
  activator, 327, 508
  antisense, 336, 510
  complementary, 516
  editing, 278, 422, 554
  functional, 245
  functions of, 275
  genes, ribosomal, 271
  genes, transfer, 271
  guide, 279
  hn 531
  messenger (m), 538
  methylase, 46
  mic, 337
  phages, 554
  polymerase, 44, 46, 246-247, 272, 554
  polymerase I, 274
  polymerase II, 274
  polymerase III, 275
  polymerase core enzyme, 554
  polymerase holoenzyme, 554
  processing, 275, 554

recombinant, 551
replicase, 554
ribosomal (rRNA), 282, 554
-RNA hybridization, 336
small nuclear (snRNA), 557
splicing, 438, 554
splicing, alternate, 508
structure, 221, 223
transcript, 554
transfer (tRNA), 563
types of, 282
U1-U7, 565
RNase, 554
D, 554
E, 282
P, 283, 554
III, 283
Reverse transcriptase, 554
Roentgen (R), 359
rRNA operon, 282, 283
Rolling circle replication, 410, 554
Rough endoplsmic reticulum (RER), 554
s, 555
S locus, 85
Salamander, 61
Saccharomyces cerevisiae,
life cycle, 121
mating types, 121
Salivary chromosomes, 555
Salmo gairdneri, 102
Salmonella, histidine metabolism, 262
Salmonella typhimurium, 61, 213, 266, 372
Satellite, 54, 555
chromosome, 555
Saturated maps, 476
Scaffold, 555
Sea urchin, 41, 42
Second
-division segregation (SDS), 555
messenger, 555
site reversion, 555
Secondary
constriction, 54
messenger, 331

Sedimentation coefficient, 555
Segmental allopolyploid, 555
Segregation, 113, 555
first-division, 176, 177, 179
second-division, 176, 177, 179
distorter (SD), 555
ratio, 555
Segregational
load, 555
petite, 555
Selection, 555
at a single locus, 493
balance, 494
directional, 492
disruptive, 492
diversifying, 492
frequency dependent, 494
density dependent, 494
normalizing, 494
stabilizing, 494
effect of, 494
Selector gene, 555
Self-
fertilization, 491, 555
incomparability, 78, 555
pollination, 556
rejection, 85
sterility, 556
Semi-conservative replication, 410, 356
Sense
codon (sense word), 287, 556
strand, 275
Sensor gene, 556
Sequence-characterized amplified region (SCAR), 476
Sequence-tagged sites (STS), 476
Serum creatine phosphokinase, 68
Sex
cell, 556
chromatin, 556
chromosome(s), 435, 556
compound, 438
correction, 441
-determination, 435, 556
chromosomal, 435

dioecious, 435
  effect of environment on, 439
  monoecious, 435
  XO type, 435
  XY type, 436
-dominance, 73
-duction, 202, 429, 556
-factor (F), 556
  genic control of, 439
  hormonal control, 440
-influenced
-lethal, 556
-limited Characters, 69, 70-73
-linked genes, 64
-linked inheritance, 556
-pilus, 556
-plasmid, 556
-reversal, 440, 556
-trait, 556
Shine-Dalgarno sequence, 557
Short index finger, 75
Short-term regulation, 326
Shotgun experiment, 557
Shuttle vector, 557
Sibling vector, 557
Sibling species, 557
Sickle cell anemia, 101
Sickle cell hemoglobin, 101
Sickle cell trait, 557
Sigma factor, 272, 557
Signal
  hypothesis, 307, 309
  peptidase, 308
  peptide, 308
  recognition particle receptor (SRPR), 308
  recognition particle (SRP), 308
  -sequence receptors, 308
Simple
  gene, 557
  sequence repeats (SSRs), 462, 474, 476
  sequences (SSs), 474
  tandem repeats (STRs), 474
Simultaneous control of different operons, 322

Simultaneous infection, 244
Single-strand binding (ssb) protein, 557
Single-strand conformation polymorphism (SSCP), 476
Sister chromatids, 557
Site-directed mutagenesis, 557
Small nucleoriboproteins, 45
Small U RNAs, 277, 279, 280
Smooth endoplasmic reticulum (SER), 557
Snails,
  dextral, 423
  sinistral, 423
Solenoid, 557
Somatic cell
  hybridization, 455, 458, 558
  hybrids, selection procedure, 456
  mutation and recombination, 558
Somatoplasm, 3
S1 nuclease, 557
Soma, 443
SOS box, 558
SOS repair system, 558
Southern blotting, 468, 558
S phase, 558
Specialized gene, 558
Specialized nucleoriboprotein structures (snurps), 558
Specialized transduction, 558
Speciation, 558
  a model of, 498-499
  phyletic 499
  true, 499
Species, 558
Specific amplicon polymorphism (SAP), 476
Spina bifida, 74
Spindle, 558
Spindle apparatus, 32, 35
Spindle fibers, 54
Spliceosome, 277, 279, 558
Splicing, 558
Split gene, 558
Split gene, evolution of, 251
Spontaneous mutations, 484, 558

Subject Index

Stabilizing selection, 558
Stacking forces, 559
Standard
  deviation, 559
  error of difference in means, 559
  sample mean, 559
Start
  codon, 559
  point, 559
  signals, recognition of, 272
Stasipatric speciation, 559
Statistic, 559
Statistical parameters, 396
Statistics, 559
Stem-and-loop structure, 248, 559
Stem-like structures, 280
Steroid hormones, 331
Sticky ends, 429, 466, 559 (also see cohesive ends)
Stop codon(s), 277, 559 (see also termination codons)
Strain, 559
Strand
  displacement, 559
  separation, 365
Strange alleles, 80
Strong promoter, 559
Structural alleles, 559
Structural gene, 559
  changes, 497
  evolution, 497
  mutations in, 317
Struggle for existence, 483
Stuttering, 75
Submetacentric chromosome, 559
Successive mutational events, 488-489
Super
  female, 559
  male, 559
Supergene, 559
Supernumerary chromosomes, 559
Supplementary genes, 560
Suppression, 560
Suppressive petite, 560
Suppressor,
  Extragenic, 560

  gene, 560
Survival of the fittest, 483, 560
S value, 560
Swivel, 560
Symbiotic association, 428
Synapsis, 35, 160, 560
Synaptonemal complex, 48, 161, 560
  central element, 161
  lateral element, 161
Syndrome, 440, 560
Synonymous codons, 276, 560
Synteny, 459, 476, 560
Synthetic phase, 560
Synthetic theory of evolution, 560
t, 560
T_2 bacteriophage, 214, 215
T_4 bacteriophage, *rII* locus, 237
Tachytelic evolution, 560
TACTAAC box, 560
Tandem duplication, 344
TATA box, 560
Tautomeric shift, 561
Tautomerism, 351
Tautomers, 350
Taylor's experiment, 231
Tay-Sachs disease, 561
T-DNA, 561
Telocentric chromosome, 561
Telomerase, 561
Telomeres, 55, 561
Telophase, 561
Telophase I, 561
Telophase II, 561
Temperate phage, 128, 561
Template (strand), 272, 275, 561
Terminal joints of fingers, enlargement of, 75
Termination codon(s), 276, 286, 288, 561 (see also stop codons)
Termination factors (TF), 551, 561
Terminator stem, 561
Test cross, 14, 561
Test of independence, 506
Test system, 561
Testing of hypothesis, 561
Tetrad(s), 561
  analysis, 176, 178, 179, 561

nonparental ditype (NPD), 176
parental ditype (PD), 176
tetratype, 176
Tetraploid(s), 347, 561
Tetrasomic, 347, 562
Tetratype (TT), 562
T even phages, circular map of, 206-207
Thalassemias, 562
Theories of evolution, 481-496
Theory of
  inheritance of acquired traits, 562
  natural selection, 482-484
    components of, 483
Theta-structure, 562
Terminators, 248
Terminator sequence, 248
Thermoregulation of a gene, 562
Thermus aquaticus, 474
Thiogalactoside transacetylase, 315
Three-point test cross, 562
3'-OH terminus, 562
Thymidine, 562
Thymine (T), 216, 562
  dimers, 358, 562
Ti plasmid, 562
Tissue compatibility, 80
Tissue culture, 562
Tobacco mosaic virus (TMV)
  infectivity, 216
Tolerable mutations, 562
Topoisomerase, 562
Topoisomerase, type I, 274
Topoisomerase, type II, 274
Totipotency, 100, 562
Train, 246, 248-249
Trailer segment, 562
Traits, 1
  acquired 3
  asymmetric, 103
  holandric, 68
  multiple allelic, 389
  qualitative, 4
  quantitative, 4, 395, 396
  sex-influenced, 73
  sex-influenced, 73-75
  sex-controlled, 73

Trans-
  acting, 562
  arrangement, 317
  heterozygote, 236, 563
  position, 235, 236
  sexual, 564
  specific evolution, 564
Transalkylation, 371
Transcribed spacer, 562
Transcribing enzyme, 246
Transcript, 275
Transcription, 247, 270, 562
  factors, 562
  eukaryotes, 274-275
  initiation, 61
  initiation reaction in, 246
  prokaryotes, 272-274
  promotion site, 61
  startise (startpoint), 247
  termination, 61, 273
  terminator sequence, 272
Transcriptional unit, 246, 562
Transduction, 194, 201, 213-214, 563
  generalized, 194, 201, 213
  mapping, 563
Transfection, 563
Transfer
  operon (*tra*), 563
  reaction, 563
  RNAs, 280
Transformant(s), 563
  single, 190
  double, 191
Transformation, 190, 211, 563
  -competent viruses, 563
  eukaryotic, 563
  in competent viruses, 563
Transformation mapping, 563
Transformed cells, 563
Transformer, 563
Transforming principle, 212, 563
Transgenic, 563
Transgressive segregation, 563
Transient polymorphism, 563
Transitions, 342
Translation, 306, 563

Subject Index 623

components of, 295
control at the level of, 334
eukaryotes, 305
prokaryotes, 295
termination signal, 300
Translational level, control at, 324
Translocase (EF-G), 564
Translocation(s), 344
 intercalary, 344
 simple, 344
 types of, 346
 reciprocal, 344
 of chromosome, 564
 of gene, 564
Transposable genetic elements, 431, 475, 564
Transposase, 475, 564
Transposition, 564
Transposition immunity, 475
Transposon(s), 475-476 (*Also see* transposable genetic elements)
 tagging, 476, 564
 -mediated gene transfer, 564
Transvection, 564
Transversion(s), 342, 564
Trihybrid, 564
Triple X, 564
Triplet code, 564
Triplet binding assay, 564
Triplicate genes, 564
Triploid(s), 347, 565
Trisomic, 347, 565
Trisomy 21, 565
Tritiated (^{3}H)-thymidine, 25, 231
Triticum aestivum, 348
Trivalent, 565
tRNAs, 280 (*see also* transfer RNAs)
 charging of, 295, 297
 deacylase, 565
 nucleotidal transferase, 565
Tryptophan synthetase, 266
t-test, 565
Tumor
 -inducing (Ti) plasmid, 565
 suppressor genes, 565
Turner's syndrome, 565
Twisting number, 565

Two-point cross, 136
Type I error, 504, 565
Type II error, 504, 565
Tyrosine metabolism, 257
Tyrosinosis, 565
Ubiquitin, 565
Unambiguous code, 287
Unassigned reading frames (URFs), 421
Underdominance, 565
Underwinding, 565
Unequal pairing, 235
Unicistronic mRNAs, 276
Unidentified reading frame (URF), 565
Unidirectional replication, 565
Uninemic chromosome, 566
Unisite mutant allele, 566
Unisite mutation, 487
Unit factor, 566
Univalent, 566
Universal
 codon, 566
 donor, 78, 566
 acceptor, 78
 code, 287, 288
 recipient, 566
Unordered tetrads, 121
Unstable mutation, 566
Unusual bases, 566
Unusual chromosomes, 566
Unwinding proteins, 566
Upper lateral incisor, absence of, 75
Up promoter mutations, 566
Upstream, 566
Upstream activating sequence (UAS), 566
Upstream sequences, 247
Uracil, 566
Uridine, 566
U-RNP, 566
Use and disuse doctrine, 481, 566
U sn RNPs, 566
Vacuole, 567
Variable number of tandem repeats (VNTRs), 462, 476, 567
Variable region, 567

Variance (σ), 567
  genetic, 529, 567
  genotypic, 5
  environmental, 5, 567
  phenotypic, 5, 567
  components of genetic, 404
  additive component of, 404, 508
  dominance component of, 404, 522
Variation,
  continuous, 564
  definition, 1
  discontinuous, 520
  environmental, 523
  genetic, 486, 487
  genotypic, 5, 529
  phenotypic, 5, 545
  somaclonal, 557
  hereditary, 483
Vector, 469, 567
Vehicle, 567
Vehicle plasmid, 567
Vesicle, 306
Vicia faba, 366
Viral coat proteins, 128
Viral genome, circular, 429
Viral infection experiment, 214
Virion, 567
Viroids, 567
Virulent phage, 128, 567
Virus(es), 567
  recombination in, 203-207
  phenotypic differences, 203
  plaques, 203
  host range, 203
  phenotypic mixing, 203
  mixed particles, 203
  gene mapping, 204
  linkage groups, 205
  independent assortment, 205
  recombination products, 205
  non-recombination products, 205
Vitamins, 125
Waxy pollen, 132
Watson-Crick
  base pairing, 567
  model, advantages of, 219
  model, DNA structure, 221
W chromosome, 567

Western blotting, 567
Wheat, grain colour in, 402
White forelock, 74
White-Hastings model, 159-160
Width of pelvis, 73
Wild-type, 567
Wobble base pairing, 568
Wobble hypothesis, 568
Wobble's rules, 288
x, 568
X_1, X_2, X_3, etc., 568
X/A ratio, 568
X chromosome, 568
X-rays, 358-359
Xenia effect, 568
Xenopus laevis,
  rDNA, 47, 48
  rDNA cistrons, 47
  rRNA, 47
Xeroderma pigmeltosum, 357, 568
X inactivation, 568
X-linkage, 568
XO condition, 568
X-rays,
  hard, 356
  soft, 356
X syndrome, 568
XXY syndrome, 568
Yate's correction, 568
  term, 506
Y chromosome, 568
Yeast, 121
  detection of linkage in, 186
  estimation of linkage in, 186
  gene mapping in, 186
  killer strain of, 416
  artificial chromosome (YAC), 568
Y-linkage, 568
Y-linked disorders, 452
Z chromosome, 569
Zaprionus paravittiger, 487
Zinc finger, 569
Zipper region, 338
Zygote, 569
Zygotene, 569
Z-DNA, 569
Z-test, 569
z values, 144, 146-148

AUTHOR INDEX

Aalalen, 62
Abelson, J.N., 576, 589
Adams, R.L.P., 232, 570
Adelberg, E.A, 130, 207
Adhya, S., 292, 570
Adleman, 275, 292
Adolph, K.W., 62
Aggarwal, S., 339, 570
Air, G.M., 571
Ajioka, J.W., 578
Alan R., 572
Allard, R.W., 406, 570
Allen, E.F., 444
Allendorp, F.W., 583
Altenberg, 355, 381, 570
Altman, S., 293, 570
Ames, B.N., 381, 570
Ammal, E.K.J., 62, 575
Ampy, F.R., 585
Anderson, W.F., 464, 479, 570
Andreason, G.L., 570
Angenon, G., 578
Anvel, P., 579
Anraku, 208
Antherton, l., 487, 594
Apgar, J., 580
Arnheim, N., 479, 570
Ashburner, M., 8, 570
Attardi, G., 433
Avery, O.T., 212, 232, 570
Auerbach, C., 8, 355, 381, 570
Ayala, F.J., 500, 571, 576
Bachmann, B.J., 207, 571
Baglioni, C., 571
Baird, J.P., 571
Bairoch, A., 578
Bajeman, C.J., 463, 571

Baker, T.A., 232, 445, 571, 573, 582
Balbiani, E.G., 58
Baltimore, D., 571
Barker, R.F., 586
Barkley, M.D., 339, 571
Barnes, W.M., 576
Barnett, L., 572, 575
Barondes, 310
Barr, M., 441
Barratt, R.W., 186, 571
Barrell, B.G., 252, 253, 571
Bass, L.W., 232, 583
Bateman, A.J., 85, 571
Bateson, W., 1, 22, 88, 136
Baltimore, D., 572
Baylor, M.B., 208, 571
Beadle, G.W., 259, 262, 268, 381, 571, 591
Beale, G.H., 130, 433, 571
Beckwith, J.R., 339, 571
Beijerinck, 128
Beisson, J., 433, 571
Bell, P.B., 571
Belling, J., 171
Benjumea, F.J.D., 583
Benzer, S., 237, 243, 252, 571, 572, 574, 577
Berchtold, 310
Berg, P., 475, 478, 478, 572, 590
Bergström, S., 587
Berlyn, M.B., 208, 572
Bertram, E., 441
Betlach, M.C., 572
Beyreuther, K., 339, 572
Bielaga, B.A., 581
Biemont, C., 104, 572
Birchler, J.A., 572

Birge, E.A., 208, 572
Blakeslee, A.F., 42
Blobel, G., 307
Blow, 232
Blum, B., 279
Boggs, S.S., 590
Bolivar, F., 478, 572
Bolle, A., 589
Bonierbale, M.W., 591
Bonner, D.M., 268, 572
Bonnet, 2
Borgaonkar, D.S., 463, 572
Bottino, P.J., 8, 85, 573
Bourgeosis, S., 339, 571
Boveri, T., 38, 41, 43, 50, 234, 572
Boyer, H.W., 572
Boycott, A.E., 433, 572
Brachet, J., 588
Brewbaker, J.L., 8, 572
Brewster, T., 463, 572
Brennan, C.A., 293, 572
Brenner, S., 268, 293, 307, 381, 572, 575, 589
Bridges, C.B., 59, 75, 381, 572, 573, 585
Brimacombe, R., 310, 584
Britten, R.D., 339, 573
Britten, R.J., 573, 575
Brock, T.D., 37, 207, 573
Broda, P., 478, 573
Brougton, W.J., 578
Brown, D.D., 573
Brown, P., 591
Brown, T.A., 8, 85, 292, 573
Brown, W.V., 573
Bruce, V., 207, 591
Brusick, D., 381, 573
Bryson, V., 590
Buc, H., 293, 573, 575
Büchi, H., 581
Buckman, C., 207, 583
Bull, J.J., 444, 573
Burd, C.G., 576
Burdette, W.J., 8, 463, 573, 581, 585, 586
Burgoyne, P.S., 75, 573

Burke, D.T., 578
Burnet, F.M., 583
Burnham, C.R., 8, 573
Burns, G.W., 8, 85, 130, 573
Burton, G.W., 406, 573
Busby, S., 339, 573, 575
Byers, R.E., 444, 573
Byme, J., 592
Cairns, J., 207, 208, 381, 573
Calendar, R., 208, 591
Callan, H.G., 62, 573
Campbell, A.M., 433, 573
Canki, N., 592
Capecchi, M.R., 462, 464, 573
Carlson, E.A., 455, 464, 406, 573
Carlson, S.S., 593
Carlson, P.S., 573
Carlton, B.C., 594
Carr, K., 573
Cartron, J.-P., 586
Caruthers, M.H., 574
Caskey, C.T., 463, 574
Castle, W.E., 105, 574
Cazenave, C., 339, 574
Cech, T.R., 574
Chambon, P., 253, 574
Chapeville, F., 293, 574
Chao, L., 500, 574
Chargaff, E., 232, 574
Charles, D.R., 406, 574
Chase, M., 214, 215, 232, 580
Cheng, S.M., 310
Cherif-Zahar, B., 586
Cheverud, J.M., 406, 574
Chilton, M.D., 479, 574
Chovnick, A., 236, 252, 574
Church, G.M., 279
Clark, A.G., 589
Clarke, A.E., 584
Clausen, H., 594
Clayton, D.A., 433, 574
Cleland, R.E., 574
Clerl, P., 579
Clewell, D.B., 130, 574
Clong, 232
Cochran, W.G., 507, 590

Cocking, E.C., 478, 574
Coe, Jr., E.H., 130, 433, 586, 592
Cohen, J.S., 588
Cohen, S.N., 232, 475, 479, 572
Cole-Turner, R., 479
Colin, Y., 586
Conner, B.N., 576
Cooper, K.W., 479
Corcos, A.F., 75, 574
Corey, R.B., 232, 587
Correns, C., 22, 411, 574
Cox, E.C., 574
Coyne, J.A., 500, 574
Crew, F.W., 440
Creighton, H.B., 154, 171, 574
Crick, F.H.C., 6, 216, 224, 232, 253, 270, 310, 571, 572, 574, 575, 592
Crock, 293
Crosa, J.H., 572
Crothers, 40
Crow, J.F., 393, 575
Cuenot, L., 105, 575
Culliton, 463
Culver, M., 586
Curuzzi, G.M., 577
Daniel, A., 575
Danner, D.B., 590
Darling, S.M., 578
Darlington, C.D., 8, 62, 575
Darnell, 339
Darwin, C., 2, 395, 482, 484, 500, 575
Das, A., 293, 575
Dausset, J., 589
Davenport, C.B., 406
Davey, M.R., 574
Davidson, E.H., 232, 573, 575
Davidson, J.N., 574, 575
Davidson, N., 589
Davis, R.W., 572
Dawid, I., 339, 584.
Day, R.R., 186, 577
Debenham, P.G., 463, 575
DeCrombrugghe, B., 575
Deich, R.O., 590
Delbruck, M., 207, 592

Demerec, M., 130, 381, 575
Denhardt, G.H., 591
Denniston, K.J., 575
DePamphilis, M.L., 479, 576
DeSerres, F.J., 381, 580
DeVicente, M.C., 591
DeVit, M.J., 593
DeVries, H., 350, 381, 484, 498, 500, 576
Dickerson, R., 576
Dickson, R.C., 339, 576
Diecumakos, E.G., 478, 570
Dilley, D.R., 573
Dillon, L.S., 252, 576
Diver, C., 433, 572
Dixon, W.J., 507, 576
Dobzhansky, Th., 8, 105, 485, 500, 576, 590
Doermann, A.H., 207, 576
Dombroski, A.J., 572
Donachoe, P.K., 579
Doolite, W.F., 501, 588
Douglas, L., 576
Dover, G.S., 576
Dower, W.J., 479, 589
Drake, J.W., 381, 576
Drapeau, G.R., 594
Dressler, D., 171, 230, 232, 576, 578
Drew, H.R., 576
Dreyfuss, G., 292, 576
Drlica, K., 207, 576
Drohan, W.H., 592
Dulbecco, R., 8
Dunn, L.C., 8, 576, 590
DuPraw, E.J., 62
Eanes, W.F., 102, 576
Earnshaw, 62
East, E.M., 22, 396, 406, 576
Eaton, G.J., 105, 576
Ebright, R.H., 339, 573
Echols, H., 208, 576
Edgar, R.S., 591
Edwards, J.W., 577
Eguchi, Y., 339, 577
Ehling, U.H., 381, 577
Eissenberg, 62

Elisson, 433
Ellis, J.R., 577
Emerson, S., 171, 186, 396, 500, 577
Emerson, S.H., 577
Englesberg, E., 339, 577
Enquist, I.W., 478, 576
Ephrussi, B., 259, 268, 433, 571, 577
Epstein, C.J., 463, 577
Erich, H., 479, 570
Erskine, A.G., 577
Evans, G.A., 570
Everett, G.A., 580
Falconer, D.S., 406, 577
Falkow, S., 572
Fahring, R., 377, 577
Falsaff, S., 579
Farnsworth, W.W., 8, 85, 577
Feigon, J., 590
Fellay, R., 578
Ferre, F., 586
Fincham, J.R.S., 186, 252, 577
Firtsch, E.F., 584
Fischer, G., 572
Fisher, R.A., 22, 500, 507, 577
Flanders, P.H., 171, 577
Flavell, R.B., 576
Flemming, W. 60
Fogel, S., 186, 577
Foster, 381
Fox, M.S., 207, 293, 478, 577, 589
Fox, T.D., 577
Fraenkel-Conrat, H., 216, 217, 232, 577
Fraley, R.T., 578
Fratini, A.W., 576
Frederick, J.E., 578
Freeling, M., 500, 578
Freese, E., 381, 577, 578
Freifelder, D., 8, 578
Frelbery, C., 578
Fujimoto, Em., 586
Fulton, T.M., 591
Galizzi, A., 589
Gallagher, 500
Galton, F., 2, 395
Ganal, M.W., 591

Garmeraad, S., 478, 578
Garniobst, L., 571
Garrod, A.E., 257, 268, 446, 578
Garza, D., 62, 578
Gasser, C.S., 479, 578
Geerts, S.J., 572
Geiger, J.H., 581
Gerald, P., 463, 572
Gesterland, 293
Gheysen, G., 479, 578
Ghosh, H., 581
Gibbs, R.A., 586
Gibson, J.B., 500, 592
Gilbert, W., 230, 232, 339, 578
Gill, K.S., 22, 104, 105, 578, 592
Gillespie, J.H., 500, 578
Gillham, N.W., 433, 578
Gillis, A.M., 445, 578
Gilman, M., 592
Giovannoni, J.J., 591
Glass, B., 588
Glover, J., 578
Golbus, M.S., 463, 577
Gold, L., 339, 578
Goldberger, R.F., 578, 589
Goldstein, A., 507, 578
Goldstein, L., 580, 588
Goodenough, U.M., 433, 583
Goodfellow, P.J., 578
Goodfellow, P.N., 578
Gorman, M., 445, 571, 579
Gottlieb, 500
Gottesman, M., 292, 570
Gould-Fogerite, S., 479, 584
Gowen, M.S., 579
Gowen, J.W., 22, 171, 579
Graley, 479
Grandillo, S., 591
Grant, V., 489, 579
Graur, D., 501, 583
Graziano, 62
Green, M.C., 579
Green, E.L., 579
Green, M.M., 105, 171, 236, 252, 456, 576, 579
Green, K.C., 579

Author Index 629

Greene, P.J., 572
Gregg, R.G., 590
Griffith, F., 579
Griffiths, A.J.F., 62, 232, 579, 591
Grivell, L.A., 579
Grossman, L., 594
Grouchy, 458
Grun, P., 433, 579
Gubbay, 445
Guest, J.R., 594
Gupta, N., 581
Gupta, P.K., 479, 579
Gustafson, A., 379
Guthrie, 280
Gyllensten, U., 433, 579
Hagan, D., 580
Hahn, S., 581
Hakomori, S., 594
Haldane, J.B.S., 500, 579
Hall, T.C., 186, 586
Haller, 463
Haqq, C.M., 579
Haqq, T.N., 579
Hamerton, 463
Hammerling, J., 433, 579
Hanna, G.C., 444, 588
Hardy, G.H., 383, 393, 433, 579
Hardy, K., 579
Harley, 445
Harris, H., 463, 579
Hartl, D.L., 463, 578, 579
Hassold, T.J., 463, 579
Hastings, P.J., 171, 593
Hayes, W., 208, 579, 593
Heard, E., 445, 579
Hecht, L.I., 580
Hélène, C., 339, 574
Henning, U., 594
Herbers, K., 479, 579
Herner, R.C., 573
Hersh, A.H., 580
Hershey, A.D., 207, 208, 214, 215, 232, 310, 580
Hershey, J.W.B., 580
Herskowitz, I., 8, 580
Heslot, H., 381, 580
Hessler, A.Y., 571
Heyneker, H.L., 572
Hill, M.B., 207, 576
Hilton, 463
Hoagland, M.B., 293, 580
Hodgkin, J., 445, 580
Hoff, M., 586
Hoffman, M., 464, 580
Hogan, 479
Hogness, D.S., 572
Hohn, B., 478
Hollaender, A., 381, 578, 580
Holley, R.W., 580
Holliday, R., 160, 186, 580
Holm, T.H., 586
Hood, L., 268, 580
Holzman, D., 580
Hopkins, N.H., 592
Horikoshi, M., 292
Horn, V., 594
Horne, R.W., 130, 580
Hotchkiss, R.D., 207, 580
Hsu, M., 589
Hurst, L.D., 500, 581
Hutchison, C.A., 571
Huxley, J., 500, 581
Ikoda, 85
Ilan, J., 581
Iltis, H., 9, 22
Immer, F.R., 171, 581
Ingram, V.M., 268, 581,
Ippen-Ihler, K.A., 207, 581
Itano, H.A., 587
Itoh, T., 336, 577
Iwanowski, 128
Jackson, I.J., 105, 581
Jackson, M., 589
Jackson, P.K., 582
Jackson, S.P., 593
Jacob, F., 207, 314, 339, 572, 581, 593
Jacob, T.M., 581
Jacobs, P.A., 463, 579
Jaenisch, R., 579
Jähner, D., 579
Jaurin, B., 587

Jeffreys, A.J., 463
Jenkins, J.B., 8, 85, 171, 581
Jeuthen, 62
Jinks, J.L., 433, 581
Johannsen, W., 5
Johansson, 488
Johnson, I.S., 581
Johnson, P.F., 479, 582
Johnston, S.A., 593
Jones, R.N., 62, 581
Josefsson, A., 579
Ju, Q., 582
Judd, B.H., 581
Jurand, A., 588
Kamada, 232
Kao, 464
Kaplan, S., 572
Kainz, M., 581
Kandasamy, M.K., 586
Kaufman, T.C., 130, 575, 581
Kavenoff, R., 62
Kemp, J.D., 586
Khorana, H.G., 581
Kibley, 381
Kiesselbach, T.A., 130
Kilbey, B.J., 581
Kilbane, J.J. II, 479, 581
Kim, Y., 339, 581
Kimura, M., 581
King, C.-Y., 579
King, M.C., 582
King, R.C., 8, 37, 582, 586
King, R.W., 582
Kirby, L.T., 463, 582
Kirch, 85
Kirschner, M.W., 37, 582, 586
Kleckner, N., 339, 589
Klug, A., 62, 582
Klukas, C.K., 339
Knowler, J.T., 433, 570, 571
Knudsen, K.L., 583
Kohne, D.E., 339
Kolata, 463
Kopka, L.M., 576
Koralesvski, M.A., 590
Kornberg, A., 232, 62, 582, 588, 589

Kornberg, R.D., 582
Koshland, D., 37, 582
Kössel, H., 581
Krafka, Jr., J., 582
Krontiris, T.G., 463, 582
Kucherlapati, R.S., 590
Kumlin, E., 586
Kumar, Ashok, 582
Kumar, Ashwni, 582
Kumar, R., 104, 105, 487, 500, 501, 582
Kusmierak, J.T., 589
Laemmi, U.K., 62
Lamarck, J.-B., 3, 481
Landauer, W., 293, 582
Lande, R., 590
Lander, E.S., 171
Landis, W., 433, 582
Landman, O.E., 500, 582
Landschulz, W.H., 337, 582
Landsteiner, K., 77, 78
Lang, W.H., 582
Lasic, D., 479, 582
Laskey, 62, 232
Lawn, R.M., 253, 582
Lawrence, W.J.C., 105, 257, 582
Leach, D., 589
Leader, D.P., 570
Lea, D.E., 381, 582
Leary, R.F., 102, 105, 583
Leder, P., 293, 586
Lederberg, E.M., 583, 585
Lederberg, J., 207, 214, 232, 381, 433, 583, 585, 594
Legator, M., 377, 581
Lengyel, L.H., 587
Lennox, E.S., 207, 583
Lepag, T., 583
Leppert, M., 586
Lerner, I.M., 102, 105, 500, 583
Letovsky, S., 208, 572
LeVan, K.C., 586
Levan, A., 463
Levene, P.A., 232, 583
Levine, P., 583
Levine, R.P., 171, 433, 463, 583

Author Index 631

Levinthal, C., 207, 583
Lewin, B., 8, 583
Lewis, E.B., 235, 252, 583
Lewontin, R,S., 393
L'Heritier, Ph., 433, 584
Li, W., 583
Liebman, S.W., 589
Lilley, 339
Lindberg, F.P., 587
Linder, 186
Lindergren, C.C., 186, 583
Lindsley, D.L., 171, 583
Lipmann, F., 574
Little, C.C., 105, 339, 574
Lively, E.R., 583
Lloyd, R.G., 207, 583
Logadon, J.M., 583
Long, E., 339, 584
Low, K.B., 207, 571
Lowentin, R.C., 591
Lubon, H., 592
Luce, W.M., 584
Luria, S., 207
Lyon, M.F., 445, 584
Lush, J.L., 393
Mackean, 8
MacLeod, C.M., 212, 570
Maddin, 37
Madison, J.T., 580
Madhani, 280
Maguire, M.P., 592
Makino, S., 62, 584
Maniatis, T., 470, 478, 584
Mann, 355. 464
Mannino, R.J., 479, 584
Manuelidis, L., 62, 584
Marin, I., 571
Marken, J., 594
Marquisee, M., 580
Martin, C., 586
Martin, G.B., 591
Marx, J., 479, 584
Massey, Jr., F.J., 507, 576
Matalova, 22
Mather, K., 171, 393, 402, 406, 584
Matthaei, J.H., 293, 587

Matton, D.P., 85, 584
Matunis, M.J., 576
Maxon, L.R., 593
Mayer, W., 500, 584
Mayloy, S.R., 207, 584
Mayr, E., 500, 584
Mays, L.L., 584
McBridge, O.A., 479, 479, 584
McCann, J., 570
McCarthy, J.E.G., 310, 584
McCarty, M., 212, 570
McClinctock, B., 54, 154, 171, 236, 475, 478, 574, 584
McClung, C.E., 444, 584
McCoubrey, W.K., 479, 584
McElligott, S.C., 593
McElroy, W.D., 588
McKeown, M., 584
McKnight, S.L., 337, 582, 584
McKusick, V.A., 463, 584, 585
McLaren, A., 3, 585
McPeek, M.S., 171, 585
Meenley, P.M., 445, 587
Mendel, G.J., 6, 9, 10, 11, 12, 22, 40, 41, 87, 484
Meneely, P.L., 584
Merlo, D.J., 586
Merrill, S.H., 580
Meselson, M., 171, 208, 226, 572, 585
Messeguer, R., 591
Meyer, 445
Miescher, F., 232, 585
Migeon, 457
Miglani, G.S., 8, 381, 501, 582, 585
Miller, J.C., 591
Miller, J.H., 207, 464, 571, 572, 585
Miller, L., 591
Milstein, C., 307
Minkley, Jr., E.G., 207, 581
Minkoff, E.C., 500, 585
Mirsky, A.E., 588
Mitchell, H.K., 268, 592
Mizutani, S., 270, 293, 591
Moazed, D., 310, 585
Mol, J.N.M., 574

Moldave, K., 310, 585, 594
Monaghan, F., 22, 574
Monod, J., 314, 339, 581
Moore, D.S., 507, 585
Montgomery, 40
Morgan, R., 581
Morgan, T.H., 8, 50, 64, 75, 234, 252, 355, 585
Morison, 232
Morrow, B.E., 582
Morse, M.L., 585
Mortimer, R.K., 186, 577
Morton, N.E., 171, 463, 585, 586
Motulsky, A., 479, 586
Mouro, I., 586
Muari, F., 479
Muller, H.J., 62, 171, 235. 355, 381, 463, 585, 586
Müller-Hill, B., 339, 578
Mullis, K.B., 474, 479, 586
Murai, N., 586
Murialdo, H., 576
Murray, A.W., 586
Murray, J.A.H., 37, 63, 478, 578
Murray, M.G., 586
Murray, R.K. 586
Nadal-Ginard, B., 590
Nakamura, Y., 463, 586
Nanninga, N., 310, 586
Narang, S.A., 581
Nasrallah, J.B., 85, 586
Nasrallah, M.E., 586
Nass, N., 584
Nathans, D., 572
Neel, J.V., 463, 586
Neuffer, M.G., 130, 586
Nevins, J.R., 293, 586
Newbigin, E., 584
Newman, H.H., 463
Newmeyer, D., 571
Nichols, W., 581
Nicklas, R.B., 37, 586
Nilsson-Ehle, H., 379, 396, 399, 406, 586
Nirenberg, M.W., 293, 310, 586, 587
Noll, M., 62, 587

Noller, H.F., 293, 310, 585, 587
Nomura, M., 587
Nordstrom, K.D., 584
Normark, S., 252, 253, 587
Notani, N.K., 587
Novitski, E., 22, 576, 589
Nowark, R., 501
Nurialdo, 208
Nurse, 37
Ochoa, S., 293, 587
O'Connell, P., 586
Ogawa, T., 232, 587
Ohno, 293, 445, 486, 487
Ohtsuka, E., 581
Okazaki, T., 232, 587
Okazaki, R., 587
Olby, R., 22, 232, 587
Old, R.W., 587
Olins, D.E., 62
Olins, R.M., 62
Olive, 130
Olsson, O., 587
O'Malley, B.W., 339, 587
O'Neill, 445
Orel, 22
Orgel, O.L., 22, 572
Osborn, 463
Oster, I.I., 381, 586
Ou-Lee, T., 479, 587
Owen, A.R.G., 171, 587
Palmer, J.D., 445, 583
Parkhurst, S.M., 445, 583, 587
Parmeggiani, A., 310, 592
Paterson, A.H., 591
Patterson, D., 463, 587
Patton, J.G., 590
Pauling, L., 232, 463, 587
Peacock, W.J., 171, 587
Pederson, 63
Pees, H., 62, 581
Pennisi, E., 479, 587
Pental, D., 574
Penwick, J.R., 580
Perkins, D.D., 571
Perret, X., 578
Perry, P.R., 292, 587

Author Index

Peters, J.A., 8, 22, 587
Peterson, J.L., 479, 584
Picard, 433
Pilgrim, I., 22, 587
Pimentel, D., 590
Pineda, O., 591
Pinol-Roma, S., 576
Platt, T., 572
Pontecorvo, G., 8, 236, 588
Porin, 337
Portugal, F.H., 232, 588
Power, J.B., 574
Potter, H., 171, 576
Preer, J., 433, 588,
Preiss, 85
Prescott, D.M., 580, 588
Primrose, S.B., 478, 587
Prince, J.P., 591
Puck, T.T., 464
Punnett, R.C., 22, 88, 136, 588
Race, R.R., 463, 588
Radding, C.M., 171, 585
Rajam, M.V., 582
Ramanis, Z., 433, 589, 590
Ramel, C., 581
Randolph, L.F., 171
Rathenberg, R., 463, 592
Reeder, H.R., 582
Reichert, N.A., 586
Rendel, J.M., 488
Rensch, B., 500
Revel, H.R., 208, 593
Reznikoff, S.S., 571
Reznikoff, W.S., 572
Rhoades, M.M., 37, 433, 588
Ribisson, 8
Richards, F.M., 463
Richardson, R.J., 339, 588
Rick, C.M., 444, 588
Riedy, M., 593
Rieger, R.A., 8
Riley, M., 207, 576
Ris, H., 62, 588
Risch, N., 171, 588
Roberts, J.W., 581, 592
Robertson, M., 292, 339, 588

Roblin, R., 572
Robson, J.M., 381, 570
Röder, M.S., 591
Rodgers, J., 22, 588
Rodriguez, R.L., 572
Rogers, H.J., 588
Roman, H., 186, 588
Rose, M.R., 501, 588
Rosenthal, A., 578
Rothenbuhler, 444
Rothman, J.L., 208, 588
Rotman, R., 207, 580
Routman E., 406, 574
Rowen, L., 588
Roy, R.P., 588
Roy, P.M., 588
Roy, Jr., W.J., 574
Roznikoff, W.S., 576
Ruckert, 60
Rudman, B., 588
Ruhland, W., 588
Russell, L.B., 445, 588
Ryder, L.P., 589
Ruzinski, M.M., 186, 588
Saenger, W., 589
Sager, R., 412, 433, 589, 590
Sakabe, K., 587
Saks, M.E., 310, 589
Sambrook, J., 584
Sampson, J.R., 589
Sandler, L., 589
Sanford, J.C., 593
Sanger, R., 232, 463, 588
Santos, 310
Sarabhai, A., 589
Sarich, V.M., 593
Sarin C., 8, 85, 589
Sarnow, 310
Sasavage, 308
Satyanarayana, 500
Savageau, M.A., 339, 589
Scaife, J., 207, 589
Schaffner, W., 594
Schultz, R., 339
Schulz-Schaeffer, J. 8, 589
Schwartz, R.J., 587

Schweinyruber, A.M., 589
Scott, J.F., 580
Searle, A.G., 589
Selander, R.K., 501, 589
Sell, H.M., 573
Sengupta-Gopalan, C., 586
Shahi, V.K., 104, 105, 589
Sharp, P.A., 130, 292, 339, 589
Shekman, R., 589
Shen, M.W., 581
Shermann, F., 589
Sherwood, 8, 22
Shigekawa, K., 479, 589
Siciliano, 458
Siddiqui, O., 207, 589
Sigler. P.B., 581
Silvers, W.K., 105, 589
Simons, R.W., 339, 589
Simpson, G.G., 445
Sinclair, 445
Singer, B., 577, 589, 590
Singer, M., 590
Singer, S.J., 216, 217, 232, 433, 587
Singh, 487
Singleton, W.R., 381, 590
Sinnott, E.W., 8, 590
Sinsheimer, R.L., 232, 590
Skleuar, V., 590
Slinghtom, J.L., 586
Slilaty, 339
Slizynski, B.M., 171
Smith, H.H., 406, 574
Smith, H.O., 590
Smith, C.W.J., 590
Smithies, O., 479, 590
Snedecor, G.W., 507, 590
Soans, A.B., 590
Soans, J.S., 590
Socha, W.W., 577
Somani, L.L., 8
Sonneborn, T.M., 293, 433, 571, 590
Southern, 479
Spear, B.B., 574
Speed, T.P., 171, 585
Srb, A.M., 433, 590
Stadler, D.R., 355

Stahl, F.W., 171, 208, 226, 590
Stallings, R.L., 458, 592
Stanley, W.M., 583
Stansfield, W.D., 8, 582, 590
Stasiak, A., 577
Stebbins, G.L., 500, 576, 590
Steel, R.G.D., 507, 590
Steele, 482
Stein, J.C., 339, 586
Steinberg, A.G., 355
Steitz, J.A., 339, 59
Stent, G.S., 208, 573, 5910
Stephen, M.C., 583
Stephenson, M.L., 580
Stern, C., 8, 22, 463, 591
Stewart, J.W., 589
Stillman, B., 571
Stormont, C., 487
Strange, 37
Streisinger, G., 207, 591
Stretton, A.O.W., 572
Stock, C.A., 586
Stokes, A.R., 593
Strasburger, E., 50
Stretton, A.D.W., 589
Strickberger, M.W., 8, 85, 130, 171, 591
Stubbe, H., 8, 591
Sturgess, V.C., 105, 582
Sturtevant, A.H., 8, 234, 381, 433, 585, 591
Sugimoto, K., 587
Suino, A., 587
Sutcliffe, 478
Sutton, D.W., 38, 43, 50, 62, 234, 586,
Suzuki, D.T., 591
Svejgaard, A., 589
Swaminathan, M.S., 8
Swammerdem, 2
Swanson, C.P., 37, 591
Szostak, J.W., 63, 586
Tamarin, R.H., 8, 85, 591
Tamin, 270
Taniguchi, T., 62
Tanksley, S.D., 406, 476, 479, 591

Author Index 635

Tartof, K.D., 591
Tatum, E.L., 262, 268, 571
Taylor, A.L., 591
Taylor, J.H., 8, 207, 231, 571, 577, 591
Temin, H.M., 293, 591
Thali, M., 594
Therman, S.E., 463
Thoday, J.M., 500, 592
Thom, 355
Thoman, M.S., 207, 591
Thomas, J.O., 582
Thomas, N.S., 62, 578
Thompson, D.H., 85
Thorpe, D., 594
Tissieres, A.N., 587
Tjian, R., 592
Tjio, J.H., 463
Tomazawa, J.-I., 577
Tomizawa, 208, 336
Tooze, J., 479, 592
Torrie, J.H., 507, 590
Touchette, N., 63
Towle, H.C., 587
Trehan, K.S., 22, 104, 105, 592
Tschermak, E. von, 22
Tuite, 310
Tukayama, 62
Turgeon, R., 587
Tycowski, K.T., 590
Uklyama, E., 579
Valander, W.H., 479, 592
Valentine, J.W., 576
Van Montagu, M., 578
vander Krol, R.R., 574
Vargas, C., 574
Vehar, A., 253, 582
Verma, R., 62, 582
Villeneuve, 445
Visconti, N., 207, 592
Vogel, F., 592
Vogel, H.J., 463, 590
Vogt, R.C., 444, 573
von Ehrenstein, G., 310, 574, 577
von Weisblum, B., 577
Wagner, R.P., 63, 268, 592,

Walbot, V., 433, 592
Waldeyer, W., 50
Waldman, 464
Wallace, A.R., 483, 484, 500, 575
Walters, M.S., 464
Wand, 274
Wang, T.S.-F., 232, 592
Warburton, D., 463, 592
Warmke, H.E., 592
Warner, J.R., 582
Watkins, P.C., 463, 592
Watson, J.D., 6, 216, 224, 232, 479, 572, 573, 592
Watts-Tobin, R.J., 575
Weigle, J.J., 208, 585
Weijland, A., 310, 592
Weiling, F., 22, 592
Weinberg, W., 383, 393, 592
Weiner, A.M., 78, 589, 592
Weinstein, A., 171, 592
Weintraub, H.M., 339, 593
Weisblum, B., 574
Weiss, E., 580
Weiss, M.A., 207, 456, 579
Weismann, A., 3, 484
Weissman, S., 572
Wells, I.C., 587
Wells, R.D., 581
West, S.C., 171, 577, 593
Wharton, D., 579
White, M.J.D., 500, 593
White, R.J., 586, 593
White, T.J., 593, 594
Whitehouse, H.L.K., 171, 593
Whittam, T.S., 589
Whittinghill, M., 171
Wichterman, R., 130, 593
Wihelmi, 85
Wilcox, G., 339, 577
Wilkie, D., 433, 593
Wilkins, M.H.F., 232, 593
Williams, N., 445, 478, 479, 593
Williams, R.S., 593
Wilson, A.C., 479, 501, 579, 582, 593
Wilson, H.R., 593
Wilson, M.R., 593

Wing, R.A., 591
Wing, R.M., 576
Winge, O., 62
Witkin, E.M., 463
Witkowski, J., 592
Woese, C.R., 293, 593
Wolfe, 37
Wollf, R., 586
Wollman, E.L., 207, 581, 593
Wood, W.B., 208, 593
Woods, P.S., 268
Woolf, C., 594
Wright, J.E., 594
Wright, R.E., 594
Wright, T.R., 8, 393, 487, 570
Wu, R., 479, 587, 594
Wu, W., 591
Xu, L., 292, 594

Yamamoto, F., 85, 594
Yamasaki, E., 570
Yanofsky, C., 251, 253, 594
Yates, F., 507, 577
Young, N.D., 292, 591
Young, R.A., 594
Yule, G.U., 399, 406, 594
Zamecnik, P.C., 580
Zamir, A., 580
Zawel, 292
Zeuthen, J., 571
Zimm, G.G., 171, 583
Zimmerman, A., 37, 594
Zinder, N.D., 214, 232, 572, 583, 594
Zipser, D., 339, 571
Zoller, M., 592
Zubay, G., 8, 85, 594